ELEMENTS
OF ACOUSTICS

SAMUEL TEMKIN
RUTGERS UNIVERSITY

JOHN WILEY & SONS
New York ● *Chichester* ● *Brisbane* ● *Toronto*

TO MY SONS

DAVID AND MICHAEL

Library of Congress Cataloging in Publication Data:

Temkin, Samuel, 1936–
   Elements of Acoustics.

   Bibliography: p.
   Includes index.
   1. Sound-waves. I. Title

QC243.T46   1981   534   80-24416
ISBN 0-471-05990-0

Printed in the United States of America

10 9 8 7 6 5 4 3 2 1

PREFACE

This book is an outgrowth of a course in acoustics I have taught for a number of years at Rutgers University. The main reason for adding one more book to an already long list of books on this subject is the lack of modern introductory texts that treat acoustics as a branch of fluid mechanics. In my view, this is the most natural approach, at least for those areas of acoustics dealing with the most common media for sound propagation, namely, air and water. This approach is, of course, not new. It was used by the authors of many of the books now considered classical, including Rayleigh, Lamb, and others. In recent times, however, many of the acoustics texts that have appeared treat the subject as a branch of electrical engineering. There are indeed many instances in which acoustic oscillations are analogous to some phenomena discussed in electrical engineering courses and the analogies are clearly advantageous to those students whose background is in that discipline. For others, the analogies may be a drawback; to them, both the acoustic equations and their electrical analogues are new.

The main subjects discussed in this book are: propagation in uniform fluids at rest; transmission and reflection phenomena; attenuation and dispersion; and emission. These are only some of the main topics in acoustics. To have attempted to cover all of them would have been presumptuous on my part. Nevertheless, there are several topics that, by some, may be considered basic enough to warrant their inclusion in a text of this nature, but that have been omitted. These include aerodynamic sound, diffraction, and propagation in nonuniform media. Some of these are mentioned in the text, but all too briefly in relation to their importance. The reasons are that some of these topics are either outside my areas of competence or are too advanced compared to the general level of the book. In any event, most of them are fully treated in one or more specialized books that have appeared recently, so that their detailed discussion in this book is unnecessary. On the other hand, sound absorption is discussed in more detail than is usual in books on acoustics. To a certain extent, this reflects my personal interest in that subject, but it is also intended to qualify the strongly held notion that dissipation effects in sound waves are unimportant.

The material given here is intended primarily for a beginning graduate course in acoustics, but includes portions suitable for more advanced courses. In writing this book, I have assumed that the student's background includes the usual preparation in undergraduate physics and mathematics, as well as a course in advanced calculus and a course in basic thermodynamics. Prior acquaintance with fluid mechanics is desirable, but not required. The required material on that subject is developed in Chapter 1. Chapter 1 also includes a summary of basic thermodynamics. To make the book self-sufficient, both of these subjects are developed to a greater degree than is needed in an introductory course.

The book contains more material than is possible to cover in one semester. By deleting some of the more advanced material, it can be used in a one-semester course in basic acoustics for students in engineering or in the physical sciences. On the other hand, with some additional material, it may be used in a one-year sequence covering both basics and applications.

Because of the basic nature of the subject of this book, I have attempted to derive each result from basic principles. However, the emphasis throughout is on the physical meaning of the results, and not on the mathematical techniques that were used to derive them. On the other hand, in some of the derivations I have included more detail than customary, since all too often the student's main effort is spent in trying to fill in the mathematical steps missing between main results. Of course, this has some pedagogical value but, more often than not, it merely improves the ability of the student to manipulate equations. In my view, a better way of learning is by doing. To this end, a number of problems have been included in the text.

Each chapter contains a brief list of suggested references. A more complete list is given in the Bibliography at the end of the book. The lists are not exhaustive; their purpose is merely to direct the interested student to other general sources, or to recent articles touching on some of the material discussed in the text.

Although I have included the results of some of my own investigations, the bulk of the material presented may be considered classical. It is therefore difficult to acknowledge the sources of many of the results that are presented. I have, however, profited much from Chapter 8 of *Fluid Mechanics* by L. D. Landau and E. M. Lifshitz and from Chapters 1–3 of *An Introduction to Fluid Dynamics* by G. K. Batchelor. Other books that have influenced this work are *The Theory of Sound* by Lord Rayleigh, *Theoretical Acoustics* by P. Morse and U. Ingard, *Fundamentals of Acoustics* by L. E. Kinsler and A. R. Frey, and *The Foundations of Acoustics* by E. Skudrzyk.

A major portion of this book was written during 1974–1975 while I was on leave at the Technion-Israel Institute of Technology. I wish to thank Rutgers University and The Lady Davis Fellowship Trust for making this leave possible. I also owe much to the faculty of the Department of Mechanical Engineering at the Technion for their kind hospitality.

I would like to express my gratitude to Professor R. A. Dobbins of Brown University, who introduced me to the subject of this book; to my colleagues at Rutgers University for their continued encouragement; to many of my students for their valuable comments and observations; and to Mrs. Rosemarie Boysen, who typed an earlier version of this book. The final manuscript was typed by Mrs. Erma Sutton, to whom I am also indebted for improving the clarity of many passages.

To conclude, I wish to express my gratitude to my wife Judy and to my sons David and Michael, who patiently endured the writing of this book.

S. TEMKIN

NOTATION AND
MAIN SYMBOLS

Boldface letters represent vectors.

Quantities with a tilde represent complex variables.

I	intensity
p_0	mean pressure
p'	acoustic pressure
$\mathbf{u}$	fluid velocity
$\mathbf{x}, \mathbf{x}'$	position vectors; $|\mathbf{x}| = r$
Π	acoustic power
ρ'	density fluctuation
ρ_0	mean density
ϕ	velocity potential

CONTENTS

CHAPTER THREE

REFLECTION AND TRANSMISSION PHENOMENA   77

CHAPTER FOUR

CHAPTER FIVE

CHAPTER SIX

APPENDIXES

INDEX

CHAPTER ONE
BASIC FLUID MECHANICS AND THERMODYNAMICS

Acoustics is the science that studies the emission, transmission, and reception of sound waves. It touches on disciplines as disparate as psychology and meteorology, and includes many subdisciplines such as architectural acoustics, bioacoustics, environmental acoustics, and musical acoustics.

This book deals only with some of the physical properties of sound waves in fluids. This topic, although limited in scope, covers one of the most important applications of acoustics, namely, the study of sound waves in air, and provides the basis for other branches of acoustics.

Most of the properties of acoustic waves in fluids may be obtained by means of the wave equation, and this can be derived from approximate conservation principles without having to resort to the far more complicated equations that describe general fluid motions. However, in doing so, one is forced to ignore, from the beginning, effects that may sometimes be important, such as dissipation and nonlinear distortion. To be sure, these effects could be included at a later stage when the more common aspects of acoustics have been studied, but this procedure is more useful when one is aware of the degree of approximation used in the initial description of the waves. This awareness can best be achieved by first deriving the general equations of fluid mechanics, an approach that has the additional advantage of introducing the concepts and symbols needed to describe acoustic fields at a relatively slower pace. We will therefore take this more complete approach, and begin by presenting a short derivation of those equations. More detailed derivations can be found in textbooks dealing with that subject, such as those listed at the end of the chapter.

1.1 INDICIAL NOTATION

We will often be interested in describing some physical property at a given point in space. The coordinates of this point will be denoted by the components x_1, x_2, and x_3, with respect to a cartesian system of coordinates, of a position vector $\mathbf{x}$. Thus, if $\mathbf{e}_1$, $\mathbf{e}_2$, and $\mathbf{e}_3$ are unit vectors along these coordinate axes, then

$$\mathbf{x} = \mathbf{e}_1 x_1 + \mathbf{e}_2 x_2 + \mathbf{e}_3 x_3 \qquad (1.1.1)$$

The magnitude of the position vector will be denoted by r, where

$$r = |\mathbf{x}| = \sqrt{x_1^2 + x_2^2 + x_3^2} \qquad (1.1.2)$$

The unit vectors $\mathbf{e}_1$, $\mathbf{e}_2$, and $\mathbf{e}_3$ are orthonormal; that is, they are mutually orthogonal and have unit length. They therefore satisfy the following conditions:

$$\mathbf{e}_1 \cdot \mathbf{e}_2 = \mathbf{e}_1 \cdot \mathbf{e}_3 = \mathbf{e}_2 \cdot \mathbf{e}_3 = 0$$

$$\mathbf{e}_1 \cdot \mathbf{e}_1 = \mathbf{e}_2 \cdot \mathbf{e}_2 = \mathbf{e}_3 \cdot \mathbf{e}_3 = 1$$

These relationships can be written more succinctly by using the so-called indicial notation. In this notation, any component of a vector $\mathbf{A} = (A_1, A_2, A_3)$, for example, may be represented by the symbol A_i, where the index i runs through the values 1 to 3. Therefore, $\mathbf{A}$ may be expressed as

$$\mathbf{A} = \sum_{i=1}^{3} \mathbf{e}_i A_i \qquad (1.1.3)$$

This can be simplified even further by adopting the convention that if in an expression where indices are used, an index is repeated twice, a summation over the range of that index is implied. Thus, the scalar product between vectors $\mathbf{A}$ and $\mathbf{B}$,

$$\mathbf{A} \cdot \mathbf{B} = A_1 B_1 + A_2 B_2 + A_3 B_3 \qquad (1.1.4)$$

is simply represented by

$$\mathbf{A} \cdot \mathbf{B} = A_i B_i \qquad (1.1.5)$$

Of course, since the result of summing over a repeated index is independent of the symbol used for that index, we could as well have written $A_k B_k$. Indices that can thus be replaced are called *dummy* indices, and are useful in writing proper indicial expressions.

Using this notation, we can write the six relationships between the unit vectors
e simply as

$$e_i e_j = \delta_{ij}, \quad i, j = 1, 2, 3 \tag{1.1.6}$$

where the quantity δ_{ij}, known as Kronecker's delta, is defined as

$$\delta_{ij} = \begin{cases} 1, & i = j \\ 0, & i \neq j \end{cases} \tag{1.1.7}$$

The quantity δ_{ij} is an example of a type of quantity known as a second-order
tensor. To specify such a quantity, one requires $3 \times 3 = 9$ components, which
may be arranged in matrix form. Thus, the components of some tensor t_{ij} may
be represented by

$$\begin{bmatrix} t_{11} & t_{12} & t_{13} \\ t_{21} & t_{22} & t_{23} \\ t_{31} & t_{32} & t_{33} \end{bmatrix}$$

The sum of the diagonal elements of this matrix, known as the trace of the
matrix, may be obtained by setting $i = j$ in t_{ij}, that is,

$$t_{11} + t_{22} + t_{33} = t_{ii} \tag{1.1.8}$$

The operation of setting one index equal to another in an indicial expression is
known as *contraction*.

In some cases, the elements of a matrix that are symmetrically located with
respect to the diagonal are equal. In such cases, the matrix (and the tensor it
represents) is said to be *symmetric*. Symmetric tensors of second order satisfy the
symmetry condition

$$S_{ij} = S_{ji} \tag{1.1.9}$$

On the other hand, a tensor ξ_{ij} for which

$$\xi_{ij} = -\xi_{ji} \tag{1.1.10}$$

is said to be *antisymmetric*.

An arbitrary second-order tensor t_{ij} can be represented in terms of a symmetric part s_{ij} and an antisymmetric part a_{ij} as follows:

$$t_{ij} = s_{ij} + a_{ij} \tag{1.1.11}$$

where

$$s_{ij} = \tfrac{1}{2}(t_{ij} + t_{ji}) \tag{1.1.12}$$

$$a_{ij} = \tfrac{1}{2}(t_{ij} - t_{ji}) \tag{1.1.13}$$

Tensors of higher order may also be needed. For example, a tensor of third order is required to represent the cross product between two vectors. This is the alternating tensor ε_{ijk}, which is equal to zero unless $i, j,$ and k are all different, in which case its value is either $+1$ or -1 depending on whether $i, j,$ and k are in cyclic order. Thus, if $e_{123} = 1$, then

$$e_{123} = e_{231} = e_{312} = +1$$

$$e_{213} = e_{132} = e_{321} = -1$$

with the remaining 21 components all being equal to zero. In terms of e_{ijk}, the ith component of the cross product between vectors **A** and **B** can be written as

$$(\mathbf{A} \times \mathbf{B})_i = e_{ijk} A_j B_k \tag{1.1.14}$$

The following relations involving e_{ijk} are often useful:

$$e_{ijk} e_{ilm} = \delta_{jl}\delta_{km} - \delta_{jm}\delta_{kl}$$

$$e_{ijk} e_{ijm} = 2\delta_{km} \tag{1.1.15}$$

$$e_{ijk} e_{ijk} = 6$$

As an example of the usefulness of these expressions, let us derive the following vector identity:

$$\mathbf{A} \times (\mathbf{B} \times \mathbf{C}) = (\mathbf{A} \cdot \mathbf{C})\mathbf{B} - (\mathbf{A} \cdot \mathbf{B})\mathbf{C} \tag{1.1.16}$$

First, denote the left-hand side of this identity by **D**. Then, using the representation given by (1.1.14), the ith component of **D** may be expressed as

$$D_i = e_{ijk} A_j (\mathbf{B} \times \mathbf{C})_k \tag{1.1.17}$$

Again, the kth component of **B** × **C** may be written as $e_{kij} B_i C_j$, but if this quantity were to be used in the above equation, it would result in an indicial expression having indices repeated more than twice. Such expressions are not admissible, as they cannot be evaluated properly. To avoid this difficulty, we write, instead, the equivalent expression

$$(\mathbf{B} \times \mathbf{C})_k = e_{klm} B_l C_m \tag{1.1.18}$$

so that

$$D_i = e_{ijk} e_{klm} A_j B_l C_m \tag{1.1.19}$$

Because of its definition, the alternating-symbol tensor is not affected by an even number of permutations of its indices. Therefore, the first of these symbols on the right-hand side of (1.1.19) can be written as e_{kij}, so that using the first of the identities given by (1.1.15) we obtain

$$D_i = (\delta_{il}\delta_{jm} - \delta_{im}\delta_{jl}) A_j B_l C_m \tag{1.1.20}$$

Remembering the properties of the Kronecker delta, we can write this as

$$D_i = A_j B_i C_j - A_j B_j C_i \tag{1.1.21}$$

This is the indicial equivalent of the identity that was to be derived.

We should notice that each of the terms on the right-hand side of the last expression for D_i above contains a repeated index, and must therefore be summed over it. The other index, i, is not repeated, and is therefore a "free" index. If an indicial expression is properly written, the free indices in each term of the expression must be the same.

Finally, we may also use the indicial notation to simplify expressions involving spatial derivatives of various quantities. Thus, the ith component of the operator ∇ defined by

$$\nabla = \mathbf{e}_1 \frac{\partial}{\partial x_1} + \mathbf{e}_2 \frac{\partial}{\partial x_2} + \mathbf{e}_3 \frac{\partial}{\partial x_3} \tag{1.1.22}$$

is simply denoted by

$$(\nabla)_i = \frac{\partial}{\partial x_i} \tag{1.1.23}$$

Therefore, the ith component of the gradient of a scalar ϕ is

$$(\nabla\phi)_i = \frac{\partial\phi}{\partial x_i} \tag{1.1.24}$$

whereas the divergence of a vector $\mathbf{f}$ becomes

$$\nabla \cdot \mathbf{f} = \frac{\partial f_i}{\partial x_i} \tag{1.1.25}$$

and the ith component of **curl f** is given by

$$(\nabla \times \mathbf{f})_i = e_{ijk} \frac{\partial f_k}{\partial x_j} \tag{1.1.26}$$

PROBLEMS

1.1.1 Use the indicial notation to derive the following identities:

(a) $\nabla \times (\nabla \times \mathbf{u}) = -\nabla^2 \mathbf{u} + \nabla(\nabla \cdot \mathbf{u})$

(b) $\nabla \times (\mathbf{u} \times \mathbf{v}) = \mathbf{u}(\nabla \cdot \mathbf{v}) - \mathbf{v}(\nabla \cdot \mathbf{u}) + (\mathbf{v} \cdot \nabla)\mathbf{u} - (\mathbf{u} \cdot \nabla)\mathbf{v}$

(c) $\nabla \cdot (\mathbf{u} \times \mathbf{v}) = \mathbf{v} \cdot (\nabla \times \mathbf{u}) - \mathbf{u} \cdot (\nabla \times \mathbf{v})$

1.2 CONTINUUM MODEL

The basic equations of fluid mechanics are mathematical statements of the physical laws of conservation of mass, momentum, and energy. These statements are made in terms of macroscopic quantities such as density and velocity. It is possible to write the conservation laws in terms of suitable averages over molecular distribution functions, but for a wide variety of applications it is simpler to adopt the continuum description of matter whereby the molecular structure of the fluid is ignored.

An obvious limitation of the continuum model is that the smallest fluid elements to which it can be applied must contain a large number of molecules so that the matter inside it appears to be uniformly distributed. Stated differently, the size of the volume element cannot be arbitrarily small, for then the molecular nature of matter could not be ignored. Thus, when we mention the fluid density at a point $\mathbf{x}$, we actually mean the density of a small volume of fluid which on a macroscopic scale appears as a point, but which on a microscopic scale would be sufficiently large so as to contain many molecules. If such a volume is denoted by δV_0, then the density at $\mathbf{x}$, at time t, would be defined as

$$\rho(\mathbf{x}, t) = \lim_{\delta V \to \delta V_0} \frac{\delta m}{\delta V}$$

where δm is the fluid mass inside δV. Similar interpretation is implied when we talk about a fluid particle.

The conditions for the applicability of the continuum model may also be expressed by saying that the average distance between molecules must be very small compared to the macroscopic length scales of interest. In gases, a measure of this average distance is provided by the *mean free path*, which is the distance that, on the average, a molecule travels between collisions with other molecules. The ratio between the mean free path Λ and the length of interest l is called the Knudsen number:

$$Kn = \Lambda / l$$

Situations for which $Kn \ll 1$ are within the limits of continuum. For example, in air at standard temperature and pressure, the mean free path is of the order of 6×10^{-6} cm. A sound wave with a frequency equal to 20,000 Hz would have, under the same conditions, a wavelength equal to 1.7 cm, so that the Knudsen number would be $Kn = 3.5 \times 10^{-6}$. Lower frequencies would correspond to proportionately lower Knudsen numbers. Thus, audible acoustic waves fall well within the limits of the continuum model. However, for extremely high frequency waves, the wavelength may be comparable to the mean free path. In such a case, a molecular model for matter would have to be used.

In what follows, we will adopt the continuum hypothesis, although from time to time we may use the molecular model to clarify some concepts.

1.3 MACROSCOPIC THERMODYNAMICS

In addition to the density, we need other parameters to describe the local state of the fluid. These can be other macroscopic properties such as the internal energy per unit mass E, the pressure p, and so on. If the fluid is at rest, these are related by means of equilibrium equations of state. For example, at sufficiently low pressures, all gases obey equations of the form

$$p = R\rho T \qquad (1.3.1)$$

where R is the gas constant and T is the absolute temperature. Thus, for these gases, called perfect gases, the equilibrium pressure is seen to be a function of two other thermodynamic properties, namely the density and the temperature. For a more general fluid, the relationship between thermodynamic properties may not be as simple as that given by (1.3.1). However, it is an empirical fact that for simple fluids, that is, fluids whose composition is uniform, the number of independent properties needed to specify the state of the fluid is also two. Which two is immaterial. What matters is that if two properties are specified, then all other properties have fixed values. Thus, if the density and the internal energy per unit mass are chosen as the independent variables, then the pressure would have a value prescribed by

$$p = p(\rho, E) \qquad (1.3.2)$$

The form of the function $p(\rho, E)$ depends on the nature of the fluid and, for our purposes, it may be assumed known.

First Law

Suppose that the state of a given mass of fluid changes from one equilibrium state to another owing to the addition of an amount Q of heat per unit mass, and to the performance of an amount of work W, also per unit mass. Then, the first law of thermodynamics states that its internal energy per unit mass will change by an amount ΔE given by

$$\Delta E = Q + W \qquad (1.3.3)$$

The internal energy E is a property of state, so that ΔE depends only on the initial and final equilibrium states. On the other hand, the heat added and the work done are path functions, that is, their values depend on the actual path connecting the end points.

Equation (1.3.3), which is basically an energy conservation equation for the system under consideration, applies whether or not the transition between the two equilibrium states is made so slowly that the system passes through a sequence of equilibrium states during the transition and, also, whether or not there is dissipation. However, if the transition does in fact pass through a sequence of equilibrium states and if there is no dissipation associated with the addition of heat or the performance of work (i.e., if the transition is reversible), then one can express W (and also Q, as we will see later) in terms of equilibrium thermodynamic properties. For example, in a simple fluid the only reversible work mode is that of slow compression (or expansion). Therefore, the elemental work done on a unit mass of fluid by compressing it reversibly is

$$dW = -p \, dv^* \qquad (1.3.4)$$

where p is the thermodynamic pressure and $v^* = 1/\rho$ is the specific volume. Thus, for such a process, (1.3.3) yields

$$dQ = dE + p \, dv^* \qquad (1.3.5)$$

This equation can also be written in terms of the enthalpy per unit mass H, defined as

$$H = E + pv^* \qquad (1.3.6)$$

In terms of H, (1.3.3) can be written as

$$dQ = dH - v^* dp \tag{1.3.7}$$

Specific Heats

Equations (1.3.5) and (1.3.7) can be used to calculate the amounts of heat that are transferred to a system during a variety of processes. In this regard, one can define quantities that give an indication of the ability of a system to "heat up." Thus, if to produce a temperature change equal to δT, an amount of heat δQ has to be added, then the quantity

$$\delta T / \delta Q$$

is a measure of that ability. The inverse of this quantity is called the specific heat, and clearly depends on the type of process used to add heat. In view of (1.3.5) and (1.3.7), useful choices are the constant-volume process and the constant-pressure process. For these processes the specific heats are, respectively,

$$c_v = \left(\frac{\partial E}{\partial T} \right)_{v*} \tag{1.3.8}$$

and

$$c_p = \left(\frac{\partial H}{\partial T} \right)_{p} \tag{1.3.9}$$

These are called, respectively, the specific heat at constant volume and the specific heat at constant pressure. Their ratio

$$\gamma = c_p / c_v \tag{1.3.10}$$

is also of some importance, and as is shown in books on thermodynamics, its value is never less than unity.

For the special but important case of perfect gases, the internal energy and the enthalpy are functions of the temperature alone. Therefore, in view of their definitions, c_p and c_v are, then, also functions of the temperature alone. Furthermore, in the case when they can be considered constant, the internal energy and the enthalpy are given simply by

$$E - E_0 = c_v T, \qquad H - H_0 = c_p T \tag{1.3.11}$$

Where E_0 and H_0 are the values of E and H at some reference temperature T_0. Gases that obey (1.3.11) are called calorically perfect gases.

Second Law

Just as the first law of thermodynamics introduced the internal energy E, the second law introduces another state property, namely, the entropy, by stating that if an elemental amount of heat dQ is added reversibly to a system, then dQ is given by

$$dQ = T\,dS \tag{1.3.12}$$

where S is the entropy per unit mass. On the other hand, if the transition is not reversible, then

$$dQ < T\,dS \tag{1.3.13}$$

In the particular case of an isolated system undergoing some process, $\delta Q = 0$, so that

$$dS \geqslant 0 \tag{1.3.14}$$

where the equality sign applies if the process is reversible. Equation (1.3.14) is important, for it shows that the entropy of an isolated system cannot decrease. Later, in Section 1.6, we will derive the equivalent form of this statement for a system not in a state of uniform equilibrium.

Making use of (1.3.12), we can write from (1.3.5) and (1.3.7)

$$T\,dS = dE + p\,dv^* \tag{1.3.15}$$

and

$$T\,dS = dH - v^*\,dp \tag{1.3.16}$$

These equations may be used to relate state properties. Also, they may be applied to transformations provided that these are reversible.

Maxwell Relations

In order to compute entropy changes in a variety of processes, it is necessary to have equations in which the changes are expressed in terms of measurable quantities. This requires the use of some relationships between partial derivatives of thermodynamic properties. These relationships are known as Maxwell's thermodynamic relations. We illustrate their derivation by obtaining one of them, namely,

$$\left(\frac{\partial S}{\partial p}\right)_T = -\left(\frac{\partial v^*}{\partial T}\right)_p \tag{1.3.17}$$

The need for this relation arises when the entropy changes are to be computed in terms of temperature and pressure changes. Thus, consider the entropy as a function of temperature and pressure, that is, $S = S(T, p)$. This gives

$$dS = \left(\frac{\partial S}{\partial T}\right)_p dT + \left(\frac{\partial S}{\partial p}\right)_T dp \qquad (1.3.18)$$

To express dS in terms of measurable quantities, we must eliminate S from the two derivatives on the right-hand side. The first may be expressed simply as

$$\left(\frac{\partial S}{\partial T}\right)_p = \frac{c_p}{T} \qquad (1.3.19)$$

This relation can be established by considering the definition of c_p together with (1.3.16). The second derivative appears in the relation (1.3.17) that is to be derived. To do this, we first integrate (1.3.15) around a closed, reversible path, and obtain

$$\oint T \, dS = \oint p \, dv^* \qquad (1.3.20)$$

This is just the reversible version of the well-known fact that in a transformation that returns to the original state, the heat transferred to a system is equal to the work performed by the system. If the closed paths are drawn in the T, S and in the p, v^* planes, it then follows from (1.3.20) that the areas enclosed by the contours are equal. Mathematically, this means[1] that the Jacobian of the transformation from the variables T, S to the variables p, v^* is unity, that is,

$$\frac{\partial(T, S)}{\partial(p, v^*)} = 1 \qquad (1.3.21)$$

This result, together with the rules for manipulating Jacobians, is very useful in deriving Maxwell relations. Thus, to derive (1.3.17), we first express $(\partial S/\partial p)_T$ as

---

[1] Consider a closed contour in the x, y plane. Then the area enclosed by the contour is given by $A = \int \int dx \, dy$. Now, if x and y depend on two other variables, v and w, say, then in the v, w plane the transformed contour has an enclosed area given by

$$A' = \int \int dv \, dw = \int \int \frac{\partial(v, w)}{\partial(x, y)} dx \, dy$$

Hence, if it happens that the enclosed areas are equal, then it follows that the Jacobian of the transformation must be unity.

a Jacobian:

$$\left(\frac{\partial S}{\partial p}\right)_T = \frac{\partial(S,T)}{\partial(p,T)} \tag{1.3.22}$$

Now, interchanging S and T in the Jacobian changes its sign, or

$$\left(\frac{\partial S}{\partial p}\right)_T = -\frac{\partial(T,S)}{\partial(p,T)} \tag{1.3.23}$$

Also, in view of (1.3.21), this can be written as

$$\left(\frac{\partial S}{\partial p}\right)_T = -\frac{\partial(T,S)}{\partial(p,T)} \cdot \frac{\partial(p,v^*)}{\partial(T,S)} \tag{1.3.24}$$

The symbols $\partial(T,S)$ appearing in both numerator and denominator can be "cancelled," so that

$$\left(\frac{\partial S}{\partial p}\right)_T = -\frac{\partial(v^*,p)}{\partial(T,p)} = -\left(\frac{\partial v^*}{\partial T}\right)_p \tag{1.3.25}$$

as was to be shown. The other three Maxwell relations are

$$\left(\frac{\partial S}{\partial v^*}\right)_T = \left(\frac{\partial p}{\partial T}\right)_{v^*} \tag{1.3.26}$$

$$\left(\frac{\partial S}{\partial v^*}\right)_p = \left(\frac{\partial p}{\partial T}\right)_S \tag{1.3.27}$$

and

$$\left(\frac{\partial S}{\partial p}\right)_{v^*} = -\left(\frac{\partial v^*}{\partial T}\right)_S \tag{1.3.28}$$

These, like (1.3.25), can be used to express dS in terms of quantities that can be measured. Thus, for example, making use of (1.3.17) and (1.3.19), we can write (1.3.18) as

$$dS = \frac{c_p}{T}dT - \left(\frac{\partial v^*}{\partial T}\right)_p dp \tag{1.3.29}$$

This is the desired result because T, c_p, and $(\partial v^*/\partial T)_p$ can be measured.

Sometimes, instead of $(\partial v^* / \partial T)_p$, one uses the *coefficient* of *thermal expansion* β. This is defined as

$$\beta = \frac{1}{v^*} \left(\frac{\partial v^*}{\partial T} \right)_p \qquad (1.3.30)$$

and gives the relative change of volume, at constant pressure, due to a temperature change. In terms of β and c_p, (1.3.29) can be written as

$$T\,dS = c_p\,dT - Tv^*\beta\,dp \qquad (1.3.31)$$

Similarly, using $S = S(T, v^*)$ and one of the Maxwell relations together with $c_v = T(\partial S / \partial T)_{v*}$, we obtain

$$T\,dS = c_v\,dT + T\left(\frac{\partial p}{\partial T} \right)_{v*} dv^* \qquad (1.3.32)$$

Fluids in Motion

The above results apply to fluids in equilibrium. In a fluid in motion, however, there may exist gradients of density, pressure, and so on so that a truly uniform equilibrium state does not exist. Under these conditions, equilibrium thermodynamics is no longer applicable. However, for many situations those gradients are quite small, so that they can be ignored within sufficiently small regions. That is, locally the fluid can be considered in equilibrium, so that a moving fluid particle can be assumed to pass through a succession of equilibrium states to which the concepts and relationships of equilibrium thermodynamics apply.

However, there are situations in fluid mechanics where the spatial or temporal departures from equilibrium are significant, such as in a high-frequency sound wave, and it is then necessary to define thermodynamic properties in a manner that does not depend on complete equilibrium. We have already seen that the definition of density is applicable, whether or not equilibrium exists. Consider now the internal energy per unit mass. It is a state property whose existence was first recognized in conjunction with the first law. As noted earlier, the first law, as expressed by (1.3.3), defines ΔE in terms of the heat added Q and the work done W, regardless of how these transfers of energy take place. Now, Q and W are measurable quantities not depending on equilibrium. Therefore, we can use (1.3.3) to define the internal energy in a nonequilibrium situation, provided that the instantaneous value to which E refers is obtained by suddenly isolating the fluid element and allowing it to reach equilibrium. This definition of E is, in fact, not entirely different from the classical thermodynamic definition of internal energy. The main difference is that in this manner one can use the internal energy concept in a moving fluid.

Other thermodynamic quantities can now be defined, in terms of ρ and E, for a moving element by means of equilibrium equations of state of the form given by (1.3.3), provided that E is interpreted as above.

PROBLEMS

1.3.1   Derive the three Maxwell relations given by (1.3.26)–(1.3.28).

1.3.2   Show that for a general (but simple) substance,

$$c_p - c_v = -T\left(\frac{\partial v^*}{\partial T}\right)_p^2 \left(\frac{\partial p}{\partial v^*}\right)_T.$$

1.3.3   The isentropic and isothermal compressibilities are defined by

$$\kappa_s = -\frac{1}{v^*}\left(\frac{\partial v^*}{\partial p}\right)_s, \qquad \kappa_T = -\frac{1}{v^*}\left(\frac{\partial v^*}{\partial p}\right)_T$$

  (a)   Show that $\kappa_s = \gamma \kappa_T$.

  (b)   Determine $(\rho \kappa_s)^{-1/2}$ for a perfect gas. Compute the value of this quantity for nitrogen at STP.

1.3.4   A more elementary derivation of the Maxwell relations requires the expressions for dU and dH given by (1.3.15) and (1.3.16), respectively, together with two similar equations for the differentials of the Gibbs function $G = H - TS$ and of the Helmholtz function $F = U - TS$. All of these differentials are of the form $dz = M\,dx + N\,dy$, and are exact; that is, the coefficients M and N are such that $(\partial M/\partial x)_y = (\partial N/\partial y)_x$. Use this property to derive the four Maxwell relations.

1.4   CONSERVATION OF MASS

We now proceed to derive the equations of fluid mechanics, and begin with the "continuity equation." This is merely an equation that states, mathematically, that mass is conserved, and will be derived in the simplest manner. Consider a region of space filled with fluid, and isolate mentally a finite volume V by means of a fixed control volume as shown in Figure 1.4.1. The control volume is bounded by surface A, which has an outward unit normal vector denoted by $\mathbf{n}$. At any time t, the total mass of fluid inside V is

$$\int_V \rho(\mathbf{x}, t)\, dV(\mathbf{x})$$

As time passes, this amount may change owing to a net mass flow out of V.

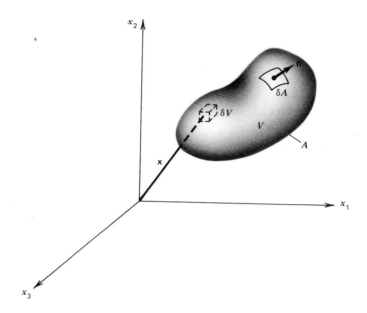

Figure 1.4.1   Control volume V fixed in space.

Thus, if $\mathbf{u}(\mathbf{x}, t)$ is the fluid velocity at a point $\mathbf{x}$, the net mass flow rate out of V is

$$\int_A \rho(\mathbf{x}, t)\mathbf{u}(\mathbf{x}, t) \cdot \mathbf{n}(\mathbf{x}) \, dA(\mathbf{x})$$

Therefore, mass conservation requires that

$$\frac{d}{dt} \int_V \rho \, dV(\mathbf{x}) = - \int_A \rho \mathbf{u} \cdot \mathbf{n} \, dA \qquad (1.4.1)$$

Now, since the volume V is fixed in space, that is, it does not depend on time, we can write

$$\frac{d}{dt} \int_V \rho \, dV(\mathbf{x}) = - \int_V \frac{\partial \rho}{\partial t} \, dV$$

Now, by the divergence theorem,

$$\int_A \rho \mathbf{u} \cdot \mathbf{n} \, dA(\mathbf{x}) = \int_V \nabla \cdot (\rho \mathbf{u}) \, dV(\mathbf{x})$$

so that our balance equation can be written as

$$\int_V \left[\frac{\partial \rho}{\partial t} + \nabla \cdot (\rho \mathbf{u}) \right] dV(\mathbf{x}) = 0$$

This result must hold for all choices of V. Therefore, the integrand must vanish identically at all points in the fluid. Hence

$$\frac{\partial \rho}{\partial t} + \nabla \cdot (\rho \mathbf{u}) = 0 \tag{1.4.2}$$

This is the required equation. It can be written in several other useful forms, but this form is sufficient for our purposes.

PROBLEMS

1.4.1 In the derivation of the continuity equation, it was implicitly assumed that there were no sources or sinks of fluid inside V. Suppose, now, that at every point $\mathbf{x}$ inside V, fluid mass is added at a rate $\rho Q(\mathbf{x}, t)$ per unit volume of fluid. Show that the appropriate continuity equation is

$$\frac{\partial \rho}{\partial t} + \nabla \cdot (\rho \mathbf{u}) = \rho Q(\mathbf{x}, t)$$

1.4.2 Show that the equation of continuity can be written as

$$\frac{\partial \rho}{\partial t} + \frac{\partial}{\partial r}(\rho u) + n\frac{\rho u}{r} = 0$$

where $n = 0$, 1, and 2 for plane flow, cylindrical flow, and spherical flow, respectively.

1.5 EQUATION OF MOTION

This equation is based on Newton's second law and is derived easily by considering a given fluid element in motion. That is, instead of considering a fixed geometrical volume V, we now consider a volume τ of fluid moving in such a manner that it always contains the same fluid particles; that is, τ is a material volume in motion. Now, the momentum of the fluid inside τ is

$$\int_{\tau(t)} \rho(\mathbf{x}, t)\, \mathbf{u}(\mathbf{x}, t)\, d\tau$$

and will change at a rate that is equal to the total forces applied to the element as it moves. These forces are of two kinds: *body forces* and *surface forces*. Body forces are long-range forces such as gravity and are proportional to the mass of fluid under consideration. Thus, if $\mathbf{F}$ denotes the body force per unit mass, the total body force acting on τ will be

$$\int_\tau \rho \mathbf{F} \, d\tau$$

On the other hand, the surface forces are short-range forces that depend on the surface of the element and on its orientation in space. Thus, if $\mathbf{\Sigma}(\mathbf{n})$ represents the local stress, the total surface force acting on τ will be

$$\int_S \mathbf{\Sigma} \, dS$$

where S is the surface bounding τ, as shown in Figure 1.5.1. As indicated in the figure, $\mathbf{\Sigma}$ is not, in general, along the unit normal.

Equating the total force on the element to the rate of change of momentum, we obtain

$$\frac{d}{dt} \int_\tau \rho \mathbf{u} \, d\tau = \int_\tau \rho \mathbf{F} \, d\tau + \int_S \mathbf{\Sigma} \, dS \qquad (1.5.1)$$

To reduce this statement to a field equation similar to (1.5.2), we must take the time derivative of the integral on the left-hand side and transform the surface integral on the right-hand side into a volume integral. The integral on the left-hand side has three scalar components. Each of these is a particular example of the more general integral

$$\frac{d}{dt} \int_{\tau(t)} \rho \theta \, d\tau$$

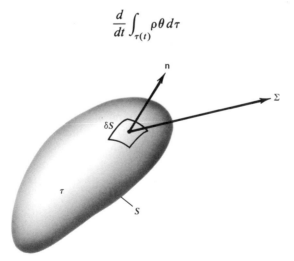

Figure 1.5.1 Surface forces acting on a moving fluid element.

where $\theta = \theta(\mathbf{x}, t)$ represents an arbitrary scalar property of the fluid. One contribution to the time derivative is due to local changes of $\rho\theta$ with time:

$$\int_\tau \frac{\partial(\rho\theta)}{\partial t} \, d\tau$$

A second contribution arises from the changes of volume τ that the fluid parcel experiences as it travels, and can be expressed as

$$\int_S \rho\theta\mathbf{u}\cdot\mathbf{n} \, dS$$

Thus, for a scalar fluid property θ, we have, using the divergence theorem,

$$\frac{d}{dt}\int \rho\theta \, d\tau = \int \left[\frac{\partial(\rho\theta)}{\partial t} + \nabla\cdot(\rho\theta\mathbf{u}) \right] d\tau \tag{1.5.2}$$

For example, if $\theta = 1$ we obtain, using the continuity equation,

$$\frac{d}{dt}\int \rho \, d\tau = \frac{dm}{dt} = 0$$

where m is the fluid mass inside τ. This simply expresses the fundamental fact that the mass of a given fluid body remains constant as it moves.

We now return to (1.5.1) and apply (1.5.2) to the ith component of the integral on the left-hand side. Thus, we let θ be equal to u_i and obtain

$$\frac{d}{dt}\int_\tau \rho u_i \, d\tau = \int_\tau \left[\frac{\partial(\rho u_i)}{\partial t} \, d\tau + \frac{\partial}{\partial x_j}(\rho u_i u_j) \right] d\tau \tag{1.5.3}$$

Consider now the integrals on the right-hand side of (1.5.1). One of them is an area integral and must be transformed into a volume integral. To do this, we first note, without proof, that the ith component of the local surface force per unit area is related to the unit normal vector by means of

$$\Sigma_i = \sigma_{ij}n_j \tag{1.5.4}$$

where σ_{ij} is the stress tensor. The meaning of a particular component of the stress tensor, say the i, j component, is that it represents the force per unit area parallel to the ith axis (first subindex) acting on a plane that is perpendicular to the jth axis (second subindex). It may also be shown that because of equilibrium conditions the stress tensor is symmetric; that is,

$$\sigma_{ij} = \sigma_{ji} \qquad (1.5.5)$$

Thus, if we consider an infinitesimal element of fluid, of rectangular shape, the surface forces acting on three of its faces would have the values shown in Figure 1.5.2. The convention used for the surface forces should be noted. In this convention, a pure tension is positive.

Returning now to the surface integral on the right-hand side of (1.5.1), we use (1.5.4) and the divergence theorem to obtain

$$\int_S \Sigma_i \, dS = \int_S \sigma_{ij} n_j \, dS = \int_\tau \frac{\partial \sigma_{ij}}{\partial x_j} \, d\tau$$

Substituting this result and that given by (1.5.3) into the ith component or our momentum-balance equation, we obtain

$$\int_\tau \left[\frac{\partial(\rho u_i)}{\partial t} + \frac{\partial}{\partial x_j}(\rho u_i u_j) - \frac{\partial \sigma_{ij}}{\partial x_j} - \rho F_i \right] d\tau = 0$$

Again, this result must hold for all τ; therefore, the integrand must vanish at

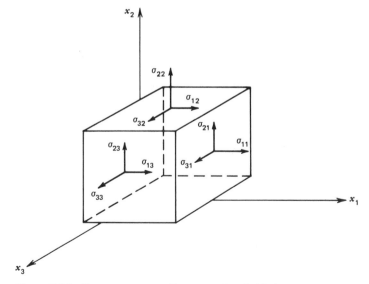

Figure 1.5.2 Stresses on a small rectangular fluid element.

every point in the fluid, or

$$\frac{\partial(\rho u_i)}{\partial t} + \frac{\partial}{\partial x_j}(\rho u_i u_j) = \rho F_i + \frac{\partial \sigma_{ij}}{\partial x_j} \qquad (1.5.6)$$

This is one form of the desired equation of motion. Another can be obtained from it by using the equation of continuity. Thus,

$$\frac{\partial u_i}{\partial t} + u_j \frac{\partial u_i}{\partial x_j} = F_i + \frac{1}{\rho}\frac{\partial \sigma_{ij}}{\partial x_j} \qquad (1.5.7)$$

We note that the two derivatives appearing on the left-hand side of (1.5.7) have a simple meaning. Their combination represents, in the field of Eulerian description, the acceleration of a fluid particle. If instead of using this description we follow the motion of a given particle, then its acceleration will be simply $d\mathbf{v}/dt$, where $\mathbf{v}$ is the velocity of the particle. In the Eulerian description, however, velocities of individual particles are not specified. Instead, one specifies the velocity at every point in space. This velocity may change from point to point and, also, it may vary with time. Therefore, the acceleration of a fluid particle is made of two parts: local changes and convective changes. These are given, respectively, by $\partial\mathbf{u}/\partial t$ and $\mathbf{u}\cdot\nabla\mathbf{u}$. The last change represents the changes of velocity in space, in the direction of $\mathbf{u}$. Therefore, the operator

$$\frac{D}{Dt} = \frac{\partial}{\partial t} + \mathbf{u}\cdot\nabla$$

gives, when applied to some fluid property, the time rate of change of that property following the motion of a fluid particle. Thus, for example, DE/Dt will represent the rate of change of the internal energy per unit mass of a fluid particle. Similar meaning can be attached to DS/DT, DT/Dt, and so on.

Stress Tensor

Equations (1.5.6) and (1.5.7) are different forms of the desired equation of motion. Neither one is in a useful form because the stress tensor has not yet been specified. As is pointed out in fluid mechanics books, it is not possible to obtain a single stress tensor that is applicable to all fluids. Here we present without derivation explicit forms of σ_{ij} applicable to two important types of fluids.

Ideal Fluids. As their name implies, these fluids represent idealizations of real-fluid behavior. Their stress tensor can be obtained from that for more general fluids. It is, however, useful to consider ideal fluids first, particularly because they play an important role in acoustics.

Basically, the stress tensor for an ideal fluid in motion has the same form as the stress tensor for a fluid at rest. That is, the forces that act on a small element of fluid are all normal to the surface of the element. No tangential stresses occur. Further, at any given point, these forces have the same magnitude independent of direction. Thus, for example, if the fluid element were a sphere, its shape would remain unchanged under the effects of such a stress. Finally, the pressure is taken to be identical to the thermodynamic pressure $p = p(\rho, E)$. In view of these assumptions, the surface force Σ acts along the normal, and its magnitude is independent of direction. Thus,

$$\Sigma = -p\mathbf{n} \tag{1.5.8}$$

where p is the pressure and where the negative sign is due to the adopted convention on the surface forces. Also, because of (1.5.4), we have for the stress tensor

$$\sigma_{ij} = -p\delta_{ij} \tag{1.5.9}$$

The components of this isotropic stress tensor are given by the matrix

$$\begin{bmatrix} -p & 0 & 0 \\ 0 & -p & 0 \\ 0 & 0 & -p \end{bmatrix}$$

We notice that one could define the mechanical pressure in terms of the stress tensor. Thus, by contracting σ_{ij} we obtain from (1.5.9)

$$p = -\tfrac{1}{3}\sigma_{ii} \tag{1.5.10}$$

where $\sigma_{ii} = \sigma_{11} + \sigma_{22} + \sigma_{33}$ is the trace of σ_{ij}. In the ideal fluid case this definition is unnecessary, as the trace elements, as with all isotropic sensors, are equal. However, this is not so for real fluids in motion, and to preserve the elementary notion of mechanical pressure in such conditions, the above definition for it is used.

Finally, substituting $\sigma_{ij} = -p\delta_{ij}$ into (1.5.7), we obtain, in vector notation,

$$\rho\left(\frac{\partial \mathbf{u}}{\partial t} + \mathbf{u}\cdot\nabla\mathbf{u}\right) = \rho\mathbf{F} - \nabla p \tag{1.5.11}$$

This is the equation of motion for an ideal fluid. It is also known as *Euler's equation* of motion and, in linearized form, plays an important role in ideal acoustics.

Newtonian Fluids. It is clear that in real fluids there will be, in addition to normal stresses, tangential stresses. That is, the stress tensor will no longer be purely isotropic, but will contain a deviatoric part that is due entirely to the appearance of velocity gradients (i.e., which vanishes when the fluid is at rest). Therefore, in a real fluid we can write

$$\sigma_{ij} = -P\delta_{ij} + d_{ij} \tag{1.5.12}$$

where the nonisotropic tensor d_{ij} is called the deviatoric stress tensor and is chosen such that $d_{ii} = 0$. The quantity P, defined as

$$P = -\tfrac{1}{3}\sigma_{ii} \tag{1.5.13}$$

is invariant under rotation of the axis of reference (because σ_{ii} is one of the three invariants of the matrix whose elements are σ_{ij}). Therefore, P does not depend on the orientation of the surface element. Also, it may be shown that the area average of the total normal force acting on a small spherical volume of fluid is just $-\tfrac{1}{3}(\sigma_{ii})$. Further, as we stated earlier, this quantity is a definition of pressure in an ideal fluid, and also in a fluid at rest. Therefore, the quantity P as given by (1.5.13) is taken as the definition of the pressure for a fluid in motion. Of course, this definition is purely mechanical so that, in general, P is not equal to the thermodynamic pressure p.

The differences between P and p, as well as the form of the *deviatoric* stress tensor, may be deduced when the velocity gradients are small, provided that the molecular structure of the fluid is isotropic. In such conditions, the deviatoric stress tensor is given by

$$d_{ij} = 2\mu\left(e_{ij} - \tfrac{1}{3}\Delta\delta_{ij}\right) \tag{1.5.14}$$

where μ is the coefficient of viscosity of the fluid (also called the shear viscosity for reasons that will be given below) and e_{ij} is the rate-of-strain tensor and is defined by

$$e_{ij} = \tfrac{1}{2}\left(\frac{\partial u_i}{\partial x_j} + \frac{\partial u_j}{\partial x_i}\right) \tag{1.5.15}$$

This definition shows that e_{ij} is a symmetric tensor. The quantity Δ is called the local rate of expansion. It is defined as $\Delta = e_{ii}$, or

$$\Delta = \nabla \cdot \mathbf{u} \tag{1.5.16}$$

Reference to (1.4.2) shows that Δ is equal to $-(1/\rho)\,D\rho/Dt$, a quantity that gives the relative change in volume of a fluid particle as it moves.

Fluids for which the deviatoric stress tensor is given by (1.5.14) are called *Newtonian*. Most gases and some simple liquids such as water appear to follow such a relationship quite well. On the other hand, liquids with complicated molecular structure do not. In this book, we consider only Newtonian fluids, in which case the stress tensor is

$$\sigma_{ij} = -P\delta_{ij} + 2\mu\left(e_{ij} - \tfrac{1}{3}\Delta\delta_{ij}\right) \tag{1.5.17}$$

Before considering the fluid-dynamic pressure, we illustrate some of the effects of a nonzero deviatoric stress tensor by means of two simple examples. In the first example, we consider a fluid that is everywhere moving steadily along the x_1 axis, say, with a velocity that varies with x_2. That is, the velocity vector **u** has components $[u_1(x_2), 0, 0]$ so that the fluid particles move in layers having different velocities, as sketched in Figure 1.5.3. For this motion, (1.5.15) shows that the rate of strain tensor has components

$$\begin{bmatrix} 0 & \tfrac{1}{2}(du_1/dx_2) & 0 \\ \tfrac{1}{2}(du_1/dx_2) & 0 & 0 \\ 0 & 0 & 0 \end{bmatrix}$$

Therefore, the only nonzero components of the deviatoric part of the stress tensor are the tangential stresses σ_{12} and $\sigma_{21} = \sigma_{12}$, where

$$\sigma_{12} = \mu\frac{du_1}{dx_2} \tag{1.5.18}$$

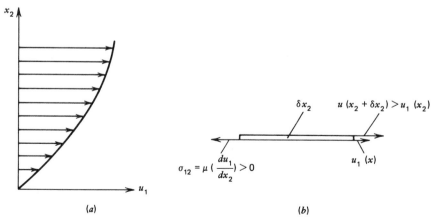

(a)   (b)

Figure 1.5.3 Simple shearing motion. (a) Velocity profile; (b) forces on a fluid layer.

Provided that μ is positive,[2] (1.5.18) represents a force that acts to eliminate velocity differences between contiguous fluid layers, that is, a frictional force. Because of its mode of action, σ_{12} is called a shearing stress. Similarly, μ is called the coefficient of shear viscosity, although it also plays a role in nonsheared motions. This may be seen in the second example, where we consider a flow in which the velocity vector is $\mathbf{u} = [u_1(x_1), 0, 0]$. In this case, the only nonzero component of e_{ij} is

$$e_{11} = du_1 / dx_1$$

so that the only force that acts on a fluid element moving in the above manner is the normal stress

$$\sigma_{11} = -P + \tfrac{4}{3}\mu\frac{du_1}{dx_1} \tag{1.5.19}$$

To fix ideas, suppose that P is identical to the equilibrium pressure. Then comparison with the ideal-fluid case shows that the first part of σ_{11} is an elastic force involving no frictional resistance. On the other hand, the second contribution, namely, $\tfrac{4}{3}\mu(du_1/dx_1)$, is frictional. At the molecular level this force arises because there is a net flow of x_1 momentum across any plane that is perpendicular to the x_1 axis and that moves with the local velocity u_1. This flow is due to random molecular motions. Thus, if the region on the left of the plane has a larger macroscopic velocity (so that a nonequilibrium situation exists in the vicinity of the plane), then molecules leaving the left side will have, on the average, larger x_1 component velocity than those entering from the right side. One of the results of this net change of momentum is a force that tends to eliminate the velocity differences, that is, that tends to restore equilibrium.

Similar arguments apply to the first example. We should point out, however, that the only molecular feature that was used to explain the origin of the frictional force was the translational motion of the molecules. In fact, we can say that it is the adjustment of the translational degrees of freedom that gives rise to the macroscopic phenomenon of shear viscosity. This type of molecular adjustment is sometimes referred to as *translational relaxation*. Now, in monatomic fluids, that is, in helium, argon, mercury, and so forth, there are no other molecular degrees of freedom so that small departures from equilibrium are dealt with by μ via the Newtonian expression for the deviatoric stress tensor.

This brings us to the question as to what happens in more complex fluids such as, for example, diatomic gases. As is well known, their molecules have not only translational degrees of freedom, but also rotational and vibrational. The energies associated with these degrees of freedom are also variable. Thus, in an

---

[2] It will be seen in Section 1.6 that the second law of thermodynamics requires that μ be positive.

equilibrium state, these energies will have equilibrium values corresponding to the macroscopic conditions of the fluid. Now, suppose that a fluid element is momentarily thrown out of equilibrium, say by means of a sudden expansion. Immediately after, the energies of the translational, rotational, and vibrational degrees of freedom will have values that differ from those corresponding to the new macroscopic conditions. Through molecular collisions, however, each one of the molecular degrees of freedom will eventually adjust to the new conditions, the number of collisions required in each case being generally different. The translational degrees require the least number to reach their equilibrium values. Therefore, by the time the translational energies have reached their new equilibrium values, the rotational and vibrational degrees are still in the process of adjustment. The net result of this lag is that the thermodynamic pressure differs from the mechanical pressure. In fact, it may be shown that during a rapid expansion, the thermodynamic pressure remains slightly larger than the mechanical pressure, and that, at least for small departures from equilibrium, the differences between P and p are proportional to the rate of expansion. Therefore, we write

$$P - p = \mu_v \Delta \qquad\qquad (1.5.20)$$

where the proportionality coefficient is called the *expansion coefficient of viscosity* and, like μ, is positive.

A simple example will illustrate the effects of a finite expansive viscosity. Suppose that we have a spherical-volume element of fluid, expanding (or contracting) in such a way that its shape is preserved. This isotropic expansion requires that $\mathbf{u} = [u_r(r), 0, 0]$ so that all the nondiagonal components of the rate-of-strain tensor are zero. Further, owing to the symmetry of the motion, the diagonal elements are all equal, so that in this case $e_{ij} = \frac{1}{3}(\Delta \delta_{ij})$. But for this value of e_{ij}, the deviatoric stress tensor vanishes identically as (1.5.14) shows. The stress tensor then becomes

$$\sigma_{rr} = -p + \mu_v \Delta$$

This shows that, in effect, the expansive viscosity also results in a frictional force, and also tends to restore equilibrium. This time, however, these effects are due to relaxation of the vibrational and rotational degrees of freedom.

One of the problems with the expansive viscosity μ_v is that very little is known about it. The reason for this is that it is very difficult to measure its effects separately from those of μ. There are, however, some experiments in which the combined effects of μ and μ_v seem to indicate that for N_2, the expansive viscosity is of the same order of magnitude as the shear viscosity. These experiments consist of sound-absorption measurements at high frequency and will be discussed in Chapter 6.

Navier–Stokes Equation

Now that we have specified the form of the stress tensor for a Newtonian fluid, we can write the equation of motion (1.5.7) in the form

$$\rho \frac{Du_i}{Dt} = \rho F_i - \frac{\partial P}{\partial x_i} + \frac{\partial}{\partial x_j} \left[2\mu \left(e_{ij} - \tfrac{1}{3}\Delta\delta_{ij} \right) \right] \qquad (1.5.21)$$

This is referred to as the *Navier–Stokes equation*. It describes the motion of Newtonian fluids in widely varied conditions. These might involve significant temperature changes, in which case μ (which depends strongly on the temperature) must be regarded as a function of position. However, in many cases, temperature variations are small so that μ can be considered constant. The case of acoustic motions is an example. For such a case, (1.5.21) becomes

$$\rho \frac{Du_i}{Dt} = \rho F_i - \frac{\partial P}{\partial x_i} + \mu \left(\frac{\partial^2 u_i}{\partial x_j \partial x_j} + \tfrac{1}{3}\frac{\partial \Delta}{\partial x_i} \right) \qquad (1.5.22)$$

Finally, using (1.5.20) and assuming that μ_v is also constant, we can write (1.5.22) as

$$\rho \frac{D\mathbf{u}}{Dt} = \rho \mathbf{F} - \nabla p + \mu \left[\nabla^2 \mathbf{u} + \left(\tfrac{1}{3} + \frac{\mu_v}{\mu} \right) \nabla(\nabla \cdot \mathbf{u}) \right] \qquad (1.5.23)$$

where we have used (1.5.16) to eliminate Δ. For an ideal fluid, this reduces to Euler's equation, which is given by (1.5.11).

PROBLEMS

1.5.1 Consider a small spherical element of fluid undergoing purely radial expansions and contractions. Show that for such a motion, the deviatoric stress tensor vanishes.

1.5.2 Show that for one-dimensional motion, Euler's equation can be written as

$$\rho \left(\frac{\partial u}{\partial t} + u\frac{\partial u}{\partial r} \right) = -\frac{\partial p}{\partial r}$$

regardless of whether the motion is plane, cylindrical, or spherical.

1.6 THE ENERGY EQUATION

We now consider the internal energy of a fluid element in motion. This energy may change owing to work done on the element by surface and body forces and to heat conduction from the surrounding fluid. The rate at which work is done on the element by body forces is

$$\int \rho F_i u_i \, d\tau$$

and that by surface forces is

$$\int_S \Sigma_i u_i \, dS = \int_S \sigma_{ij} n_j u_i \, dS = \int_\tau \frac{\partial(u_i \sigma_{ij})}{\partial x_j} \, d\tau \tag{1.6.1}$$

Therefore, the rate at which work is done on the element is

$$\int_\tau \left[\rho F_i u_i + u_i \frac{\partial \sigma_{ij}}{\partial x_j} + \sigma_{ij} \frac{\partial u_i}{\partial x_j} \right] d\tau$$

However, from (1.5.7) we see that

$$\rho F_i u_i + u_i \frac{\partial \sigma_{ij}}{\partial x_j} = \rho u_i \frac{Du_i}{Dt} \tag{1.6.2}$$

The right-hand side of this represents the increase of kinetic energy of the element. Consequently, the two terms on the left-hand side represent the fraction of the total work performed on the fluid that was used to increase the kinetic energy of the fluid element. These terms should, therefore, not appear on an equation describing the balance of the internal energy of a given body of fluid. On the other hand, the remaining term, namely,

$$\int \sigma_{ij} \frac{\partial u_i}{\partial x_j} \, d\tau$$

represents the rate at which work is done in deforming the element without changing its velocity. Therefore, the rate at which the internal energy per unit mass of the element increases due to this account is

$$\frac{1}{\rho} \sigma_{ij} \frac{\partial u_i}{\partial x_j}$$

The internal energy may also increase owing to a net gain of heat. If the only mode by which heat is gained is conduction and if the temperature gradients are small, then the heat flux will be given by Fourier's law:

$$\dot{\mathbf{q}}_h = k \nabla T \tag{1.6.3}$$

where k is the thermal conductivity of the fluid. Therefore, the rate at which heat enters the fluid element across its surface will be

$$\dot{Q} = \int_S k n_j \frac{\partial T}{\partial x_j} \, dS = \int_\tau \frac{\partial}{\partial x_j} \left(k \frac{\partial T}{\partial x_j} \right) d\tau \tag{1.6.4}$$

and the rate at which heat is gained by the fluid element per unit mass of fluid is

$$\frac{1}{\rho} \frac{\partial}{\partial x_j} \left(k \frac{\partial T}{\partial x_j} \right)$$

Finally, the rate of change of the internal energy per unit mass of the fluid element is DE/Dt, so that the balance equation is

$$\frac{DE}{Dt} = \frac{\sigma_{ij}}{\rho} \frac{\partial u_i}{\partial x_j} + \frac{1}{\rho} \frac{\partial}{\partial x_j} \left(k \frac{\partial T}{\partial x_j} \right) \tag{1.6.5}$$

Now, since the stress tensor is symmetric and since $\sigma_{ij}(\partial u_i/\partial x_j) = \sigma_{ji}(\partial u_j/\partial x_i)$ (i and j are dummy), we have

$$2\sigma_{ij} \frac{\partial u_i}{\partial x_j} = \sigma_{ij} \left(\frac{\partial u_i}{\partial x_j} + \frac{\partial u_j}{\partial x_i} \right)$$

Thus,

$$\frac{DE}{Dt} = \frac{\sigma_{ij} e_{ij}}{\rho} + \frac{1}{\rho} \frac{\partial}{\partial x_j} \left(k \frac{\partial T}{\partial x_j} \right) \tag{1.6.6}$$

where e_{ij} is given by (1.5.15).

Now, the stress tensor is given by (1.5.17). Substituting this value in (1.6.6), we obtain

$$\frac{DE}{Dt} = -\frac{P}{\rho} \Delta + \frac{2\mu}{\rho} \left(e_{ij} e_{ij} - \tfrac{1}{3} \Delta^2 \right) + \frac{1}{\rho} \frac{\partial}{\partial x_j} \left(k \frac{\partial T}{\partial x_j} \right) \tag{1.6.7}$$

Furthermore, from (1.5.20) $P = p - \mu_v \Delta$, where the expansion rate $\Delta = \nabla \cdot \mathbf{u}$ may be expressed as

$$\Delta = \rho \frac{D(1/\rho)}{Dt} \tag{1.6.8}$$

Therefore, instead of (1.6.7) we can write

$$\frac{DE}{Dt} + p\frac{D(1/\rho)}{Dt} = \frac{\mu_v}{\rho}\Delta^2 + \Phi + \frac{1}{\rho}\frac{\partial}{\partial x_j}\left(k\frac{\partial T}{\partial x_j}\right) \tag{1.6.9}$$

where the nonnegative quantity Φ is given by

$$\Phi = \frac{2\mu}{\rho}\left(e_{ij}e_{ij} - \tfrac{1}{3}\Delta^2\right) \tag{1.6.10}$$

Because of the role it plays, this quantity is called the *viscous dissipation function*. Now, by (1.3.15), the left-hand side of (1.6.9) is just $T(DS/Dt)$. Hence

$$T\frac{DS}{Dt} = \frac{\mu_v}{\rho}\Delta^2 + \Phi + \frac{1}{\rho}\nabla\cdot(k\nabla T) \tag{1.6.11}$$

This is the required equation. Its left-hand side can also be written by means of (1.3.29) as

$$T\frac{DS}{Dt} = c_p\frac{DT}{Dt} - \frac{\beta T}{\rho}\frac{Dp}{Dt} \tag{1.6.12}$$

where the coefficient of thermal expansion β is defined by (1.3.30).

The heat-conduction term on the right-hand side of (1.6.9) can be expressed differently by means of the following identity:

$$\frac{1}{T}\nabla\cdot(k\nabla T) = \frac{k}{T^2}(\nabla T)^2 + \nabla\cdot\left(\frac{k\nabla T}{T}\right) \tag{1.6.13}$$

If this is used in (1.6.11), then we obtain

$$T\left[\frac{DS}{Dt} - \frac{1}{\rho}\nabla\cdot\left(\frac{k}{T}\nabla T\right)\right] = \Phi + \frac{\mu_v}{\rho}\Delta^2 + \frac{k}{\rho}\frac{(\nabla T)^2}{T} \tag{1.6.14}$$

This equation has a form that makes it suitable to comparison with the second law of thermodynamics.

Second Law for a Continuum

For a continuum, the second law can be derived in the following manner. Consider a fixed control volume V inside a continuum material such as a fluid. The entropy inside V changes with time at a rate given by

$$\frac{d}{dt} \int \rho S \, dV$$

This rate must be balanced by (a) the net flow of entropy out of V and (b) the rate at which entropy is produced in V. Thus, if we denote by $\mathbf{J}_s$ the *entropy flux* (entropy flow out per unit area and unit time) and by $\rho\sigma$ the *entropy production* in V per unit volume in unit time, then we have

$$\frac{d}{dt} \int_V \rho S \, dV = -\int_A \mathbf{J}_s \cdot \mathbf{n} \, dA + \int_V \rho\sigma \, dV \qquad (1.6.15)$$

Therefore, transforming the area integral into a volume integral and taking the time derivative inside the volume integral, we obtain

$$\int_V \left[\frac{\partial(\rho S)}{\partial t} + \nabla \cdot \mathbf{J}_s - \rho\sigma \right] dV = 0 \qquad (1.6.16)$$

or

$$\frac{\partial(\rho S)}{\partial t} + \nabla \cdot \mathbf{J}_s - \rho\sigma = 0 \qquad (1.6.17)$$

The second law of thermodynamics states that the entropy production in V cannot be negative, that is, we must have

$$\sigma = \frac{1}{\rho}\left[\frac{\partial(\rho S)}{\partial t} + \nabla \cdot \mathbf{J}_s \right] \geq 0 \qquad (1.6.18)$$

Now, using the equation of continuity to evaluate the time derivative, we obtain

$$\frac{\partial(\rho S)}{\partial t} = \rho\frac{DS}{Dt} - \nabla \cdot (\rho\mathbf{u}S) \qquad (1.6.19)$$

and so

$$\sigma = \frac{DS}{Dt} + \frac{1}{\rho}\nabla \cdot (\mathbf{J}_s - \rho\mathbf{u}S) \geq 0 \qquad (1.6.20)$$

However, for the fluid continuum we are considering, the only manner in which energy can leave a given region is by direct convection and by heat conduction. Therefore, the entropy flux vector is given by

$$\mathbf{J}_s = \rho \mathbf{u} S - \frac{k}{T} \nabla T \qquad (1.6.21)$$

A negative sign in front of the heat conduction flux appears because of the sign convention used earlier [see (1.6.3)]. Therefore, (1.6.20) becomes

$$\sigma = \frac{DS}{Dt} - \frac{1}{\rho} \nabla \cdot \left(\frac{k}{T} \nabla T \right) \geq 0 \qquad (1.6.22)$$

Comparing this equation with (1.6.14) shows that the left-hand side of that equation is just $T\sigma$. Therefore, we find that

$$\sigma = \frac{1}{T} \Phi + \frac{1}{T} \frac{\mu_v}{\rho} \Delta^2 + \frac{k}{\rho} \frac{(\nabla T)^2}{T^2} \geq 0 \qquad (1.6.23)$$

This is an important result for at least two reasons. First, since Φ, Δ^2, and $(\nabla T)^2/T^2$ are nonnegative, it shows that μ, μ_v, and k must be positive. This result was anticipated in Section 1.5. Second, as we know from thermodynamics, a net entropy increase implies that some useful energy has been dissipated. Thus, since (1.6.23) gives in fact the entropy increase for our system, we can compute from it energy losses in various situations of interest.

Related to the second result, we observe from (1.6.22) that if μ_v, μ, and k are all effectively zero, the entropy of a fluid particle is a constant. Such flows are known as *isentropic* and play an important role in acoustics. A related concept is that of a *homentropic* flow. Here the entropy is uniform throughout the fluid. Most of what is covered in Chapters 3 to 5 of this book refers to fluids whose entropy is uniform throughout.

PROBLEMS

1.6.1 Show that the viscous dissipation function Φ is nonnegative. For simplicity, consider a two-dimensional flow $\mathbf{u} = [u_1(x_1, x_2), u_2(x_1, x_2), 0]$.

1.7 COMPLETE SYSTEM OF EQUATIONS

We now rewrite the equations presented in Sections 1.3–1.6.

(1) Continuity

$$\frac{\partial \rho}{\partial t} + \nabla \cdot (\rho \mathbf{u}) = 0$$

(2) Navier–Stokes

$$\rho \frac{\partial \mathbf{u}}{\partial t} + \mathbf{u} \cdot \nabla \mathbf{u} = \rho \mathbf{F} - \nabla p + \mu \left[\nabla^2 \mathbf{u} + \left(\frac{1}{3} + \frac{\mu_v}{\mu} \right) \nabla (\nabla \cdot \mathbf{u}) \right]$$

(3) Energy

$$T \frac{DS}{Dt} = c_p \frac{DT}{Dt} - \frac{\beta T}{\rho} \frac{DP}{Dt} = \Phi + \frac{1}{\rho} \nabla \cdot (k \nabla T) + \frac{\mu_v}{\rho} \Delta^2$$

(4) State

$$p = p(\rho, S)$$

We thus have a system of six equations for the six scalar unknowns: ρ, u_1, u_2, u_3, p, and S. The system must be complemented by suitable initial and boundary conditions. Here, we will only mention that a real fluid "sticks" to solid boundaries. This "no-slip" condition is usually stated as

$$\mathbf{u} = \mathbf{U}_b$$

on boundaries where $\mathbf{U}_b$ is the velocity of the boundary. On the other hand, for an ideal fluid we can only impose the fact that the fluid cannot penetrate an impermeable boundary. Thus

$$\mathbf{u} \cdot \mathbf{n} = \mathbf{U}_b \cdot \mathbf{n}$$

where $\mathbf{n}$ is a unit vector normal to the boundary. Other conditions may also be stated, but these will be developed when needed.

In the next few chapters we shall obtain the basic physical properties of acoustic waves by applying the ideal version of the above system of equations to a variety of simple conditions. The problem of dissipation, which requires the more complete system of equations, will not be taken up until Chapter 6.

SUGGESTED REFERENCES[3]

Batchelor, G. K. *An Introduction to Fluid Dynamics*. The first three chapters of this advanced text present a very clear and complete exposition of basic fluid dynamics.

Currie, I. G. *Fundamental Mechanics of Fluids*. The first two chapters of this intermediate level text have a concise derivation of the basic conservation laws.

Kestin, J. *A Course in Thermodynamics*. The first volume of this work contains a very thorough presentation of the methods and concepts of classical thermodynamics.

---

[3] Full bibliographical description of the works listed at the end of each chapter and in the text will be found in the bibliography at the end of the book.

CHAPTER TWO
BASIC PROPERTIES OF ACOUSTIC WAVES

We now begin our study of acoustic waves in fluids. This will be done by applying the system of equations given in Chapter 1 to situations in which the departures from equilibrium are small. As a first step, we consider fluids that are ideal in the sense that all of the irreversible mechanisms that result in entropy production are absent. This requires, as (1.6.21) shows, that μ, μ_v, and k be zero. In real fluids, these transport quantities are finite. However, there are fluids for which they are very small, so that one may expect that under some conditions such fluids will behave as ideal. These conditions will be considered in some detail in Chapter 6. In this chapter, as well as in the next three chapters, we deal primarily with ideal fluids.

2.1 IDEAL FLUIDS

For an ideal fluid, the transport coefficients are all zero. Thus, with $\mu = \mu_v = k = 0$, we obtain from Section 1.7 the following system of specialized equations:

Continuity

$$\frac{D\rho}{Dt} + \rho \nabla \cdot \mathbf{u} = 0 \tag{2.1.1}$$

Momentum

$$\rho \frac{D\mathbf{u}}{Dt} = \rho \mathbf{g} - \nabla p \tag{2.1.2}$$

Energy

$$DS/Dt = 0 \tag{2.1.3}$$

State

$$p = p(\rho, S) \tag{2.1.4}$$

In (2.1.2) we have assumed that gravity is the only body force acting on the fluid. Now, from (2.1.4) we obtain for an infinitesimal change of pressure

$$dp = \left(\frac{\partial p}{\partial \rho}\right)_S d\rho + \left(\frac{\partial p}{\partial S}\right)_\rho dS \tag{2.1.5}$$

or, for a fluid element in motion,

$$\frac{Dp}{Dt} = \left(\frac{\partial p}{\partial \rho}\right)_S \frac{D\rho}{Dt} + \left(\frac{\partial p}{\partial S}\right)_\rho \frac{DS}{Dt} \tag{2.1.6}$$

Therefore, from the isentropic relation (2.1.3), we have

$$\frac{Dp}{Dt} = c^2 \frac{D\rho}{Dt} \tag{2.1.7}$$

where

$$c^2 = \left(\frac{\partial p}{\partial \rho}\right)_{S_0} \tag{2.1.8}$$

is a nonnegative quantity[1] that is a function of the density, and where S_0 is the constant value of the entropy for that element. The quantity c will later be identified with the "local" speed of sound. Using this notation, we can rewrite the continuity equation as

$$\frac{Dp}{Dt} + \rho c^2 \nabla \cdot \mathbf{u} = 0 \tag{2.1.9}$$

or

$$\frac{\partial p}{\partial t} + \mathbf{u} \cdot \nabla p + \rho c^2 \nabla \cdot \mathbf{u} = 0 \tag{2.1.10}$$

Expanding $D\mathbf{u}/Dt$ in (2.1.2), we write that equation as

$$\rho\left(\frac{\partial \mathbf{u}}{\partial t} + \mathbf{u} \cdot \nabla \mathbf{u}\right) + \nabla p = \rho \mathbf{g} \tag{2.1.11}$$

---

[1] It is shown in books on thermodynamics that equilibrium conditions require that $(\partial p/\partial \rho)_S > 0$.

So far no assumption other than ideal-fluid behavior has been made, so that (2.1.10) and (2.1.11) apply to all motions of ideal fluids in the presence of gravity. In the next section, it is shown that for most applications of acoustics, it is possible to neglect the effects of gravity. One of these effects results in spatial variations of the mean values of pressure, density, and temperature. Such variations are significant in ocean and atmospheric acoustics. In this book, we are concerned mainly with fluids that, at rest, are spatially uniform. The acoustic equations for such fluids are derived in Section 2.3.

2.2 LINEARIZATION

Consider now a body of fluid at rest under the influence of gravity. The equilibrium pressure p_0 and density ρ_0 satisfy the equilibrium condition

$$\nabla p_0 = \rho_0 \mathbf{g} \qquad (2.2.1)$$

which shows that p_0 (and also possible ρ_0) must vary with position. We assume that the spatial gradients associated with p_0 and ρ_0 are small. Suppose now that the state of equilibrium is disturbed slightly, say, by means of an isentropic compression, so that the density is now given by

$$\rho = \rho_0(\mathbf{x}) + \rho'(\mathbf{x}, t) \qquad (2.2.2)$$

where the deviation from equilibrium ρ' is such that $|\rho'| \ll \rho_0$ throughout the fluid. Corresponding to this density deviation, we also have a pressure variation $p'(\mathbf{x}, t)$ such that

$$p = p_0(\mathbf{x}) + p'(\mathbf{x}, t) \qquad (2.2.3)$$

with $|p'| \ll p_0$. Now, if the external forces producing the deviation from equilibrium are removed, we may expect the fluid to oscillate about equilibrium with a phase that, in general, will vary from point to point in the fluid. The oscillatory fluid velocity $\mathbf{u}$ of the fluid can also be expected to be small. This assumption, together with the restrictions on ρ' and p', may be used to simplify (2.1.10) and (2.1.11)

Consider the derivative appearing as a coefficient in the isentropic relation (2.1.7). As mentioned earlier, it is, for a given S_0, a function of ρ. Since the departures from equilibrium density are small, we can expand that derivative in Taylor series about ρ_0. Thus,

$$c^2 = \left(\frac{\partial p}{\partial \rho}\right)_{S_0} = \left[\left(\frac{\partial p}{\partial \rho}\right)_{S_0}\right]_{\rho = \rho_0} + \left[\left(\frac{\partial^2 p}{\partial \rho^2}\right)_{S_0}\right]_{\rho = \rho_0} (\rho - \rho_0) + \cdots$$

or

$$c^2 = c_0^2(\mathbf{x}) + d_0(\mathbf{x})\rho' + \cdots$$

where

$$c_0^2 = \left[\left(\frac{\partial p}{\partial \rho} \right)_{S_0} \right]_{\rho = \rho_0}, \qquad d_0 = \left[\left(\frac{\partial^2 p}{\partial \rho^2} \right)_{S_0} \right]_{\rho = \rho_0} \qquad (2.2.4)$$

Making use of this expansion, and of (2.2.2) and (2.2.3), the continuity equation can be written as

$$\frac{\partial p'}{\partial t} + \mathbf{u} \cdot \nabla p_0 + \mathbf{u} \cdot \nabla p' + (\rho_0 + \rho')(c_0^2 + d_0 \rho' + \cdots)\nabla \cdot \mathbf{u} = 0 \qquad (2.2.5)$$

Now, by assumption, ρ' and $\mathbf{u}$ are small. Therefore, the terms $\mathbf{u} \cdot \nabla p'$, $d_0 \rho'^2$, and $\rho_0 d_0 \rho' \nabla \cdot \mathbf{u}$ are of second order. Retaining only first-order terms, we obtain

$$\frac{\partial p'}{\partial t} + \rho_0(\mathbf{x}) c_0^2(\mathbf{x}) \nabla \cdot \mathbf{u} = -\mathbf{u} \cdot \nabla p_0 \qquad (2.2.6)$$

Similarly, using the equilibrium equation (2.2.1), the momentum equation can be written as

$$(\rho_0 + \rho')\left(\frac{\partial \mathbf{u}}{\partial t} + \mathbf{u} \cdot \nabla \mathbf{u} \right) + \nabla p' = \rho' \mathbf{g} \qquad (2.2.7)$$

Again, retaining only first-order quantities, this becomes

$$\rho_0 \frac{\partial \mathbf{u}}{\partial t} + \nabla p' = \frac{1}{\rho_0} \rho' \nabla p_0 \qquad (2.2.8)$$

The quantities on the right-hand sides of (2.2.6) and (2.2.8) are, in most cases of interest, small compared to those on the left-hand side of those equations. For example, the magnitude of $\mathbf{u} \cdot \nabla p_0$ in (2.2.6), on using (2.2.1), is

$$|\mathbf{u} \cdot \nabla p_0| \sim U g \rho_0 \qquad (2.2.9)$$

where U is some representative fluid velocity. On the other hand, the magnitude of $\nabla \cdot \mathbf{u}$ is

$$|\nabla \cdot \mathbf{u}| \sim U/L \qquad (2.2.10)$$

where L is length scale for the variations of velocity. Hence,

$$\frac{|\mathbf{u}\cdot\nabla p_0|}{\rho_0 c_0^2 |\nabla\cdot\mathbf{u}|} \sim \frac{L}{c_0^2/g} \tag{2.2.11}$$

Now, in air at standard temperature and pressure, $c_0^2/g \simeq 12$ km. However, the length scales of interest in acoustics are usually much smaller than this. Therefore, the term on the right-hand side of (2.2.6) may then be neglected compared to those on the left, so that the equation becomes

$$\frac{1}{\rho_0(\mathbf{x})c_0^2(\mathbf{x})}\frac{\partial p'}{\partial t} + \nabla\cdot\mathbf{u} = 0 \tag{2.2.12}$$

Similar arguments show that the right-hand side of (2.2.8) is also of the order $L/(c_0^2/g)$ relative to those on the left, so that in the same approximation we may write that equation as

$$\frac{\partial\mathbf{u}}{\partial t} + \frac{1}{\rho_0}\nabla p' = 0 \tag{2.2.13}$$

We now combine (2.2.12) and (2.2.13) by taking cross derivatives. Thus, the time derivative of (2.2.12) is

$$\frac{1}{\rho_0(\mathbf{x})c_0^2(\mathbf{x})}\frac{\partial^2 p'}{\partial t^2} + \nabla\cdot\frac{\partial\mathbf{u}}{\partial t} = 0 \tag{2.2.14}$$

and the spatial derivative of (2.2.13) is

$$\nabla\cdot\frac{\partial\mathbf{u}}{\partial t} + \frac{1}{\rho_0}\nabla^2 p' = \frac{1}{\rho_0^2}\nabla p'\cdot\nabla\rho_0 \tag{2.2.15}$$

The magnitude of the term on the right-hand side is also of order $L/(c_0^2/g)$ relative to those on the left, and may therefore be neglected. Finally, we eliminate the terms containing $\mathbf{u}$ and obtain

$$\frac{\partial^2 p'}{\partial t^2} = c_0^2(\mathbf{x})\nabla^2 p' \tag{2.2.16}$$

This is the wave equation for a fluid whose equilibrium state is not uniform. It is the same as the classical wave equation, except that the quantity c_0 is a function of position.

PROBLEMS

2.2.1 Show that in a fluid at rest in a gravitational field, the constant-pressure and constant-density surfaces coincide.

2.3 UNIFORM FLUIDS

Equation (2.2.16) describes the propagation of acoustic waves in a medium that is not uniform, and therefore relates to sound waves propagating in the oceans and in the atmosphere. However, in many situations, the mean properties of the medium are nearly uniform throughout. For example, the characteristics of sound waves in a room can be obtained quite accurately under the assumption of uniform mean density and pressure. In what follows, we will be concerned mainly with sound waves in a uniform medium.

For a uniform fluid, the pressure deviation from equilibrium satisfies (2.2.16), with c_0 being a constant throughout. However, it is convenient to derive this equation from the linearized equations of continuity and momentum that are applicable in the case of uniform mean density, pressure, and entropy. These equations are (see Problem 2.3.1)

$$\frac{\partial p'}{\partial t} + \rho_0 c_0^2 \, \nabla \cdot \mathbf{u} = 0 \tag{2.3.1}$$

$$\rho_0 \frac{\partial \mathbf{u}}{\partial t} + \nabla p' = 0 \tag{2.3.2}$$

These may be combined in the manner of (2.2.6) and (2.2.8). However, before doing so, it is convenient to introduce another important quantity: the velocity potential ϕ. To do this, we take the curl of the second equation and obtain

$$\frac{\partial}{\partial t}(\nabla \times \mathbf{u}) = 0 \tag{2.3.3}$$

The quantity $\nabla \times \mathbf{u}$ is called the *vorticity* ω of the fluid, and is proportional to the local angular velocity of the fluid. From above, we see that

$$\omega = \nabla \times \mathbf{u} = \mathbf{f}(\mathbf{x})$$

for all t. However, since the fluid was initially at rest, the vorticity was originally zero. We therefore conclude that in an acoustic wave in an ideal, uniform fluid,

$$\omega = 0 \tag{2.3.4}$$

Flows for which $\omega = 0$ are called *irrotational*. They play an important role in idealized acoustics. Since for them $\nabla \times \mathbf{u} = 0$, we can write

$$\mathbf{u} = \nabla \phi \qquad (2.3.5)$$

where $\phi = \phi(\mathbf{x}, t)$ is known as the *velocity potential*. This substitution enables us to obtain, from a single scalar quantity, all components of the fluid velocity. Further, the pressure can also be obtained from ϕ, as can be seen by substituting $\mathbf{u} = \nabla \phi$ in the linearized momentum equation. This yields

$$\nabla \left(\rho_0 \frac{\partial \phi}{\partial t} + p' \right) = 0$$

Thus,

$$\rho_0 \frac{\partial \phi}{\partial t} + p' = g(t) \qquad (2.3.6)$$

The function $g(t)$ can be set equal to zero without loss of generality. This can be seen by defining another potential ϕ', say, such that

$$\rho_0 \phi = \rho_0 \phi' + \int^t g(t')\, dt'$$

Clearly,

$$\nabla \phi = \nabla \phi'$$

and so the velocity field is not affected. Substitution in (2.3.6) yields

$$\rho_0 \frac{\partial \phi'}{\partial t} + p' = 0$$

We therefore can put $g(t) = 0$ in (2.3.6) so that the pressure may be obtained from ϕ by means of

$$p' = -\rho_0 \frac{\partial \phi}{\partial t} \qquad (2.3.7)$$

The velocity potential satisfies the wave equation, as can be seen by substituting $\mathbf{u} = \nabla \phi$ and $p' = -\rho_0 (\partial \phi / \partial t)$ into the continuity equation. The result is

$$\frac{\partial^2 \phi}{\partial t^2} = c_0^2 \nabla^2 \phi \qquad (2.3.8)$$

Also, since the vorticity $\nabla \times \mathbf{u}$ is identically zero, we have $\nabla(\nabla \cdot \mathbf{u}) = \nabla^2 \mathbf{u}$. Hence, we may eliminate p' from (2.3.1) and (2.3.2) to obtain

$$\frac{\partial^2 \mathbf{u}}{\partial t^2} = c_0^2 \nabla^2 \mathbf{u} \tag{2.3.9}$$

This is a vector wave equation and is, therefore, generally less convenient than the scalar equations for p', ρ', or ϕ.

Associated with the changes of density and pressure in the fluid, there will be variations of temperature. These can be obtained from the isentropic condition together with (1.6.11), that is, by

$$\frac{DT}{Dt} = \frac{\beta T}{\rho c_p} \frac{Dp}{Dt}$$

In linearized form, the changes of temperature $T' = T - T_0$ are then given by

$$T' = \frac{\beta_0 T_0}{\rho_0 c_{p0}} p' = -\frac{\beta_0 T_0}{c_{p0}} \frac{\partial \phi}{\partial t} \tag{2.3.10}$$

For a perfect gas, $\beta_0 = 1/T_0$, so that $T' = -(1/c_{p0})(\partial \phi / \partial t)$.

PROBLEMS

2.3.1 Consider (2.1.10) and (2.1.11) for the case when $\mathbf{g} = 0$. Show that for small departures for the equilibrium state (p_0, ρ_0), those equations may be approximated by (2.3.1) and (2.3.2), respectively.

2.3.2 Show that for one-dimensional cases, the wave equation for ϕ can be written as

$$\frac{\partial^2 \phi}{\partial t^2} = c_0^2 \frac{\partial^2 \phi}{\partial r^2} + \frac{n}{r} \frac{\partial \phi}{\partial r}$$

where $n = 0$, 1, 2 for plane, cylindrical, or spherical flow, respectively.

2.3.3 Show that if $\phi_1, \phi_2, \ldots, \phi_n$ are n different solutions of the wave equation (2.3.8), then $\phi = \phi_1 + \phi_2 + \cdots + \phi_n$ is also a solution. This is an example of superposition for linear motions.

2.4 ONE-DIMENSIONAL PLANE WAVES

To understand the meaning of the quantity c_0, and to obtain a basic solution of the wave equation, we consider the case where ϕ depends only on one cartesian coordinate, say x, and on time t. That is, the pressure, density, and other quantities are assumed to be constant on a plane perpendicular to the x axis. Thus, with $\phi = \phi(x,t)$ we obtain from (2.3.8)

$$\frac{\partial^2 \phi}{\partial t^2} = c_0^2 \frac{\partial^2 \phi}{\partial x^2} \tag{2.4.1}$$

The general solution of this equation can be obtained by introducing the new independent variables ξ and η defined by

$$\xi = x - c_0 t$$

$$\eta = x + c_0 t \tag{2.4.2}$$

Since in terms of ξ and η, $d\phi$ is given by

$$d\phi = \frac{\partial \phi}{\partial \xi} d\xi + \frac{\partial \phi}{\partial \eta} d\eta$$

we obtain

$$\frac{\partial \phi}{\partial x} = \frac{\partial \phi}{\partial \xi}\frac{\partial \xi}{\partial x} + \frac{\partial \phi}{\partial \eta}\frac{\partial \eta}{\partial x}$$

or, on using (2.4.2),

$$\frac{\partial \phi}{\partial x} = \frac{\partial \phi}{\partial \xi} + \frac{\partial \phi}{\partial \eta}$$

Also,

$$\frac{\partial^2 \phi}{\partial x^2} = \frac{\partial}{\partial \xi}\left(\frac{\partial \phi}{\partial \xi} + \frac{\partial \phi}{\partial \eta} \right) + \frac{\partial}{\partial \eta}\left(\frac{\partial \phi}{\partial \xi} + \frac{\partial \phi}{\partial \eta} \right)$$

$$= \frac{\partial^2 \phi}{\partial \xi^2} + 2\frac{\partial^2 \phi}{\partial \xi \partial \eta} + \frac{\partial^2 \phi}{\partial \eta^2}$$

Similarly,

$$\frac{\partial \phi}{\partial t} = c_0\left(\frac{\partial \phi}{\partial \eta} - \frac{\partial \phi}{\partial \xi} \right)$$

$$\frac{\partial^2 \phi}{\partial t^2} = c_0^2\left(\frac{\partial^2 \phi}{\partial \xi^2} + \frac{\partial^2 \phi}{\partial \eta^2} \right) - 2c_0^2 \frac{\partial^2 \phi}{\partial \xi \partial \eta}$$

Substitution into (2.4.1) yields

$$\frac{\partial^2 \phi}{\partial \xi \partial \eta} = 0$$

Integrating with respect to η gives

$$\frac{\partial \phi}{\partial \xi} = F(\xi)$$

A second integration yields

$$\phi = f(\xi) + g(\eta) \tag{2.4.3}$$

where $f(\xi) = \int^{\xi} F(\xi') d\xi'$ is another function of ξ.

Equation (2.4.3) is the general solution of the one-dimensional, plane wave equation. In terms of x and t, it can be written as

$$\phi(x,t) = f(x - c_0 t) + g(x + c_0 t) \tag{2.4.4}$$

To understand the meaning of the general solution, we consider the more easily understood quantities p' and ρ'. Since these quantities also satisfy the one-dimensional, plane wave equation, it is clear that they will also be given for the one-dimensional case by general solutions of the type given by (2.4.4). However, we can use, directly, the general solution for the potential, together with $p' = -\rho_0(\partial \phi / \partial t)$, $p' = c_0^2 \rho'$, and $\mathbf{u} = \nabla \phi$. For example, if ϕ is given by (2.4.4), then the pressure fluctuation will be given by

$$p'(x,t) = \rho_0 c_0 f'(x - c_0 t) - \rho_0 c_0 g'(x + c_0 t) \tag{2.4.5}$$

where the primes represent derivatives with respect to the arguments. Thus, $f'(x - c_0 t) = df/d\xi$, where $\xi = x - c_0 t$. Consider the special case when $g' = 0$. The fluctuating part of the pressure will then be given by

$$p'(x,t) = \rho_0 c_0 f'(x - c_0 t) \tag{2.4.6}$$

Whatever the form of f might be, it is clear that at a given time p' is a function of position, and that at a fixed location it varies with time. However, x and t appear in the combination $x - c_0 t$, so that f' will be a constant for constant values of $x - c_0 t$. In particular, suppose that at time $t = t_0$ and at location $x = x_0$, that is, when $x_0 - c_0 t_0 = \psi_0$, the pressure deviation p' has the value $P_0'(x_0, t_0)$ as shown schematically in Figure 2.4.1. Then, the same value of p' will be found

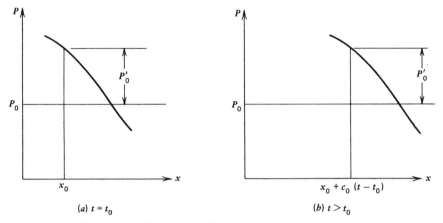

(a) $t = t_0$ (b) $t > t_0$

Figure 2.4.1 Pressure profiles at two different times.

later, at a distance

$$x = \psi_0 + c_0 t$$

Thus, we find that the pressure disturbance moves toward increasing values of x with a speed c_0. This is, of course, a wave, and c_0 is its speed.

Similarly, the function $g(x + c_0 t)$ represents a plane, one-dimensional wave traveling with speed c_0 toward decreasing values of x. Both types of waves are called *progressive waves*, and (2.4.4) shows that the general solution of the one-dimensional, plane wave equation is given by the superposition of two opposite-traveling waves. Now, in a plane, progressive wave, the velocity vector is $\mathbf{u} = \{u(x, t), 0, 0\}$, where $u = \partial \phi / \partial x$. Thus, the fluid velocity is along the direction of propagation. This type of wave is therefore known as *longitudinal*.

Compressional longitudinal waves can also be sustained by elastic solids. These waves have much in common with sound waves in fluids. In fact, many of the results derived for sound waves in fluids also apply to longitudinal waves in solids. These are, however, outside the scope of this book.

Speed of Sound in a Perfect Gas

From its definition, it is clear that the speed of sound c_0 is a thermodynamic property depending only on the equilibrium state of the fluid. Accordingly, its value can be computed from equilibrium equations of state if these are available. In the important case of perfect gases, of which dry, clean air is a good example, the equation of state is of the form given by (1.3.1), that is,

$$p = R\rho T$$

This is to be supplemented with the statement that for such gases, the internal energy is a function of temperature alone, so that

$$dE = c_v\, dT$$

Combining this with (1.3.15), we obtain for an isentropic change

$$c_v(T)\, dT = \frac{p}{\rho^2}\, d\rho, \qquad S = \text{constant} \tag{2.4.7}$$

Now, from (1.3.1), we have

$$dp = R(T\, d\rho + \rho\, dT) \tag{2.4.8}$$

Therefore, (2.2.24) gives

$$c_0^2 = RT_0 + R\rho_0 \left(\frac{\partial T}{\partial \rho} \right)_{S_0}$$

Using (2.4.7) to evaluate the partial derivative, we obtain

$$c_0^2 = RT_0 \left(1 + \frac{R}{c_{v0}} \right)$$

Finally, since for a perfect gas $R = c_p - c_v$, we obtain

$$c_0^2 = \gamma RT_0 \tag{2.4.9}$$

where, as before, $\gamma = c_p/c_v$. Thus, in a perfect gas, c_0 is a function of temperature alone. In air, at 273 K for example, $c_0 = 331$ m/sec, so that at any temperature and pressure (for which air behaves as a perfect gas), we have $c = 331(T/273)^{1/2}$.

One of the most surprising facts about (2.4.9) is that it agrees with the results of careful measurements of the speed of sound in *real gases*. This is surprising because the equation, first derived by Laplace in 1816, assumes that there is no heat conduction into or out of a fluid element during the passage of a sound wave. In ideal fluids this is, of course, true because they have zero thermal conductivity. However, since real fluids have finite thermal conductivity, one might have expected that heat is evolved during the passage of a wave, and that this heat should flow quickly out of a given element, thereby keeping its temperature constant. This was, in fact, the argument used by Newton in deriving what is now known as the isothermal speed of sound c_{T0} (see Problem 2.4.1). The reasons that explain why Laplace's equation is applicable will be

Table 2.4.1 Sound Speed in Selected Gases at 273 K, 1 Atm

Gas	γ	$\rho \ (\text{kg}/\text{m}^3)$	$c_0 \ (\text{m}/\text{sec})$
Air, dry	1.40	1.293	331
Argon	1.667	1.783	319
Carbon dioxide	1.40	1.977	259
Helium	1.667	0.178	965
Hydrogen	1.40	0.0899	1284
Nitrogen	1.40	1.251	334
Oxygen	1.40	1.429	316

Source: *Handbook of Chemistry and Physics*, Forty-eighth Ed. (Chemical Rubber Company, Cleveland, Ohio, 1968).

studied in Chapter 6. Here we mention only that unless the frequency of the waves is exceedingly large, sound propagation in gases is basically an adiabatic phenomenon, as assumed in deriving (2.4.9).

Table 2.4.1 gives values of the speed of sound in various gases at low pressures, where they obey the perfect-gas equation of state. The data are applicable provided the frequency of the waves is not so large that the molecular degrees of freedom of the gases exhibit relaxation effects (see Section 1.5). In CO_2, for example, the vibrational degrees of freedom are not excited at ambient temperature. However, for frequencies of the order of 10^{10} sec^{-1}, they take a significant portion of their full share of the internal energy. The net result of this is a lowering of the effective specific heat ratio and therefore of the speed of sound. Diatomic gases display similar effects, at somewhat higher frequencies. Such frequencies are, however, beyond the range of applicability of our basic conservation equations, mainly because for them the continuum model ceases to be valid. This matter is discussed in more detail in Chapter 6.

Speed of Sound in Other Fluids

The definition of c_0 given by (2.1.8) also applies to more general fluids, so that, in principle, all that is required to compute their speed of sound is their equation of state. However, except for some results of restricted validity, these equations are not available, and one has to resort to experiment to obtain thermodynamic data. These data may, of course, be used to determine the speed of sound. In fact, however, the opposite is true; that is, experimentally determined values of the speed of sound are often used to determine thermodynamic data. The reasons for this are that the speed of sound is related to the isentropic compressibility of a pure substance (see Problem 2.4.2), and that accurate measurements of the speed of sound are relatively easy to perform.

Table 2.4.2   Sound Speed in Selected Liquids and Solids at 298 K

Substance	$\rho(\text{kg}/\text{m}^3)$	$c_0(\text{m}/\text{sec})$
Liquids		
Benzene	870	1295
Castor oil	969	1477
Kerosene	810	1324
Mercury	1350	1450
Water, distilled	998	1498
Solids		
Aluminum	2700	6420
Brass	8600	4700
Gold	19,700	3240
Lead	11,400	1960
Silver	10,400	3650
Steel, stainless	7900	5790

Source: Handbook of Physics and Chemistry.

Table 2.4.2 gives measured values of the speed of sound in various liquids. It also includes speeds of propagation of longitudinal waves in some elastic solids. As stated earlier, this type of wave in solids has much in common with acoustic waves in ideal fluids, and is therefore called acoustic. Solids, however, can sustain other types of waves. For an account of these, the reader is referred to the book by Kolsky.

Finally, Table 2.4.3 gives speed-of-sound data as a function of temperature for the important case of pure water. The data are taken from Del Grosso (1969),

Table 2.4.3   Sound Speed in Pure Water

Temperature[a] (C)	Speed of sound[a] (m/sec)
0	1402.4
5	1426.1
10	1447.2
20	1482.1
25	1496.6
50	1542.5
60	1551.0
74	1555.1

[a]All data rounded to the nearest one-tenth.
Source: V. A. Del Grosso, *J. Acoust. Soc. Amer.*,
47, 947 (1969).

who also gives data for seawater. The pure-water data are fairly well fitted by means of

$$c_0 = 1402.40 + 5.01\theta - 0.055\theta^2 + 0.000022\theta^3$$

where θ is the temperature in degrees Celsius. This equation was adapted from a more accurate one given by Del Grosso (1974).

Relationships between Acoustic Quantities

We now return to (2.3.5) and derive some relationships between pressure, density, velocity, and so on that hold for plane progressive waves. Thus, if the wave is moving toward increasing values of x (i.e., to the right in a right-handed coordinate system), the acoustic density and velocity are given by

$$\rho' = (\rho_0/c_0)f'(x - c_0t) \tag{2.4.10}$$

$$u = f'(x - c_0t) \tag{2.4.11}$$

Thus, the acoustic density and velocity in a plane wave are related by

$$\rho' = \rho_0 u/c_0 \tag{2.4.12}$$

Also, since $p' = c_0^2\rho'$, we obtain

$$p' = \rho_0 c_0 u \tag{2.4.13}$$

Finally, since $T' = (\beta_0 T_0/\rho_0 c_{p0})p'$, we have

$$T' = (\beta_0 T_0 c_0/c_{p0})u \tag{2.4.14}$$

For a wave moving toward decreasing values of x, the right-hand sides of (2.4.12)–(2.4.14) change sign. Thus, if the general solution of ϕ is

$$\phi = f(x - c_0t) + g(x + c_0t)$$

the fluid velocity in the acoustic wave will be given by

$$u(x,t) = f'(x - c_0t) + g'(x + c_0t) \tag{2.4.15}$$

and the acoustic pressure by

$$p'(x,t) = \rho_0 c_0 f'(x - c_0t) - \rho_0 c_0 g'(x + c_0t) \tag{2.4.16}$$

These equations give the pressure and velocity in an arbitrary plane, one-dimensional wave. The solution for a particular problem may be obtained from it by applying the initial or boundary conditions suitable to that specific problem. That is, these conditions may be used to obtain the, so far, arbitrary functions f and g, as is done in the simple example treated below. These equations may also be used to arrive at criteria for the relative smallness of u and p'. For example, (2.4.12) gives the criterion for u. This equation can be written as

$$\frac{\rho'}{\rho_0} = \frac{u}{c_0} \tag{2.4.17}$$

Thus, since by assumption $|\rho'| \ll \rho_0$, we see that we must also have $|u| \ll c_0$. Similarly, (2.4.13) shows that the pressure fluctuation p' must be such that $|p'| \ll \rho_0 c_0^2$. For a perfect gas, this condition can be written as $|p'| \ll \gamma p_0$.

We also note from (2.4.13) that in a plane wave the pressure and fluid velocity are related by

$$p'/u = \rho_0 c_0 \tag{2.4.18}$$

Thus, for a plane, one-dimensional propagating wave, the ratio of acoustic pressure to velocity depends only on the properties of the medium.

In some works on acoustics, the ratio of pressure to velocity is called the specific impedance because under some restricted conditions, that ratio is analogous to the electrical impedance of an alternating-current circuit. For a plane, one-dimensional wave, this ratio is, as we have seen, a constant depending only on the properties of the medium. The quantity $\rho_0 c_0$ is therefore called the *characteristic impedance* of the medium.

For a more general type of wave, the ratio p'/u depends not only on the properties of the medium, but also on the properties of the wave.

EXAMPLE

Although (2.4.15) and (2.4.16) will be studied later in some detail, we use them now to study a simple problem. Suppose that a very long tube is divided into two parts by means of a membrane located at $x=0$. The pressure in each of the two parts is initially uniform, and has values p_1 on the right, say, and $p_2 > p_1$ on the left. If the pressure difference $p_2 - p_1$ is small compared to p_1, we may expect our linearized equations to be applicable, and may therefore use them to describe the effects that take place when the diaphragm is removed at $t=0$. Since the initial pressure difference is $p_2 - p_1$, the initial distribution of excess pressure is as shown in Figure 2.4.2. As stated above, we take this difference to be small so that we may identify the above distribution of overpressure with the

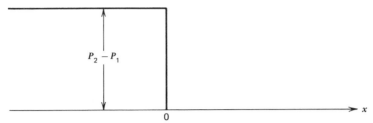

Figure 2.4.2 Initial distribution of excess pressure.

initial value of p'. Thus,

$$p'(x,0)= \begin{cases} P_2-P_1, & x<0 \\ 0, & x>0 \end{cases} \tag{2.4.19}$$

But p' is given by (2.4.16) so that

$$f'(x)-g'(x)= \begin{cases} \dfrac{P_2-P_1}{\rho_0 c_0}, & x<0 \\ 0, & x>0 \end{cases} \tag{2.4.20}$$

A second relationship between f' and g' may be obtained from the initial distribution of fluid velocity. Thus, since initially $u=0$, we have from (2.4.15)

$$f'(x)=-g'(x) \tag{2.4.21}$$

Combining this with (2.4.20), we obtain

$$f'(\xi)= \begin{cases} \dfrac{P_2-P_1}{2\rho_0 c_0}, & \xi<0 \\ 0, & \xi>0 \end{cases} \tag{2.4.22}$$

Thus,

$$f'(x-c_0 t)= \begin{cases} \dfrac{P_2-P_1}{2\rho_0 c_0}, & x-c_0 t \leqslant 0 \\ 0, & x-c_0 t \geqslant 0 \end{cases} \tag{2.4.23}$$

Similarly,

$$g'(x+c_0 t)= \begin{cases} -\dfrac{P_2-P_1}{2\rho_0 c_0}, & x+c_0 t \leqslant 0 \\ 0, & x+c_0 t \geqslant 0 \end{cases} \tag{2.4.24}$$

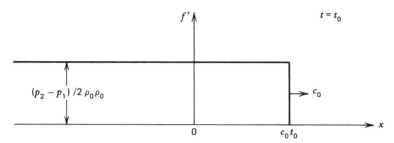

Figure 2.4.3 Function f' for $t > t_0$ (compressive wave).

To see the meaning of these results more clearly, we plot separately the two functions f' and g' at some fixed time $t = t_0$. Thus, our result above shows that if $x > c_0 t$, f' is zero, whereas if $x < c_0 t_0$ it has a constant value equal to $(p_2 - p_1)/2\rho_0 c_0$, as shown in Figure 2.4.3. The arrow placed at $x = c_0 t_0$ implies that an instant later the wave would move to the right. Since in the range where it does not vanish, $f' > 0$, it follows that it represents a *compression* wave moving to the right and having a pressure amplitude equal to one-half of the initial overpressure.

On the other hand, in the range where it does not vanish, $g' < 0$, as shown schematically in Figure 2.4.4. This is an *expansion* wave, as it lowers the pressure of the fluid into which it moves.

The complete pressure distribution at the same instant $t = t_0$ may be obtained simply by superposition of f' and g'. Thus, from (2.4.16), with f' and g' given by (2.4.23) and (2.4.24), we find that the total pressure distribution has the profile. shown in Figure 2.4.5.

The velocity can be obtained from f' and g' by means of (2.4.15). These results show that in the region where it does not vanish, the fluid velocity is constant and positive; that is, the flow is uniform and to the right.

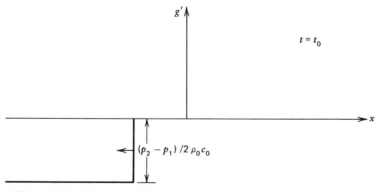

Figure 2.4.4 Function g' for $t > t_0$ (expansive wave).

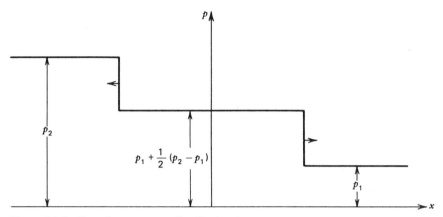

Figure 2.4.5 Complete pressure distribution for $t > t_0$.

It should be noted that in this example, the two traveling waves $f'(x - c_0 t)$ and $g'(x + c_0 t)$ represented either a compression or an expansion wave. Later we will see that in more general cases, each of these waves may contain compressions and expansions.

PROBLEMS

2.4.1 The isothermal speed of sound for a pure substance is defined by $c_{T0}^2 = (\partial p / \partial \rho)_T$. Show that $c_0^2 = \gamma c_{T0}^2$ so that the error of c_{T0} relative to c_0 is

$$|\varepsilon| = \frac{\sqrt{\gamma} - 1}{\sqrt{\gamma}}$$

Compare this error for the case of air to your estimate of the experimental accuracy for the measurements of the speed of sound in Newton's days.

2.4.2 For liquids, it is customary to specify their thermal equation of state in terms of their coefficient of thermal expansion β and of their isentropic compressibility κ_s. Show that the isentropic compressibility is related to the speed of sound in a pure substance by means of

$$\kappa_s = 1/\rho_0 c_0^2$$

2.5 MONOCHROMATIC WAVES

A monochromatic wave is a wave in which the pressure, velocity, and so on depend on time only through periodic functions of time of a single (circular) frequency ω. These waves are very important because waves of more general time dependence can be represented as a sum of their monochromatic components. Also, many situations of practical importance relate to monochromatic waves. In a monochromatic wave, the velocity potential will depend on time through the elementary functions $\sin \omega t$, $\cos \omega t$. However, it is often more convenient to use, instead, the simple complex exponential $\exp(-i\omega t)$. Using this notation, the velocity potential may be written as

$$\phi(\mathbf{x},t) = \mathrm{Re}\{\tilde{\Phi}(\mathbf{x})e^{-i\omega t}\} \tag{2.5.1}$$

where $\tilde{\Phi}$ may be complex. The symbol $\mathrm{Re}\{\ \}$ is used to remind us that ϕ is given by the real part of the quantity in brackets. However, so long as we deal only with linear operations, the symbol may be omitted, provided, of course, that we remember that only the real part of such an expression is physically meaningful. With this restriction in mind, we substitute (2.5.1) into (2.3.8) and obtain

$$\nabla^2\tilde{\Phi} + (\omega/c_0)^2\tilde{\Phi} = 0 \tag{2.5.2}$$

This type of equation is known as a *Helmholtz equation* and, also, as a time-independent wave equation. It is sometimes written as

$$\nabla^2\tilde{\Phi} + k^2\tilde{\Phi} = 0 \tag{2.5.3}$$

where the quantity k, defined by

$$k = \omega/c_0 \tag{2.5.4}$$

is called the *wave number*. Its meaning will be made clear by considering the special case of plane, monochromatic waves.

Plane, One-Dimensional Monochromatic Waves

These are, by far, the simplest type. Suppose that the direction of propagation is along the x axis of a coordinate system. Then, since the acoustic quantities are constant on a plane perpendicular to the x axis, the time-independent wave equation reduces to

$$\frac{d^2\tilde{\Phi}}{dx^2} + k^2\tilde{\Phi} = 0 \tag{2.5.5}$$

The general solution of this equation can be written as

$$\tilde{\Phi} = \tilde{A}e^{ikx} + \tilde{B}e^{-ikx} \tag{2.5.6}$$

and so

$$\tilde{\phi}(x,t)=\tilde{A}e^{i(kx-\omega t)}+\tilde{B}e^{-i(kx+\omega t)} \qquad (2.5.7)$$

where $\tilde{A}$ and $\tilde{B}$ are, in general, complex quantities, that is, $\tilde{A}$ is of the form $|\tilde{A}|e^{ia}$, where $|\tilde{A}|$ and a are constants. The two terms in (2.5.7) are the monochromatic version of the functions $f(x-c_0t)$ and $g(x+c_0t)$ in (2.4.4). Consider the case $\tilde{B}=0$ so that only the positive-going wave remains. Then, ϕ is given by

$$\phi=A\cos(kx-\omega t+a) \qquad (2.5.8)$$

where the quantity $A=|\tilde{A}|$ is called the *amplitude* of the wave and where the quantity

$$\psi=kx-\omega t+a \qquad (2.5.9)$$

is called the *phase*. We note that the phase is constant for an observer moving with velocity

$$\frac{dx}{dt}=\frac{\omega}{k}=c_f \qquad (2.5.10)$$

The quantity ω/k is called the *phase velocity*, and in this simple case is equal to c_0. Now, we already know that the acoustic variables in the wave are sinusoidal functions of time with period T, that is,

$$\cos(kx-\omega t+a)=\cos[kx-\omega(t+T)+a]$$

Thus, we must have

$$\omega=2\pi/T \qquad (2.5.11)$$

Similarly, the solution is also a periodic function of x. The spatial period is called the *wavelength* and will be denoted by the symbol λ. Its relationship to the quantity k may be obtained from the condition $\cos(kx)=\cos[k(x+\lambda)]$. This yields

$$k=2\pi/\lambda \qquad (2.5.12)$$

Therefore, the wave number is the spatial counterpart to ω. That is, while ω gives the number of oscillations in time 2π, k gives the number of wave crests in a distance 2π.

These relationships, together with $k = \omega/c_0$, yield the following important relationship between wavelength, frequency, and speed of sound in a plane wave:

$$c_0 = f\lambda \tag{2.5.13}$$

Plane, Monochromatic Waves in Three Dimensions

In some applications, we require the velocity potential for a plane, monochromatic wave that is traveling along a direction which does not corespond with one of the axes of a system of coordinates. In this case, the solution may be obtained, by rotation, from the solution given by (2.5.7). However, it is instructive to derive it from Helmholtz equation

$$\nabla^2\tilde{\Phi} + k^2\tilde{\Phi} = 0 \tag{2.5.14}$$

Now, since the wave is plane and its direction does not change, we may assume that the wave is propagating along some axis $s(\mathbf{x})$. We then have

$$\frac{d^2\tilde{\Phi}}{ds^2} + k^2\tilde{\Phi} = 0 \tag{2.5.15}$$

so that the solution for the complex potential $\tilde{\Phi}$ is of the form

$$\tilde{\Phi} = e^{\pm iks(\mathbf{x})} \tag{2.5.16}$$

or

$$\tilde{\phi}(x,t) = \tilde{A}e^{i[ks(\mathbf{x}) - \omega t]} + \tilde{B}^{-i[ks(\mathbf{x}) + \omega t]}$$

We now express s in terms of the position vector $\mathbf{x}$. Since the wave is one dimensional, we may take the s axis to pass through any given point on an infinite plane perpendicular to s. At that point, we may place the origin of a cartesian system of coordinates. Then it is clear that for any point along s, the position vector $\mathbf{x}$ can be expressed as

$$\mathbf{x} = \mathbf{n}s \tag{2.5.17}$$

where $\mathbf{n}$ is a unit vector along s, that is, along the direction of propagation. Thus,

$$s(\mathbf{x}) = \mathbf{n}\cdot\mathbf{x} \tag{2.5.18}$$

Therefore, the solution for $\tilde{\Phi}$ is

$$\tilde{\Phi} = \tilde{A}e^{i(k\mathbf{n}\cdot\mathbf{x})} \tag{2.5.19}$$

or, if we define a *wave vector* **k** by means of

$$\mathbf{k} = k\mathbf{n} = \frac{\omega}{c_0}\mathbf{n} \qquad (2.5.20)$$

we have

$$\tilde{\Phi}(x) = \tilde{A}e^{i\mathbf{k}\cdot\mathbf{x}} \qquad (2.5.21)$$

We can convince ourselves that this is, in fact, a solution of (2.5.14) by writing that equation as

$$\frac{\partial^2 \tilde{\Phi}}{\partial x_i \, \partial x_i} + k^2 \tilde{\Phi} = 0$$

Thus, with

$$\tilde{\Phi} = \tilde{A}e^{ik_j x_j} \qquad (2.5.22)$$

we obtain, since $\partial x_j / \partial x_i = \delta_{ij}$,

$$\frac{\partial \tilde{\Phi}}{\partial x_i} = ik_i \tilde{\Phi} \qquad (2.5.23)$$

Similarly,

$$\frac{\partial^2 \tilde{\Phi}}{\partial x_i \, \partial x_i} = \nabla^2 \tilde{\Phi} = -k^2 \tilde{\Phi}$$

so that (2.5.14) is identically satisfied. Therefore, for plane, monochromatic, three-dimensional waves, we can write

$$\tilde{\phi}(x, t) = \tilde{A}e^{i(\mathbf{k}\cdot\mathbf{x} - \omega t)} + \tilde{B}e^{-i(\mathbf{k}\cdot\mathbf{x} + \omega t)} \qquad (2.5.24)$$

This solution is very useful in problems dealing with plane sound waves over boundaries that are neither parallel nor perpendicular to the direction of propagation. Some of these problems will be treated in Chapter 3.

Relation between Variables in Monochromatic Wave

In the case of monochromatic waves, whether plane or not, the relationships between different variables such as pressure and velocity can be expressed in a simple manner. Consider, first, the acoustic pressure p'. Since p' also satisfies the wave equation, it can be expressed as the real part of

$$\tilde{p}' = \tilde{P}'(\mathbf{x})e^{-i\omega t} \tag{2.5.25}$$

Similarly, the velocity vector $\boldsymbol{u}$ is giver by

$$\tilde{\mathbf{u}} = \tilde{\mathbf{U}}(\mathbf{x})e^{-i\omega t} \tag{2.5.26}$$

But p' and $\boldsymbol{u}$ are related by

$$\frac{\partial \mathbf{u}}{\partial t} = -\frac{1}{\rho_0}\nabla p' \tag{2.5.27}$$

Hence,

$$\tilde{\mathbf{U}}(x) = \frac{1}{i\rho_0\omega}\nabla\tilde{P}'(\mathbf{x}) \tag{2.5.28}$$

For plane waves, $\nabla\tilde{P}(\mathbf{x}) = ik\tilde{P}'(\mathbf{x})$, and (2.5.28) reduces to a particular example of the simple-wave relationship between $\mathbf{u}$ and P' given by (2.4.13).

Time Averages

In some applications of acoustics, one is interested in obtaining time averages of the acoustic pressure, velocity and so on. By definition, the average, over a time τ, of a time-dependent quantity $f(t)$ is given by

$$\langle f(t)\rangle = \frac{1}{\tau}\int_0^\tau f(t)\,dt \tag{2.5.29}$$

For monochromatic waves, we may average over one or more complete periods. The result would be the same in either case because of the periodic nature of those waves. However, the time averages of the acoustic variables will be zero, and to obtain a measure of the average magnitude of the variations, one uses the so-called root-mean-squared average defined by

$$f_{\text{rms}} = \langle f^2\rangle^{1/2} \tag{2.5.30}$$

For example, if we have a monochromatic function $p' = A\cos(\omega t)$, then $p'_{\text{rms}} = A/\sqrt{2}$. Now, in many of the equations in this section, we have represented the acoustic variables by means of complex quantities on the understanding that at the end of the calculations only the real part should be used. This requirement is particularly important when dealing with products such as p'^2, u^2, and so on because for any such quantity, $f^2 = (\text{Re}\tilde{f})^2 \neq \text{Re}(\tilde{f}^2)$. Thus, if a variable is expressed in complex form, then in order to obtain its rms value, we must first extract its real part, square it, and finally integrate the result over one period. In many situations, such as in the plane, monochromatic wave case, this procedure is very simple. However, on some occasions, it may require considerably more effort because the complex functions appearing then are not as simple as $\exp(-i\omega t)$. Fortunately, we may, even then, easily obtain time averages of squared monochromatic quantities by writing for the real part of any complex quantity $\tilde{F}$ the following identity:

$$\text{Re}\,\tilde{F} = \tfrac{1}{2}(\tilde{F} + \tilde{F}^*) \qquad (2.5.31)$$

where $\tilde{F}^*$ is the complex conjugate of $\tilde{F}$. That is, if $\tilde{F} = a + ib$, with a and b real, then $\tilde{F}^* = a - ib$. This identity gives

$$(\text{Re}\,\tilde{F})^2 = \tfrac{1}{4}(\tilde{F}^2 + 2\tilde{F}\tilde{F}^* + \tilde{F}^{*2}) \qquad (2.5.32)$$

We are interested in the time average of this quantity. Substituting (2.5.32) into (2.5.29), and remembering that the average of harmonic functions of frequency 2ω also vanishes, we obtain

$$\langle(\text{Re}\,\tilde{F})^2\rangle = \tfrac{1}{2}(\tilde{F}\tilde{F}^*) \qquad (2.5.33)$$

Similarly, if we want to obtain the average of the product between two variables represented in complex form by $\tilde{F}$ and $\tilde{G}$, say, then

$$\langle(\text{Re}\,\tilde{F})(\text{Re}\,\tilde{G})\rangle = \tfrac{1}{4}(\tilde{F}\tilde{G}^* + \tilde{F}^*\tilde{G}) \qquad (2.5.34)$$

but since $(\tilde{F}\tilde{G}^*)^* = (\tilde{F}^*\tilde{G})$, it then follows that

$$(\tilde{F}\tilde{G}^* + \tilde{F}^*\tilde{G}) = \left[\tilde{F}\tilde{G}^* + (\tilde{F}\tilde{G}^*)^*\right] = 2\,\text{Re}(\tilde{F}\tilde{G}^*) \qquad (2.5.35)$$

Thus,

$$\langle(\text{Re}\,\tilde{F})(\text{Re}\,\tilde{G})\rangle = \tfrac{1}{2}\text{Re}(\tilde{F}\tilde{G}^*) = \tfrac{1}{2}\text{Re}(\tilde{F}^*\tilde{G}) \qquad (2.5.36)$$

These expressions considerably simplify the computations of time averages, as will be seen in later sections of this book.

PROBLEMS

2.5.1 Let

$$\tilde{\phi} = u_0 \frac{R^2}{ikR-1} \frac{1}{r} e^{[ik(r-R)-\omega t]}$$

where r, R, and u_0 are real. Determine $\langle p'u \rangle$, where $p' = -\rho_0(\partial\phi/\partial t)$ and $u = \partial\phi/\partial r$.

2.5.2 Show that if $\tilde{\phi} = \tilde{\Phi}(\mathbf{x})e^{-i\omega t}$ satisfies the wave equation, then

$$\langle \mathbf{u}\cdot\mathbf{u} \rangle = \tfrac{1}{2}k^2|\tilde{\Phi}|^2 + \tfrac{1}{2}\nabla\cdot(\tilde{\Phi}\nabla\tilde{\Phi})$$

where $k = \omega/c_0$.

2.6 FOURIER ANALYSIS

Although no specific problems were solved in the last section, it should be clear that monochromatic waves are the simplest type of oscillatory wave that may be found in nature. However, this simplicity is not shared by most waves that actually occur. Fortunately, it is possible to decompose particular types of actual waves into monochromatic components having different frequencies. This procedure, called Fourier's decomposition in honor of its originator, is briefly outlined below.

Periodic Waveforms—Fourier Series

We first consider waves that produce, at a fixed point in space, periodic changes of pressure, velocity, and so on. That is, if these quantities are represented by some function of time and position $f(\mathbf{x}, t)$, then

$$f(\mathbf{x}, t+T) = f(\mathbf{x}, t)$$

where T is the period of the induced changes, as shown schematically in Figure 2.6.1. If the function f and its derivatives are mathematically well behaved (in the sense that they are continuous functions except for a finite number of jump discontinuities), then Fourier's theorem states that f may be represented as

$$f(\mathbf{x}, t) = \tfrac{1}{2}a_0(\mathbf{x}) + \sum_{n=1}^{\infty} \left[a_n(\mathbf{x})\cos(n\omega t) + b_n(\mathbf{x})\sin(n\omega t) \right] \qquad (2.6.1)$$

where the quantity

$$\omega = 2\pi/T \qquad (2.6.2)$$

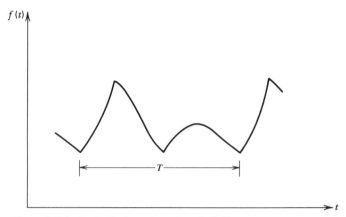

Figure 2.6.1 Periodic function of period T.

is known as the fundamental frequency of the wave. The constant term as $a_0/2$ is basically the average value of f. The terms in the infinite series represent the monochromatic components of the wave. Thus, the term with $n=1$ represents the fundamental component, whereas terms with frequencies $2\omega, 3\omega$, and so on represent the *harmonic* components. All periodic waves actually occurring meet the mathematical requirements stated above so that our problem is reduced to finding the Fourier coefficients a_n and b_n. These may be obtained from (2.6.1), together with some of the properties of the trigonometric functions. Thus, the coefficient $a_0(x)$ can be obtained by integrating (2.6.1), term by term, in the range $-T/2 \leqslant t \leqslant T/2$. This gives

$$a_0(\mathbf{x}) = \frac{2}{T} \int_{-T/2}^{T/2} f(\mathbf{x}, t)\, dt \tag{2.6.3}$$

To obtain the other coefficients, we take advantage of the orthogonality conditions of the trigonometric functions. These are, for any nonzero integers m and n,

$$\int_{-T/2}^{T/2} \cos(m\omega t)\cos(n\omega t)\, dt = \frac{T}{2}\delta_{nm} \tag{2.6.4}$$

$$\int_{-T/2}^{T/2} \sin(m\omega t)\sin(n\omega t)\, dt = \frac{T}{2}\delta_{nm} \tag{2.6.5}$$

$$\int_{-T/2}^{T/2} \sin(n\omega t)\cos(n\omega t)\, dt = 0 \tag{2.6.6}$$

where δ_{nm} is Kronecker's delta. Thus, if we multiply (2.6.1) by $\cos m\omega t$ and integrate the result between $-T/2$ and $T/2$, the only term on the right-hand side that does not vanish upon integration is

$$a_m \int_{-T/2}^{T/2} \cos^2 m\omega t \, dt = \frac{T}{2} a_m \qquad (2.6.7)$$

Therefore,

$$a_n(\mathbf{x}) = \frac{2}{T} \int_{-T/2}^{T/2} f(\mathbf{x}, t) \cos n\omega t \, dt \qquad (2.6.8)$$

Similarly,

$$b_n(\mathbf{x}) = \frac{2}{T} \int_{-T/2}^{T/2} f(\mathbf{x}, t) \sin n\omega t \, dt \qquad (2.6.9)$$

Hence, if $f(\mathbf{x}, t)$ is given, (2.6.3), (2.6.8), and (2.6.9) may be used to obtain the coefficients in the Fourier series for $f(\mathbf{x}, t)$. It is clear that some terms of this series may sometimes be absent. For example, if the function $f(\mathbf{x}, t)$ is even in t, that is, if $f(\mathbf{x}, -t) = f(\mathbf{x}, t)$, then $b_n(\mathbf{x}) = 0$ for all n so that the series will not contain any sine terms. Similarly, no cosine terms will appear in the series for the special case of odd functions, that is, when $f(\mathbf{x}, -t) = -f(\mathbf{x}, t)$. However, since most functions are neither even nor odd, a typical Fourier series will contain both trigonometric functions.

A more compact, but equivalent representation of $f(\mathbf{x}, t)$ is

$$f(\mathbf{x}, t) = \sum_{n=-\infty}^{\infty} \tilde{c}_n(\mathbf{x}) e^{-in\omega t} \qquad (2.6.10)$$

where

$$\tilde{c}_n = \frac{1}{T} \int_{-T/2}^{T/2} f(\mathbf{x}, t) e^{in\omega t} \, dt \qquad (2.6.11)$$

To obtain a relationship between the complex coefficients $\tilde{c}_n$ and the real coefficients a_n and b_n, we write (2.6.10) as

$$f(\mathbf{x}, t) = \tilde{c}_0 + \sum_{n=1}^{\infty} (\tilde{c}_{-n} + \tilde{c}_n) \cos(n\omega t) + \sum_{n=1}^{\infty} i(\tilde{c}_{-n} - \tilde{c}_n) \sin(n\omega t) \qquad (2.6.12)$$

Thus, $a_n = \tilde{c}_{-n} + \tilde{c}_n$ and $b_n = i(\tilde{c}_{-n} - \tilde{c}_n)$. But since a_n and b_n are real, we must have $\tilde{c}_{-n} = \tilde{c}^*_n$. Therefore,

$$\tilde{c}_n = \tfrac{1}{2}(a_n + ib_n) \qquad (2.6.13)$$

The expansion (2.6.10) is well suited to represent our acoustic variables, but for this to be so, the coefficients $\tilde{c}_n(\mathbf{x})$ must be such that $f(\mathbf{x}, t)$ satisfies the wave equation. To obtain the conditions on $\tilde{c}_n(\mathbf{x})$, we substitute (2.6.10) in the wave equation and obtain

$$\sum_{n=-\infty}^{\infty} \left[\nabla^2 \tilde{c}_n(\mathbf{x}) + k_n^2 \tilde{c}_n(\mathbf{x}) \right] e^{-in\omega t} = 0 \qquad (2.6.14)$$

where

$$k_n = n\omega/c_0 \qquad (2.6.15)$$

Now, the functions $\sin n\omega t, \cos n\omega t$, with $n = 0, \pm 1, \ldots$ are linearly independent. That is, no one of them may be expressed as a linear combination of the others. Therefore, the vanishing of the series (2.6.14) requires that for each value of n, the quantity in brackets by zero. Thus,

$$\nabla^2 \tilde{c}_n(\mathbf{x}) + k_n^2 \tilde{c}_n(\mathbf{x}) = 0 \qquad (2.6.16)$$

This is a Helmholtz equation, and reference to (2.5.1) and (2.5.2) will show that each term of the series (2.6.10) simply represents a monochromatic wave of frequency $n\omega$.

The plane case is particularly simple, for then (2.6.16) has solutions of the form

$$\tilde{c}_n = \tilde{A}_n e^{inkx} + \tilde{B}_n e^{-inkx} \qquad (2.6.17)$$

so that

$$f(x, t) = \sum_{n=-\infty}^{\infty} \left[\tilde{A}_n e^{in(kx-\omega t)} + \tilde{B}_n e^{-in(kx+\omega t)} \right] \qquad (2.6.18)$$

The constants $\tilde{A}_n$ and $\tilde{B}_n$ may be obtained from boundary conditions or from given wave profiles. Thus, for example, suppose f and $\partial f/\partial x$ are specified at planes $x = x_1$ and $x = x_2$, respectively, by $f(x_1, t) = F(t)$ and $\partial F(x_2, t)/\partial x = G(t)$. Then, if $F(t)$ and $G(t)$ can be expanded in Fourier series,

$$F(t) = \sum_{n=-\infty}^{\infty} \tilde{F}_n e^{-in\omega t}, \qquad G(t) = \sum_{n=-\infty}^{\infty} \tilde{G}_n e^{-in\omega t} \qquad (2.6.19)$$

where

$$\tilde{F}_n = \frac{1}{T} \int_{-T/2}^{T/2} F(t) e^{in\omega t}\, dt, \qquad \tilde{G}_n = \frac{1}{T} \int_{-T/2}^{T/2} G(t) e^{in\omega t}\, dt \qquad (2.6.20)$$

Then, it follows that

$$\tilde{A}_n e^{inkx_1} + \tilde{B}_n e^{-inkx_1} = \tilde{F}_n$$

$$\tilde{A}_n e^{inkx_2} - \tilde{B}_n e^{-inkx_2} = \tilde{G}_n / ink \qquad (2.6.21)$$

from which $\tilde{A}_n$ and $\tilde{B}_n$ may be obtained.

Expansions obtained in the above form are most useful in obtaining solutions of the wave equation that satisfy prescribed boundary conditions. Another application of Fourier series is the original one: decomposition of a *given* wave into its monochromatic components. Here, instead of studying the time changes of some acoustic property at some fixed point in space, one considers the instantaneous spatial variations of that property in the wave. In general, these variations are not exactly periodic, even if the wave is monochromatic, so that the above Fourier series expansion cannot be used. However, in the case of plane, one-dimensional waves in ideal fluids, the spatial variations are periodic if the temporal variations are periodic. Consider, for example, the first term in (2.6.18). Then, if we put $x = x + \lambda$, we obtain

$$f(x+\lambda, t) = \sum_{n=-\infty}^{\infty} \tilde{A}_n e^{ink\lambda + in(kx-\omega t)} \qquad (2.6.22)$$

But since for this wave $k\lambda = 2\pi$, it follows that $f(x, t)$ is also periodic in x. Therefore, we can use the method given above to obtain the Fourier components of a wave that at some time $t = t_0$, say, has a given profile along the x axis. The working equation is still (2.6.18), but now the coefficients are determined by the shape of the wave at $t = t_0$. For example, suppose that $F(x)$ represents the acoustic pressure at $t = t_0$ in a wave traveling to the right. Then, because of the periodicity of $F(x)$, we may express it as

$$F(x) = \sum_{n=-\infty}^{\infty} \tilde{F}_n e^{inkx} \qquad (2.6.23)$$

where

$$\tilde{F}_n = \frac{1}{\lambda} \int_{-\lambda/2}^{\lambda/2} F(x) e^{-inkx} \qquad (2.6.24)$$

Therefore,

$$\tilde{A}_n = \tilde{F}_n e^{i\omega n t_0} \qquad (2.6.25)$$

so that at any time t, the positive-going wave may be expressed as

$$f(x,t)= \sum_{n=-\infty}^{\infty} \tilde{F}_n e^{in[kx-\omega(t-t_0)]} \tag{2.6.26}$$

As an example, consider the sawtooth profile shown in Figure 2.6.2. The location of the origin of the x axis has been chosen to simplify the calculations. Any other location will give the same results. Now, in the region between the discontinuities of $F(x)$, we have

$$F(x)= \frac{2H}{\lambda} x, \quad -\frac{\lambda}{2} \leqslant x \leqslant \frac{\lambda}{2}$$

Therefore,

$$\tilde{F}_n = \frac{2H}{\lambda^2} \int_{-\lambda/2}^{\lambda/2} xe^{-inkx} dx$$

For $n=0$, this gives $\tilde{F}_0=0$. For nonzero n, we integrate by parts and obtain

$$\tilde{F}_n = - \frac{H \cos n\pi}{\pi i n}, \quad n \neq 0$$

Therefore,

$$f(x,t)= \frac{H}{\pi} \sum_{n=-\infty}^{\infty} \frac{(-1)^{n+1}}{in} e^{in(kx-\omega t)}, \quad n\neq 0$$

$$= \frac{H}{\pi} \sum_{n=1}^{\infty} \frac{(-1)^{n+1}}{in} \left[e^{in(kx-\omega t)} - e^{-in(kx-\omega t)} \right]$$

$$f(x,t)= \frac{2H}{\pi} \sum_{n=1}^{\infty} \frac{(-1)^{n+1}}{n} \sin n(kx-\omega t) \tag{2.6.27}$$

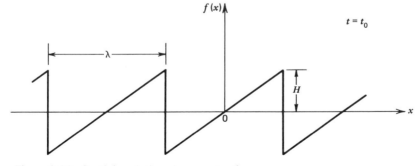

Figure 2.6.2 Spatial variations in a sawtooth wave.

This is the desired result, as it gives the acoustic variable represented by f as an infinite series of positive-going waves having frequencies ω, 2ω, 3ω, and so on. It should be noted that each of the monochromatic components travels at the same speed $c_0 = \omega/k$. Stated differently, the speed of propagation of each component is independent of the frequency. If this were not the case, the profile of the wave would not remain constant. Waves for which the speed of propagation depends on the frequency are known as *dispersive* and will be considered later.

The result given by (2.6.27) may also be used to obtain the frequency distribution at a fixed location. Thus, at $x = 0$, (2.6.27) gives

$$f(t) = \frac{2H}{\pi} \sum_{n=1}^{\infty} \frac{(-1)^n}{n} \sin n\omega t \qquad (2.6.28)$$

As consideration of Figure 2.6.2 will show, this is also a sawtooth, but the "teeth" now have a negative slope.

Nonperiodic Functions—Fourier Transform

Functions that are not periodic cannot be expressed as an infinite series of harmonic frequency components. Instead, their frequency spectrum will contain frequencies of all values. In other words, whereas the spectrum of a periodic function of fundamental frequency ω contains components of frequencies 2ω, 3ω, 4ω, and so on, the spectrum of a nonperiodic function will be a continuous function of ω.

The mathematical techniques required to represent nonperiodic functions by their various frequency components are usually derived by considering the limit when the basic interval in the Fourier series expansion, $-T/2 \leqslant t \leqslant T/2$, is allowed to cover the whole real axis. That is, the Fourier series representation of a function is considered in the limit $T \to \infty$. Now, as T increases, the frequency difference between two consecutive terms in the series decreases so that in the limit as $T \to \infty$ the series becomes an integral over all frequencies. The derivation may be found in books on the subject. (See, for example, I. Sneddon, *Fourier Transforms*, McGraw-Hill, New York.) Here we merely give the final results. These are that under some conditions, the nonperiodic function $f(t)$ may be expressed as

$$f(t) = \frac{1}{\sqrt{2\pi}} \int_{-\infty}^{\infty} F(\omega) e^{-i\omega t} d\omega \qquad (2.6.29)$$

where

$$F(\omega) = \frac{1}{\sqrt{2\pi}} \int_{-\infty}^{\infty} f(t) e^{i\omega t} dt \qquad (2.6.30)$$

is called the *Fourier Transform* of $f(t)$. The conditions that a function $f(t)$ must satisfy so that it may be represented in the above manner are that the function and its first derivative be piecewise continuous, and that

$$\int_{-\infty}^{\infty} |f(t)|\, dt < \infty$$

Except for the case when $f(t)$ represents a stationary time function, these conditions are satisfied by all acoustic variables. In the case of a stationary time signal, say $f(t) = a \sin \Omega t$, the above integral is not bounded so that $F(\omega)$ does not exist. What is usually done in such a case is to evaluate $F(\omega)$ over a finite time interval, instead of the infinite time interval specified by (2.6.30). The resulting spectrum will then contain other frequencies. Such a procedure is, in fact, adopted when a frequency analyzer is used to measure the spectrum of a time-dependent signal.

Returning now to (2.6.29), we substitute $F(\omega)$ from (2.6.30) and obtain the Fourier integral representation

$$f(t) = \frac{1}{2\pi} \int_{-\infty}^{\infty} e^{-i\omega t} \int_{-\infty}^{\infty} F(u) e^{i\omega u}\, du\, d\omega$$

$$= \frac{1}{2\pi} \int_{-\infty}^{\infty} \int_{-\infty}^{\infty} f(u) e^{i\omega(u-t)}\, du\, d\omega \qquad (2.6.31)$$

If the function $f(t)$ is real, the imaginary part of the right-hand side must be zero so that we may write

$$f(t) = \frac{1}{\pi} \int_{0}^{\infty} \int_{-\infty}^{\infty} f(u) \cos \omega(u-t)\, du\, d\omega \qquad (2.6.32)$$

For the special cases of even or odd functions, this may be simplified further. Thus, since

$$\cos \omega(u-t) = \cos \omega u \cos \omega t + \sin \omega u \sin \omega t$$

it follows that if $f(t)$ is even, then

$$f(t) = \sqrt{\frac{2}{\pi}} \int_{0}^{\infty} F(\omega) \cos \omega t\, d\omega \qquad (2.6.33)$$

where

$$F(\omega) = \sqrt{\frac{2}{\pi}} \int_{0}^{\infty} f(u) \cos \omega u\, du \qquad (2.6.34)$$

Similarly, when $f(t)$ is odd, we have

$$f(t) = \sqrt{\frac{2}{\pi}} \int_0^\infty F(\omega) \sin \omega t \, d\omega \qquad (2.6.35)$$

where

$$F(\omega) = \sqrt{\frac{2}{\pi}} \int_0^\infty f(u) \sin \omega u \, du \qquad (2.6.36)$$

As an example, we consider a sinusoidal wave of finite duration,

$$f(t) = \sin \Omega t, \quad -\tau \leqslant t \leqslant \tau$$

Since this function is odd, we may use (2.6.36) to obtain its transform. Thus,

$$F(\omega) = \sqrt{\frac{2}{\pi}} \int_0^\tau \sin \Omega u \sin \omega u \, du \qquad (2.6.37)$$

Now,

$$\int \sin au \sin bu \, du = \frac{1}{2} \left[\frac{\sin(a-b)u}{a-b} - \frac{\sin(a+b)u}{a+b} \right]$$

Therefore,

$$F(\omega) = \frac{1}{\sqrt{2\pi}} \left[\frac{\sin(\Omega-\omega)\tau}{\Omega-\omega} - \frac{\sin(\Omega+\omega)\tau}{\Omega+\omega} \right] \qquad (2.6.38)$$

This gives the frequency distribution of our monochromatic wave of finite duration. Figure 2.6.3 shows the variations of $F(\omega)$ with ω for one value of τ. It is seen that in addition to a main component at $\omega = \Omega$, this spectrum contains many other frequencies. The appearance of these is due to finite duration of the signal. This statement may be physically obvious, but it is nevertheless instructive to extract it directly from (2.6.38). To do this, we first note that $F(\omega)$ is controlled mainly by the first term inside the square bracket in (2.6.38). It therefore follows that the value of $F(\omega)$ at $\Omega = \omega$ is

$$F(\omega) \approx \tau / \sqrt{2\pi} \qquad (2.6.39)$$

Similarly, the most significant frequency components, that is, those with largest amplitude, have frequencies in the band

$$\Delta\omega \approx 2\pi / \tau \qquad (2.6.40)$$

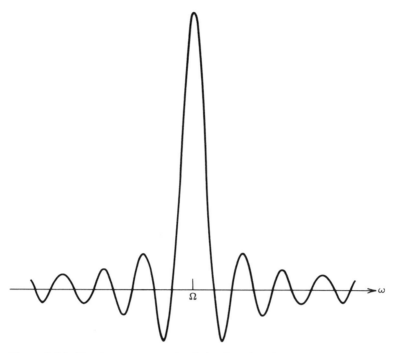

Figure 2.6.3 Fourier spectrum of a finite-duration sinusoidal oscillation.

centered at $\omega = \Omega$. Hence, the longer the duration of the wave train, the closer will a real sinusoidal pulse approximate the single-frequency ideal of a monochromatic wave.

PROBLEMS

2.6.1 Obtain the Fourier series expansion for the function sketched in the figure. By numerical addition, show that on either side of the discontinuities, the series gives a value that differs significantly from unity.

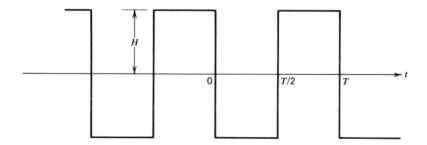

2.6.2   A "sonic boom" produces, at a given location, a pressure variation of the form shown below. Because of its similarity with the letter N, such a profile is sometimes called an N wave. If the duration of this N wave is τ, obtain the frequency distribution in the wave and plot your results versus $\omega\tau$. Show that the largest-amplitude contributions to the frequency spectrum are with a frequency band $\Delta\omega \sim 2\pi/\tau$. Compare your frequency distribution with that applicable for the periodic N wave given by (2.6.27).

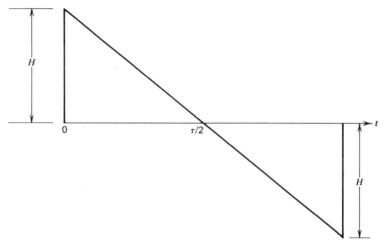

2.6.3   Consider periodic but nonmonochromatic functions of time having the same fundamental period. Then, any two such functions may be expressed in Fourier series as

$$F = \sum_{n=0}^{\infty} F_n \cos(\alpha_n - n\omega t), \qquad G = \sum_{n=0}^{\infty} G_n \cos(\beta_n - n\omega t)$$

where F_n and G_n are real. Show that

$$\langle FG \rangle = \tfrac{1}{2} \sum_n F_n G_n \cos(\alpha_n - \beta_n)$$

2.7   ACOUSTIC ENERGY

We now return to our main topic and consider the energy of a sound wave in a uniform fluid that is otherwise at rest. The important case of nonuniform fluids is not considered here. The interested reader may consult some of the works on acoustics in nonuniform media that are listed in the bibliography.

For the case of uniform fluids at rest, we may obtain an equation for the conservation of acoustic energy from the equation for the internal energy that was derived in Section 1.6. to do this, we first use (1.6.5) to obtain an equation for the kinetic and potential energy of an ideal fluid. Thus, with all transport quantities set equal to zero, (1.6.7) yields

$$\rho \frac{DE}{Dt} + p\nabla \cdot \mathbf{u} = 0 \qquad (2.7.1)$$

Now, in the absence of body forces, Euler's equation (2.1.2) reduces to

$$\rho \frac{D\mathbf{u}}{Dt} = -\nabla p \qquad (2.7.2)$$

Taking the dot product of this with $\mathbf{u}$ and substituting the result in (2.7.1), we obtain

$$\rho \frac{D}{Dt}\left(E + \tfrac{1}{2}q^2\right) = -\nabla \cdot (p\mathbf{u}) \qquad (2.7.3)$$

where $q^2 = \mathbf{u} \cdot \mathbf{u}$. Finally, the continuity equation

$$\frac{D\rho}{Dt} + \rho\nabla \cdot \mathbf{u} = 0 \qquad (2.7.4)$$

may be used to write (2.7.3) as

$$\frac{D}{Dt}\left(\rho E + \tfrac{1}{2}\rho q^2\right) + \rho\left(E + \tfrac{1}{2}q^2\right)\nabla \cdot \mathbf{u} = -\nabla \cdot (p\mathbf{u}) \qquad (2.7.5)$$

Using the definition of D/Dt, this may also be written as

$$\frac{\partial}{\partial t}\left(\rho E + \tfrac{1}{2}\rho q^2\right) + \nabla \cdot \left(\rho E + \tfrac{1}{2}\rho q^2\right)\mathbf{u} = -\nabla \cdot (p\mathbf{u}) \qquad (2.7.6)$$

This is the desired equation. It applies to general motions of ideal fluids. Presently, however, we are interested in sound wave motions, that is, in small departures from the equilibrium state $\rho = \rho_0, p = p_0, \mathbf{u} = 0$. For these, (2.7.6) may be simplified considerably. Let us consider the energy per unit volume of fluid, $\rho E + \tfrac{1}{2}(\rho q^2)$, in conditions near to complete equilibrium. Remembering that the entropy of the fluid is constant, we may expand this quantity as a function of the density fluctuation $(\rho - \rho_0)$. Clearly, such an expansion must contain at least second-order terms. Otherwise, for example, the kinetic energy of the sound wave would have to be taken as zero. Therefore,

$$\rho E + \tfrac{1}{2}\rho q^2 = \rho_0 E_0 + \left[\left(\frac{\partial(\rho E)}{\partial \rho}\right)_S\right]_{\rho=\rho_0} (\rho - \rho_0)$$

$$+ \frac{1}{2}\left[\left(\frac{\partial^2 \rho E}{\partial \rho^2}\right)_S\right]_{\rho=\rho_0} (\rho - \rho_0)^2 + \cdots + \tfrac{1}{2}\rho_0 q^2 \qquad (2.7.7)$$

Consider the first thermodynamic derivative. It can be written as

$$\left[\frac{\partial(\rho E)}{\partial \rho}\right]_S = \rho\left(\frac{\partial E}{\partial \rho}\right)_S + E \qquad (2.7.8)$$

Now, from (1.3.15) we know that

$$T\,dS = dE - \frac{p}{\rho^2}\,d\rho$$

Hence,

$$\left(\frac{\partial E}{\partial \rho}\right)_S = \frac{p}{\rho^2} \qquad (2.7.9)$$

Therefore,

$$\left[\frac{\partial(\rho E)}{\partial \rho}\right]_S = E + \frac{p}{\rho} \qquad (2.7.10)$$

but the quantity $(E + p/\rho)$ is the enthalpy per unit mass H. Hence,

$$\left[\frac{\partial(\rho E)}{\partial \rho}\right]_S = H \qquad (2.7.11)$$

The second derivative appearing in Equation (2.7.7) is therefore

$$\left[\frac{\partial^2(\rho E)}{\partial \rho^2}\right]_S = \left(\frac{\partial H}{\partial \rho}\right)_S \qquad (2.7.12)$$

But, since $T\,dS = dH - (1/\rho)\,dp$, we obtain

$$\left(\frac{\partial H}{\partial \rho}\right)_S = \frac{1}{\rho}\left(\frac{\partial p}{\partial \rho}\right)_S = \frac{c^2}{\rho} \qquad (2.7.13)$$

We now substitute these derivatives, evaluated at equilibrium conditions, in the equation for the energy of the fluid and obtain, with $\rho - \rho_0 = \rho'$,

$$\rho E + \tfrac{1}{2}\rho q^2 = \rho_0 E_0 + H_0 \rho' + \frac{c_0^2}{2\rho_0}\rho'^2 + \tfrac{1}{2}\rho_0 q^2 + \cdots \qquad (2.7.14)$$

Substituting this into (2.7.6) yields

$$\frac{\partial}{\partial t}\left(\tfrac{1}{2}\rho_0 q^2 + \frac{1}{2}\frac{c_0^2}{\rho_0}\rho'^2\right) + H_0\left(\frac{\partial \rho'}{\partial t} + \rho'\nabla\cdot\mathbf{u} + \mathbf{u}\cdot\nabla\rho'\right)$$

$$+ \rho_0\left(E_0 + \frac{p_0}{\rho_0}\right)\nabla\cdot\mathbf{u} = -\nabla\cdot(p'\mathbf{u})$$

$$(2.7.15)$$

The second and third bracketed terms on the left-hand side of this equation cancel out because of the continuity equation, so that (2.7.15) becomes

$$\frac{\partial}{\partial t}\left(\tfrac{1}{2}\rho_0 q^2 + \frac{1}{2}\frac{c_0^2}{\rho_0}\rho'^2\right) = -\nabla\cdot(p'\mathbf{u}) \qquad (2.7.16)$$

This is an acoustic energy balance equation. Its meaning may become clearer by integrating the equation over a *fixed* volume of fluid. The result is

$$\frac{\partial}{\partial t}\int_V\left(\tfrac{1}{2}\rho_0 q^2 + \frac{1}{2}\frac{c_0^2}{\rho_0}\rho'^2\right)dV = -\int_A p'\mathbf{u}\cdot\mathbf{n}\,dA \qquad (2.7.17)$$

This shows that the quantity within the time derivative is the total acoustic energy in V, that is,

$$E_a = \int\left(\tfrac{1}{2}\rho_0 q^2 + \frac{1}{2}\frac{c_0^2}{\rho_0}\rho'^2\right)dV \qquad (2.7.18)$$

and that the quantity on the right-hand side is the rate at which acoustic energy, that is, acoustic power, leaves V.

Energy Density

The quantity in the integrand on the left-hand side of the acoustic energy balance equation has the dimensions of energy per unit volume. It can therefore be regarded as an *acoustic energy density* and will be denoted by the symbol ε:

$$\varepsilon = \tfrac{1}{2}\rho_0 q^2 + \frac{1}{2}\frac{c_0^2}{\rho_0}\rho'^2 \tag{2.7.19}$$

Since, for our case, the pressure and density are related by $p' = c_0^2 \rho'$, we can also write ε as

$$\varepsilon = \tfrac{1}{2}\rho_0 q^2 + \frac{1}{2\rho_0 c_0^2}p'^2 \tag{2.7.20}$$

The two terms on the right-hand side are, respectively, the kinetic and potential energy densities of the sound wave.

Plane Waves. The kinetic energy density $\tfrac{1}{2}(\rho_0 q^2)$ and the potential energy density $\tfrac{1}{2}(c_0^2\rho'^2/\rho_0)$ are, in general, different. However, in the case of plane waves they are identical. This can be seen from (2.4.16), which gives $u = \pm(c_0/\rho_0)\rho'$. Therefore,

$$q^2 = (c_0/\rho_0)^2\rho'^2 \tag{2.7.21}$$

so that $\tfrac{1}{2}(\rho_0 q^2) = \tfrac{1}{2}(c_0^2\rho'^2/\rho_0)$. Hence, for a plane propagating wave, we simply have

$$\varepsilon = \rho_0 q^2 = \frac{p'^2}{\rho_0 c_0^2} \tag{2.7.22}$$

Monochromatic Waves. Another instance where similar simplification can be made in the computation of the acoustic energy is when the waves are monochromatic. In that special case, one is interested in the time average of the total energy; thus,

$$\langle E_a \rangle = \int_V \left[\tfrac{1}{2}\rho_0\langle q^2 \rangle + \frac{c_0^2}{2\rho_0}\langle \rho'^2 \rangle \right] dV \tag{2.7.23}$$

where V represents the total volume occupied by the fluid. Now, $q^2 = \mathbf{u} \cdot \mathbf{u}$ and $\mathbf{u} = \nabla\phi$, where $\phi = \Phi(\mathbf{x})\cos(\omega t - a)$. However, it is simpler to work with complex quantities. Hence, if $\phi = \tilde{\Phi}(\mathbf{x})e^{-i\omega t}$, then we can write

$$\langle q^2 \rangle = \tfrac{1}{2}\left[(\nabla\tilde{\Phi}) \cdot (\nabla\tilde{\Phi})^* \right] \tag{2.7.24}$$

Similarly, since $\rho' = -(\rho_0/c_0^2)(\partial\phi/\partial t)$, so that $\tilde{\rho}' = (i\omega\rho_0/c_0^2)\tilde{\Phi}(x)e^{-i\omega t}$, we obtain

$$\langle \rho'^2 \rangle = \frac{\rho_0^2}{2c_0^4}\omega^2(\tilde{\Phi}\tilde{\Phi}^*) \tag{2.7.25}$$

Now consider the first term in $\langle E_a \rangle$, that is,

$$\langle E_a \rangle_K = \int \tfrac{1}{2}\rho_0\langle q^2 \rangle \, dV = \tfrac{1}{4}\rho_0 \int \nabla\tilde{\Phi}\cdot\nabla\tilde{\Phi}^* \, dV \tag{2.7.26}$$

But

$$\nabla\tilde{\Phi}^*\cdot\nabla\tilde{\Phi} = \nabla\cdot(\tilde{\Phi}\nabla\tilde{\Phi}^*) - \tilde{\Phi}\nabla^2\tilde{\Phi}^* = \nabla\cdot(\tilde{\Phi}\nabla\tilde{\Phi}^*) + k^2\tilde{\Phi}\tilde{\Phi}^*$$

Hence,

$$\langle E_a \rangle_K = \tfrac{1}{4}\rho_0 k^2 \int_V \tilde{\Phi}\tilde{\Phi}^* \, dV + \tfrac{1}{4}\rho_0 \int_A \tilde{\Phi}\mathbf{u}\cdot\mathbf{n} \, dA \tag{2.7.27}$$

where V is the volume occupied by the fluid and A is the surface area enclosing V. Since on A we must have $\mathbf{u}\cdot\mathbf{n} = 0$, it follows that

$$\langle E_a \rangle_K = \tfrac{1}{4}\rho_0 k^2 \int \tilde{\Phi}\tilde{\Phi}^* \, dV$$

But since $\omega = kc_0$, it then follows from (2.7.25) that, provided the waves are monochromatic, the average total kinetic and potential energies are equal. Therefore,

$$\langle E_a \rangle = \rho_0 \int \langle q^2 \rangle \, dV \tag{2.7.28}$$

Further, if a nonmonochromatic wave is expressed in terms of its monochromatic components, then it may be shown (see Problem 2.7.1) that the mean energy will be given by the sum of the mean energies of the monochromatic components.

Acoustic Intensity

Consider now a region of volume V in a fluid where a sound wave is being propagated. Then, the right-hand side of (2.7.17)

$$\int_A p'\mathbf{u}\cdot\mathbf{n} \, dS$$

gives the net acoustic energy flow out of the region. The quantity

$$\mathbf{q}_a = p'\mathbf{u} \tag{2.7.29}$$

represents the instantaneous energy flow per unit area out of V, and $\mathbf{q}_a$ is therefore called the *energy-flux vector*. The time average of this vector is the *acoustic intensity* $\mathbf{I}$

$$\mathbf{I} = \langle p'\mathbf{u} \rangle \tag{2.7.30}$$

and has the dimensions of power per unit area, for example watt/m². The integral of $\mathbf{I}$ over a surface with unit normal $\mathbf{n}$ gives the *acoustic power* crossing the surface

$$\Pi = \int_A \langle p'\mathbf{u} \rangle \cdot \mathbf{n}\, dA \tag{2.7.31}$$

Of special interest are the energy flux and intensity for a plane, traveling wave. The energy flux is $\mathbf{q}_a = p'\mathbf{u}$, but for a plane, traveling wave $p' = \rho_0 c_0 u$, so that if $\mathbf{n}$ is a unit vector along the direction of propagation, then

$$\mathbf{q}_a = \rho_0 c_0 u^2 \mathbf{n} \tag{2.7.32}$$

However, the quantity $\rho_0 u^2$ is simply ε, the acoustic energy density. Thus,

$$\mathbf{q}_a = c_0 \varepsilon \mathbf{n} \tag{2.7.33}$$

That is, the energy flux is simply the product of the energy density times the speed of the wave. This result was to be expected, since the wave whose energy density is ε is moving along $\mathbf{n}$ with speed c_0. This fact is sometimes used to define the energy flux. Finally, the power per unit area in a plane wave is simply

$$\Pi/\text{Area} = \rho_0 c_0 \langle u^2 \rangle \tag{2.7.34}$$

Reference Levels

Acoustic waves have intensities that vary over a very wide range. For example, the lowest intensity level that can be heard by an individual whose hearing is normal is of the order of 10^{-12} watt/m² at 1000 Hz. This level is customarily used as the reference level in acoustics, that is,

$$I_{\text{ref}} = 10^{-12}\, \text{watt/m}^2 \tag{2.7.35}$$

On the other hand, the intensities associated with loud sounds can be of the order of 10^2 watt/m^2. In view of this wide range, and because the human ear seems to have a logarithmic response with respect to changes of intensity, an intensity level scale is defined as

$$IL = 10 \log_{10} \frac{I}{I_{ref}} \qquad (2.7.36)$$

The resulting nondimensional level is said to be in decibels (dB).

Similarly, since the intensity is proportional to the square of the root-mean-square pressure, one may introduce a sound pressure level (SPL) scale by means of

$$SPL = 20 \log_{10} \frac{P_{rms}}{P_{ref}} \qquad (2.7.37)$$

Here, P_{ref} is a root-mean-square reference pressure. For the special case of sound waves in air, it is usually taken as $P_{ref} = 2.04 \times 10^{-5}$ N/m^2 because this value nearly corresponds to I_{ref} in a plane wave (see Problem 2.7.1).

In terms of the sound pressure level, one can write

$$\frac{P_{rms}}{P_{ref}} = 10^{(SPL/20)} \qquad (2.7.38)$$

Sometimes it is more convenient to obtain the ratio between P_{rms} and the mean pressure p_0. Thus,

$$SPL = 20 \log_{10} \frac{p_0}{P_{ref}} + 20 \log_{10} \frac{P_{rms}}{p_0} \qquad (2.7.39)$$

Also,

$$\frac{P_{rms}}{p_0} = \frac{P_{ref}}{p_0} 10^{(SPL/20)} \qquad (2.7.40)$$

This can be used to obtain an idea of the magnitudes involved. For example, consider a sound wave with a sound pressure level of 100 dB in air at 1 Atm ($\approx 10^5$ N/m^2). This gives $P_{rms}/P_0 = 2 \times 10^{-5}$. Thus, even at that large sound pressure level, the pressure fluctuation is indeed very small.

PROBLEMS

2.7.1 Determine the rms velocity in a plane sound wave traveling in air at standard temperature and pressure, and having an intensity equal to I_{ref}.

2.7.2 Consider a nonmonochromatic but periodic wave. Show that the average of the total energy is equal to the sum of the averages of its monochromatic components, that is,

$$\langle E_a \rangle = \langle E_1 \rangle + \langle E_2 \rangle + \cdots + \langle E_n \rangle$$

2.7.3 In 1878, Thomas A. Edison was awarded patent no. 210,767 for a "vocal engine." This device would convert acoustical energy into useful work. According to Edison, this engine can act as a prime motor for various light mechanisms. Assuming that the device has an input area of 1 m^2, and converts 75 percent of all of the incident energy into work, determine the acoustic intensity level required to light a 15-watt light bulb.

2.7.4 Consider two sound sources, each producing at some location an intensity level equal to IL$_1$. What is the combined intensity level?

SUGGESTED REFERENCES

Lindsay, R. B. "The Story of Acoustics." This article presents a very enjoyable historical account of the development of mathematical and experimental tools for studying the production, propagation, and reception of sound waves.

Westfall, R. S. "Newton and the Fudge Factor." A significant portion of this article deals with Newton's efforts to make his formula for the speed of sound agree with experimental observation.

CHAPTER THREE
REFLECTION AND TRANSMISSION PHENOMENA

In the last chapter, we studied some of the properties of sound waves traveling in unbounded media. Such waves do not exist, of course, as there are always boundaries limiting the regions in which the waves propagate.

When a wave meets a boundary, there will appear *reflected* and *transmitted* waves. The nature of these waves depends on several factors, which include the properties of the incident wave, the angle of incidence, the nature of the boundary, and so forth. For example, when a wave meets a rigid wall, we expect that it will be reflected almost completely, whereas if the boundary is free to move, some transmission is to be expected.

Closely related to these problems is the problem of acoustic-wave transmission in tubes. Here we find that changes in the geometry of the tube also result in reflected waves. Further, even for tubes of constant cross-sectional area, it is found that not all types of acoustic waves can be propagated in them.

Both of these situations are of some practical importance in determining the amount of noise that will be propagated through a wall, or that will be transmitted in a silencing duct.

In this chapter, we consider some simple, one- and two-dimensional problems related to the reflection and transmission of plane waves. Included here are some examples of transmision and excitation of waves in tubes of constant and varying cross-sectional area. Other types of waves will be considered in the next chapter.

3.1 NORMAL INCIDENCE

Reflection at a Rigid Surface

By far the simplest of these problems is that of a plane, monochromatic wave falling on an infinite, *rigid* plane at a normal angle of incidence, as shown

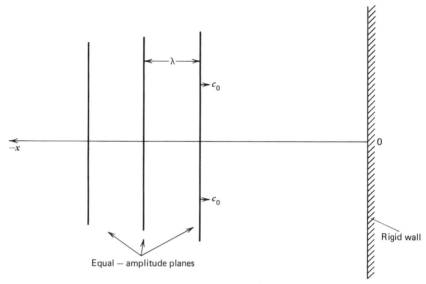

Figure 3.1.1 Plane, monochromatic wave falling on a rigid, plane wall at a normal angle of incidence.

schematically in Figure 3.1.1. Some of the aspects of the problem can be solved with arbitrary plane waves, but the simple monochromatic case is important and will be studied first. We take the direction of propagation to coincide with the x axis, and denote the potential of the incident wave by ϕ_i. Corresponding to this potential, there is an incident pressure p_i' and an incident velocity u_i. Since the boundary condition is in terms of the velocity, namely, $u = 0$, we consider this problem in terms of u. This quantity can be expressed in terms of incident and reflected waves as

$$\tilde{u} = \tilde{u}_i + \tilde{u}_r = A e^{i(kx - \omega t)} + \tilde{B} e^{-i(kx + \omega t)} \tag{3.1.1}$$

where the phase of A has been set equal to zero. Applying the boundary condition at $x = 0$, we obtain $\tilde{B} = -A$. Several interesting results follow from this simple result. First, the magnitude of the reflected-wave amplitude is equal to that of the incident. This implies that the intensities of these two waves are the same, so that no losses occur at the boundary. Second, since $B = A \exp(i\pi)$, it follows that the reflected wave has a phase angle of 180 degrees relative to the incident wave. That is, when any portion of the incident wave passes a given location, the corresponding portion in the reflected wave will not appear there until a time $\Delta t = \pi/\omega$ has elapsed. This phase lag may be understood easily by considering the general solution to the one-dimensional wave equation (see Problem 3.1.1). That solution shows that in order to satisfy the conditon $u = 0$ at a rigid boundary, the reflected wave must be of the same character as the incident wave. Thus, for example, if the incident wave is a compression pulse, the reflected wave will also be compressive.

Returning now to (3.1.1), we have, with $\tilde{B} = -A$,

$$\tilde{u}(x, t) = Ae^{-i\omega t}(e^{ikx} - e^{-ikx}) = 2iAe^{-i\omega t} \sin kx \qquad (3.1.2)$$

Taking the real part of this, we obtain

$$u(x, t) = 2A \sin \omega t \sin kx \qquad (3.1.3)$$

This velocity field is stationary in the sense that at any point x the field simply fluctuates harmonically in time. Thus, for example, if u is zero somewhere, at some time, it will remain zero at all other times. Solutions of the type give by (3.1.3) are known as *standing-wave* solutions. Similarly, (3.1.3) is said to represent a standing wave. It should be clear, however, that we may make use of the traveling-wave solution to represent a standing wave. Physically, both representations are equivalent.

The acoustic pressure corresponding to the velocity field given by (3.1.3) is

$$p'(x, t) = 2\rho_0 c_0 A \cos \omega t \cos kx \qquad (3.1.4)$$

Figure 3.1.2 shows the spatial variations of the root-mean-square (rms) pressure and velocity. The points where the pressure vanishes are called *nodes*, whereas the points where it is a maximum are called *antinodes*. Clearly, pressure and velocity nodes are interchanged. We also note that the first pressure node occurs

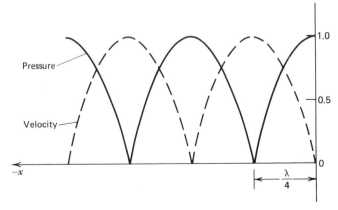

Figure 3.1.2 Standing-wave pattern in front of a rigid reflector for norma incidence.

at a distance x_0 from the wall given by

$$kx_0 = \pi/2 \qquad (3.1.5)$$

that is, at a distance $x_0 = \lambda/4$. Since $\lambda = c_0/f$, measurements of x_0 and of the frequency can be used to determine the speed of sound.

Finally, it should be noted that if (3.1.3) and (3.1.4) are used to compute the intensity at any point in the fluid, then we would obtain $I = 0$. This is as it should be, because as pointed out earlier, the intensities of the incident and reflected waves are equal so that they simply cancel each other.

Reflection at the Interface between Two Media

We now consider a plane wave falling at normal angle of incidence on the interface that separates two elastic media. These may be two immiscible fluids, a fluid in contact with an elastic solid, or two elastic solids. Since, in this case, the second medium is elastic, some of the incident-wave energy will be transmitted. Therefore, the velocity does not, in general, vanish at the interface, as was the case with the rigid reflector. The boundary conditions that apply at the interface are that the pressure and velocity be continuous across the interface. These conditions may be easily understood physically. For example, if the pressures were not the same, the interface would move until the pressures would equalize. However, it is instructive to derive them from our equations of motion. We will do this only for the case of two immiscible fluids.

Consider a volume τ of fluid enclosing part of the interface. The equation of motion for this fluid mass is given by (1.5.1). In the case of ideal fluids under no body forces, the equation reduces to

$$\frac{d}{dt} \int_\tau p\mathbf{u} \, d\tau = - \int_S p\mathbf{n} \, dA \qquad (3.1.6)$$

where S is the surface bounding the volume τ. In the acoustic approximation, this becomes

$$\frac{d}{dt} \int_{\tau^-} \rho\mathbf{u} \, d\tau = - \int_S p'\mathbf{n} \, dS \qquad (3.1.7)$$

where ρ is the mean equilibrium value of the density, which may, of course, not be the same for both media.

Now, let the volume of integration be the pillbox shown schematically in Figure 3.1.3, and let the thickness of the pillbox decrease to zero. Then we find

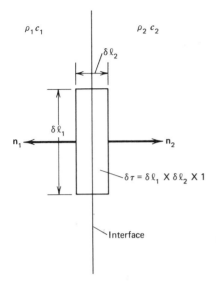

Figure 3.1.3 Control volume of integration on interface.

the the volume integral vanishes more rapidly than the area integral, so that

$$\int_S p'\mathbf{n}\, dS = 0$$

Now the contributions to this from the top and bottom of the pillbox cancel out. Therefore, if the outward normals through the lateral sides are $\mathbf{n}_1$ and $\mathbf{n}_2$, we obtain

$$p'_1\mathbf{n}_1 + p'_2\mathbf{n}_2 = 0$$

But since $\mathbf{n}_1 = -\mathbf{n}_2$, we obtain the boundary conditon that at the interface[1]

$$p'_1 = p'_2 \tag{3.1.8}$$

---

[1] It should be added that if the interface were curved, surface tension effects would produce a pressure discontinuity across the interface of magnitude

$$\Delta p = \sigma\left(\frac{1}{R_1} + \frac{1}{R_2}\right)$$

where σ is the coefficient of surface tension and R_1 and R_2 are the principal radii of curvature. The higher value of the pressure would exist in that side of the interface where the center of curvature is located.

In writing (3.1.7), we have implicitly assumed that the two media do not separate. However, since the media may move, the motion must be such that on the interface

$$u_1 = u_2 \qquad (3.1.9)$$

This boundary condition may also be obtained from the linearized continuity equation, by integration over the volume element of Figure 3.1.3, and by applying the same limits as with (3.1.7) above.

Let us now return to the transmission problem, and note that the speeds of sound in the two fluids may be different. Within each fluid, however, they have constant values, so that the problem requires the solution of two one-dimensional-wave equations:

$$\frac{\partial^2 \phi}{\partial t^2} = c_1^2 \frac{\partial^2 \phi}{\partial x^2}, \quad x < 0 \qquad (3.1.10)$$

$$\frac{\partial^2 \phi}{\partial t^2} = c_2^2 \frac{\partial^2 \phi}{\partial x^2}, \quad x > 0 \qquad (3.1.11)$$

The problem is usually solved by writing down monochromatic-wave solutions for these equations. We will, however, use the D'Alambert solution. Thus, if the incident wave is traveling from medium 1 to medium 2, we have

$$\phi = f(x - c_1 t) + g(x + c_1 t), \quad x < 0 \qquad (3.1.12)$$

$$\phi = F(x - c_2 t), \quad x > 0 \qquad (3.1.13)$$

subject to (3.1.8) and (3.1.9). In (3.1.13), only the positive-going wave has been included because medium 2, like medium 1, is taken to be of infinite extent. Now, the pressure and velocity in medium 1 are

$$\left. \begin{array}{l} p' = \rho_1 c_1 [\, f'(x - c_1 t) - g'(x + c_1 t)] \\ u = f'(x - c_1 t) + g'(x + c_1 t) \end{array} \right\} \quad x \leqslant 0 \qquad (3.1.14)$$

whereas in medium 2, the transmitted pressure and velocity are

$$\left. \begin{array}{l} p_t' = \rho_2 c_2 \, F'(x - c_2 t) \\ u_t = F'(x - c_2 t) \end{array} \right\} \quad x \geqslant 0 \qquad (3.1.15)$$

In view of the boundary conditions (3.1.8) and (3.1.9), these give

$$\rho_1 c_1 \left[f'(-c_1 t) - g'(c_1 t) \right] = \rho_2 c_2 F'(-c_2 t) \tag{3.1.16}$$

$$f'(-c_1 t) + g'(c_1 t) = F'(-c_2 t) \tag{3.1.17}$$

These represent two differential equations from which the functions g and F can be obtained. However, we are interested mainly in the intensity of the transmitted wave, and this is given in terms of the function F'. We therefore eliminate $g'(c_1 t)$ from the above equations and obtain

$$F'(-c_2 t) = \frac{2\rho_1 c_1}{\rho_1 c_1 + \rho_2 c_2} f'(-c_1 t) \tag{3.1.18}$$

This may also be written as

$$u_t(0, t) = \frac{2\rho_1 c_1}{\rho_1 c_1 + \rho_2 c_2} u_i(0, t) \tag{3.1.19}$$

where u_i is the velocity in the incident wave. Now, in many transmission problems, one of the quantities of interest is the *transmission coefficient*. This is defined as the ratio of transmitted to incident energy fluxes

$$\alpha_t = \frac{p'_t \mathbf{u}_t \cdot \mathbf{n}_t S_t}{p'_i \mathbf{u}_i \cdot \mathbf{n}_i S_i} \tag{3.1.20}$$

where S represents the area perpendicular to the direction of propagation, $\mathbf{n}$ represents a unit vector along the direction of propagation, and the substripts i and t refer to the incident and the transmitted waves, respectively. For a plane, one-dimensional case, and for $S_i = S_t$, (3.1.20) reduces to

$$\alpha_t = \frac{\rho_2 c_2}{\rho_1 c_1} \frac{u_t^2}{u_i^2} \tag{3.1.21}$$

For a monochromatic wave, one uses, instead of (3.1.20),

$$\alpha_t = \frac{\langle p'_t \mathbf{u}_t \rangle \cdot \mathbf{n} S_t}{\langle p'_i \mathbf{u}_i \rangle \cdot \mathbf{n} S_i} \tag{3.1.22}$$

In the present situation, we may use (3.1.21). Thus, with (3.1.19), be obtain

$$\alpha_t = \frac{4\rho_1 c_1 \rho_2 c_2}{(\rho_1 c_1 + \rho_2 c_2)^2} \tag{3.1.23}$$

A reflection coefficient α_r, defined as the ratio of the reflected to incident energy fluxes, may be obtained in terms of the function g'. However, since the total acoustic energy is conserved, we must have

$$\alpha_r = 1 - \alpha_t \qquad (3.1.24)$$

or

$$\alpha_r = \left(\frac{\rho_2 c_2 - \rho_1 c_1}{\rho_2 c_2 + \rho_1 c_1} \right)^2 \qquad (3.1.25)$$

Thus, the transmission and reflection coefficients depend on the characteristic impedances $\rho_1 c_1$ and $\rho_2 c_2$, the particular form of the wave being of no importance.

Several special results follow from (3.1.23) and (3.1.25). First, the incident wave is completely reflected in either of the limiting cases $\rho_2 c_2 \gg \rho_1 c_1$ or $\rho_1 c_1 \gg \rho_2 c_2$. Second, when $\rho_2 c_2 = \rho_1 c_1$, $\alpha_r = 0$, so that the wave is totally transmitted. For other combinations we obtain partial transmission.

The air-water interface case is of special interest. Thus, if fluid 2 represents water, it follows that $\rho_2 c_2 \gg \rho_1 c_1$ and

$$\alpha_r \approx 1 - 4 \frac{\rho_1 c_1}{\rho_2 c_2} = 0.9988$$

Therefore, when a sound wave traveling in air impinges perpendicularly on a water surface, most of its energy is reflected. However, the pressure amplitude of the transmitted wave is nearly twice as large as that of the incident wave, as shown by (3.1.15) and (3.1.18). On the other hand, the transmitted fluid velocity is then nearly negligible.

Acoustic Impedance at a Boundary

In some reflection and transmission problems, it is necessary to consider reflecting surfaces that are neither rigid nor elastic. For example, some sound-"absorbing" materials are made of loosely packed fibers, others of rubber foam, and so on. The surfaces of these materials will move in response to the pressure fluctuations of the incident sound waves. However, this surface motion cannot be determined because dynamic equations for the material boundary are not generally available. Therefore, the conditions that should be imposed on the acoustic variables at the boundary are not known, and in order to determine the field in front of the boundary, one must resort to experiments. In many practical situations, however, one is not interested in the field in front of an absorber. Instead, one may be interested in determining, for example, how efficient a given material is to absorb sounds of given frequencies. However, because of the lack of information about absorbing materials, such data are usually obtained from measurements of the sound pressure field in front of the absorber when a monochromatic wave falls on it. The method requires an assumption about the surface motion that is based on the intuitive idea that the local displacement of the boundary should be proportional to the pressure fluctuation of the incoming wave, but having, in general, a time lag with respect to it. That is, if the incoming wave is monochromatic, and the pressure fluctuation at the boundary is given by $\tilde{p}'$, then the velocity of the boundary, along the local normal, is given by

$$\tilde{u}_n = \tilde{p}'/\tilde{z} \qquad (3.1.26)$$

where the complex quantity $\tilde{z}$ is called the *specific acoustic impedance* of the boundary. In general, $\tilde{z}$ will depend on the properties of the boundary, the properties of the fluid in front of it, the frequency of the wave, and so on, but not on time. Also, it may vary from point to point on the boundary.

The impedance $\tilde{z}$ introduced in (3.1.26) is called specific because the term "acoustic impedance" is usually reserved for the ratio of pressure to volume flow rate Su. Such an impedance will be introduced later.

It should be clear that the assumption that (3.1.26) holds is not generally valid. For example, the boundary may be such that its response to the waves is nonlinear. However, in those instances where the assumption has some validity, (3.1.26) may be used to relate the properties of the acoustic field in front of the boundary to those properties of the boundary described by $\tilde{z}$. This may be done by reconsidering the simple reflection problem treated at the beginning of the section, but without assuming a perfectly rigid reflector. The pressure field in front of the absorber can still be represented in terms of an incident and a reflected wave as

$$\tilde{p}' = Ae^{i(kx-\omega t)} + \tilde{B}e^{-i(kx+\omega t)} \qquad (3.1.27)$$

where A and $\tilde{B}$ now represent pressure magnitudes of the incident and reflected waves, respectively. Denoting their ratio by $\tilde{R}$, that is,

$$\tilde{R} = \tilde{B}/A \qquad (3.1.28)$$

we can write $\tilde{p}'$ as

$$\tilde{p}'_1 = Ae^{-i\omega t}(e^{ikx} + \tilde{R}e^{-ikx}) \qquad (3.1.29)$$

The corresponding fluid velocity is

$$\tilde{u}_1 = \frac{A}{\rho_0 c_0} e^{-i\omega t}(e^{ikx} - \tilde{R}e^{-ikx}) \qquad (3.1.30)$$

The pressure-amplitude reflection coefficient $\tilde{R}$ can be determined if $\tilde{z}$ is known. Thus, since at $x=0$, $\tilde{p}'/\tilde{u}=\tilde{z}$, we have, using (3.1.29) and (3.1.30),

$$\tilde{z} = \frac{1+\tilde{R}}{1-\tilde{R}} \rho_0 c_0 \tag{3.1.31}$$

Therefore,

$$\tilde{R} = \frac{(\tilde{z}/\rho_0 c_0) - 1}{(\tilde{z}/\rho_0 c_0) + 1} \tag{3.1.32}$$

If we introduce the ratio

$$\tilde{\zeta} = \frac{\tilde{z}}{\rho_0 c_0} \tag{3.1.33}$$

we can write (3.1.32) as

$$\tilde{R} = \frac{\tilde{\zeta} - 1}{\tilde{\zeta} + 1} \tag{3.1.34}$$

This defines $\tilde{R}$ as a function of the specific impedance of the boundary, and of the characteristic impedance of the medium in front of it.

The standing-wave field in front of the reflector may be expressed in several useful forms. One of them is in terms of circular functions of complex argument, as these will enable us to arrive at relatively simple expressions from which $\tilde{z}$ can be determined. To obtain these expressions, we first write the quantity $\exp(ikx) + \tilde{R}\exp(-ikx)$ in (3.1.29) as a trigonometric function. Thus, if we anticipate that upon reflection the amplitude and the phase of the wave may change, we can express $\tilde{R}$ as

$$\tilde{R} = e^{-2\psi - i\pi\sigma} \tag{3.1.35}$$

where

$$e^{-2\psi} = R \tag{3.1.36}$$

This gives

$$e^{ikx} + \tilde{R}e^{-ikx} = e^{-\psi - i\pi\sigma/2}\left[e^{i(kx - i\psi + \pi\sigma/2)} + e^{-i(kx - i\psi + \pi\sigma/2)} \right] \tag{3.1.37}$$

so that (3.1.29) can be written as

$$\tilde{p}' = 2Ae^{-i(\omega t + \pi\sigma/2 - i\psi)} \cos(kx + \pi\sigma/2 - i\psi) \tag{3.1.38}$$

The corresponding fluid velocity is

$$\tilde{u} = \frac{2A}{i\rho_0 c_0} e^{-i(\omega t + \pi\sigma/2 - i\psi)} \sin(kx + \pi\sigma/2 - i\psi) \qquad (3.1.39)$$

At the boundary, the fluid velocity is equal to the velocity of the boundary, so that on using (3.1.26) together with the above equations, we obtain

$$\tilde{\zeta} = i \cot(\pi\sigma/2 - i\psi) \qquad (3.1.40)$$

Thus, $\tilde{\zeta}$ may be obtained if σ and ψ are known. These quantities, in turn, can be determined from measurements of the rms pressure field in front of the reflector. The basis for the method is Equation (3.1.29). This gives, for the rms pressure, $P(x) = (\tilde{p}'\tilde{p}'^*)^{1/2}/\sqrt{2}$,

$$P(x) = \frac{1}{\sqrt{2}} A \left[1 + R^2 + \tilde{R}e^{ikx} + \tilde{R}^* e^{-ikx} \right]^{1/2} \qquad (3.1.41)$$

Using (3.1.35) and (3.1.36), this may be written as

$$P(x) = \frac{1}{\sqrt{2}} A \left[1 + R^2 + 2R\cos 2(kx + \pi\sigma/2) \right]^{1/2} \qquad (3.1.42)$$

Thus, $P(x)$ varies with distance from the reflector in a manner different from that in front of a rigid reflector. Of course, if $R = 1$ and $\sigma = 0$, we obtain the rigid-reflector result

$$P(x) = \sqrt{2} A |\cos kx| \qquad (3.1.43)$$

This shows that in that case $P(x)/(\sqrt{2} A)$ varies between 0 and 1 and has a first zero at a distance $\lambda/4$ from the reflector. For $\tilde{R} \neq 1$, we see from (3.1.42) that the minimum value of $P(x)/\sqrt{2} A$ is now

$$\frac{P_{min}(x)}{\sqrt{2} A} = \frac{1 - R}{2} \geqslant 0 \qquad (3.1.44)$$

and that the maximum value of that ratio is now

$$\frac{P_{max}(x)}{\sqrt{2} A} = \frac{1 + R}{2} \leqslant 1 \qquad (3.1.45)$$

Furthermore, at $x=0$, the pressure no longer has a maximum value. Thus,

$$\frac{P(0)}{\sqrt{2}\,A} = \tfrac{1}{2}(1+R^2+2R\cos\pi\sigma)^{1/2} \leqslant \frac{1+R}{2} \qquad (3.1.46)$$

Finally, the pressure minima occur for values of x determined by

$$2(kx_n+\pi\sigma/2) = -n\pi \qquad (3.1.47)$$

Therefore, the first minimum is now at a distance from the reflector given by

$$|x_0| = \frac{\pi}{2k}(1+\sigma) = \frac{\lambda}{4}(1+\sigma) \qquad (3.1.48)$$

Returning now to the determination of σ and ψ, it is clear that a measurement of the distance between the reflector and the first minimum and use of (3.1.48) with $\lambda = c_0/f$ yield the value of σ. The quantity ψ, on the other hand, may be determined by measuring the ratio of pressure maxima to pressure minima. This ratio, called the *standing-wave ratio* and denoted by SWR, is from (3.1.44) and (3.1.4) given by

$$\text{SWR} = \frac{1+R}{1-R} \qquad (3.1.49)$$

Thus, if this quantity is known, the magnitude of $\tilde{R}$ is given by

$$R = \frac{\text{SWR}-1}{\text{SWR}+1} \qquad (3.1.50)$$

This result, together with (3.1.36), defines ψ. Once σ and ψ are known, the nondimensional specific impedance $\tilde{z}$ may be determined by means of (3.1.40).

Another procedure that can also be used to determine $\tilde{\zeta}$ is to plot lines having constant values of R and of σ in a plane where the coordinates are the real and imaginary parts of $\tilde{\zeta}$ (see Problem 3.1.3).

Acoustical Elements Having Complex Impedances

As the preceding discussion shows, it is sometimes necessary to consider acoustical elements or acoustical materials whose mechanical properties are not known. For instance, one may be interested in the amplitudes of the waves that are transmitted into the absorbing material of the example treated above. From that example, it follows that we can represent the mechanical properties of the material in terms of a specific acoustic impedance $\tilde{z}_a$. Thus, for any point where the acoustic pressure is p' and where the acoustic velocity is u, we put

$$\tilde{p}' = \tilde{z}_a \tilde{u} \tag{3.1.51}$$

The complex impedance may be expressed as

$$\tilde{z}_a = x + iy \tag{3.1.52}$$

where x and y may be functions of position. The question that now arises is: What do the real and imaginary parts of $\tilde{z}_a$ represent? To answer this, we consider a material element having sufficiently small thickness so that it may be considered to move as a lump, and study its response to the fluctuating forces in the wave. For small pressure amplitudes, we expect that the dynamic response of this element is linear. Thus, if the element moves a small distance ξ in response to the waves, we expect that the medium on the other side of the element will exert a restoring force on it proportional to ξ and to the area of the element. Similarly, we expect that a resistance proportional to the velocity of the element and to its area will also exist. Therefore, the element can be written as

$$M' \frac{d^2\xi}{dt^2} + R' \frac{d\xi}{dt} + K'\xi = p' \tag{3.1.53}$$

when M' is the mass per unit area of the element, and R' and K' are the frictional resistance and stiffness of the medium, respectively, also per unit area. Now, we are considering monochromatic waves, and for these, p' may be expressed by the real part of $\tilde{p}' = a\exp[-i(\omega t - kx)]$. Thus, in the steady state, the displacement of the element will also be proportional to $\exp(-i\omega t)$, so that the complex pressure may be expressed as

$$\tilde{p}' = [R' + i(K'/\omega - M'\omega)]\tilde{u} \tag{3.1.54}$$

where $\tilde{u} = -i\omega\xi$ is the complex velocity of the element. Comparing this result with (3.1.51) we obtain

$$\tilde{z}_a = R' + i(K'/\omega - M'\omega) \tag{3.1.55}$$

so that

$$x = R' \tag{3.1.56}$$

$$y = (K'/\omega - M'\omega) \tag{3.1.57}$$

Hence, the real part of $\tilde{z}_a$ represents a frictional force per unit area that tends to resist the motion. Because of its definition, this force leads to a net expenditure of energy. This fact may also be deduced as follows: The work done by the waves in displacing a unit area of the element is, per unit time,

$$\dot{W} = \frac{1}{T} \int_0^T p' \frac{d\xi}{dt} dt$$

$$= \langle p'u \rangle = \tfrac{1}{2} x |u|^2 \tag{3.1.58}$$

As it may be checked by direct computation, this power is identical to the rate at which energy is spent to overcome the frictional resistance.

The imaginary part of $\tilde{z}_a$ also represents a force, but, as the above calculation shows, this force does not lead to energy expenditure. Rather, it represents the reaction of the medium on the waves, and is caused by the medium stiffness and inertia. In view of these interpretations, it is customary to refer to the real part of the impedance as the *resistive* part, and to the imaginary part as the *reactive* part.

Helmholtz Resonators

As an example of an acoustic element whose mechanical behavior may be described by a complex impedance, we consider the device shown in Figure 3.1.4. It consists of a rigid-wall cavity of volume V, and having at least one short and narrow orifice through which the fluid filling it communicates with the external medium. For simplicity we will only consider the case when the internal and external fluids are the same perfect gas.

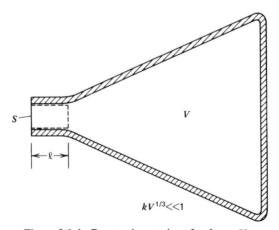

Figure 3.1.4 Resonating cavity of volume V.

To obtain the device's specific impedance, we study the action of a monochromatic wave on the device, and assume that the lateral dimensions of the cavity V are small compared with the wavelength of the incident wave. Since the neck is short and narrow, the fluid inside it will act as a solid mass, and will oscillate in response to the waves. Thus, the mass of the element, per unit area is

$$M' = \rho_0 l_e \tag{3.1.59}$$

where ρ_0 is the density of the gas in the neck and l_e is the effective length of the moving element. In general this length is not equal to the neck's length l, but for simplicity we will use the value.[2] Thus, $M' = \rho_0 l$. The next element that is needed to compute $\tilde{z}_a$ is the frictional resistance per unit area R'. In an actual device, there are at least two distinct contributions to R'. The first is due to viscous effects at the walls of the neck. The second is due to the energy that is radiated as sound due to the oscillatory motion. As far as the vibrating mass is concerned, such energy is "lost" so that we may express such losses in terms of a resistance force due to radiation. It is shown in Chapters 5 and 6 that both of these effects produce an energy expenditure proportional to $|\tilde{u}|^2$ and may therefore be symbolically represented by some resistance per unit area R'. In ideal fluids, however, R' is entirely due to radiation losses.

The last quantity that is required is the stiffness per unit area K'. By definition, this is equal to the restoring force per unit area that results due to a unit displacement. The restoring force per unit area is of course due to the pressure difference that exists across the neck due to a small displacement of the moving mass. Thus, if the mass moves a distance x' into the cavity, then

$$K' = (p_{\text{inside}} - p_{\text{outside}})/x'$$

The outside pressure is p_0. If the cavity's dimensions are small compared with the wavelength (as already assumed), the pressure inside the cavity will be uniform throughout the cavity and will be larger than the pressure outside by an amount ΔP that is due to the reduction of the cavity volume from V to $V - Sx'$, where S is the cross-sectional area of the neck. Now, this reduction in volume may be accompanied by changes of the other properties of the gas in the cavity. For ideal fluids, those changes occur at constant entropy, so that the relationship between pressure and volume is the familiar isentropic equation of state,

$$pV^\gamma = \text{constant}$$

Hence

$$(p_0 + \Delta p) = p_0 \left(\frac{V}{V - Sx'} \right)^\gamma$$

---

[2]A correction to this approximation is obtained in Chapter 5.

Since $Sx \ll V$, this yields

$$\Delta p = \gamma p_0 S x' / V$$

Therefore

$$K' = \frac{\rho_0 c_0{}^2 S}{V} \tag{3.1.60}$$

where we have used the perfect gas relationship $\gamma p_0 = \rho_0 c_0{}^2$.

Substitution of the computed values of M' and K' into (3.1.55) yields

$$\tilde{z}_a = R' + i\rho_0 \left(\frac{c_0{}^2 S}{V\omega} - \omega l \right) \tag{3.1.61}$$

Because of the definition of $\tilde{z}_a$, it follows that for a given pressure amplitude, the resulting velocity will be maximum at the frequency for which $|\tilde{z}|$ attains its minimum value. For the important case when dissipation is very small, this condition approximately occurs when the reactive part of the specific impedance vanishes, or

$$\omega_H = c_0 \sqrt{S/lV} \tag{3.1.62}$$

This is the natural frequency of oscillation of the device as may be verified by computing the value of the natural frequency, $\sqrt{K/M}$, of a harmonic oscillator. At input frequencies near this natural frequency the amplitude of oscillation would become very large. This is of course the well-known phenomenon of resonance. As we move away from resonance, $|\tilde{z}_a|$ increases rapidly so that the amplitude is greatly reduced.

Devices of the type discussed above are known as Helmholtz resonators, in honor of Herman L. F. Helmholtz (1821–1894), who used them extensively in his pioneering work on the physiology of hearing. Some of the resonators Helmholtz used had shapes of the type shown in Figure 3.1.5. The funnel-shaped opening was adapted for insertion in the ear by covering it with a layer of wax that was molded to fit the inner surface of the ear's canal thus providing an air-tight seal. Used in this manner, the resonator will provide the ear amplified sounds having a frequency equal to its natural frequency. Using a series of properly tuned resonators, Helmholtz was able to experiment with low-amplitude tones in the presence of much louder ones.

At present Helmholtz resonators are used in a manner opposite to that described above; that is, they are used to eliminate unwanted sound. This is

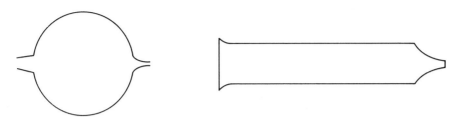

Figure 3.1.5 Helmholtz resonators.

accomplished by increasing the amount of viscous damping, or by placing the resonators in such a manner that sound is sent back to the source. We will later illustrate the procedure by means of simple examples. The reader interested in more detailed information of the application of Helmholtz resonators should read some of the works on the subject cited in the bibliography.

Electrical Analogies

The specific acoustic impedance, $\tilde{z}_a$, derives its name from the electrical impedance of a simple, lumped-system circuit, to which it is analogous. The circuit is the basic series circuit shown schematicaly in Figure 3.1.6. It consists of a resistance R_e, an inductance L, and a capacitance C connected in series to an emf source. It is shown in texts dealing with electrical circuits (see, for example, J. Brophy, *Basic Electronics for Scientists*, McGraw-Hill, New York, 1972) that if the applied emf is $E' = E_0 \cos \omega t$, the differential equation satisfied by the charge q stored in the condenser is

$$L\frac{d^2q}{dt^2} + R_e\frac{dq}{dt} + \frac{1}{C}q = E' \qquad (3.1.63)$$

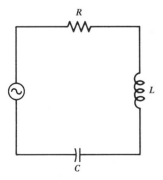

Figure 3.1.6 Series *RLC* circuit.

To establish the correspondence between lumped electrical and acoustical quantities, we write (3.1.53) as

$$\frac{M'}{S}\frac{d^2\Xi}{dt^2} + \frac{R'}{S}\frac{d\Xi}{dt} + \frac{K'}{S}\Xi = p' \qquad (3.1.64)$$

where

$$\Xi = S\xi \qquad (3.1.65)$$

is the volume flow displaced by the element. A third type of system also analogous to the systems under consideration is the mechanical harmonic oscillator, for which we simply have

$$M\frac{d^2\xi}{dt^2} + R\frac{d\xi}{dt} + K\xi = F' \qquad (3.1.66)$$

where F' is a sinusoidally varying force. It is clear that the dynamical behavior of these three systems is similar so that the three variables may be said to be analogous. If this is done, then it is also clear that there will be a definite correspondence between the mechanical, electrical, and acoustical elements that appear in these dynamical equations. Direct comparison of the quantities appearing in these equations result in the analogies shown in Table 3.1.1

We are now in a position to introduce the electrical impedance $\tilde{Z}_e$, and define, by analogy, mechanical and acoustical impedances. Thus, if we use complex notation, the steady-state solution to (3.1.63) may be readily obtained. This solution may be expressed in terms of the complex current $\tilde{j} = \tilde{q} = -i\omega\tilde{q}$ as

$$\tilde{j} = \frac{1}{\tilde{Z}_e}\tilde{E}' \qquad (3.1.67)$$

Table 3.1.1 Mechanical, Electrical, and Acoustical Analogies

Mechanical		Electrical		Acoustical	
Displacement	ξ	Charge	q	Volume flow	$S\xi$
Velocity	$\dot{\xi}$	Current	$\dot{q}$	Volume flow rate	Su
Mass	M	Inductance	L	Mass/squared area	MS^{-2}
Force	F'	emf	E'	Pressure	p'
Stiffness	K	Inverse capacitance	C^{-1}	Stiffness/squared area	KS^{-2}
Resistance	R	Resistance	R_e	Resistance/squared area	RS^{-2}

where

$$\tilde{Z}_e = R_e + i(\omega L - 1/\omega C) \qquad (3.1.68)$$

is the electrical impedance of the circuit of Figure 3.1.6; it is defined as the ratio of a complex, harmonic emf to a complex current. Since the mechanical analogue of an emf is a force, and since that of a current is a velocity, it follows that an analogous *mechanical* impedance may be defined by

$$\tilde{Z}_m = \tilde{F}'/\tilde{u} \qquad (3.1.69)$$

Similarly, since the acoustical analogue of an emf is the acoustic pressure, and that of a current is the volume flow rate, we define an analogous *acoustical* impedance by means of

$$\tilde{Z}_a = \tilde{p}'/S\tilde{u} \qquad (3.1.70)$$

The impedance thus defined is related to the specific acoustic impedance $\tilde{z}_a$ by means of

$$\tilde{Z}_a = \frac{1}{S}\tilde{z}_a \qquad (3.1.71)$$

Therefore, we may express $\tilde{z}_a$ as

$$\tilde{Z}_a = X + iY = \frac{R'}{S} + i\left(\frac{K'}{\omega S} - \frac{M'\omega}{S}\right) \qquad (3.1.72)$$

PROBLEMS

3.1.1 Consider a wave with potential $\phi = f(x - c_0 t)$ impinging normally on a rigid, infinite plane located at $x = 0$. Show that the reflected wave, given by $g(x + c_0 t)$, will be of the same nature (i.e., a compression or an expansion) as that of the incident wave. Show by graphical construction that the reflected wave may be thought of as the mirror image of the incident wave (i.e., as a wave traveling toward $x = 0$ from the other side of the plane).

3.1.2 Use the monochromatic-wave solution to derive the transmission and reflection coefficients at an interface.

3.1.3 Let $\tilde{\zeta} = x + iy$, and $\tilde{R} = R\exp(-\pi i\sigma)$. Show that on the x, y plane the constant R and constant σ curves are circles. Sketch some of these curves.

3.1.4 The effectiveness of a sound absorber may be measured by an absorption coefficient a, defined as the ratio of absorbed to incident powers. Thus, $a = 1 - R^2$. Show that if $\theta = \tan^{-1}(y/x)$, and $\tilde{\zeta} = x + iy$, then

$$a = \frac{1}{\dfrac{1}{2} + \dfrac{1}{4\cos\theta}\left(\zeta + \dfrac{1}{\zeta}\right)}$$

where $\zeta = |\tilde{\zeta}|$. Assuming that $\theta = $ constant, show that a is maximum for $\zeta = 1$. Sketch a versus ζ for a few values of $\cos\theta$.

3.1.5 Consider a plane, monochromatic wave falling normally on an interface between two media. The specific impedances of the two media have values $\tilde{z}_1$ and $\tilde{z}_2$, respectively. At the boundary, the velocity and pressure are continuous, that is, $u_1(0) = u_2(0)$, $p'_1(0) = p'_2(0)$. Show that the complex amplitude of the transmitted pressure at the interface has value given by

$$\tilde{C}(0) = \frac{2\tilde{z}_2(0)}{\tilde{z}_2(0) + \tilde{z}_1(0)} A$$

where $z_2(0)$ and $\tilde{z}_1(0)$ are the specific impedances evaluated at the interface.

3.1.6 Consider plane-wave propagation in a medium for which the linearized momentum equation is

$$\rho_0 \frac{\partial u}{\partial t} + \frac{\partial p'}{\partial x} = \frac{4}{3}\mu_0 \frac{\partial^2 u}{\partial x^2}$$

[Compare with equation (2.3.2).] Assume that p' and u may be expressed as the real parts of

$$\tilde{p}' = P_0 e^{-i(\omega t - Kx)}, \qquad \tilde{u} = U_0 e^{-i(\omega t - Kx)}$$

where $K = k + i\alpha$ is a complex wave number. Determine the specific impedance for such a medium and compute the intensity in the wave.

3.1.7 Consider the device shown in the figure. It consists of a small circular bottle having a massless but rigid partition separating two perfect gases. The gas in the first partition communicates with the exterior medium (whose composition is the same as that of gas 1) through a short and narrow cavity of cross sectional area S and length l Determine:

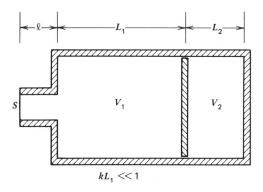

$$kL_1 \ll 1$$

(1) The acoustical impedance of the device

(2) The devices's natural frequency of oscillation.

3.2 CHARACTERISTIC WAVES

So far, we have considered situations in which there is only one reflecting surface. In many situations, however, other reflecting surfaces exist. These impose additional conditions, and it is of interest to see how our plane, one-dimensional solution is then affected. The more general case of oblique incidence will be considered in the following section. Here we consider one-dimensional waves in a region bounded by plane, rigid reflectors of infinite extent, whose normals are along the direction of propagation. Other conditions may be considered, but this is perhaps the simplest. Let the planes be located at $x=0$ and $x=L$. Then, since the reflectors are rigid, the boundary conditions are $u(0,t)=u(L,t)=0$. To obtain the field in the region $0 \leqslant x \leqslant L$, we consider the general solution for u:

$$u(x,t)=f(x-c_0 t)+g(x+c_0 t) \tag{3.2.1}$$

Because of the boundary conditions at $x=0$, we have

$$f(-c_0 t)=-g(c_0 t) \tag{3.2.2}$$

that is, $f(-\xi)=-g(\xi)$. Therefore,

$$f(x-c_0 t)=-g(c_0 t-x) \tag{3.2.3}$$

so that

$$u(x, t) = g(c_0 t + x) - g(c_0 t - x) \tag{3.2.4}$$

The second boundary condition, namely, $u(L, t) = 0$, gives

$$g(c_0 t + L) = g(c_0 t - L) \tag{3.2.5}$$

or

$$g(\xi) = g(\xi + 2L) \tag{3.2.6}$$

In other words, the function g is a periodic function, with period $T = 2L$. Therefore, we may expand it in a Fourier series having a fundamental period equal to $2L$, that is,

$$g(\xi) = \tfrac{1}{2} a_0 + \sum_{n=1}^{\infty} \left[a_n \cos \frac{n \pi \xi}{L} + b_n \sin \frac{n \pi \xi}{L} \right] \tag{3.2.7}$$

Hence,

$$g(c_0 t \pm x) = \tfrac{1}{2} a_0 + \sum_{n=1}^{\infty} \left[a_n \cos \frac{n \pi}{L} (c_0 t \pm x) + b_n \sin \frac{n \pi}{L} (c_0 t \pm x) \right] \tag{3.2.8}$$

Substituting these expansions into (3.2.4), we obtain

$$u(x, t) = \sum_{n=1}^{\infty} \sin \frac{n \pi x}{L} \left[A_n \cos \frac{n \pi c_0}{L} t + B_n \sin \frac{n \pi c_0}{L} t \right] \tag{3.2.9}$$

where $A_n = -2b_n$ and $B_n = -2a_n$. The corresponding acoustic pressure is given by

$$\frac{p'(x, t)}{\rho_0 c_0} = \tfrac{1}{2} A_0 + \sum_{n=1}^{\infty} \cos \frac{n \pi x}{L} \left[B_n \cos \frac{n \pi c_0}{L} t - A_n \sin \frac{n \pi c_0}{L} t \right] \tag{3.2.10}$$

where $A_0 = 2a_0$. This completes the solution. The Fourier coefficients may be obtained from the initial distributions of pressure and velocity.

The solution represented by either of the above equations is interesting, for it shows that no matter what the initial distributions of pressure and velocity are, the result is an infinite series of standing waves, the velocity in each one being of the form

$$u_n = \left[A_n \cos \frac{n \pi c_0 t}{L} + B_n \sin \frac{n \pi c_0 t}{L} \right] \sin \frac{n \pi x}{L} \tag{3.2.11}$$

Each such term in the infinite series represents a possible *mode* of oscillation within the region, the frequency of oscillation of the nth mode being, in this case,

$$\omega_n = n\pi c_0 / L \tag{3.2.12}$$

These modes and frequencies of oscillation are characteristic of the system. They are therefore called the characteristic modes (or eigenmodes) and the characteristic frequencies (or eigenfrequencies) of the system. Figure 3.2.1 shows schematically the spatial variations of the first few velocity modes in the region. The characteristic frequencies, as well as the spatial distribution of the characteristic modes, depend only on the geometry of the region, the boundary conditions, and the properties of the medium. They do not depend on the distribution of the initial disturbance. It may therefore appear that an arbitrary disturbance will split itself into Fourier components whose frequencies correspond to the characteristic frequencies for the region. In reality, however, the disturbance splits itself into positive-going and negative-going waves, which may be expressed mathematically by a Fourier series having a fundamental period specified by the boundary conditions. However, if we use a frequency analyzer to observe our field, it will be found that the spatial variations of velocity, say, for a given characteristic frequency are as specified by the corresponding mode. Thus, we may regard these modes as being physical entities, and not mere mathematical symbols. We have therefore two ways of looking at a given acoustic field. The first and more basic one is in terms of waves traveling in opposite directions; the second is in terms of an infinite number of modes of oscillation combining together to give the instantaneous distribution of pressure and velocity in the acoustic field.

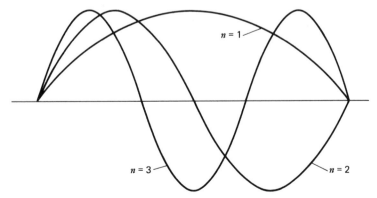

Figure 3.2.1 Spatial distribution of velocity in first three modes in the region between two rigid walls.

The characteristic frequencies of oscillation may be obtained more directly by considering the monochromatic-wave solution. This is so because as we have seen, any periodic function may be expressed in Fourier series. To illustrate the procedure, we reconsider the above region between two rigid reflectors. Reference to the beginning of Section 3.1 will show that in order that our monochromatic-wave solution be applicable at all for that region, the spatial variations of the velocity must be given by a sine and not by a cosine (otherwise the condition at $x=0$ would never by met by a monochromatic wave). In fact, the velocity is given by (3.1.3):

$$u = 2A \sin \omega t \sin kx \qquad (3.2.13)$$

but this time k must be such that the second boundary condition $u(L,t)=0$ be met. Thus, we require that

$$\sin kL = 0 \qquad (3.2.14)$$

or

$$k = n\pi/L, \quad n=1,2,3,\dots \qquad (3.2.15)$$

The allowed values of ω are therefore $\omega_n = n\pi c_0/L$. Corresponding to each possible value of n, there is a function

$$u_n(x,t) = 2A \sin \frac{n\pi c_0 t}{L} \sin \frac{n\pi x}{L} \qquad (3.2.16)$$

that satisfies the wave equation and meets the boundary condition. Therefore, the most general solution will be a linear combination of all such functions, that is,

$$u(x,t) = \sum_{n=1}^{\infty} c_n \sin \frac{n\pi c_0 t}{L} \sin \frac{n\pi x}{L} \qquad (3.2.17)$$

This corresponds to (3.2.9).

Linearized Shock Tube

As an application, we consider the acoustic approximation to the flow field in a shock tube. Basically, this consists of a tube closed at both ends and divided into high and low pressures by means of a membrane, as depicted in Figure 3.2.2. At some time t, the membrane is removed. As we will see, this results, among many other effects, in a compressive wave with a very steep front, that is, a shock wave. The situation does not normally fall within the scope of acoustics,

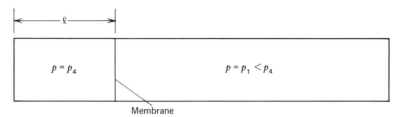

Figure 3.2.2 Initial distribution of pressure in a shock tube.

but if the initial overpressure is small, we may use our equations to describe the sequence of events after the diaphragm is removed. The initial distribution of pressure is similar to that studies in Section 2.4. This time, however, the region if finite, being in fact limited by rigid walls at $x=0$ and $x=L$. Therefore, the initial pressure distribution is given by

$$p'(x,0) = \begin{cases} \Delta, & 0<x<l \\ 0, & l<x<L \end{cases} \qquad (3.2.18)$$

and the initial velocity is $u(x,0)=0$. The problem may be solved easily by using (3.2.9) and (3.2.10). These are applicable here because the boundary conditions are equal to those used to derive them. Thus, (3.2.9) gives, at $t=0$,

$$\sum_{n=1}^{\infty} A_n \sin\frac{n\pi x}{L} = 0 \qquad (3.2.19)$$

This can be so only if $A_n=0$ for all n. Therefore,

$$p'(x,t) = \rho_0 c_0 \left[\tfrac{1}{2}A_0 + \sum_{n=1}^{\infty} B_n \cos\frac{n\pi x}{L} \cos\frac{n\pi c_0 t}{L} \right] \qquad (3.2.20)$$

The coefficients A_0 and B_n may be obtained for the half-range $0, L$ as

$$\rho_0 c_0 A_0 = \frac{2}{L} \int_0^L p'(x,0)\, dx \qquad (3.3.21)$$

Substitution from (3.2.18) gives

$$\rho_0 c_0 A_0 = 2\Delta l/L \qquad (3.2.22)$$

Similarly,

$$\rho_0 c_0 B_n = \frac{2}{L} \int_0^L p'(x,0) \cos\frac{n\pi x}{L}\, dx \qquad (3.2.23)$$

or

$$\rho_0 c_0 B_n = \frac{2\Delta}{\pi n} \sin n\pi \frac{l}{L} \tag{3.2.24}$$

Therefore, the acoustic pressure and velocity in the linearized shock tube are given by

$$p'(x, t) = \Delta\left[\frac{\ell}{L} + \frac{2}{\pi} \sum_{n=1}^{\infty} \frac{1}{n} \sin n\pi \frac{l}{L} \cos \frac{n\pi x}{L} \cos \frac{n\pi c_0 t}{L}\right] \tag{3.2.25}$$

$$u(x, t) = \frac{2\Delta}{\pi \rho_0 c_0} \sum_{n=1}^{\infty} \frac{1}{n} \sin n\pi \frac{l}{L} \sin \frac{n\pi x}{L} \sin \frac{n\pi c_0 t}{L} \tag{3.2.26}$$

Equations (3.2.25) and (3.2.26) complete the solution of the problem, and may be used to compute the acoustic energy (see Problem 3.2.1). It is clear, however,

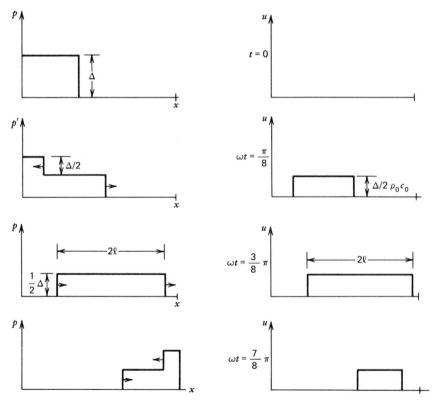

Figure 3.2.3 Pressure and velocity distributions at various instances in a linearized shock tube.

that their form is not very convenient to visualize the variations of the pressure or of the velocity inside the tube. Better suited for this purpose is the general solution of the wave equation. In fact, following a procedure similar to that used in Section 2.4 (see Problem 3.2.2), D'Alambert's solution can be used to obtain the pressure and velocity profiles sketched in Figure 3.2.3 at various instants. These can, of course, be obtained also from the Fourier-series representation. The arrows in the figure indicate the direction of motion of the wave fronts. It should be noticed from the figure that during a finite time interval, both the pressure and the velocity have constant values. This time is called the "test time," and in the linear approximation has a maximum duration given by

$$t_S = 2l/c_0 \qquad (3.2.27)$$

PROBLEMS

3.2.1 Show that the instantaneous acoustic energy in the linearized shock tube is (per unit cross-sectional area)

$$E = \frac{L}{4\rho_0 c_0^2} \left[\left(\frac{2}{\pi}\right)^2 \sum_{n=1}^{\infty} \frac{1}{n^2} \sin^2 n\pi \frac{l}{L} + 2\left(\frac{l}{L}\right)^2 \right] \Delta^2$$

Evaluate E for the special case $l = (1/2)L$. How is this energy related to the initial distribution of overpressure? [For odd n, $\sum_{n=1}^{\infty} (1/n^2) = \pi^2/8$.]

3.2.2 Determine the pressure and velocity fields in the shock tube using D'Alambert's solution. Compare your results with those shown graphically in Figure 3.2.3.

3.2.3 Consider the region between two planes. One of the planes is rigid, and the other, at a distance L, is such that $p' = 0$. This would correspond approximately to the boundary condition at the open end of a tube.

   (1) Determine the characteristic frequency for the region between the planes.

   (2) Sketch the spatial variations of the first three pressure and velocity modes.

3.3 TRANSMISSION THROUGH A WALL

An important problem in acoustics is that of sound transmission through a solid wall separating two fluid media. An example would be the problem of sound transmission between two adjacent rooms. The problem can be solved

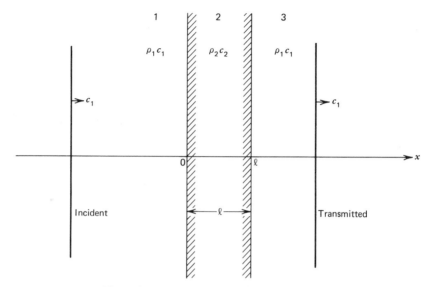

Figure 3.3.1   Transmission through solid wall.

when the media on both sides of the wall are different, but here we will consider only the special but important case when they are equal, as sketched in Figure 3.3.1. Further, we will consider only one-dimensional monochromatic waves, and assume that the field is fully developed. That is, no transient effects that may have existed initially are considered. Media 1 and 3 extended to $-\infty$ and $+\infty$, respectively. In these conditions, the pressures in the three media are given by

$$\tilde{p}_1' = A_1 e^{i(k_1 x - \omega t)} + \tilde{B}_1 e^{-i(k_1 x + \omega t)}$$

$$\tilde{p}_2' = \tilde{A}_2 e^{i(k_2 x - \omega t)} + \tilde{B}_2 e^{-i(k_2 x + \omega t)} \qquad (3.3.1)$$

$$\tilde{p}_3' = \tilde{A}_3 e^{i[k_1(x - l) - \omega t]}$$

The quantities we are interested in are $\tilde{B}_1 / A_1$ and $\tilde{A}_3 / A_1$, as they give the amplitudes of the reflected and transmitted waves, respectively. We obtain these quantities by means of the boundary conditions at $x = 0$ and $x = l$. These conditions are

$$p_1'(0, t) = p_2'(0, t)$$

$$u_1(0, t) = u_2(0, t) \qquad (3.3.2)$$

and

$$p_2'(l, t) = p_3'(l, t)$$

$$u_2(l, t) = u_3(l, t) \tag{3.3.3}$$

The velocities can be obtained from the pressure, either by integration of the linearized momentum equation or, more simply, by remembering that in a plane wave, the pressure and velocity are related by $p' = \pm \rho_0 c_0 u$, where the positive sign refers to waves traveling toward increasing values of x. Thus, the boundary conditions at $x = 0$ yield

$$A_1 + \tilde{B}_1 = \tilde{A}_2 + \tilde{B}_2 \tag{3.3.4}$$

$$\frac{1}{\rho_1 c_1}(A_1 - \tilde{B}_1) = \frac{1}{\rho_2 c_2}(\tilde{A}_2 - \tilde{B}_2) \tag{3.3.5}$$

Similarly, at $x = l$, we obtain

$$\tilde{A}_2 e^{ik_2 l} + \tilde{B}_2 e^{-ik_2 l} = \tilde{A}_3 \tag{3.3.6}$$

and

$$\frac{1}{\rho_2 c_2}(\tilde{A}_2 e^{ik_2 l} - \tilde{B}_2 e^{-ik_2 l}) = \frac{1}{\rho_1 c_1}\tilde{A}_3 \tag{3.3.7}$$

These four equations for $\tilde{B}_1$, $\tilde{A}_2$, $\tilde{B}_2$, and $\tilde{A}_3$ can be arranged conveniently in matrix form as

$$\begin{bmatrix} \rho_2 c_2 & \rho_1 c_1 e^{ik_2 l} & -\rho_1 c_1 e^{ik_2 l} & 0 \\ 1 & -e^{ik_2 l} & -e^{ik_2 l} & 0 \\ 0 & -\rho_1 c_1 & \rho_1 c_1 & \rho_2 c_2 \\ 0 & 1 & 1 & -1 \end{bmatrix} \begin{bmatrix} \tilde{A}_3 \\ \tilde{B}_2 \\ \tilde{A}_2 \\ \tilde{B}_1 \end{bmatrix} = A_1 \begin{bmatrix} 0 \\ 0 \\ \rho_2 c_2 \\ 1 \end{bmatrix} \tag{3.3.8}$$

To solve for any of the four unknowns, we require the inverse of the coefficient matrix $\overline{\overline{A}}$. This is given by

$$\overline{\overline{A}}^{-1} = \frac{\overline{\overline{C}}}{\det \overline{\overline{A}}} \tag{3.3.9}$$

where $\overline{\overline{C}}$ is a matrix whose j, i element is the cofactor of the i, j element in $\overline{\overline{A}}$, and $\det \overline{\overline{A}}$ is the determinant of $\overline{\overline{A}}$. Thus,

$$\det \overline{\overline{A}} = (\rho_2 c_1 + \rho_2 c_2)^2 e^{-ik_2 l} - (\rho_1 c_1 - \rho_2 c_2)^2 e^{ik_2 l} \qquad (3.3.10)$$

This can also be written as

$$\det \overline{\overline{A}} = -4\rho_1 c_1 \rho_2 c_2 \cos k_2 l - 2i \left[(\rho_1 c_1)^2 + (\rho_2 c_2)^2 \right] \sin k_2 l \qquad (3.3.11)$$

In terms of $\overline{\overline{C}}$ and $\det \overline{\overline{A}}$, the unknown amplitudes are given by

$$
\begin{bmatrix} \tilde{A}_3 \\ \tilde{B}_2 \\ \tilde{A}_2 \\ \tilde{B}_1 \end{bmatrix}
= \frac{A_1}{\det \overline{\overline{A}}}
\begin{bmatrix} c_{11} & c_{12} & c_{13} & c_{14} \\ c_{21} & c_{22} & c_{23} & c_{24} \\ c_{31} & c_{32} & c_{33} & c_{34} \\ c_{41} & c_{42} & c_{43} & c_{44} \end{bmatrix}
\begin{bmatrix} 0 \\ 0 \\ \rho_2 c_2 \\ 1 \end{bmatrix}
\qquad (3.3.12)
$$

To obtain the transmitted intensity, we require A_3. This is from the above equation given by

$$A_3 = \frac{A_1}{\det \overline{\overline{A}}} (\rho_2 c_2 c_{13} + c_{14}) \qquad (3.3.13)$$

where

$$c_{13} =
\begin{vmatrix}
\rho_1 c_1 e^{-eik_2 l} & -\rho_1 c_1 e^{ik_2 l} & 0 \\
-e^{-ik_2 l} & -e^{ik_2 l} & 0 \\
1 & 1 & -1
\end{vmatrix}
\qquad (3.3.14)$$

and

$$c_{14} = -
\begin{vmatrix}
\rho_1 c_1 e^{-ik_2 l} & -\rho_1 c_1 e^{ik_2 l} & 0 \\
-e^{-k_2 l} & -e^{ik_2 l} & 0 \\
-\rho_1 c_1 & \rho_1 c_1 & \rho_2 c_2
\end{vmatrix}
\qquad (3.3.15)$$

Expanding these determinants, we obtain

$$c_{13} = 2\rho_1 c_1 \qquad (3.3.16)$$

and

$$c_{14} = 2\rho_1 c_1 \rho_2 c_2 \qquad (3.3.17)$$

Therefore,

$$\frac{\tilde{A}_3}{A_1} = -\frac{2\rho_1 c_1 \rho_2 c_2}{2\rho_2 c_2 \rho_1 c_1 \cos k_2 l + i\left[(\rho_1 c_1)^2 + (\rho_2 c_2)^2\right]\sin k_2 l} \tag{3.3.18}$$

The appearance of a complex amplitude A_3 implies that the transmitted and incident waves are not in phase. Now, we are interested in the ratio of transmitted-to-incident intensities. Since media 1 and 3 are the same, this ratio is given simply by

$$\alpha_t = \frac{|\tilde{A}_3|^2}{A_1^2} \tag{3.3.19}$$

Therefore, using $|\tilde{A}_3|^2 = \tilde{A}_3 \tilde{A}_3^*$, we obtain

$$\alpha_t = \frac{4}{4\cos^2 k_2 l + \left(\dfrac{\rho_1 c_1}{\rho_2 c_2} + \dfrac{\rho_2 c_2}{\rho_1 c_1}\right)^2 \sin^2 k_2 l} \tag{3.3.20}$$

This equation can be applied to several particular situations. The first one, when $\rho_1 = \rho_2, c_2 = c_1$, gives the trivial result that when there is no change in the properties of the media, $\alpha_t = 1$. Less trivial is the result that when the characteristic impedances of the two media are the same, that is, when $\rho_1 c_1 = \rho_2 c_2 (= \rho_3 c_3)$, we also obtain $\alpha_t = 1$. In that condition, the material dividing regions 1 and 3 is "transparent" to the incident wave.

In general, however, this situation does not arise, and then α_t becomes a strong function of $k_2 l = (\omega l / c_2)$. Figure 3.3.2 shows the variations of α_t with $\omega l / c_2$. For a given wall, that is, given values of l and c_2, the figure shows the effects of frequency on transmission. We see that for most frequencies, the wall acts as an effective sound barrier; the larger the difference between the two characteristic impedances, the smaller the transmission coefficient. However, we also see that there are many frequencies for which $\alpha_t = 1$. These are given by

$$\frac{\omega l}{c_2} = n\pi, \quad n = 0, 1, 2, \ldots \tag{3.3.21}$$

and correspond to the characteristic frequencies of the region $0 \leqslant x \leqslant l$. The reason, then, for the wall to act as a perfect transmitter is *resonance*. That is, the matching of the incoming-wave frequency with one of the characteristic frequencies of the wall results in a more efficient energy transfer to the wall and, therefore, higher amplitudes of oscillation in the wall; that is, the amplitudes $\tilde{A}_2$

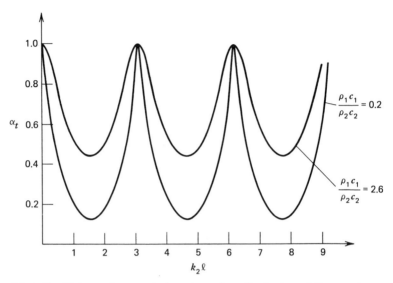

Figure 3.3.2 Coefficient of transmission through walls having different characteristic impedances.

and $\tilde{B}_2$ are then larger (see Problem 3.3.1). These larger amplitudes in the wall are, in turn, responsible for the large amplitude of the transmitted wave.

In some situations of practical interest, (3.3.20) acquires a simpler form. Thus, for example, if medium 1 is air and medium 2 is a solid wall, then $\rho_2 c_2 \gg \rho_1 c_1$, so that

$$\alpha_t = \frac{1}{\cos^2 k_2 l + (\rho_2 c_2/2\rho_1 c_1)^2 \sin^2 k_2 l} \tag{3.3.22}$$

Furthermore, if the wall thickness is very small and the frequency is not very high, then $k_2 l \ll 1$, and our equation becomes

$$\alpha_t = \frac{1}{1 + (\rho_2 c_2 k_2 l/2\rho_1 c_1)^2} \tag{3.3.23}$$

In this limit, our wall has become a thin plate oscillating in response to the incident waves, thereby producing waves in medium 3. Now, the quantity $\rho_2 l$ is the mass per unit area of the plate. Denoting it by σ, we have

$$\alpha_t = \frac{1}{1 + \left(\dfrac{\omega\sigma}{2\rho_1 c_1}\right)^2} \tag{3.3.24}$$

Another measure of the effectiveness of a wall to prevent sound transmision is given by the *transmission-loss index*. This is defined as

$$TL = 10 \log_{10} \frac{I_i}{I_t} = 10 \log_{10} \frac{1}{\alpha_t} \tag{3.3.25}$$

Thus, for a plate we have

$$TL = 10 \log_{10} \left[1 + \left(\frac{\omega \sigma}{2 \rho_1 c_1} \right)^2 \right] \tag{3.3.26}$$

In the particular case of a very massive plate, that is, when $\sigma \gg 2 \rho_1 c_1 / \omega$, this becomes

$$TL = 20 \log_{10} \frac{\omega \sigma}{2 \rho_1 c_1} \tag{3.3.27}$$

Therefore, if the mass per unit area of the plate is doubled, the transmission-loss index increases by $20 \log_{10} 2 \approx 6 \, \text{dB}$. This increase of TL with σ is known as the "mass law," and is approximately correct provided that the incident sound waves do not excite flexural waves in the plate. If this occurs the flexural motion of the plate would necessarily excite waves in medium 3, and therefore would result in an increase of α_t.

The thin-plate results may also be obtained by considering the motion of a plate under given applied forces. The procedure is more direct than that used above, and clearly shows how the motion of the plate excites the waves in medium 3.

Consider, then, a very large plate of mass σ per unit area, freely suspended in a fluid, and dividing it into sections 1 and 3, respectively. Let the plate fall on a plane that is perpendicular to the direction of propagation of the incident wave, as sketched in Figure 3.3.3. The fluid pressures on the sides of the plate are $p_1 = p_0 + p_1'(0)$ and $p_2 = p_0 + p_2'(0)$, respectively, so that the net force acting on an area A of the plate is

$$(p_1 - p_2)_{x=0} A = \left[p_1'(0) - p_2'(0) \right] A \tag{3.3.28}$$

Therefore, the plate's equation of motion is

$$\sigma \frac{du}{dt} = p_1'(0) - p_2'(0) \tag{3.3.29}$$

where u is the plate's velocity. Now, since in medium 3 there are no negative-going waves, we have

$$\tilde{p}_2'(0) = \tilde{A}_3 = \rho_1 c_1 \tilde{u}_2 \tag{3.3.30}$$

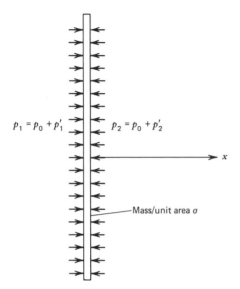

$p_1 = p_0 + p_1'$ $p_2 = p_0 + p_2'$

Mass/unit area σ

Figure 3.3.3 Thin plate in a sound field.

On the other hand, $p_1'(0)$ is due to both incident and reflected waves; thus,

$$\tilde{p}_1'(0) = A_1 + \tilde{B}_1 \tag{3.3.31}$$

But if the fluids are to remain in contact with the plate, we require that $u_1(0) = u_2(0) = u$. Therefore,

$$A_1 - \tilde{B}_1 = \tilde{A}_3 \tag{3.3.32}$$

so that $\tilde{p}_1'(0) = 2A_1 - \tilde{A}_3$. Substituting this into (3.3.29), we obtain

$$A_1 - \tilde{A}_3 = -i\omega\sigma\tilde{u}/2 \tag{3.3.33}$$

Thus, with $\tilde{u} = \tilde{u}_2 = \tilde{A}_3/\rho_1 c_1$, we obtain

$$\frac{\tilde{A}_3}{A_1} = \frac{1}{1 - i\dfrac{\omega\sigma}{2\rho_1 c_1}} \tag{3.3.34}$$

from which (3.3.24) follows directly.

PROBLEMS

3.3.1 From (3.3.1) and (3.3.12), determine the acoustic pressure in a solid wall under normal incidence. Discuss the case $\omega = n\pi c_0/l$.

3.3.2 Evaluate the transmission coefficient through a wall when $k_2 l = (2n - 1)\pi/2$. Discuss the meaning of this result.

3.3.3 Consider a large plate of mass per unit area σ. A monochromatic wave of frequency ω falls normally on one side of the plate. The other side of the plate is evacuated, and the plate is held in position by means of springs located in the evacuated side that exert on the plate, elastic and dissipative forces. Determine the reflection coefficient.

3.3.4 Determine the acoustic impedance of the thin plate treated in this section.

3.4 OBLIQUE INCIDENCE

We now consider a plane, one-dimensional wave falling on a plane fluid interface at an arbitrary angle of incidence θ_1, as shown in Figure 3.4.1. In view of our previous results, we expect that, in general, reflected and transmitted waves will appear. The amplitude, frequency, and angle of incidence of the incident wave are prescribed. We would like to determine some of the properties of the reflected and transmitted waves. Of special interest are their directions and the acoustic power transmitted by them.

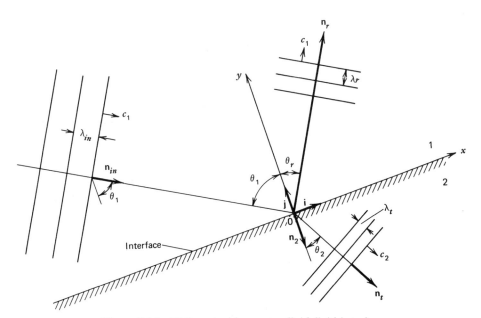

Figure 3.4.1 Oblique incidence at a fluid-fluid interface.

It is convenient to analyze the problem in terms of velocity potentials. Thus, the incident wave will have a velocity potential given by [see Equation (2.5.24)]

$$\tilde{\phi}_{in} = A e^{i\omega(\mathbf{n}_{in} \cdot \mathbf{x}/c_1 - t)} \tag{3.4.1}$$

Here $\mathbf{n}_{in}$ is a unit vector normal to the direction of propagation and c_1 is the speed of sound in fluid medium 1. The reflected and transmitted waves will also have potentials of the same general form as $\tilde{\phi}_{in}$. Now, at the interface the boundary conditions are independent of time. This can be so only if the frequencies of the transmitted and reflected waves are the same as the frequency of the incident wave. Therefore, their potentials can be written as

$$\tilde{\phi}_r = \tilde{B} e^{i\omega(\mathbf{n}_r \cdot \mathbf{x}/c_1 - t)} \tag{3.4.2}$$

and

$$\tilde{\phi}_t = \tilde{C} e^{i\omega(\mathbf{n}_t \cdot \mathbf{x}/c_2 - t)} \tag{3.4.3}$$

where c_2 is the speed of sound in medium 2. In terms of the system of coordinates shown in Figure 3.4.1, the unit normal vectors are given by

$$\mathbf{n}_{in} = \mathbf{i} \sin\theta_1 - \mathbf{j}\cos\theta_1$$

$$\mathbf{n}_r = \mathbf{i} \sin\theta_r + \mathbf{j}\cos\theta_r \tag{3.4.4}$$

$$\mathbf{n}_t = \mathbf{i} \sin\theta_2 - \mathbf{j}\cos\theta_2$$

Here we have also assumed that the incident, reflected, and transmitted waves are coplanar. This is physically obvious, for if $\mathbf{n}_r$ and $\mathbf{n}_t$ had a z component, then the boundary conditions would require that the three potentials depend on z in the same manner. However, the incident potential does not depend on z; that is, $\mathbf{n}_{in}$ does not have a z component. Hence, we conclude that $\mathbf{n}_r$ and $\mathbf{n}_t$ are also on the x–y plane. Now, since the two-dimensional position vector is $\mathbf{x} = \mathbf{i}x + \mathbf{j}y$, the potentials can be written as

$$\tilde{\phi}_{in} = A \exp\left[i\omega\left(\frac{x}{c_1}\sin\theta_1 - \frac{y}{c_1}\cos\theta_1 - t \right) \right]$$

$$\tilde{\phi}_r = \tilde{B} \exp\left[i\omega\left(\frac{x}{c_1}\sin\theta_r + \frac{y}{c_1}\cos\theta_r - t \right) \right] \tag{3.4.5}$$

$$\tilde{\phi}_t = \tilde{C} \exp\left[i\omega\left(\frac{x}{c_2}\sin\theta_2 - \frac{y}{c_2}\cos\theta_2 - t \right) \right]$$

The boundary conditions are that the pressure and normal component of velocity are continuous across the interface, that is,

$$p'_1(0, t) = p'_2(0, t)$$

$$\mathbf{u}_1(0, x, t) \cdot \mathbf{n}_1 + \mathbf{u}_2(0, x, t) \cdot \mathbf{n}_2 = 0$$

where $\mathbf{n}_1 = \mathbf{j}$ and $\mathbf{n}_2 = -\mathbf{j}$ are unit vectors perpendicular to the interface plane. We use the second condition first. Thus, since

$$\tilde{\mathbf{u}}_1 \cdot \mathbf{n}_1 = \frac{\partial \tilde{\phi}_{in}}{\partial y} + \frac{\partial \tilde{\phi}_r}{\partial y}$$

$$\tilde{\mathbf{u}}_2 \cdot \mathbf{n}_2 = -\frac{\partial \tilde{\phi}_t}{\partial y}$$

we obtain, evaluating the derivatives at $y = 0$,

$$-\frac{i\omega \cos \theta_1}{c_1} A e^{i(\omega x/c_1)\sin \theta_1} + \frac{i\omega \cos \theta_r}{c_1} \tilde{B} e^{i(\omega x/c_1)\sin \theta_r}$$

$$= -\frac{i\omega \cos \theta_2}{c_2} \tilde{C} e^{i(\omega x/c_2)\sin \theta_2} \tag{3.4.6}$$

This equation must hold for all x. Therefore, we must have the following two equalities:

$$\frac{\sin \theta_1}{c_1} = \frac{\sin \theta_r}{c_1} = \frac{\sin \theta_2}{c_2}$$

From the first one, it follows that

$$\theta_1 = \theta_r \tag{3.4.7}$$

Thus, the angle of incidence is equal to the angle of reflection. Similarly, the second equality yields the well-known *Snell's law* of refraction

$$\frac{\sin \theta_1}{\sin \theta_2} = \frac{c_1}{c_2} \tag{3.4.8}$$

Using these results in the boundary condition for the velocity yields

$$\frac{\cos \theta_1}{c_1}(\tilde{B} - A) = -\frac{\cos \theta_2}{c_2}\tilde{C} \tag{3.4.9}$$

Similarly, since the pressure in medium 1 is

$$\tilde{p}_1' = i\omega\rho_1\left(\tilde{\phi}_{in} + \tilde{\phi}_r\right) \tag{3.4.10}$$

and the transmitted pressure is

$$p_t' = i\omega\rho_2\tilde{\phi}_t \tag{3.4.11}$$

the boundary condition on the pressure becomes

$$\rho_1(A + \tilde{B}) = \rho_2\tilde{C} \tag{3.4.12}$$

Using Snell's law, the boundary condition on the velocity can be written as

$$\frac{\tilde{B} - A}{\tan\theta_1} = -\frac{\tilde{C}}{\tan\theta_2} \tag{3.4.13}$$

This can be combined with (3.4.12) to yield

$$\frac{\tilde{B} - A}{\rho_1\tan\theta_1} = -\frac{A + \tilde{B}}{\rho_2\tan\theta_2} \tag{3.4.14}$$

or

$$\frac{\tilde{B}}{A} = \frac{\rho_2\tan\theta_2 - \rho_1\tan\theta_1}{\rho_2\tan\theta_2 + \rho_1\tan\theta_1} \tag{3.4.15}$$

But from Snell's law, $\theta_2 = \sin^{-1}[(c_2/c_1)\sin\theta_1]$, so that

$$\tan\theta_2 = \frac{c_2\sin\theta_1}{\sqrt{c_1^2 - c_2^2\sin^2\theta_1}} \tag{3.4.16}$$

Therefore,

$$\frac{\tilde{B}}{A} = \frac{\rho_2 c_2\cos\theta_1 - \rho_1\sqrt{c_1^2 - c_2^2\sin^2\theta_1}}{\rho_2 c_2\cos\theta_1 + \rho_1\sqrt{c_1^2 - c_2^2\sin^2\theta_1}} \tag{3.4.17}$$

This expression is more convenient than (3.4.15) because it gives $\tilde{B}/A$ as a function of the characteristic impedances of the two media and of the angle of incidence. Now, the reflection coefficient can be expressed as

$$\alpha_r = \frac{\langle q_r^2 \rangle}{\langle q_{in}^2 \rangle} \tag{3.4.18}$$

where

$$\langle q^2 \rangle = \tfrac{1}{2}\nabla\tilde{\phi}\cdot(\nabla\tilde{\phi})^* \tag{3.4.19}$$

But from (2.5.23), $\nabla\tilde{\phi} = i\bar{k}\tilde{\phi}$. Hence,

$$\langle q^2 \rangle = \tfrac{1}{2}k^2(\tilde{\phi}\tilde{\phi}^*) \tag{3.4.20}$$

Using this and (3.4.5) in (3.4.18) yields

$$\alpha_r = \tilde{B}\tilde{B}^*/A^2 \tag{3.4.21}$$

Finally, substitution from (3.4.19) gives

$$\alpha_r = \left[\frac{\rho_2 c_2 \cos\theta_1 - \rho_1\sqrt{c_1^2 - c_2^2\sin^2\theta_1}}{\rho_2 c_2 \cos\theta_1 + \rho_1\sqrt{c_1^2 - c_2^2\sin^2\theta_1}} \right]^2 \tag{3.4.22}$$

For normal incidence, this equation reduces to the corresponding result given in Section 3.1. There we found that if $\rho_2 c_2 = \rho_1 c_1$, the reflection coefficient would be zero. As (3.4.22) shows, this is not true for all angles of incidence. In the oblique-incidence case, the reflection coefficient is zero only for an angle of incidence θ_1^*, given by [see Equation (3.4.15)]

$$\rho_2 \tan\theta_2 = \rho_1 \tan\theta_1^* \tag{3.4.23}$$

Using (3.4.16), this can be expressed as

$$\tan^2\theta_1^* = \frac{\rho_2^2 c_2^2 \sin^2\theta_1^*}{\rho_1^2(c_1^2 - c_2^2\sin^2\theta_1^*)} = \frac{\rho_2^2 c_2^2 \sin^2\theta_1^*}{\rho_1^2(c_1^2 - c_2^2)\sin^2\theta_1^* + \rho_1^2 c_1^2 \cos^2\theta_1^*} \tag{3.4.24}$$

or

$$\tan^2\theta_1^* = \frac{(\rho_2 c_2)^2 - (\rho_1 c_1)^2}{\rho_1^2(c_1^2 - c_2^2)} \tag{3.4.25}$$

Thus, total transmission can occur if the angle of incidence is equal to θ_1^*. However, since $\tan^2 \theta_1^*$ is a real positive quantity, this angle exists only for certain combinations of the parameters involved. Thus, if $\rho_2 c_2$ is larger than $\rho_1 c_1$, it is necessary that c_1 be larger than c_2. Conversely, if $\rho_2 c_2 < \rho_1 c_1$, then we must have $c_2 > c_1$. For example, for the case of a sound wave reaching an air-water interface from either side, there are no possible solutions to (3.4.25); that is, complete transmission is impossible.

Also of interest is the total-reflection (or critical) angle of incidence θ_{1c}. From (3.4.22) we see that $\alpha_r = 1$ for an angle $\theta_1 = \theta_{1c}$, such that

$$\sin \theta_{1c} = c_1/c_2 \tag{3.4.26}$$

For this angle to exist, we must clearly have $c_1 < c_2$. Now, for $\theta_1 = \theta_{1c}$, the angle of refraction (θ_2) is $\pi/2$, that is, along the surface of the interface. However, there is no energy flow across the surface, as may be seen by considering the transmission coefficient α_t. Although this can be derived from α_r, it is instructive to derive it from the more primitive definition given by (3.1.22). Thus,

$$\alpha_t = \frac{\langle p_t' u_t \rangle S_t}{\langle p_{in}' u_{in} \rangle S_i} \tag{3.4.27}$$

where the areas S_i and S_t are perpendicular to the direction of propagation. Of course, since we are considering unbounded media, these areas are infinite. However, their ratio is finite and is given by (see Figure 3.4.2)

$$\frac{S_t}{S_i} = \frac{\cos \theta_2}{\cos \theta_1} \tag{3.4.28}$$

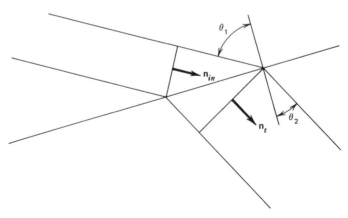

Figure 3.4.2   Change of wave-front area due to refraction at an interface.

Using Snell's law, this can be written as

$$\frac{S_t}{S_i} = \frac{\sqrt{1-(c_2/c_1)^2 \sin^2 \theta_1}}{\cos \theta_1} \tag{3.4.29}$$

Therefore,

$$\alpha_t = \frac{\rho_2 c_1}{\rho_1 c_2} \frac{\tilde{C}\tilde{C}^*}{A^2} \frac{\sqrt{1-(c_2/c_1)^2 \sin^2 \theta_1}}{\cos \theta_1} \tag{3.4.30}$$

But from (3.4.12),

$$\frac{\tilde{C}}{A} = \frac{\rho_1}{\rho_2}\left(1+\frac{\tilde{B}}{A}\right) \tag{3.4.31}$$

Using (3.4.17), this can be written as

$$\frac{\tilde{C}}{A} = \frac{2\rho_1 c_2 \cos \theta_1}{\rho_2 c_2 \cos \theta_1 + \rho_1 c_1 \sqrt{1-(c_2/c_1)^2 \sin^2 \theta_1}} \tag{3.4.32}$$

Finally, substitution of this into (3.4.30) yields

$$\alpha_t = \frac{4\rho_1 c_1 \rho_2 c_2 \cos \theta_1 \sqrt{1-(c_2/c_1)^2 \sin^2 \theta_1}}{\left[\rho_2 c_2 \cos \theta_1 + \rho_1 c_1 \sqrt{1-(c_2/c_1)^2 \sin^2 \theta_1}\right]^2} \tag{3.4.33}$$

Thus, if $\theta_1 = \theta_{1c}$, $\alpha_t = 0$, so that no energy originally in medium 1 penetrates medium 2.

What happens when $\theta_1 > \theta_{1c}$ (but still with $c_1 < c_2$, for otherwise θ_{1c} would not exist)? To answer this, we first note that since $\sin \theta_1 > \sin \theta_{1c} = c_1/c_2$, then

$$1-(c_2/c_1)^2 \sin^2 \theta_1 < 0$$

Thus, the square root in (3.4.30) is purely imaginary; that is,

$$\sqrt{1-(c_2/c_1)^2 \sin^2 \theta_1} = i\sqrt{(c_2/c_1)^2 \sin^2 \theta_1 - 1} \tag{3.4.34}$$

But the radical on the left hand side is equal to $\cos \theta_2$. Therefore, the equation

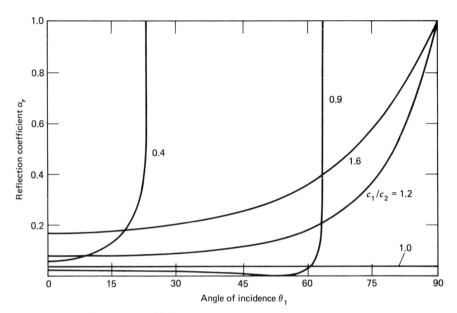

Figure 3.4.3 Reflection coefficient for oblique incidence. All curves correspond to $\rho_1/\rho_2 = 1.5$.

for the transmitted-wave potential, (3.4.5), becomes

$$\phi_t = \tilde{C}\exp\left[-k_2\sqrt{(c_2/c_1)^2\sin^2\theta_1 - 1}\,|y| + i\omega(x\sin\theta_2/c_2 - t)\right] \quad (3.4.35)$$

where $k_2 = \omega/c_2$ and where we have used the fact that in medium 2, y is always negative. We can then see that if $\theta_1 > \theta_{1c}$, the amplitude of the transmitted wave decreases rapidly beyond the interface. Of course, this surface field does not dissipate any energy, so that all of the incoming energy is reflected as indicated by (3.4.22), which gives $\alpha_r = 1$. Figure 3.4.3 shows the variations of α_r with θ_1 for various combinations of sound speeds.

Field in Front of a Rigid Reflector

We now consider the field in front of the interface, that is, in the medium where the incident wave travels, for the case when medium 2 is much denser than medium 1 (i.e., $\rho_2 \gg \rho_1$). In this case we obtain, from (3.4.17),

$$\tilde{B} = A \quad (3.4.36)$$

Therefore, the pressure field in medium 1, which is given by (3.4.10), becomes

$$\tilde{p}_1' = i2\omega\rho_1 A e^{i(k_x x - \omega t)} \cos k_y y \qquad (3.4.37)$$

where

$$k_y = k\cos\theta = 2\pi/\lambda_y$$

$$k_x = k\sin\theta = 2\pi/\lambda_x \qquad (3.4.38)$$

and

$$k = \omega/c_1 = \sqrt{k_x^2 + k_y^2} \qquad (3.4.39)$$

To understand the meaning of (3.4.37), suppose, first, that x is held constant. Then, we see that the variable term represents a standing wave in front of the reflector, which has a nodal pattern similar to that obtained for normal incidence. Here, the first node occurs at a distance y_0 from the reflector equal to

$$y_0 = \lambda_y/4 \qquad (3.4.40)$$

where $\lambda_y = \lambda/\cos\theta_1$. On the other hand, the exponential factor represents a traveling wave moving toward increasing values of x. Therefore, we can say that (3.4.37) represents a nodal pattern that is being propagated to the right. It is of interest to determine some of the characteristics of this propagation. For this purpose, we consider Figure 3.4.4, which depicts the nodal pattern at some fixed instant. In particular, let us consider some specific feature of the pattern, say a line of antinodes that falls on the plane $x = x_0$. An instant later, this line will be located at

$$x = x_0 + c_{fx}(t - t_0) \qquad (3.4.41)$$

where c_{fx} is the velocity with which points of equal phase are moving to the right, and is given by

$$c_{fx} = \frac{c_1}{\sin\theta_1} \qquad (3.4.42)$$

In view of (3.4.38) and (3.4.39), this can also be writen as

$$c_{fx} = c_1(k/k_x) \qquad (3.4.43)$$

The velocity thus defined is called the *phase velocity*, to distinguish it from the

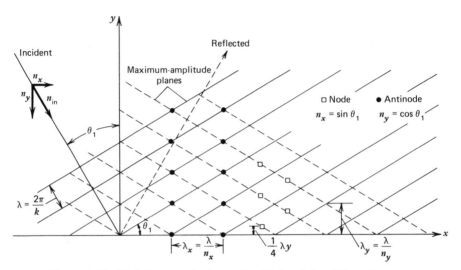

Figure 3.4.4 Nodal pattern in front of a rigid wall for oblique incidence.

velocity with which the waves are moving. Thus, since we are considering motion along the x axis, the speed of the waves is merely the "x component" of c_1, or

$$c_{gx} = n_x c_1 = \frac{k_x}{k} c_1 \qquad (3.4.44)$$

This is called the *group velocity*. In this example, the differences between c_{fx} and c_{gx} arise because the x axis is not along the direction of propagation. It will be shown later that these velocities may be defined from a relation between ω and k by means of

$$c_g = d\omega/dk \qquad (3.4.45)$$

$$c_f = \omega(k)/k \qquad (3.4.46)$$

Thus, if we put $k = k_x$ and use $\omega = c_1\sqrt{k_x^2 + k_y^2}$ in these, we recover the above results. It should be noted that as $\theta_1 \rightarrow 0$, the group velocity (along x) becomes smaller and smaller and vanishes when $\theta_1 = 0$. This implies that there are no waves traveling along x, so that, in particular, there is no energy being propagated along that direction. On the other hand, the phase velocity becomes infinite in that limit. This does not imply that there is propagation with infinite velocity because propagation is associated only with the group velocity c_g. What it implies is simply that as $\theta_1 \to 0$, the distance λ_x between any two consecutive nodes becomes very large, so that k_x becomes very small and vanishes in the limit $\theta_1 = 0$.

Dispersion

To see the significance of the group velocity c_g, introduced earlier, we consider, briefly, dispersive waves. These are waves whose phase velocity $c_f = \omega/k$ is not constant. Their name originated from the fact that light waves, whose speed of propagation depends on the frequency, are "dispersed" by a prism into a spectrum. Whether a given wave is dispersive depends on the functional relationship between the frequency and the wave number, that is, by the form of

$$\omega = \omega(k) \qquad (3.4.47)$$

This relationship is termed a *dispersion equation*. In the nondispersive case, $\omega(k)$ is of the form

$$\omega(k) = ck, \quad c = \text{constant} \qquad (3.4.48)$$

In this case, the phase velocity $c_f = \omega(k)/k$ and the group velocity $c_g = d\omega/dk$ are identical. In general, acoustic waves are nondispersive.

We will derive the relationship defining c_g by means of the well-known phenomenon of *beats* between tones having nearly the same frequency. Thus, we consider two sinusoidal acoustic waves having the same amplitude, but having slightly different frequencies and wave numbers. The combined field is given simply by the sum of each wave. Hence,

$$\tilde{\phi} = A\left[e^{i(k_1 x - \omega_1 t)} + e^{-i(k_2 x - \omega_2 t)} \right] \qquad (3.4.49)$$

Thus, if we put $\omega_2 = \omega_1 + \Delta\omega$ and $k_2 = k_1 + \Delta k$, we can write for $\tilde{\phi}$

$$\tilde{\phi} = 2A \cos\tfrac{1}{2}(\Delta k x - \Delta\omega t) e^{i[(k_1 + \frac{1}{2}\Delta k)x - (\omega + \frac{1}{2}\Delta\omega)t]} \qquad (3.4.50)$$

This represents waves of frequency $(\omega_1 + \Delta\omega/2)$, having a phase velocity $[(\omega_1 + \frac{1}{2}\Delta\omega)/(k_1 + \frac{1}{2}\Delta k)]$, and having a varying amplitude a given by

$$a = 2A \cos\tfrac{1}{2}(\Delta k x - \Delta\omega t) \qquad (3.4.51)$$

Figure 3.4.5 shows the spatial variations of ϕ at some time t_0. The slowly-varying curves in the figure represent the envelope of the amplitude function.

While the curves represent variations of the field due to the interactions between two pure tones of nearly equal frequency, it is possible to interpret each

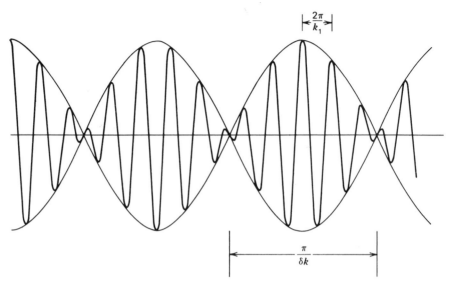

Figure 3.4.5 Beats due to combination of two waves having nearly equal frequencies.

"hump" in the curves as a group of waves; namely, a group of waves having a wavelength $\sim 2\pi/k_1$. It is clear that the velocities of a group of waves and of an individual wave within a group are different. Thus, as (3.4.50) shows, the velocity of the phase of an individual wave—the phase velocity—is $c_f \doteq \omega_1/k_1$, whereas (3.4.51) shows that the velocity with which each group is moving is given by

$$c_g = \Delta\omega/\Delta k \qquad (3.4.52)$$

Although this argument is rather specialized, we might expect that if we had a group of waves having nearly equal wavelengths instead of two beating sinusoids, then the velocity of the group would be given by

$$c_g = d\omega/dk \qquad (3.4.53)$$

as used earlier.

PROBLEMS

3.4.1 Consider a combination of densities and speeds of sound such that θ_1^* exists [see Equation (3.4.25)]. Compute α_r for $\theta_1 > \theta_1^*$ using (3.4.22). Explain your results.

3.4.2 Consider (3.4.17) for $\tilde{B}/A$. Reduce the equation for the case $\rho_1 c_1 / \rho_2 c_2 \ll 1$. Discuss the validity of your result for the case $\theta_1 \to 0$.

3.4.3 Construct the nodal pattern in front of a rigid reflector for the case $\theta_1 = \pi/6$.

3.5   PROPAGATION IN A TWO-DIMENSIONAL CHANNEL

In this section we use some of the results obtained earlier to study the simple but important problem of propagation in a two-dimensional channel. The channel's height is l, as shown in Figure 3.5.1. Its walls, which extend to $\pm \infty$ in a direction perpendicular to the plane of the figure, are assumed to be rigid. We also assume that at a sufficiently long distance to the left, a sound source exists that produces acoustic waves of frequency ω. The pressure field in the channel may then be obtained by solving (2.5.2) and using (2.3.7). Equivalently, it can be obtained from

$$\frac{\partial^2 \tilde{p}'}{\partial x^2} + \frac{\partial^2 \tilde{p}'}{\partial y^2} + k^2 \tilde{p}' = 0 \tag{3.5.1}$$

where $k = \omega/c_0$. The solution to this equation that satisfies the condition that at $y = 0$ the normal velocity component vanishes, is given by

$$\tilde{p}' = i 2 \rho_0 \omega A e^{i(k_x x - \omega t)} \cos(k_y y) \tag{3.5.2}$$

where

$$k_x^2 + k_y^2 = k^2 \tag{3.5.3}$$

Corresponding to $\tilde{p}'$ as given above, there exists a velocity potential given by

$$\tilde{\phi} = 2 A e^{i(k_x x - \omega t)} \cos(k_y y) \tag{3.5.4}$$

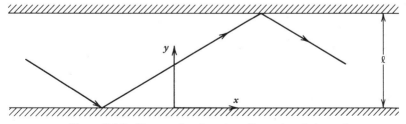

Figure 3.5.1   Propagation in a two-dimensional channel.

Now, at $y = l$ the boundary condition is also $\mathbf{u} \cdot \mathbf{n} = 0$. Thus, since $\tilde{\mathbf{u}} = \nabla \tilde{\phi}$, and $\mathbf{n} = -\mathbf{j}$, we obtain

$$2Ak_y \sin k_y l = 0 \qquad (3.5.5)$$

Thus, for $A \neq 0$, the y component of the wave vector must be such that

$$k_y l = n\pi, \quad n = 0, 1, 2, \ldots \qquad (3.5.6)$$

The root $k_y = 0$ is allowed because it gives the nontrivial result $k_x = k$. The x component of the wave vector can then be obtained from (3.5.3), and is

$$k_x = k\sqrt{1 - (n\pi c_0 / \omega l)^2} \qquad (3.5.7)$$

Denoting this quantity by k_{xn}, we can write the solution for $\tilde{p}'$ as

$$\tilde{p}' = \sum_{n=0}^{\infty} \tilde{B}_n e^{i(k_{xn}x - \omega t)} \cos \frac{n\pi y}{l} \qquad (3.5.8)$$

where $B_n = 2\rho_0 \omega i A_n$ and where the quantities A_n can be obtained in terms of the conditions at the source. The meaning of this solution can be understood easily by writing out the series. Thus,

$$\tilde{p}' = \tilde{B}_0 e^{i(k_{x0}x - \omega t)} + \tilde{B}_1 e^{i(k_{x1}x - \omega t)} \cos \frac{\pi y}{l} + \cdots \qquad (3.5.9)$$

Each term in this series represents a different *mode* of oscillation, and the quantities $|\tilde{B}_n|$ are the amplitudes of each one of these modes. Now, for $n = 0$, $k_{xn} = k$. Therefore, the first term of this series, namely,

$$\tilde{p}'_0 = \tilde{B}_0 e^{i(kx - \omega t)} \qquad (3.5.10)$$

represents a purely one-dimensional, plane wave traveling along the axis of the channel. This mode is sometimes denoted a *longitudinal* mode to differentiate it from the other modes, which are of the form

$$\tilde{p}'_n = \tilde{B}_n e^{i(k_{xn}x - \omega t)} \cos \frac{n\pi y}{l}, \quad n \neq 0 \qquad (3.5.11)$$

These modes are called *transverse* because they propagate along a direction that does not coincide with the x axis, but that is determined by the values of k_{xn} and k_{yn}. The quantity that determines whether or not these modes are propagated along the channel is k_{xn}. Thus, in view of (3.5.7), we have the following possibilities:

(1) $\omega > \omega_n = n\pi c_0/l$. In this case, the imposed frequency is larger than the nth characteristic frequency for lateral oscillations. Here, k_{xn} is real and positive, so that (3.5.11) represents a *transverse* acoustic mode that is progressing in time toward increasing values of x.

(2) $\omega < \omega_n$. Here,

$$k_{xn} = i\sqrt{(\omega_n/\omega)^2 - 1} \qquad (3.5.12)$$

so that

$$e^{ik_{xn}x} = e^{-\sqrt{(\omega_n/\omega)^2 - 1}\,x} \qquad (3.5.13)$$

Therefore, in this case the nth mode decays exponentially with distance along the channel so that at sufficiently long distances from the source its amplitude is negligible.

(3) $\omega = \omega_n$. This is known as the cutoff condition, and separates cases (1) and (2) above. Here $k_{xn} = 0$ so that (3.5.11) becomes

$$\tilde{p}'_n = \tilde{B}_n \cos n\pi y/l \qquad (3.5.14)$$

This is a purely transverse oscillation, independent of x, and with frequency ω_n, which is the nth characteristic frequency of oscillation for the region $0 \leqslant y \leqslant l$. Here we have a simple resonance; that is, the frequency of the source matches, exactly, one of the characteristic frequencies of the region between the walls.

We therefore arrive at the following important result. A given mode, that is, a given term of the series (3.5.8), will be propagated along the channel according to whether or not the frequency ω of the imposed waves is greater or smaller, respectively, than $\omega_n = n\pi c_0/l$. If $\omega > \omega_n$, then the nth mode will be propagated, and if $\omega < \omega_n$, it will be damped. Therefore, as ω is decreased from a large value, fewer and fewer transverse modes will be propagated along the channel. In particular, none of these transverse modes will be propagated if the frequency ω is lower than the lowest characteristic frequency for transverse oscillations, that is, lower than the cutoff frequency for the $n=1$ mode. In these conditions, only the longitudinal mode ($n=0$) will be propagated without attenuation:

$$\tilde{p}' = B_0 e^{i(kx-\omega t)} + B_1 e^{-k\sqrt{(\omega_1/\omega)^2-1}\,x} + \cdots \qquad (3.5.15)$$

Thus, for sufficiently long distances x, only the first term will survive, and the pressure field will be given by

$$\tilde{p}' = B_0 e^{i(kx - \omega t)} \qquad (3.5.16)$$

The above conclusion on whether a given mode will be propagated may also be arrived at by using the concept of group velocity introduced earlier, and determining for what frequency this quantity becomes zero. Thus, since

$$\omega = c_0 \sqrt{k_{xn}^2 + k_{yn}^2} \qquad (3.5.17)$$

we have for $c_{gx} = \partial \omega / \partial k_{xn}$

$$c_{gx} = c_0 \sqrt{1 - (\omega_n / \omega)^2} \qquad (3.5.18)$$

where $\omega_n = n\pi c_0 / l$. Thus, c_{gx} is real and positive only if $\omega > \omega_n$. Figure 3.5.2 depicts the variations of c_{gx} and c_{fx} with frequency. It is clear that different modes, when propagated, are propagated with different speeds. Therefore, a given profile will be distorted as the wave is propagated along the channel.

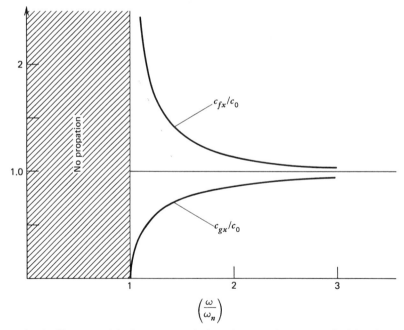

Figure 3.5.2 Changes with frequency of the phase and group velocities in a two-dimensional channel.

Similar results hold for other geometries. In particular, for a circular tube, the frequency below which only the plane wave mode is possible is given by

$$\omega_0 = 0.5861\pi c_0/R \qquad (3.5.19)$$

where R is the tube radius. This result will be derived later. Together with the results derived here, it shows that, in general, this cutoff frequency is of the order c_0/d, where d is the maximum lateral dimension of the tube.

We should mention, also, that in order to have only longitudinal modes propagating in a tube, there is, in addition to the above restriction on the frequency, a restriction on the length of the tube. This additional restriction arises because the only manner in which the transverse modes can become negligible is if x can take sufficiently large values. However, in some instances this is not the case. Suppose, for example, that we have sound waves being excited at one end of a tube of length L. Then, in order to have a fairly plane sound wave inside the tube, the transverse modes must have, at $x=L$, amplitudes that are negligible small. Otherwise they will be reflected back to the source with a possible buildup that cannot be ignored. Now, a minimum length that insures purely plane-wave modes cannot be specified exactly. However, a useful criterion for that purpose could be that L be such that

$$kL\sqrt{(\omega_1/\omega)^2 - 1} \geqslant \pi \qquad (3.5.20)$$

because then the amplitude of the first transverse mode at $x=L$ will be a small fraction $(1/e^\pi)$ of the amplitude in front of the source.

Excitation of Transverse Modes

In addition to satisfying the condition that the excitation frequency ω must be larger than the cutoff frequency of the nth mode for that mode to be propagated, it is necessary that the mode be excited at the source. To see how this excitation can take place, consider a channel fitted at $x=0$ by a two-dimensional, movable membrane, Suppose now that the membrane executes harmonic oscillations about $x=0$. The amplitude of these oscillations will be assumed small, but will be allowed to vary from point to point of the membrane's surface. Thus, the membrane velocity will have only an x component, and this is given by

$$V = U(y)\cos\omega t \qquad (3.5.21)$$

or, also, by the real part of $\tilde{V} = U(y)\exp(-i\omega t)$. The potential in the channel is given by

$$\tilde{\phi}(x, y, t) = \sum_{n=0}^{\infty} \tilde{A}_n e^{i(k_{xn}x - \omega t)} \cos\frac{n\pi y}{l} \qquad (3.5.22)$$

The amplitudes A_n may be obtained from the condition that the velocity of the fluid in the channel in front of the membrane be equal to the velocity of the membrane, Now, since the membrane is moving, the condition should, in principle, be applied at the instantaneous position of the membrane. However, since we have assumed that the membrane's displacement amplitude is very small, we may apply the condition at the mean position $x=0$. Therefore, since the membrane's only motion is along the x axis, we require that

$$\left(\frac{\partial \tilde{\phi}}{\partial x}\right)_{x=0} = U(y) \tag{3.5.23}$$

Therefore,

$$\sum i\tilde{A}_n k_{xn} \cos \frac{n\pi y}{l} = U(y) \tag{3.5.24}$$

We evaluate the coefficients $\tilde{A}_n$ in the usual manner, taking advantage of the orthogonality of the functions $\cos(n\pi y/l)$. Thus, multiplying both sides of (3.5.24) by $\cos(m\pi y/l)$ and integrating between 0 and l, we obtain

$$\sum_{n=0}^{\infty} i\tilde{A}_n k_{xn} \int_0^l \cos \frac{m\pi y}{l} \cos \frac{n\pi y}{l} \, dy = \int_0^l U(y) \cos \frac{m\pi y}{l} \, dy \tag{3.5.25}$$

But

$$\int_0^l \cos \frac{m\pi y}{l} \cos \frac{n\pi y}{l} \, dy = \frac{l}{2}\delta_{nm}, \quad m,n>0 \tag{3.5.26}$$

where δ_{nm} is the Kronecker delta. Thus, for $n>0$, we have

$$ik_{xn}\tilde{A}_n = \frac{2}{l}\int_0^l U(y) \cos \frac{n\pi y}{l} \, dy \tag{3.5.27}$$

whereas if $n=0$, we obtain, from (3.5.24)

$$ik\tilde{A}_0 = \frac{1}{l}\int_0^l U(y) \, dy \tag{3.5.28}$$

Hence, if the membrane's velocity is uniform, then all modes with $n\neq 0$ will have zero amplitude regardless of the frequency of oscillation. In this case, we merely have a rigid surface; for example, a piston, moving back and forth and generating a plane, one-dimensional wave. On the other hand, if $U(y)$ is not uniform, then some transverse modes will be excited. If $U(y)$ is specified, then

(3.5.27) and (3.5.28) determine the coefficients $\tilde{A}_n$, and this completes the solution. It is nevertheless instructive to express these coefficients in a slightly different manner. Thus, in view of (3.5.24) we may express $U(y)$ as

$$U(y) = \tfrac{1}{2}U_0 + \sum_{n=1}^{\infty} U_n \cos \frac{n\pi y}{l} \qquad (3.5.29)$$

where the coefficients U_n are obtained as usual. In fact, (3.5.27) and (3.5.28) show that $U_0 = 2ikA_0$, and that $U_n = ik_{xn}\tilde{A}_n$. It then follows that for all n, the amplitude of the nth mode is given by

$$\tilde{A}_n = U_n / ik_{xn} \qquad (3.5.30)$$

Therefore,

$$\tilde{\phi} = \sum_{n=0}^{\infty} (U_n / ik_{xn}) e^{i(k_{xn}x - \omega t)} \cos \frac{n\pi y}{l} \qquad (3.5.31)$$

where k_{xn} is given by (3.5.7). The point of this simple calculation is to show that if some U_n's are zero, then the corresponding modes do not exist. Also, for n sufficiently large, the frequency ω will be smaller than ω_n, the cutoff frequency corresponding to that value of n. Accordingly, the nth mode will not be propagated. Therefore, small-scale spatial variations of the membrane's motion, which are represented by the high-order Fourier components in (3.5.29), will not be transmitted in the channel. One could then say that the channel is acting as a filter because its output does not contain all the information about the membrane's motion.

Many problems of practical importance involve the propagation of acoustic waves in tubes. These include, for example, acoustic waves propagating in "wind" instruments, in air-conditioning ducts, and so forth. In general, the waves propagating in these and in other examples are three-dimensional, and are quite difficult to handle. However, we have seen that below the lowest cutoff frequency, only the plane, longitudinal mode can be propagated along the tube, so that below that frequency, the field in the tube will be made of plane, one-dimensional waves.

The exact value of the cutoff frequency depends on the shape of the tube's cross-section area, and can be obtained analytically only in a few simple cases such as the rectangular and circular shapes. Other procedures can be used for more complicated geometries, but it is well to remember that the lowest cutoff frequency for a given tube is of the order of the speed of sound, divided by the widest dimension in a cross section of the tube. In the following sections, we will study some simple examples of wave propagation in tubes below cutoff.

PROBLEMS

3.5.1 Consider a two-dimensional channel that is driven at $x=0$ by a membrane executing harmonic oscillations about $x=0$, with a velocity amplitude given by

$$U(y) = \frac{4U_m y}{l}\left(1 - \frac{y}{l}\right)$$

where l is the channel's height and U_m is the maximum velocity amplitude.

(1) Determine the velocity potential in the channel if the channel extends to infinity in the x direction.

(2) Plot the number of modes that are propagated in the channel versus the nondimensional quantity $\omega l / c_0$.

(3) Compute the variations with y of the x component of the fluid velocity. Plot these variations for the particular case $\omega l / c_0 = 10\pi$ at a distance x_0 from the driver equal to $x_0 = 2\pi c_0/\omega$ and at a time $t_0 = 2\pi/\omega$. Compare this velocity with that of the membrane.

3.5.2 Consider a tube of rectangular cross section. The height and width of the tube are a and b, respectively. The tube is fitted at $x=0$ with a membrane executing harmonic oscillations about x with a velocity

$$U = U_e(y, x)\cos \omega t$$

Determine the velocity potential in the tube if the tube extends to infinity in the x direction. Discuss the case $a = b$.

3.6 ACOUSTIC FIELD IN A PISTON-DRIVEN TUBE

As an extension of the results presented in the last section, we consider here the sound field in a rigid tube that is closed at one end by a rigid wall and fitted with a rigid, frictionless, movable piston at the other, as shown in Figure 3.6.1. The piston moves back and forth along the axis of the tube according to

$$x_p = x_0 + l\sin \omega t \tag{3.6.1}$$

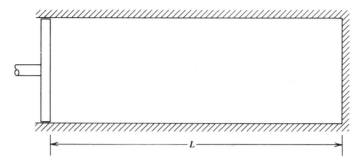

Figure 3.6.1 Piston-driven tube.

Here $x_0 = 0$ is the mean position of the piston, l is the displacement amplitude, and ω is the frequency of the motion. From our previous discussion, it follows that the only mode that can be generated is the plane-wave mode, so that the waves in the tube can be considered purely longitudinal. [For $\omega < 0.586(\pi c_0 / R)$, this would be so even if the piston motion were not exactly perpendicular to the axis of the tube.] We are presently interested in the field after the steady-state condition has been reached. Initially, a transient field will be excited whose frequency corresponds to the characteristic frequency for the region between the piston and the rigid wall. After some time, this field will die out and the remaining field will have a frequency equal to the imposed frequency ω. Thus, the field in the tube will be assumed to be of the form

$$\tilde{u}(x, t) = A e^{i(kx - \omega t)} + \tilde{B} e^{-i(kx + \omega t)} \tag{3.6.2}$$

The boundary conditions are

$$u(x_p, t) = U_p(t)$$

$$u(L, t) = 0 \tag{3.6.3}$$

where U_p, the piston velocity, is equal to $l\omega \cos \omega t$. The first boundary condition is exact, as written, and expresses the fact that the fluid velocity at the piston face is equal to the velocity of the piston. As we will see, this boundary condition must be approximated if (3.6.2) is to be applicable. First, we make use of the second boundary condition and obtain

$$\tilde{B} = -A e^{2ikL} \tag{3.6.4}$$

This gives

$$\tilde{u}(x, t) = -2iA e^{-i(\omega t - kL)} \sin k(L - x) \tag{3.6.5}$$

The first boundary condition, written as

$$\tilde{u}(x_p, t) = \omega l e^{-i\omega t} \qquad (3.6.6)$$

gives

$$\frac{i\omega l}{2A} = e^{ikL} \sin kL \left[1 - \frac{x_p(t)}{L} \right]$$

Thus, we find that A is a function of time. Therefore, our assumed solution is incorrect. However, if for all t, $|x_p|/L \ll 1$, then our solution works well. This is equivalent to applying the boundary at $x = 0$ instead of at $x = x_p$. Thus, if $u(0, t) = U_p$, then

$$-2iAe^{ikL} = \frac{\omega L}{\sin kL} \qquad (3.6.7)$$

so that

$$\tilde{u}(x, t) = \frac{\omega l e^{-i\omega t}}{\sin kL} \sin k(L - x) \qquad (3.6.8)$$

or

$$u(x, t) = \frac{\omega l}{\sin kL} \cos \omega t \sin k(L - x) \qquad (3.6.9)$$

The corresponding acoustic pressure is

$$\tilde{p}'(x, t) = \rho_0 c_0 \frac{\omega l}{\sin kL} e^{-i(\omega t - \pi/2)} \cos k(L - x) \qquad (3.6.10)$$

or

$$p'(x, t) = \rho_0 c_0 \frac{\omega l}{\sin kL} \sin \omega t \cos k(L - x) \qquad (3.6.11)$$

The total acoustic energy in the tube

$$E_a = S \int_0^L \varepsilon \, dx \qquad (3.6.12)$$

where S is the cross-sectional area of the tube and ε, the acoustic-energy density, is given by

$$\varepsilon = \tfrac{1}{2}\rho_0 q^2 + \frac{1}{2\rho_0 c_0^2} p'^2 \tag{3.6.13}$$

The acoustic energy per unit area of the tube is, therefore,

$$e_a = \tfrac{1}{2}\frac{\rho_0}{k}\left(\frac{\omega l}{\sin kL}\right)^2\left[\cos^2\omega t\int_0^{kL}\sin^2 z\,dz + \sin^2\omega t\int_0^{kL}\cos^2 z\,dz\right] \tag{3.6.14}$$

Now, the integrals appearing above are

$$\int_0^{kL}\sin^2 z\,dz = \frac{kL}{2}\left(1 - \frac{\sin 2kL}{2kL}\right)$$

$$\int_0^{kL}\cos^2 z\,dz = \frac{kL}{2}\left(1 + \frac{\sin 2kL}{2kL}\right)$$

Therefore, the instantaneous acoustic energy in the tube is

$$e_a = \frac{\rho_0 L}{4}\left(\frac{\omega l}{\sin kL}\right)^2\left[1 - \frac{\sin 2kL}{2kL}(\cos^2\omega t - \sin^2\omega t)\right]$$

or

$$e_a(t) = \frac{\rho_0 L}{4}\left(\frac{\omega l}{\sin kL}\right)^2\left[1 - \frac{\sin 2kL}{2kL}\cos 2\omega t\right] \tag{3.6.15}$$

The time average of this is

$$\langle e_a(t)\rangle = \frac{\rho_0 L}{4}\left(\frac{\omega l}{\sin kL}\right)^2 \tag{3.6.16}$$

Also of interest is the rate at which the piston does work on the fluid. This is simple compression work, that is,

$$\dot{W}_p = p'(0, t)(dx_p/dt)S_p \tag{3.6.17}$$

where $S_p = S$ is the area of the piston. In view of (3.6.1), this can also be written as

$$\dot{W}_p = p'(0, t)U_p = p'(0, t)u(0, t) \tag{3.6.18}$$

The average of this quantity if the net power given by the piston to the fluid and is

$$\dot{W}_p = \tfrac{1}{2}\langle \sin 2\omega t \rangle = 0 \qquad (3.6.19)$$

This is as it should be since for an ideal fluid, once the steady-state condition has been established, no more energy is required to maintain the motion.

We observe from the solution that the field is a standing wave with nodes and antinodes. In this sense, the field in the tube is very similar to the field in front of a rigid reflector. Now, however, the energy becomes infinitely large for those values of k that satisfy the condition

$$kL = n\pi, \quad n = 1, 2, 3, \ldots \qquad (3.6.20)$$

This is resonance, for then the frequency of the piston $\omega = kc_0$ is equal to one of the characteristic frequencies for longitudinal waves in the tube. Thus, if an experiment were performed in a piston-driven tube by varying the frequency of the driving source, one would expect that the acoustic pressure, for example, would become very large at those frequencies. Consider the pressure at the closed end. Its magnitude may, from (3.6.11), be written as

$$\frac{|p'(L, t)|}{\rho_0 c_0 \omega l} = \frac{1}{\sin \omega L/c_0} \qquad (3.6.21)$$

In the vicinity of the resonance condition, we can put

$$\omega L/c_0 = n\pi(1 \pm \eta) \qquad (3.6.22)$$

where $\eta \ll 1$. Therefore, in the neighborhood of resonance, the nondimensional pressure magnitude $|p'|/\rho_0 c_0^2$ is given by

$$\frac{|p'(L, t)|}{\rho_0 c_0^2} = \frac{l/L}{\eta} \qquad (3.6.23)$$

Thus, even for small l/L, our results predict values of p' incompatible with the restriction that $|p'|$ be small. These results are due in part to our neglecting the effects of dissipation. We will discuss these effects later, using the complete energy equation. Presently, we give a heuristic description of those effects on the present problem. For that purpose, we resort to experimental observation, which shows that if a plane, one-dimensional wave traveling in a fluid has an amplitude A_0 at position x_0, then its amplitude at position x will be

$$A_0 e^{-\alpha(x - x_0)} \qquad (3.6.24)$$

The quantity α is known as the *attenuation coefficient* or as the spatial damping coefficient. It should be clear that such a behavior cannot be obtained from our ideal-wave equation, and that α must be obtained experimentally, or from a more complete description of the physical phenomena involved. In any event, one can obtain the attenuated behavior from our original, nonattenuated solution by letting k be complex. Thus, if in $A_0 \exp(ikx)$ we set

$$k \rightarrow k + i\alpha \qquad (3.6.25)$$

we obtain $A_0 \exp(ikx - \alpha x)$, as required.

Consider, now, the acoustic pressure in the piston-driven problem. At $x = L$, we obtain from (3.6.10)

$$\tilde{p}'(L, t) = \frac{\rho_0 \omega c_0 l}{\sin kL} e^{-i(\omega t - \pi/2)} \qquad (3.6.26)$$

Thus, if $k \rightarrow k + i\alpha$, this gives

$$\tilde{p}'(L, t) = \frac{\rho_0 \omega c_0 l e^{-i(\omega t - \pi/2)}}{\sin kL \cosh \alpha L + i \cos kL \sinh \alpha L} \qquad (3.6.27)$$

In many cases, $\alpha L \ll 1$ so that $\sinh \alpha L \simeq \alpha L$ and $\cosh \alpha L \simeq 1$. Therefore,

$$\tilde{p}'(L, t) = \rho_0 c_0 \frac{\omega l e^{-i(\omega t - \pi/2)}}{\sin kL + i\alpha L \cos kL} \qquad (3.6.28)$$

Thus, if we let $\tan \theta = \alpha L / \tan kL$, then

$$\sin kL + i\alpha L \cos kL = \sqrt{\sin^2 kL + (\alpha L)^2 \cos^2 kL} \; e^{i\theta}$$

so that

$$\tilde{p}'(L, t) = \frac{\rho_0 c_0 \omega l e^{-i(\omega t - \pi/2 + \theta)}}{\sqrt{\sin^2 kL + (\alpha L)^2 \cos^2 kL}} \qquad (3.6.29)$$

Taking the real part of this result, we obtain

$$p'(L, t) = \rho_0 c_0 \omega l \frac{\cos(\omega t + \theta - \pi/2)}{\sqrt{\sin^2 kL + (\alpha L)^2 \cos^2 kL}} \qquad (3.6.30)$$

At resonance, $\omega = n\pi c_0 / L$ so that $\sin kL = 0$ as before. But now, instead of an

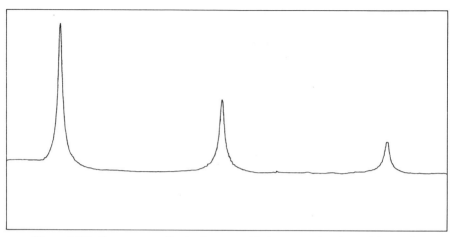

Figure 3.6.2 Pressure amplitude changes with frequency in piston-driven tube.

infinite value for p', we find that its magnitude at $x = L$ is

$$|p'(L, t)|_R = \rho_0 c_0 \omega l / \alpha L \qquad (3.6.31)$$

This value increases with piston amplitude and decreases with attenuation. Figure 3.6.2 displays actual variations of $|p'(L, t)|$ with frequency in a tube. It is seen that the resonance peaks are different. The reasons for this are that α increases with frequency and that, for a given driving force, the piston displacement l decreases with frequency. Nevertheless, it is seen that the values of $|p'|$ are quite large at resonance, especially at the lowest resonance frequency. Under those conditions, the above results may be questionable, because they were derived using the linearized equations of motion.

Experimental Determination of α

There are several techniques that can be used to determine experimentally the attenuation coefficient α. Most of them use a tube, driven by a membrane (e.g., a loudspeaker) or a piston. The attenuation may be due to a variety of factors such as, for example, viscosity. Here we will outline three experimental methods that can be used with a driven tube, using some of the results derived earlier in this section.

Resonance-Peak Method. The first method depends on the shape of the resonance curves. Consider (3.6.30) for p'. At resonance, it gives

$$\langle p'^2 \rangle = \tfrac{1}{2}(\rho_0 c_0 \omega l / \alpha L)^2 \qquad (3.6.32)$$

Therefore, at some other frequency, the relative amplitude squared

$$\frac{\langle p'^2 \rangle}{\frac{1}{2}(\rho_0 c_0 \omega l/\alpha L)^2} = \frac{1}{\sin^2 kL/(\alpha L)^2 + \cos^2 kL} \qquad (3.6.33)$$

has maxima, each of value 1, at $kL = n\pi$. One such maximum is sketched in Figure 3.6.3. The width of the curve depends on α, being greater for larger values of that parameter. This is shown in Figure 3.6.4, which displays two

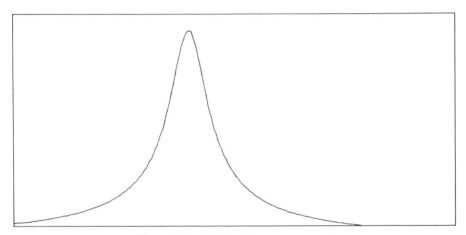

Figure 3.6.3 Width of a resonance peak.

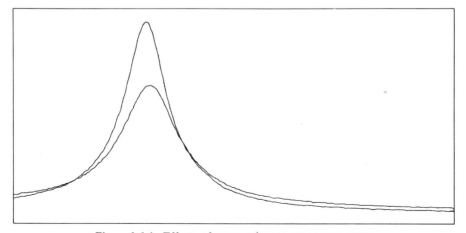

Figure 3.6.4 Effects of attenuation on resonance curve.

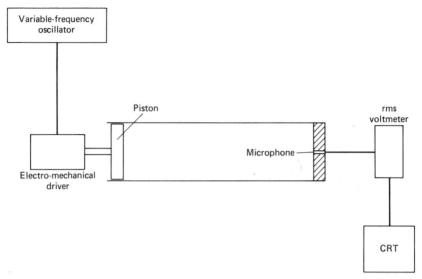

Figure 3.6.5   Apparatus used to study resonance in a tube.

resonance curves having the same resonance frequency but having different amounts of damping. The curves were obtained with the arrangement shown schematically in Figure 3.6.5 by changing the input frequency at a very slow rate.

Let us obtain from (3.6.30) a relation that will yield α in terms of the width of the curve at some predetermined amplitude. For this purpose, we consider one resonance peak, with maxima at some frequency $\omega = \omega_0$, say. As Figures 3.6.3 and 3.6.4 show, the peak is not symmetric about ω_0. However, in the vicinity of that frequency it is nearly so. Now, define the width of the resonance peak by the difference between the frequencies for, ω_1 and ω_2, for which the squared amplitude is one-half of its value at maxima, that is, by

$$\delta = (\omega_2 - \omega_1)/2\pi \tag{3.6.34}$$

Assuming symmetry, we have

$$\omega_1 = \omega_0 - \pi\delta \tag{3.6.35}$$

where $\omega_0 = n\pi c_0/L$. But from the above condition determining ω_1, we have

$$\frac{1}{\sin^2(\omega_1 L/c_0)/(\alpha L)^2 + \cos^2(\omega_1 L/c_0)} = \frac{1}{2} \tag{3.6.36}$$

and, in view of (3.6.35),

$$\sin(\omega_1 L/c_0) = -\sin(\pi\delta L/c_0) \tag{3.6.37}$$

Further, if in addition to symmetry, we assume that $\delta \ll c_0/L$, then $\sin(\omega_1 L/c_0)$ $\approx -(\pi\delta L/c_0)$ and $\cos(\omega_1 L/c_0) \approx 1$. Therefore, (3.6.36) becomes

$$\frac{1}{1+(\pi\delta/\alpha c_0)^2} = \frac{1}{2} \tag{3.6.38}$$

so that

$$\alpha = \pi\delta/c_0 \tag{3.6.39}$$

Hence, a measurements of δ is sufficient to determine α.

Although the method is relatively simple to use, it suffers from a variety of problems. For example, if (3.6.39) is to be applicable, we require that $\alpha L \ll 1$. But if this is satisfied, then p' may be very large, as (3.6.31) shows. Under these conditions, nonlinear effects may not be negligible. Further, if α is small, then δ is also small, so that it may be very difficult to measure accurately.

When αL is not very small, these difficulties do not arise, but then (3.6.39) is not applicable. Also, the curve may not be symmetrical, as it was assumed above. For such cases, one can still obtain α experimentally, but the complete equation (3.6.27) must be used instead of (3.6.28). Otherwise the method remains the same, with frequencies f_1 and f_2 now being asymmetrical with respect to f_0. Of course, the simplicity of (3.6.39) is now lost.

Decay Method. Another method that can be used to determine α experimentally is based on the fact that if the sound source is turned off, the pressure amplitude at any location in the tube decays exponentially with time. Thus, if at time $t=t_0$ the pressure amplitude at some fixed point $x=x_0$ is P_0, and if the sound source is turned off, then the pressure amplitude after an elapsed time t will be given by

$$P = P_0 e^{-\beta t} \tag{3.6.40}$$

when β is called the temporal damping coefficient. The reasons for the decay are basically the same as those that produce spatial attenuation, for example viscosity. Therefore, β should be related to α. The relationship between these two quantities may be established simply by noting that the standing wave in the tube is made of traveling waves having an attenuation coefficient α. Consider the decay in one cycle. In terms of β, the amplitude decays from P_0 to

$$P_0 e^{-\beta T} = P_0 e^{-2L\beta/c_0} \tag{3.6.41}$$

where T is the period. Similarly, in terms of traveling wave, we know that it travels a distance $2L$ in a time T. Therefore, after traveling this distance, its amplitude decays from P_0 to

$$P_0 e^{-2\alpha L} \tag{3.6.42}$$

Comparing with (3.6.40), we obtain

$$\alpha = \beta/c_0 \tag{3.6.43}$$

so that if β is known, α can be obtained.

The simplest technique available to measure β is to measure two or more consecutive peaks in a pressure-decay record, as shown in Figure 3.6.6. Thus, if A_1 and A_2 represent two consecutive peaks of the instantaneous pressure, and if the oscillations are sinusoidal, A_1 and A_2 are related by $A_2 = A_1 \exp(-\beta T)$, and so

$$\beta = f \, ln(A_1/A_2) \tag{3.6.44}$$

Another technique is to plot the decay envelope on a semilogarithmic graph and then measure the slope of the resulting line.

It should be pointed out, however, that the decay method measures an overall damping coefficient in the system. That is, it will contain contributions from

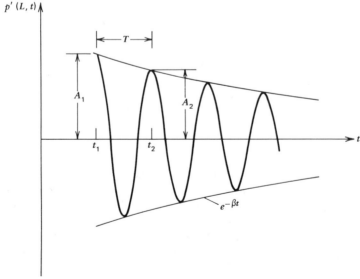

Figure 3.6.6 Temporal decay of pressure at the closed end of the tube.

whatever dissipative mechanisms exist in the tube, and not just those accounted for by α. For example, if the tube is being driven by a piston, friction effects between the piston and the walls of the tube will result in increased values of β. similarly, any "leaks" in the tube will produce higher values of that quantity.

Standing-Wave Method. A more accurate method than the two methods described above is based on the interference pattern that is formed in front of the closed end of the tube. This is so mainly because the pattern exists at all frequencies, so that one can avoid resonance, and therefore eliminate the problems associated with the resonance-peak method. A second reason for its high accuracy is that the dissipation effects that are measured are due to whatever absorption exists between the closed end and the point of measurement. Stated differently, dissipation caused by nonideal conditions at the source does not affect the final result.

This method relies on the fact that the difference between the ideal-case pattern and that for the actual case may be related to α. In fact, the method is similar to that used in Section 3.1 to study the absorption properties of a nonrigid reflector. There, absorption occurred only at the reflector, whereas now it occurs also in the main body of the fluid filling the tube.

To obtain α, we must first relate it to easily measurable quantities in the tube. The most suitable quantity for this purpose is the acoustic pressure in front of the tube's closed end because it is from this end that measurements can be made easily by means of a movable microphone probe, as shown in Figure 3.6.7. The probe is hollow and has a small orifice near its tip that is used to communicate the pressure at some point in the tube to the externally placed microphone. Now, the pressure in front of the reflector may be obtained from (3.6.10), but it is simpler to obtain it in terms of incident and reflected waves. Thus, if the origin is placed at the closed end of the tube (so that x' measures the distance to the closed end), the pressure in the tube may be expressed as

$$\tilde{p}'(x', t) = \rho_0 c_0 A e^{-i\omega t}(e^{-ikx'} + \tilde{R}e^{ikx'}) \tag{3.6.45}$$

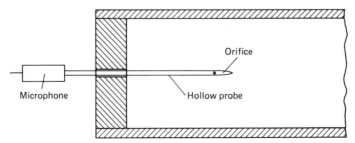

Figure 3.6.7 Microphone probe used to study the standing-wave pattern inside the tube.

where $\tilde{R}$ is the pressure-amplitude ratio. Using (3.1.35), this becomes

$$\tilde{p}' = \rho_0 c_0 A e^{-\psi - i(\omega + \pi\sigma/2)} \left[e^{-i(kx' + i\psi - \pi\sigma/2)} + e^{i(kx' + i\psi - \pi\sigma/2)} \right] \quad (3.6.46)$$

So far we have allowed for absorption only at the reflecting end. To account for absorption in the fluid, we let $k \to k + i\alpha$, as in (3.6.25), so that

$$\tilde{p}' = \rho_0 c_0 A e^{-\psi - i(\omega t + \pi\sigma/2)} \left[e^{-i(kx' - \pi\sigma/2) + (\alpha x' + \psi)} + e^{i(kx' - \pi\sigma/2) - (\alpha x' + \psi)} \right]$$

$$(3.6.47)$$

This equation may be adapted, in a manner similar to that shown below, to obtain α for a wide range of values of the quantity. Of special importance, however, is the case of small attenuation because this requires the highest accuracy. For such a case, (3.6.47) takes on a specially simple form. Thus, we consider the case where $\alpha x' \ll 1$. Further, we take the closed end to be a rigid reflector so that we may safely put $\sigma = 0$. On the other hand, the quantity ψ is not zero, even for a rigid reflector, because there may be some losses associated with heat conduction there. However, we take these losses to be small so that ψ is also small. Therefore,

$$e^{\pm(\alpha x' + \psi)} = 1 \pm (\alpha x' + \psi) + \cdots \quad (3.6.48)$$

and so (with $\sigma = 0$) we have

$$\tilde{p}' = 2\rho_0 c_0 A e^{-\psi - i\omega t} \left[\cos kx' - i(\alpha x' + \psi) \sin kx' \right] \quad (3.6.49)$$

The magnitude of this quantity is

$$|\tilde{p}'| = 2\rho_0 c_0 A e^{-\psi} \left[\cos^2 kx' + (\alpha x' + \psi)^2 \sin^2 kx' \right]^{1/2} \quad (3.6.50)$$

The quantity in the radical oscillates between a value near zero and a value near unity. The values at maxima are approximately given by

$$|\tilde{p}'|_{\max} = 2\rho_0 c_0 A e^{-\psi} \quad (3.6.51)$$

and occur at distances from $x' = 0$ given by

$$D_n = \tfrac{1}{2} n\lambda, \quad n = 0, 1, \ldots \quad (3.6.52)$$

On the other hand, $|\tilde{p}'|$ has minima given by

$$|\tilde{p}'|_{\min} = 2\rho_0 c_0 A e^{-\psi} |\alpha d_n + \psi| \quad (3.6.53)$$

where d_n is their distance to the reflector and is given by

$$d_n = \tfrac{1}{4}n\lambda \qquad (3.6.54)$$

Therefore, the ratio of $|\tilde{p}'|_{min}$ to $|\tilde{p}'|_{max}$ is given simply by

$$\frac{|\tilde{p}'|_{min}}{|\tilde{p}'|_{max}} = \alpha d_n + \psi \qquad (3.6.55)$$

where we have used the fact that α, d_n, and ψ are positive. Figure 3.6.8a displays the variations of $|\tilde{p}'|$ with distance from the reflector, and Figure 3.6.8b shows the corresponding variation of the ratio of pressure magnitudes.

When the above equations are applicable, α can be determined by measuring the amplitude at one maximum and measuring the amplitudes and locations of

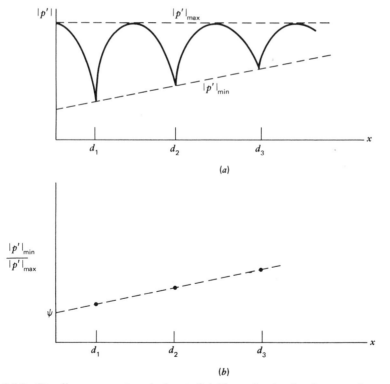

Figure 3.6.8 Standing-wave pattern in front of rigid termination for the case of very small attenuation. (a) Pressure variations; (b) variations of standing-wave ratio.

several minima. When the pressure-amplitude ratios are plotted versus distance, and the line connecting the data points are extrapolated to $x' = 0$, the value of ψ is obtained. The value of α is then obtained from the slope of the line, or from

$$\alpha = \frac{1}{d_n}\left[-\psi + \frac{|p'|_{\min,\,n}}{|p'|_{\max}} \right] \tag{3.6.56}$$

Amplitude Growth at Resonance

We now return to our discussion of the acoustic field in the tube during resonance conditions. Earlier, we showed that if we let the wave number become complex, and then take the limit as ω approaches one of the characteristic frequencies of the tube, the resulting field is found to have finite amplitudes. While the procedure is correct, it is desirable to use a different argument to show how the finite-amplitude condition arises. The basic idea is simply that waves whose amplitudes grow indefinitely are not physically possible. Therefore, a limiting-amplitude state must eventually be reached where a balance exists between the energy-input rate and the energy-dissipation rate.

We begin by studying the manner in which the field grows at resonance. In the absence of dissipation, our results for p' and u may be written as

$$p' = \frac{\rho_0 c_0 \omega l}{\sin \omega L/c_0}\sin \omega t \cos \frac{\omega(L-x)}{c_0} \tag{3.6.57}$$

$$u = \frac{\omega l}{\sin \omega L/c_0}\cos \omega t \sin \frac{\omega(L-x)}{c_0} \tag{3.6.58}$$

Suppose that at some time t the piston frequency is set equal to $\omega_0 = n\pi c_0/L$. Then our equations predict that p' and u become infinitely large. However, this cannot happen instantaneously, and at least during some interval, p' and u have finite amplitudes. The manner in which these quantities grow in time may be obtained from (3.6.57) and (3.6.58) by taking the limit $\omega \to \omega_0$ in these expressions. Thus, in the limit $\omega \to \omega_0$, the terms in p' and u whose amplitudes grow more rapidly are found to be

$$p' = \frac{\rho_0 c_0 \omega_0 l t}{(-1)^n (L/c_0)}\cos \omega_0 t \cos \frac{\omega_0(L-x)}{c_0} \tag{3.6.59}$$

and

$$u = \frac{\omega_0 l t}{(-1)^{n+1}(L/c_0)}\sin \omega_0 t \cos \frac{\omega_0(L-x)}{c_0} \tag{3.6.60}$$

The factors in front of the trigonometric functions represent the instantaneous amplitudes. Thus, the pressure amplitude is given by

$$P(t) = \rho_0 c_0^2 (\omega_0 l / L) t \tag{3.6.61}$$

so that at some fixed location in the tube the pressure grows linearly in time, as sketched in Figure 3.6.9. The velocity also grows linearly in time. Therefore, the energy grows with time as t^2. In fact, making use of (3.6.59) and (3.6.60), we obtain

$$E(t) = \frac{L}{4} \rho_0 c_0^2 \left(\frac{\omega_0 l}{L} t \right)^2 S \tag{3.6.62}$$

where S is the tube's cross-sectional area. Therefore, the energy grows at a rate given by

$$\frac{dE}{dt} = \frac{L}{2} \omega_0 c_0^2 \left(\frac{\omega_0 l}{L} \right)^2 S t \tag{3.6.63}$$

The reason for this growth is simply that the energy-input rate is growing. This may be seen by computing the rate at which the piston is doing work on the fluid. Thus, using (3.6.18), (3.6.59), and (3.6.60), we obtain

$$\dot{E}_{in} = \frac{L}{2} \rho_0 c_0^2 \left(\frac{\omega_0 l}{L} \right)^2 S t \tag{3.6.64}$$

This is identical to the energy-growth rate. Thus, as long as no dissipation exists, all of the energy input goes to increase the amplitude of the waves in the tube.

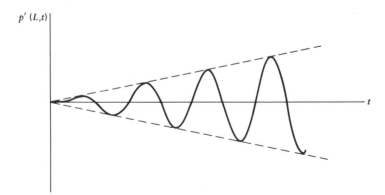

Figure 3.6.9 Pressure-amplitude growth at resonance.

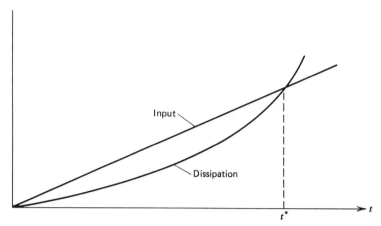

Figure 3.6.10 Increases in time of the energy-input rate and of the energy-dissipation rate.

This is, of course, not the case when dissipative effects are included. In fact, as we will see below, the rate at which the energy dissipation grows is proportional to t^2. Hence, for small t, the energy-input rate is larger than the energy-dissipation rate so that the wave's amplitude grows. At a later time, however, the situation is reversed. Therefore, at some intermediate time $t=t^*$, the two rates become equal, as shown in Figure 3.6.10. Beyond this time, no further growth occurs (except, perhaps, for a momentary overshoot).

To determine t^*, we first show that if dissipative effects are taken into account, the instantaneous dissipation rate $\dot{E}_{out}$ is given by

$$\dot{E}_{out} = 2\beta E(t) \tag{3.6.65}$$

To derive this, we make use of the fact that under the effects of dissipation alone, the pressure and velocity would decay exponentially with time, that is,

$$p' = P_0 e^{-\beta t} \sin \omega t \cos k(L-x) \tag{3.6.66}$$

$$u = (P_0/\rho_0 c_0) e^{-\beta t} \cos \omega t \sin k(L-x) \tag{3.6.67}$$

These yield

$$E(t) = \frac{L}{4\rho_0 c_0^2} S P_0^2 e^{-2\beta t} \tag{3.6.68}$$

so that the rate at which energy is lost by dissipation is

$$\dot{E}_{\text{out}} = 2\beta \frac{L}{4\rho_0 c_0^2} S\left(P_0 e^{-\beta t}\right)^2 \tag{3.6.69}$$

Since $P_0 \exp(-\beta t)$ is the instantaneous pressure amplitude, we may take (3.6.69) to apply at all times and not just during decay. Therefore, in view of (3.6.68), we obtain (3.6.65). During the growth stage, at resonance, $E(t)$ is given by (3.6.62) so that (3.6.65) becomes

$$\dot{E}_{\text{out}} = 2\beta \frac{L}{4} \rho_0 c_0^2 \left(\frac{\omega_0 l}{L}\right)^2 St^2 \tag{3.6.70}$$

When this is equated to the energy-input rate $\dot{E}_{\text{in}}$, as given by (3.6.64), we obtain the time t^* beyond which no further growth occurs. This is

$$t^* = 1/\beta \tag{3.6.71}$$

Thus, from (3.6.61) the pressure amplitude in the limiting condition is

$$P(t^*) = \rho_0 c_0^2 \frac{\omega_0 l}{\beta L} \tag{3.6.72}$$

Or, on using $\beta = \alpha c_0$,

$$P(t^*) = \rho_0 c_0 \frac{\omega_0 l}{\alpha L} \tag{3.6.73}$$

in agreement with (3.6.31).

PROBLEMS

3.6.1 Consider a tube of length L, open at one end and driven by a flat piston at the other end. The motion of the piston is sinusoidal and has an amplitude l. Determine the velocity amplitude at the open end under resonance conditions. Assume that the boundary condition there is $p' = 0$.

3.6.2 In our derivation of (3.6.49), we assumed that $\alpha x' + \psi$ was small. For sufficiently large values of x', this is, of course, not true. Determine $|\tilde{p}'|$ when approximation (3.6.48) is not used. Sketch the variations of pressure maxima and minima with distance.

3.6.3 The quality factor Q of a resonance peak is defined by the ratio of resonance frequency to the half-power width. Determine the Q of a resonance peak for the tube of Problem 3.6.1. Assume that a very small amount of damping is present.

3.7 SOME NONLINEAR EFFECTS

In the previous section, we mentioned several times that at resonance, nonlinear effects might not be negligible so that our linear solutions may not be applicable. An idea of what happens when significant nonlinear effects are present may be obtained from Figure 3.7.1, which displays the pressure oscillations in time at the closed end of a tube driven by a piston at frequencies near its lowest characteristic frequency. These records were obtained with an experimental setup of the type shown in Figure 3.6.5, in which both αL and l/L were small and had values in the vicinity of 10^{-2}. It is seen that away from resonance, the oscillations appear to be sinusoidal as theory predicts, but that as the resonance condition is approached, the profiles become quite distorted. Clearly, our sinusoidal solution has then ceased to be applicable. Another departure from linear theory may be seen by plotting the pressure amplitude at the closed end, $|p'(L, t)|$ versus the parameter l/L. Our previous result for this quantity is given by (3.6.31). This predicts that $|p'(L, t)|$ increases linearly with that parameter. In fact, this trend actually occurs when l/L is very small. Figure 3.7.2 displays some results of measurements at resonance that clearly show the departures from a linear dependence. The pressure profiles corresponding to these measurements show increased amounts of distortion when l/L increases. As we know, this implies that the Fourier components of the wave are increasing both in number and in amplitude. Figure 3.7.3 displays the growth of the second and third harmonics relative to that of the fundamental. That a nonsinusoidal periodic wave may be decomposed in Fourier components is, of course, not surprising. What is surprising, at least from the linear point of view, is that the energy was fed to the gas in the tube at a frequency equal to the lowest characteristic frequency of the tube. This transfer of energy from the fundamental mode to its harmonics cannot be accounted for in linear acoustics.

The above effects are examples of phenomena that occur when the amplitudes of oscillation in a wave are large. These phenomena are the subject matter of nonlinear acoustics and, to a certain extent, gas dynamics, and are therefore beyond the scope of this book. Nevertheless, we study in this section some simple situations in which nonlinear effects are important.

Distortion of a Progressive Wave

To see how energy is transferred from one monochromatic component to its harmonics, we must return to the basic, nonlinear equations of motion. For

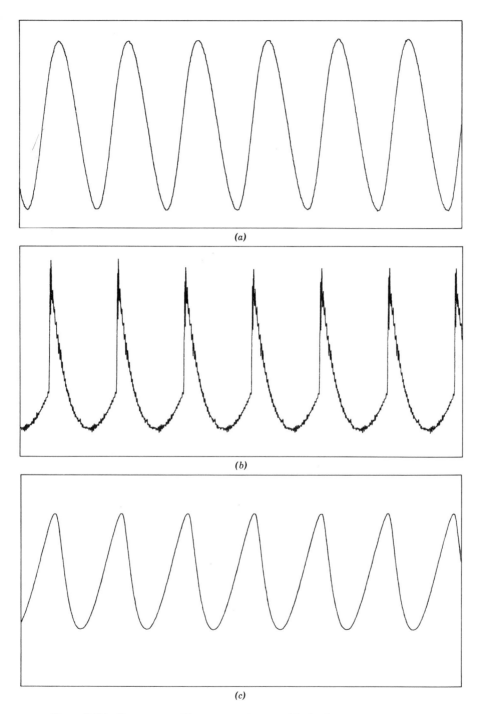

Figure 3.7.1 Pressure profiles near resonance· (a), $f < f_0$; (b), $f = f_0$; (c), $f > f_0$.

149

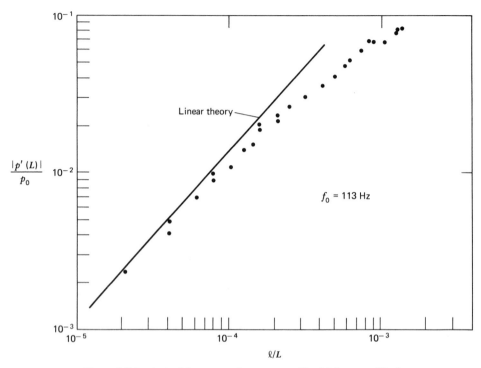

Figure 3.7.2 Actual increase of pressure with driving amplitude.

one-dimensional motions, these are

$$\frac{\partial \rho}{\partial t} + \frac{\partial}{\partial x}(\rho u) = 0 \qquad (3.7.1)$$

$$\frac{\partial(\rho u)}{\partial t} + \frac{\partial}{\partial x}(\rho u^2) = -\frac{\partial p}{\partial x} \qquad (3.7.2)$$

The last equation is Euler's equation written in the form of (1.5.6). If the fluid is ideal, we may still consider the motion to be isentropic, so that

$$p = p(\rho) \qquad (3.7.3)$$

As we will see later, this does not apply if there are shock waves in the fluid. The reason is that in the shocks themselves, the effects of viscosity and thermal conductivity cannot be neglected.

Now, to obtain a working equation suitable for studying nonlinear effects in acoustics, we first multiply (3.7.1) by c_0^2 and then add $\partial p/\partial t$ to both sides of the resulting equation. Next, we multiply (3.7.2) by c_0^2 and then eliminate, by

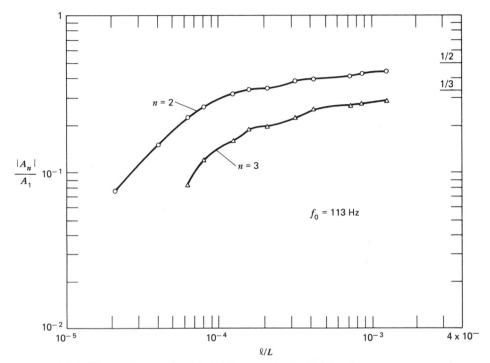

Figure 3.7.3 Harmonic growth with driving amplitude. Driving frequency was equal to frequency of first mode.

cross-differentiation from both equations, the terms containing derivatives of ρu. The result of these operations is

$$\frac{\partial^2 p}{\partial t^2} - c_0^2 \frac{\partial^2 p}{\partial x^2} = \frac{\partial^2}{\partial t^2}\left(p - \rho c_0^2\right) + c_0^2 \frac{\partial^2}{\partial x^2}\rho u^2 \qquad (3.7.4)$$

So far, no approximation has been made. Therefore, (3.7.4) describes all isentropic flows of fluids. Presently, we would like to use it only to study progressive distortion in a simple monochromatic wave. To do this, we adopt a perturbation procedure in which the dependent variables in (3.7.4) are approximated by the following expansions:

$$p = p_0 + \varepsilon p_1 + \varepsilon^2 p_2 + \cdots \qquad (3.7.5)$$

$$\rho = \rho_0 + \varepsilon \rho_1 + \varepsilon^2 \rho_2 + \cdots \qquad (3.7.6)$$

$$u = \varepsilon u_1 + \varepsilon^2 u_2 + \cdots \qquad (3.7.7)$$

where ε is taken to be much smaller than unity. In this scheme, consecutive terms in the expansions are assumed to be only small corrections to the previous solution. For example, we assume that $\varepsilon|p_2|\ll p_1$. If this is satisfied, we may truncate the series after obtaining the first correction to p_1. Now, in view of (3.7.3) and (3.7.6), p may also be expressed as

$$p=p(\rho_0)+c_0^2\left(\varepsilon\rho_1+\varepsilon^2\rho_2+\cdots\right)+\tfrac{1}{2}d_0\left(\varepsilon\rho_1+\varepsilon^2\rho_2+\cdots\right)^2+\cdots$$

$$=p_0+c_0^2\varepsilon\rho_1+\varepsilon^2\left(c_0^2\rho_2+\tfrac{1}{2}d_0\rho_1^2\right)+\cdots \tag{3.7.8}$$

where $d_0=[(\partial^2 p/\partial\rho^2)_S]_{\rho=\rho_0}$. Comparing (3.7.5) and (3.7.8) and equating terms with equal powers of ε, we obtain

$$p_1=\rho_1 c_0^2 \tag{3.7.9}$$

$$p_2=\rho_2 c_0^2+\tfrac{1}{2}d_0\rho_1^2 \tag{3.7.10}$$

Substitution of (3.7.6) and (3.7.8) into (3.7.4) yields

$$\varepsilon\left(\frac{\partial^2 p_1}{\partial t^2}-c_0^2\frac{\partial^2 p_1}{\partial x^2}\right)+\varepsilon^2\left(\frac{\partial^2 p_2}{\partial t^2}-c_0^2\frac{\partial^2 p_2}{\partial x^2}-\tfrac{1}{2}d_0\frac{\partial^2\rho_1^2}{\partial t^2}-\rho_0 c_0^2\frac{\partial^2 u_1^2}{\partial x^2}\right)+\cdots=0 \tag{3.7.11}$$

As expected, the leading term of this is a wave equation for p_1:

$$\frac{\partial^2 p_1}{\partial t^2}-c_0^2\frac{\partial^2 p_1}{\partial x^2}=0 \tag{3.7.12}$$

The correction p_2 to this is seen to satisfy

$$\frac{\partial^2 p_2}{\partial t^2}-c_0^2\frac{\partial^2 p_2}{\partial x^2}=\frac{1}{2}d_0\frac{\partial^2\rho_1^2}{\partial t^2}+\rho_0 c_0^2\frac{\partial^2 u_1^2}{\partial x^2} \tag{3.7.13}$$

Thus, the first correction to the linear solution p_1 satisfies a nonhomogeneous wave equation in which derivatives of the squares of the first-order solution are cast in the role of driving forces. Two such derivatives appear on the right-hand side of (3.7.13). The first,

$$\frac{\partial^2}{\partial t^2}\tfrac{1}{2}d_0\rho_1^2$$

has its origin in (3.7.8). That is, it is due to the nonlinear relationship between pressure and density. The results it produces are therefore said to be caused by *medium nonlinearity*. The second derivative,

$$\frac{\partial^2}{\partial x^2} \rho_0 c_0^2 u_1^2$$

has its origins in the convective term in Euler's equation, and its effects are therefore said to be caused by *convective nonlinearity*. While the mechanisms responsible for these nonlinearities are different, the effects they produce are similar. Therefore, in order to simplify our calculations we consider only the effects of one of them, and use

$$\frac{\partial^2 p_2}{\partial t^2} - c_0^2 \frac{\partial^2 p_2}{\partial x^2} = \rho_0 c_0^2 \frac{\partial^2 u_1^2}{\partial x^2} \tag{3.7.14}$$

to study the initial distortion of a sinusoidal wave in an ideal fluid. To do this, we take a situation where plane waves are being produced by a piston oscillating about $x=0$ with frequency ω. To simplify matters, we assume that the boundary condition can be applied at $x=0$, although as is shown in Problem 3.7.7, this approximation is not entirely consistent with our retaining second-order terms in our formulation. Now, if the piston displacement is given by $x_p = l \sin \omega t$, then the approximate boundary condition is $u(0, t) = \omega l \cos \omega t$. Also, since $u = \varepsilon u_1 + \cdots$, where u_1 satisfies the wave equation, that is, $u_1 = f(x - c_0 t)$, at $x=0$ we have

$$\varepsilon f(-c_0 t) = \omega l \cos \omega t \tag{3.7.15}$$

Thus, $\varepsilon f(\xi) = \omega l \cos k \xi$, and $u = \omega l \cos(kx - \omega t)$, so that

$$\varepsilon = kl \ll 1 \tag{3.7.16}$$

and

$$u_1 = c_0 \cos(kx - \omega t) \tag{3.7.17}$$

Substitution of this into (3.7.14) gives

$$\frac{\partial^2 p_2}{\partial t^2} - c_0^2 \frac{\partial^2 p_2}{\partial x^2} = -2 \rho_0 c_0^2 \omega^2 \cos 2(kx - \omega t) \tag{3.7.18}$$

This may be written as the real part of

$$\frac{\partial^2 \tilde{p}_2}{\partial t^2} - c_0^2 \frac{\partial^2 \tilde{p}_2}{\partial x^2} = -2 \rho_0 c_0^2 \omega^2 e^{2i(kx - \omega t)} \tag{3.7.19}$$

We seek solutions of this of the form

$$\tilde{p}_2 = f(x)e^{-2i\omega t} \tag{3.7.20}$$

Substitution of this into the previous equation yields

$$f''(x) + 4k^2 f(x) = 2\rho_0 c_0^2 k^2 e^{2ikx} \tag{3.7.21}$$

A particular solution of this equation is

$$f(x) = \frac{\rho_0 c_0^2 kx}{2i} e^{2ikx} \tag{3.7.22}$$

The solutions of the corresponding homogeneous equation are of the form $\exp(\pm 2ikx)$. Hence, for positive-going waves, we have

$$\tilde{p}_2 = \left(\tilde{A} + \frac{kx\rho_0 c_0^2}{2i} \right) e^{2i(kx-\omega t)} \tag{3.7.23}$$

However, since at $x = 0$ the oscillation has frequency ω, we must have $\tilde{A} = 0$ so that

$$p_2 = \tfrac{1}{2} kx\rho_0 c_0^2 \cos 2(kx - \omega t - \pi/4) \tag{3.7.24}$$

Using this result, together with $p_1 = \rho_0 c_0 u_1$ in (3.7.5), yields

$$\frac{p - p_0}{\rho_0 c_0^2} = \varepsilon \cos(\omega t - kx) + \tfrac{1}{2}\varepsilon^2 kx \cos 2(kx - \omega t - \pi/4) \tag{3.7.25}$$

Thus, the second approximation to the nonlinear equations is also a monochromatic wave, but of twice the frequency of the first approximation. Because this represents a modification of the purely monochromatic wave, arising from nonlinear effects, the process is called nonlinear harmonic distortion.

A second point of interest regarding (3.7.25) is that the amplitude of the harmonic increases linearly with distance, so that the profile of the wave becomes increasingly distorted as the waves move away from the source. Eventually, however, our main assumption, namely, that $\varepsilon|p_2| \ll |p_1|$, will cease to be valid so that the profile will not be properly described by (3.7.25). In principle, it is possible to include more terms in the approximation. However, it is clear that the procedure is too cumbersome to be of much value. Nevertheless, it is also clear that further distortion takes place. In fact, it is experimentally found that for sufficiently large initial amplitude, a monochromatic wave eventually attains a discontinuous profile of the type shown in Figure 3.7.4. The

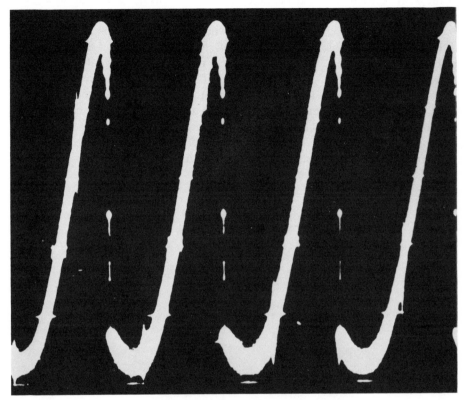

Figure 3.7.4 Profile showing discontinuities.

reasons for the formation of the discontinuity may be understood from the following discussion. Let us consider a wave that initially is purely monochromatic. In the linear approximation, every portion of the wave is propagated with speed c_0 so that the profile of the wave remains constant as it travels. When finite-amplitude effects are included, this speed of propagation is modified in a manner not uniform throughout. One of the sources of the modification is the increase of the local speed of sound $c = \sqrt{(\partial p / \partial \rho)_S}$ with temperature; that is, c is larger or smaller than c_0, depending on whether the point under consideration is in the compressive or in the expansive part of the wave. The second, and usually more important, modification arises because the speed of a wave, at a point in the fluid where the fluid velocity is u, will have a value $c + u$ with respect to a fixed observer. Therefore, due to this effect, points in the compressive part of the wave will move faster than points in the expansive part of the wave. Clearly, both effects involve modifications of equal signs at a given point. Thus, if we draw our profile at several consecutive instants, taking into account the above-mentioned effects, we would obtain curves of the type shown in Figure 3.7.5. The last profile is particularly interesting, for if it could occur, then the pressure and density at a given point in space would have three different values. This is, of course, not possible. In reality, the profile becomes discontinuous, as shown by the dotted line in the figure. In real fluids, these transitions are thin but finite regions in which the variables change continuously due to the smoothing effects of viscosity and heat conductivity. Ideal fluids are, of course, fluids for which these transport coefficients vanish. Therefore, when these transitions, or shocks, occur in a wave, they must be considered to have

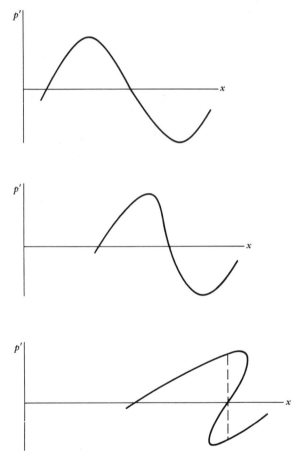

Figure 3.7.5 Progressive distortion of a finite-amplitude wave.

zero thickness. But as we will see below, discontinuous motions of ideal fluids can occur only if the entropy of the fluid moving across such a discontinuity increases; these flows are not isentropic. Further, if entropy changes occur, then some energy dissipation must take place, and this is of interest in some cases, as shown in a later example.

Entropy Changes

Let us consider a wave in which the pressure, density, and so forth suffer a small but finite change across a discontinuity, as shown in Figure 3.7.6. This may be part of a train of waves such as those sketched earlier, or may be a single pulse. We wish to compute the entropy change of the fluid that is overtaken by the shock. For simplicity, we consider the particular case where the fluid is a perfect gas with constant specific heats. In this case, an explicit equation of state is available that simplifies our calculations considerably. Thus, if we use (1.3.32) together with (1.3.1), we obtain for the change of entropy per unit mass

$$\frac{S_2 - S_1}{c_v} = \ln\left[\frac{p_2}{p_1}\left(\frac{\rho_1}{\rho_2}\right)^\gamma\right] \tag{3.7.26}$$

This equation gives the changes of entropy between any two equilibrium states in a perfect gas. However, the states on either side of the discontinuity in Figure 3.7.6 cannot be arbitrary. That is, the values of p_2/p_1 and ρ_1/ρ_2 must be such that the basic conservation equations are satisfied. To determine the values of these variables, we consider a system of coordinates moving with the shock, and *assume* that the shock is moving at a constant speed. In this coordinate system

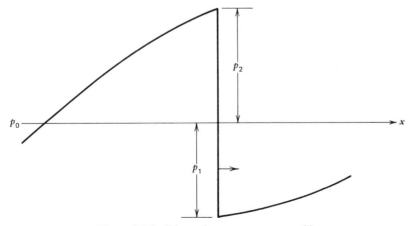

Figure 3.7.6 Discontinuous pressure profile.

the flow is steady so that the conservation equations become

$$\frac{d}{dx}(\rho U)=0 \tag{3.7.27}$$

$$\frac{d}{dx}(p+\rho U^2)=0 \tag{3.7.28}$$

and, from (2.7.6),

$$\frac{d}{dx}\rho u\left(H+\tfrac{1}{2}U^2\right)=0 \tag{3.7.29}$$

where U represents the fluid velocity as seen by an observer moving with the shock. Integrating these equations, we obtain

$$\rho_1 U_1 = \rho_2 U_2 \tag{3.7.30}$$

$$p_1 + \rho_1 U_1^2 = p_2 + \rho_2 U_2^2 \tag{3.7.31}$$

$$c_p T_1 + \tfrac{1}{2}U_1^2 = c_p T_2 + \tfrac{1}{2}U_2^2 \tag{3.7.32}$$

where we have used the fact that for a perfect gas having constant specific heats, $H_2 - H_1 = c_p(T_2 - T_1)$. The subscripts 1 and 2 refer to points that may be located anywhere in the fluid. In particular, they apply to points located on the sides of the discontinuity so that they may be used to obtain ρ_2/ρ_1 as a function of p_2/p_1. First, we use the perfect-gas equation of state to write the last equation as

$$U_1^2 - U_2^2 = \frac{2\gamma}{\gamma-1}\left(\frac{p_2}{\rho_2} - \frac{p_1}{\rho_1}\right) \tag{3.7.33}$$

This may be combined with (3.7.30) and (3.7.31) to eliminate U_1 and U_2. Eliminating, first, U_2^2 from (3.7.31) and (3.7.32) by means of (3.7.30), and then eliminating U_1^2 from the two resulting equations, we obtain

$$p_2 - p_1 = \frac{2\gamma\rho_1}{\gamma-1}\frac{\dfrac{p_2}{\rho_2} - \dfrac{p_1}{\rho_1}}{1+\rho_1/\rho_2} \tag{3.7.34}$$

Solving this for ρ_2/ρ_1 yields

$$\frac{\rho_2}{\rho_1} = \frac{2\gamma p_2 - (p_2 - p_1)(\gamma-1)}{2\gamma p_1 + (p_2 - p_1)(\gamma-1)}$$

$$= \frac{\gamma-1+(\gamma+1)(p_2/p_1)}{\gamma+1+(\gamma-1)(p_2/p_1)} \tag{3.7.35}$$

This equation can also be written in terms of a shock-strength parameter δ defined by

$$\delta = \frac{p_2 - p_1}{p_1} \qquad (3.7.36)$$

Thus,

$$\frac{\rho_2}{\rho_1} = \frac{2\gamma + (\gamma + 1)\delta}{2\gamma + (\gamma - 1)\delta} \qquad (3.7.37)$$

Figure 3.7.7 displays the variations of ρ_2/ρ_1 with p_2/p_1, and shows that according to (3.7.35) it is possible to have both compressive and expansive discontinuities. However, the second law of thermodynamics rules out the second possibility. To see this, we note that the allowed states on the second side of the discontinuity are those for which $S_2 \geqslant S_1$. Reference to (3.7.26) shows that

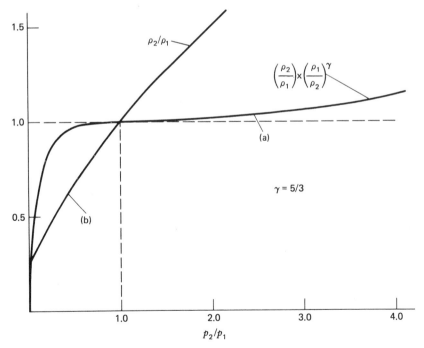

Figure 3.7.7 Variations of $(p_2/p_1)(\rho_1/\rho_2)^\gamma$ and of ρ_2/ρ_1 with pressure ratio. Curve (a) shows that for $p_2 < p_1$, the entropy of the fluid crossing the discontinuity would decrease. Curve (b) shows that if $p_2 > p_1$, then $\rho_2 > \rho_1$.

for this to be satisfied, we need

$$p_2/p_1 \geqslant (\rho_2/\rho_1)^\gamma \tag{3.7.38}$$

However, since $\gamma \geqslant 1$, it follows from Figure 3.7.7 that this condition can be satisfied only if $p_2/p_1 \geqslant 1$, $\rho_2/\rho_1 \geqslant 1$ (see Problem 3.7.8). Thus, only compressive shocks are allowed in perfect gases. For other fluids, it can be shown that the same result holds provided that $(\partial^2 v^*/\partial p^2)_S > 0$. This condition is satisfied by all common fluids. [For possible exceptions, see K. C. Lambrakis and P. A. Thompson, *Phys. Fluids* **15**, 933 (1972).]

Let us now return to our computation of the entropy change across the shock. This is given by (3.7.26), which, on using (3.7.3) and $p_2/p_1 = 1 + \delta$, may be written as

$$\frac{S_2 - S_1}{c_v} = \ln(1+\delta) + \gamma \ln\left(1 + \frac{\gamma-1}{2\gamma}\delta\right) - \gamma \ln\left(1 + \frac{\gamma-1}{2\gamma}\delta\right) \tag{3.7.39}$$

As stated earlier, this applies for all positive values of δ. At present, however, we are interested in shock waves having relatively small pressure differences. These are called *weak shock waves*, and are specified by the condition $\delta \ll 1$. For this case, we expand the logarithmic functions in power series of δ using the expansion $\ln(1+\varepsilon) \approx \varepsilon - \varepsilon^2/2 + \varepsilon^3/3 + \cdots$, as $\varepsilon \to 0$. Expanding the three logarithms in (3.7.39) and collecting terms having the same power of δ, we find that the terms of order δ and δ^2 are absent. The first nonvanishing term that appears is of order δ^3, and is given by

$$\frac{S_2 - S_1}{c_v} = \frac{\gamma^2 - 1}{12\gamma^2}\left(\frac{p_2 - p_1}{p_1}\right)^3 \tag{3.7.40}$$

a result which also shows that δ must be positive. In addition, in view of the assumed smallness of δ, the entropy changes are small compared to c_v. Nevertheless, the dissipation of energy associated with this entropy increase may be a significant fraction of the energy of a wave, so that it cannot be neglected.

The amount of energy dissipated by a weak shock wave as it moves in a fluid can be computed as follows. First, as shown in Problem 3.7.4, the speed of the waves in the weak-shock approximation is equal to the speed of sound evaluated at ambient conditions (which, by definition, do not differ much from conditions immediately ahead or behind the shock wave). Therefore, the entropy increase of the volume of fluid that crosses the shock in a unit time is

$$A\rho_1 c_1 (S_2 - S_1) \simeq A\rho_0 c_0 (S_2 - S_1) \tag{3.7.41}$$

where A is the area of the wave front. As shown in books on thermodynamics, the corresponding dissipation of energy is given by the above entropy increase multiplied by the ambient temperature. Thus,

$$\dot{E}_s = A\rho_0 c_0 T_0 (S_2 - S_1) \qquad (3.7.42)$$

For perfect gases and for weak shocks, $S_2 - S_1$ is given by (3.7.40). Hence, the energy dissipation rate is given by

$$\dot{E}_s = \frac{(\gamma+1)Ac_0}{12\gamma^2 p_0^2}(p_2 - p_1)^3 \qquad (3.7.43)$$

This small but nonnegligible energy is acoustic energy being dissipated at the shock. We should therefore expect that as the wave travels, its amplitude should decay. However, contrary to situations in which the attenuation is linear, such as those introduced in the last section, now the decay is not exponential. The reason for the difference is that $\dot{E}_s$ is proportional to the cube of the amplitude, not the square.

Attenuation of a Sawtooth Wave

As an example of nonlinear attenuation produced by discontinuities in a wave, we consider the sawtooth profile shown in Figure 3.7.8. In a system of coordinates moving with the wave, the pressure distribution between the discontinuities is given by

$$p = \frac{2P}{\lambda}x, \qquad -\frac{\lambda}{2} \leqslant x \leqslant \frac{\lambda}{2} \qquad (3.7.44)$$

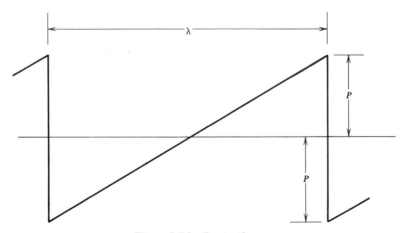

Figure 3.7.8 Sawtooth wave.

where λ is the distance between two consecutive discontinuities and P is the amplitude of the wave as shown in the figure. The acoustic energy for the volume $A\lambda$ between two discontinuities is

$$E_w = \frac{A}{\rho_0 c_0^2} \int_{-\lambda/2}^{\lambda/2} p^2 \, dx$$

$$= \tfrac{1}{3}\lambda A \frac{P^2}{\rho_0 c_0^2} \tag{3.7.45}$$

In the absence of dissipation, this energy would, of course, remain constant so that the amplitude P also would remain constant. In this case, however, there is dissipation so that P decreases as the wave moves forward. Since the volume is moving with speed c_0, the time rate of change of the acoustic energy is $\dot{E}_w = -c_0(dE_w/dx)$, or

$$\dot{E}_w = -\frac{2}{3} \frac{\lambda A}{\rho_0 c_0} P \frac{dP}{dx} \tag{3.7.46}$$

This must be equal to the rate at which energy is dissipated at the forward shock. Equating (3.7.43) and (3.7.46), we obtain, with $p_2 - p_1 = 2P$,

$$\frac{dP}{P^2} = -\frac{\gamma+1}{\gamma p_0 \lambda} dx \tag{3.7.47}$$

Integrating this yields

$$\frac{1}{P_0} - \frac{1}{P} = -\frac{\gamma+1}{\gamma p_0} \frac{x-x_0}{\lambda} \tag{3.7.48}$$

where P_0 is the amplitude at $x = x_0$. This result may be rewritten as

$$\frac{P}{P_0} = \frac{1}{1 + \dfrac{\gamma+1}{\gamma} \dfrac{P_0}{p_0} \dfrac{x-x_0}{\lambda}} \tag{3.7.49}$$

This gives the decay of the wave's amplitude with distance. The result has two interesting features. First, as discussed earlier, the decay is not exponential; second, the larger the value of P_0, the faster the wave decays. Neither of these results is, of course, applicable in the linear limit.

This concludes our brief discussion of nonlinear effects. The particular effects we have discussed are only a few of the many that appear when the amplitudes of the waves are not negligible. The interested reader is referred to books on nonlinear acoustics that have appeared recently and that are listed in the bibliography.

PROBLEMS

3.7.1 Derive the equation satisfied by the third-order approximation p_3.

3.7.2 Show that the medium nonlinearity vanishes for a perfect gas with $\gamma = 1$.

3.7.3 Suppose that a membrane is oscillating at $x=0$ with a velocity given by $U = A_1 \cos \omega_1 t + A_2 \cos \omega_2 t$. What are the frequencies of the second-order approximation p_2? (Note that the complete solution of p_2 is not required.)

3.7.4 Show that, in the weak-shock limit, the shock propagates with the ambient speed of sound.

3.7.5 Compute $(S_2 - S_1)/c_v$ across a weak shock wave in a perfect gas for $\delta = 0.001, 0.01$, and 0.1 using both the exact result given by (3.7.39) and the approximate one given by (3.7.40).

3.7.6 Compute the decay with distance of the amplitudes P_n of the first three Fourier components of a sawtooth wave. Plot P_n/P_0 versus $(x-x_0)/\lambda$.

3.7.7 The exact boundary condition for the fluid in front of a plane piston that is moving sinusoidally in a tube is $u(x_p, t) = U_0 \cos \omega t$, where $x_p = l \sin \omega t$ is the instantaneous position on the piston. Assuming that $u = f(x - c_0 t)$, determine $f(x - c_0 t)$ to order ε^2, where $\varepsilon = \omega l / c_0$.

3.7.8 In order to show that expansive shocks of the permanent type do not exist, we used a graphical argument based on (3.7.38) and Figure 3.7.7. A more satisfying proof may be derived as follows: From (3.7.40), we know that for $(p_2/p_1 - 1)$ small and positive, $(S_2 - S_1) > 0$ as required by the second law of thermodynamics. For other values of p_2/p_1, use (3.7.26) and (3.7.35) to show that

$$\frac{d}{d(p_2/p_1)}\left(\frac{S_2 - S_1}{c_v}\right) = \frac{(p_1/p_2)\left[(\gamma^2 - 1)(p_2/p_1 - 1)^2\right]}{\left\{(\gamma^2 - 1)\left[1 - (p_2/p_1)^2\right] + 2(\gamma^2 + 1)(p_2/p_1)\right\}^2}$$

Since the denominator is positive, this shows that for any possible value of γ, the slope of $(S_2 - S_1)/c_v$ versus p_2/p_1 is positive. Hence, $S_2 - S_1$ is positive only when $p_2 \geqslant p_1$.

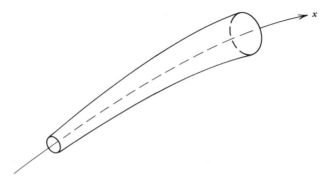

Figure 3.8.1   Tube with slowly varying cross-sectional area.

3.8   PLANE WAVES IN TUBES OF VARYING CROSS SECTION

So far, we have been studying plane-wave propagation in tubes of uniform cross-sectional area. There are, however, situations where propagation takes place in tubes whose area changes with distance. In general, propagation in such cases does not relate to plane waves, but if the variations of area are very small (e.g., Figure 3.8.1), then we may assume the waves to be plane, provided, of course, that the lateral dimensions of the tube are such that the frequencies involved are below some cutoff frequency for the propagation of transverse modes.[3]

By small variations of the cross-sectional area, we mean that the changes of area over distances comparable to the smallest length scale of interest should be small. One such length may be the wavelength. Another might be the lateral dimension of the tube, and so on. If this length is represented by x_0, then we require that $S(x+x_0)-S(x)\ll S(\tilde{x})$, where $S(\tilde{x})$ is the cross-sectional area of some intermediate location $\tilde{x}$. If these requirements are satisfied, we can assume that the propagation is still one-dimensional in the sense that properties of the wave on any plane perpendicular to the axis of the tube are uniform. However, since the cross-sectional area of the tube varies from point to point, the one-dimensional wave equation given earlier does not work here. To derive the

---

[3] Of course, since the cross-sectional area is varying, there is no unique cutoff frequency for transverse modes. However, the main concept can still be applied. Thus, for a tube such as that depicted in Figure 3.8.1, where variations are very small, one can assign cutoff frequencies to various sections of the tube. Then, for example, if propagation originates from the narrower section, we can take a portion of this section as straight and compute for it a cutoff frequency based on the lateral dimension there. If the frequency of the source is less than the computed cutoff frequency, then only the plane-wave mode will be propagated along the tube.

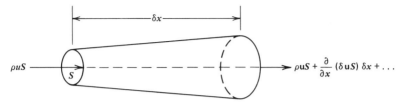

Figure 3.8.2 Element of length along tube.

proper equation, we reconsider the equation of continuity in integral form:

$$\frac{d}{dt}\int_V \rho(x,t)\,dV = -\int_S \rho\mathbf{u}\cdot\mathbf{n}\,dS \tag{3.8.1}$$

Let us apply this equation to a tube element of length δx, shown in Figure 3.8.2. For this case, $dV = S(x)\,dx$. Also, on the side walls of the tube, $\mathbf{u}\cdot\mathbf{n} = 0$, so that after integrating the area integral on the right-hand side of (3.8.1), we obtain

$$\int_x^{x+\delta x} S(x)\frac{\partial\rho}{\partial t}\,dx + (\rho uS)_{x+\delta x} - (\rho uS)_x = 0 \tag{3.8.2}$$

Thus, in the limit as $\delta x \to 0$,

$$S(x)\frac{\partial\rho}{\partial t} + \frac{\partial}{\partial x}(\rho uS) = 0 \tag{3.8.3}$$

This is the exact field form of the continuity equation for variable cross section. We require only the approximate version for small density changes. Thus, if as before we let $\rho = \rho_0 + \rho'$ and retain only first-order terms, we obtain

$$S\frac{\partial\rho'}{\partial t} + \rho_0\frac{\partial}{\partial x}(uS) = 0 \tag{3.8.4}$$

This is the required equation. The linearized momentum equation remains (see Problem 3.8.7)

$$\rho_0\frac{\partial u}{\partial t} + \frac{\partial p'}{\partial x} = 0 \tag{3.8.5}$$

As before, the isentropic relation gives $p' = c_0^2\rho'$. Therefore, the continuity equation becomes

$$\frac{\partial p'}{\partial t} + \rho_0 c_0^2\left[\frac{\partial u}{\partial x} + u\frac{d}{dx}(\ln S)\right] = 0 \tag{3.8.6}$$

We can use the momentum equation to eliminate p' or u but since the flow is irrotational, we also have $u = \partial\phi/\partial x$. Further, $p' = -\rho_0(\partial\phi/\partial t)$, so that

$$\frac{\partial^2\phi}{\partial t^2} - c_0^2 \frac{d}{dx}(\ln S)\frac{\partial\phi}{\partial x} = c_0^2\frac{\partial^2\phi}{\partial x^2} \tag{3.8.7}$$

This is the desired equation. Because of its applications, it is known as the horn equation. It differs from the usual wave equation in that it contains a term proportional to $\partial\phi/\partial x$. One of the effects of this term may be seen by considering the monochromatic time dependence. For this case, (3.8.7) gives

$$\frac{d^2\tilde{\Phi}}{dx^2} + \frac{d}{dx}(\ln S)\frac{d\tilde{\Phi}}{dx} + k^2\tilde{\Phi} = 0 \tag{3.8.8}$$

where, as before, $k = \omega/c_0$. This equation may be compared with the well-known equation for a simple harmonic oscillator, in which case the middle term appears because of dissipation. Thus, we expect that if $S(x)$ increases with distance, one of the effects of the increase will be to attenuate the waves. Similarly, if $S(x)$ decreases with x, the waves will be amplified. Both effects are physically obvious, as they are a direct consequence of the requirement that energy be conserved. However, there are other, less obvious effects that arise because of the changes of area. For example, not all frequency components of a given wave are propagated in certain cases.

Although the differential equation (3.8.8) is linear in $\tilde{\Phi}$, its solutions are not readily obtainable because the coefficient of $d\tilde{\Phi}/dx$ is a function of x. There are, however, some geometries that yield equations that may be solved without much difficulty. Some of these will be treated below.

Exponential Horn

The simplest case, other than that for which $S(x) = 0$, is a horn whose cross-sectional area varies exponentially:

$$S(x) = S(0)e^{2mx} \tag{3.8.9}$$

where $S(0)$ is the value of the cross-sectional area at $x = 0$. In this case, the coefficient of $d\tilde{\Phi}/dx$ is simply $2m$, so that we obtain an equation with constant coefficients:

$$\tilde{\Phi}''(x) + 2m\tilde{\Phi}'(x) + k^2\tilde{\Phi}(x) = 0 \tag{3.8.10}$$

Solutions for the form $\exp(Kx)$ can be obtained provided

$$K^2 + 2mK + k^2 = 0 \tag{3.8.11}$$

Thus, solutions of the form $\exp(Kx)$ exist for values of K given by

$$K_{\pm} = -m \pm \sqrt{m^2 - k^2} = -m \pm i\sqrt{k^2 - m^2} \tag{3.8.12}$$

Therefore, the complex velocity potential is given by

$$\tilde{\phi} = A \exp\left[-mx + i\left(\sqrt{k^2 - m^2}\, x - \omega t\right)\right] + \tilde{B} \exp\left[-mx - i\left(\sqrt{k^2 - m^2}\, x + \omega t\right)\right] \tag{3.8.13}$$

Consider the case when a sound source is placed in a very long horn so that no reflected waves occur and $\tilde{B} = 0$. Therefore,

$$\tilde{\phi} = A \exp\left\{-mx + i\left[(k^2 - m^2)^{1/2} x - \omega t\right]\right\} \tag{3.8.14}$$

Depending on the relative values of m and k, this may or may not represent propagation. Three separate cases can be considered.

$k > m$. The wave produced by the source is propagated along the tube with an amplitude that decays exponentially. Since the area was assumed to change only gradually, the quantity m must necessarily be small. Therefore, the decay, also, is small.

Since the effective wave number is $\sqrt{k^2 - m^2}$, the propagation is dispersive; that is, the speed of propagation depends on the frequency. The relevant speeds are the group velocity $c_g = d\omega/d\sqrt{k^2 - m^2}$ and the phase velocity $c_f = \omega/\sqrt{k^2 - m^2}$:

$$c_g = c_0\sqrt{1 - (m/k)^2} \tag{3.8.15}$$

and

$$c_f = c_0 \Big/ \sqrt{1 - (m/k)^2} \tag{3.8.16}$$

$k < m$. Here we can write

$$\tilde{\phi} = A \exp\left\{-m\left[1 - \sqrt{1 - (k/m)^2}\,\right]x - i\omega t\right\} \tag{3.8.17}$$

Since $m > k, \tilde{\phi} \to 0$ as x increases so that the wave is not propagated. In other words, if $m > \omega/c_0$, the waves of frequency ω will not be propagated.

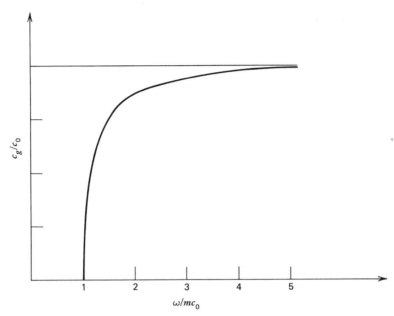

Figure 3.8.3 Group velocity in exponential tube.

$k = m$. This is a cutoff condition, and implies that the fluid in the horn oscillates with the same phase (but with amplitudes that depend on position). This cutoff condition arises because of the spreading of the cross-sectional area of the tube. It is not related to the cutoff condition mentioned in Sec. 3.5, as that condition applies to transverse oscillations that may be excited in a tube.

   Figure 3.8.3 shows the variations of the group velocity c_g with the ratio ω/mc_0. We note that the larger m is, the lower the value of ω for which propagation does not exist. Such effects are clearly important in determining the transmission properties of a horn, and will be discussed later in some more detail.

Power-Law Horns

   Exact analytical solutions can also be obtained for horns whose cross-sectional area varies with distance according to a simple power law, that is, in those cases where S is given by

$$S(x) = S_0(x/x_0)^n \qquad (3.8.18)$$

where S_0 is the value of S at $x = x_0$. For positive values of n, this gives divergent tubes, whereas for negative values, the tubes are convergent. The following

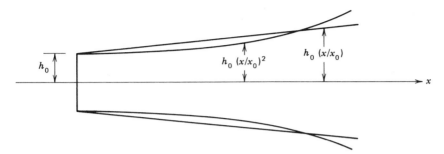

Figure 3.8.4 Conical and parabolic tube sections.

special cases should be noted:

$n=2$, straight cone

$n=4$, parabolic cone

Their names derive from the variations with distance of their lateral dimension, as shown in Figure 3.8.4.

Let us now return to our monochromatic horn equation (3.8.8). Using (3.8.18), that equation becomes

$$\tilde{\Phi}''(z)+\frac{n}{z}\tilde{\Phi}'(z)+\tilde{\Phi}(z)=0 \qquad (3.8.19)$$

Here $z=kx$ and the primes denote differentiation with respect to the argument. Solutions of (3.8.19) may be obtained in terms of Bessel functions. To do this, we first transform (3.8.19), so that it takes the form of the standard Bessel's differential equation (see Appendix C):

$$f''(z)+\frac{1}{z}f'(z)+\left[1-\frac{s^2}{z^2}\right]f(z)=0 \qquad (3.8.20)$$

This may be accomplished by the substitution into (3.8.19) of

$$\tilde{\Phi}=z^m f(z) \qquad (3.8.21)$$

and then by determining those values of m that will make the result have the form given by (3.8.20). Thus, with

$$\tilde{\Phi}'=z^m f'(z)+mz^{m-1}f(z)$$

and

$$\tilde{\Phi}''=z^m f''(z)+2mz^{m-1}f'(z)+m(m-1)z^{m-2}f(z)$$

we obtain

$$f''(z) + \frac{2m+n}{z}f'(z) + 1 + \frac{m(m-1)+mn}{z^2}f(z) = 0 \qquad (3.8.22)$$

To obtain (3.8.20) from this, we first require that the coefficient of $f'(z)$ in (3.8.22) be $1/z$. Therefore, we set $2m+n=1$ so that

$$m = (1-n)/2 \qquad (3.8.23)$$

and

$$\Phi = z^{(1-n/2)}f(z) \qquad (3.8.24)$$

where $f(z)$ represents the solution of Bessel's differential equation (3.8.20).

The next step is to find s in (3.8.20) in terms of n. Thus, in view of (3.8.23), we have

$$m(m-1) + mn = -\left(\frac{n-1}{2}\right)^2 \qquad (3.8.25)$$

Therefore, $s = \pm(n-1)/2$. As these values of s are generally not integers, the solution of (3.8.20) may be expressed as[4]

$$f(z) = AJ_s(z) + BJ_{-s}(z). \qquad (3.8.26)$$

where $J_{\pm s}(z)$ is the Bessel function of the first kind of order s.

Therefore, the general solution of (3.8.19) may be written as

$$\tilde{\Phi}(x) = (kx)^{(1-n)/2}\left[AJ_{n-1/2}(kx) + BJ_{-n+1/2}(kx)\right] \qquad (3.8.27)$$

For integral values of n, the Bessel function $J_{\pm(n-1)/2}$ may be expressed in terms of elementary functions. Thus, for $n=0$, we have

$$J_{-1/2}(z) = \sqrt{2/\pi z}\ \cos z \qquad (3.8.28)$$

$$J_{1/2}(z) = \sqrt{2/\pi z}\ \sin z \qquad (3.8.29)$$

---

[4]Except when $n=1$, in which case $s=0$ and the solution becomes

$$f(z) = AJ_0(z) + BY_0(z)$$

where $Y_0(z)$ is the Bessel function of the second kind (Neuman function) of zero order.

For $n=0$, we simply have $\tilde{\Phi}=A\cos z+B\sin z$. For $n=2$, we still require $J_{\pm 1/2}(z)$, but this time $z^m=z^{-1/2}$, so that

$$\tilde{\Phi}=\sqrt{\frac{2}{\pi}}\,\frac{1}{kx}(A\sin kx+B\cos kx)\tag{3.8.30}$$

The velocity potential corresponding to this may be expressed as

$$\tilde{\phi}=\frac{1}{kx}\left[Ae^{i(kx-\omega t)}+Be^{-i(kx+\omega t)}\right]\tag{3.8.31}$$

This form is useful for studying propagation in cones. Thus, if the cone is so long that no reflected waves exist in it, then

$$\tilde{\phi}=\frac{A}{kx}e^{i(kx-\omega t)}\tag{3.8.32}$$

This form shows that contrary to the exponential horn case, propagation in a straight cone is nondispersive so that waves of all frequencies are transmitted with the same velocity. On the other hand, the power transmitted at low frequencies is smaller than in the corresponding exponential horn, as shown later.

Other Shapes

In general, shapes other than those treated above result in differential equations for Φ that cannot be solved by elementary means. However, since (3.8.8) is linear in Φ, several mathematical techniques are available to obtain approximate solutions for Φ. These techniques are beyond the scope of this book. In addition to reading standard books on differential equations, the reader may want to consult the references listed at the end of this chapter.

Transmission Coefficient

As is well known, one of the uses of horns is to increase the acoustic power that is radiated by small sources. These may be capable of emitting sound waves having a wide range of frequencies, but their emitted power is usually small. The reasons for this may be understood by considering a simple situation where a small membrane is oscillating about a mean equilibrium position with uniform velocity $u=u_0\cos\omega t$. As we saw in Section 3.5, such motion will produce an acoustic power proportional to

$$\langle p'u\rangle S$$

where p' is the pressure fluctuation in front of the membrane and S is the membrane's area. Suppose that the membrane is placed on the center of a wall in a very large room. The pressure fluctuation p' will necessarily be very small so that the ability of the membrane to do work on the fluid, and therefore emit waves, is limited. If we take the membrane and place it at one end of a tube of uniform cross-sectional area, the pressure fluctuation remains relatively large owing to unavoidable reflections from the other end of the tube. Therefore, the membrane is able to do more work on the fluid, thereby increasing the acoustic power emitted by the membrane relative to the previous case. Another factor that affects the radiation is the area of the tube at the open end. At this location, there are other factors that are also important, but it should be clear that for a given velocity and pressure at the opening, the emitted power will increase with the area there. Thus, if we want to increase the power emitted by a small source, we may attach to it a tube whose area is larger at the open end than at the source. Unfortunately, such a device may act as a filter so that not all waves are transmitted. For example, the exponential horn will not transmit any wave whose frequency is smaller than a certain value. Other geometries may also produce similar effects on the transmission of waves inside a horn.

A quantity that may be used to give some idea about the transmission properties of sound waves in horns is the transmission coefficient τ. This is defined as the ratio of the power transmitted in any given horn to the corresponding power in a straight tube having the same cross-sectional area as the narrowest area in the horn, and having the same type of excitation; that is,

$$\tau = \frac{\Pi_{\text{horn}}}{\Pi_{\text{tube}}} \tag{3.8.33}$$

Note that this definition does not refer to the waves radiated by the horn. It refers only to the transmission properties of a given tube under the assumption that no reflected waves exist inside the tubes.

We will obtain the value of τ for two of the geometries treated earlier: the exponential cone and the straight cone. To compare these two cases against each other, we select a truncated cone and an exponential horn having the same cross-sectional area at some position $x = x_0$, as sketched in Figure 3.8.5. At this position, we place a membrane oscillating along x with uniform velocity $u = u_0 \cos \omega t$. This will produce a velocity potential given by (3.8.32) in the case of the cone, and by (3.8.14) in the case of the exponential horn. In both cases, the constants A are evaluated from the condition that at $x = x_0$, the fluid velocity be equal to that of the membrane. Applying this condition, we obtain for the exponential horn

$$A_{\text{exp}} = \frac{u_0}{-m + i\sqrt{k^2 - m^2}} \exp\left(mx_0 - i\sqrt{k^2 - m^2}\, x_0 \right) \tag{3.8.34}$$

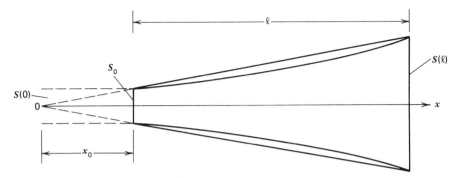

Figure 3.8.5 Conical and exponential tubes having equal lengths and equal area ratios.

Similarly, we obtain for the cone

$$A_{\text{cone}} = -u_0 x_0 e^{-ikx_0} \frac{kx_0}{1 - ikx_0} \qquad (3.8.35)$$

For the straight-tube case, we simply have

$$\tilde{\phi}_{\text{tube}} = \frac{u_0}{ik} e^{i[k(x-x_0)-\omega t]} \qquad (3.8.36)$$

The related powers may be obtained from

$$\Pi = \langle p'u \rangle S \qquad (3.8.37)$$

Using $p' = -\rho_0(\partial\phi/\partial t)$, $u = \partial\phi/\partial x$, we have

$$\Pi = \tfrac{1}{2}\rho_0\omega S(x)\,\text{Re}\!\left(-i\tilde{\phi}^*\,\partial\tilde{\phi}/\partial x\right) \qquad (3.8.38)$$

Thus, for the straight tube, this gives

$$\Pi_t = \tfrac{1}{2}\rho_0 c_0 u_0^2 S_0 \qquad (3.8.39)$$

where S_0 is the tube's cross-sectional area. To obtain the corresponding result for the exponential horn, we substitute (3.8.14) into (3.8.38). This gives

$$\Pi_{\text{exp}} = \tfrac{1}{2}\rho_0\omega|A_{\text{exp}}|^2 S(x)\sqrt{1-(m/k)^2}\,e^{-2mx} \qquad (3.8.40)$$

where A_{exp} is given by (3.8.34) and $S(x)$ is given by (3.8.9). Using these equations, we obtain

$$\Pi_{exp} = \tfrac{1}{2}\rho_0 c_0 u_0^2 S(0)\sqrt{1-(m/k)^2}\, e^{2mx_0} \qquad (3.8.41)$$

But $S(0)e^{2mx_0} = S_0$, and so

$$\Pi_{exp} = \tfrac{1}{2}\rho_0 c_0 u_0^2 S_0\sqrt{1-(m/k)^2} \qquad (3.8.42)$$

The transmission coefficient for the exponential horn is therefore given by

$$\tau_{exp} = \sqrt{1-(m/k)^2} \qquad (3.8.43)$$

Thus, the exponential horn does not transmit acoustic power at all frequencies. In fact, it does not transmit any power below the cutoff frequency $\omega = mc_0$. The reason for this is dispersion. In fact, reference to (3.8.15) will show that the variations of the transmission frequency are basically the same as those for the group velocity c_g.

Let us now consider the case of the straight cone. Proceeding as before, we obtain for the transmitted power

$$\Pi_{cone} = \tfrac{1}{2}\rho_0 k\omega|A_{cone}|^2 S(x)/(kx)^2 \qquad (3.8.44)$$

Using (3.8.35) and $S(x)=S_0(x/x_0)^2$, this may be written as

$$\Pi_{cone} = \tfrac{1}{2}\rho_0 c_0 u_0^2 S_0 \frac{(kx_0)^2}{1+(kx_0)^2} \qquad (3.8.45)$$

The corresponding transmission coefficient is

$$\tau_{cone} = \frac{(kx_0)^2}{1+(kx_0)^2} \qquad (3.8.46)$$

Therefore, transmission in a straight cone is also frequency dependent, but this time the reason is not dispersion, for propagation in the cone is not dispersive. Rather, the reason is that the area of the cone near the throat increases more rapidly than the corresponding area of the exponential horn as sketched in Figure 3.8.5. This more rapid increase results in a smaller pressure fluctuation at $x=x_0$ in the case of the cone than in the case of the exponential horn. This may be seen by considering the corresponding velocity potentials. At $x=x_0$ these give, as they should, velocity fluctuations of magnitude u_0. On the other hand, the magnitude of the pressure fluctuations there are

$$|p'(x_0, t)|_{\text{exp}} = \rho_0 c_0 u_0 \qquad (3.8.47)$$

and

$$|p'(x_0, t)|_{\text{cone}} = \rho_0 c_0 u_0 \frac{kx_0}{\sqrt{1 + k^2 x_0^2}} \qquad (3.8.48)$$

Only when $kx_0 \gg 1$ does the pressure fluctuation at $x = x_0$ in the cone become equal to that in the exponential horn. This is not surprising, for as x_0 increases (for given ω), while holding the throat area S_0 constant, our cone becomes more and more like a constant-cross-section tube. The same effect is, of course, achieved by decreasing the size of the wavelength relative to x_0, that is, by increasing the frequency.

To compare the transmission properties of the cone and of the exponential horn, we select a cone and a horn having the same entrance and exit areas, and having the same length l as sketched in Figure 3.8.5. Clearly, depending on the value of x_0, many such pairs may be found. However, if we also prescribe the area ratio $S(l)/S_0 = (l/x_0)^2$ for the cone, then only one exponential horn will have the desired properties. For this horn, the value of m may be obtained from (see Problem 3.8.4)

$$mx_0 = \frac{\ln(l/x_0)}{l/x_0 - 1} \qquad (3.8.49)$$

Now, in terms of mx_0 and kx_0, the transmission coefficient for the exponential horn may be expressed as

$$\tau_{\text{exp}} = \sqrt{1 - \frac{(mx_0)^2}{(kx_0)^2}} \qquad (3.8.50)$$

This may be used to plot the variations of τ_{exp} versus kx_0, as in Figure 3.8.6, for different values of (l/x_0). It is clear that the exponential horn is better than the conical horn to transmit acoustic power, except for frequencies below the cutoff value mc_0, where it does not transmit any.

It should be added that the results for τ were obtained by assuming no reflected waves. These will, of course, be present in an actual tube, and may result in resonances at certain frequencies. Also, although the transmission coefficient determines in part the amount of acoustic energy radiated by a horn,

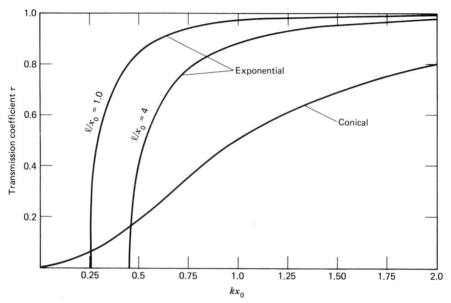

Figure 3.8.6 Transmission coefficients for two sets of conical and exponential horns.

there are other factors that also must be taken into account. Some of these will be mentioned in Chapter 5, where we discuss sound emission.

PROBLEMS

3.8.1 For integral values of m, the Bessel function $J_{m \pm 1/2}$ may be obtained from

$$J_{m+1/2}(z) = \frac{2m-1}{z} J_{m-1/2}(z) - J_{m-3/2}(z)$$

Use this relationship together with (3.8.23) and (3.8.29) to show that for the parabolic cone, the velocity potential may be written as

$$\tilde{\phi} = \frac{1-kx}{(kx)^2} \left[A e^{i(kx-\omega t)} + B e^{-i(kx+\omega t)} \right]$$

3.8.2 Show that (3.8.19) for the case $n=2$ can be written as

$$\frac{d^2(\tilde{\Phi}x)}{dx^2} + k^2(\tilde{\Phi}x) = 0$$

so that $\tilde{\Phi}$ must be given by (3.8.30).

3.8.3 Consider a truncated cone of length *l*, closed at both ends. Determine the first three characteristic frequencies of the cone, and compare them with a straight tube of the same length.

3.8.4 Derive (3.8.49).

3.8.5 Use the answer given in Problem 3.8.1 to obtain the transmission coefficient for a parabolic horn. Compare the parabolic-horn results with those obtained for the exponential horn and for the straight cone.

3.8.6 Describe in physical terms the effects on the transmission coefficient for a cone, of placing a converging "mouthpiece" at the beginning of the cone.

3.8.7 Show that (3.8.5) is applicable when the cross-sectional area changes.

3.9 SUDDEN AREA CHANGES

There are many practical situations where acoustic waves propagate in tubes or ducts that have cross-sectional areas that change rapidly, as in the examples shown in Figure 3.9.1. This figure depicts a plane wave approaching (a) a sudden enlargement, (b) a side branch, and (c) a bifurcation. In each of these cases, some of the energy of the incident beam may be reflected and some transmitted, and it is of some interest to compute the respective coefficients. In view of the comments made in the last section regarding one-dimensional propagation in tubes of varying cross-sectional area, it is clear that waves propagating in tubes such as those shown in the figure cannot possibly be one-dimensional, at least near the locations where the area changes rapidly. Nevertheless, because of the difficulties involved in solving such problems exactly, and because of their practical importance, it is customary to treat them by assuming that the reflected and transmitted waves are also plane, one-dimensional waves. This is a reasonable assumption for those waves having frequencies smaller than the cutoff frequencies of the different sections of a tube. In such cases, the transverse modes that might be excited at the transition will decay exponentially, as shown in Section 3.5, so that at sufficient distances from the transition, the waves will be plane.

In what follows, we consider this problem for the monochromatic case. To fix ideas, we use case (a) in Figure 3.9.1, and assume that the tubes on either side of the transition extend to infinity. Thus, the acoustic pressure in the incident wave is given by

$$\tilde{p}'_i = A_0 e^{i(kx - \omega t)} \tag{3.9.1}$$

where A_0, ω, and $k = \omega/c_0$ are assumed known. Due to the transition, a reflected

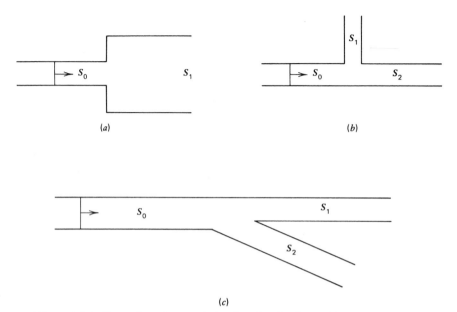

Figure 3.9.1 Examples of transmission in tubes having sudden-area transitions.

wave will exist in the tube where the incident wave is being propagated (denoted by subindex zero). Because of our plane-wave model, the pressure in the reflected wave will be given by

$$\tilde{p}'_r = \tilde{R}_0 e^{-i(kx+\omega t)}$$

so that in region zero the acoustic pressure is

$$\tilde{p}'_0 = A_0 e^{i(kx-\omega t)} + \tilde{R}_0 e^{-i(kx+\omega t)} \tag{3.9.2}$$

For region 1, we assume that only transmitted waves exist, and for these,

$$\tilde{p}'_t = \tilde{T}_1 e^{i(kx-\omega t)} \tag{3.9.3}$$

The amplitudes of the reflected and transmitted waves, $\tilde{R}_0$ and $\tilde{T}_1$, are to be determined from appropriate boundary conditions at the transition. However, because of the complicated nature of the field near the transition, it is not clear what these boundary conditions are. In fact, only for very special cases can we state conditions compatible with (3.9.1) and (3.9.3). The most important such case occurs when the frequency of the waves is very low, for here the width of the region in which the waves are not one-dimensional may be small when compared to the wavelength so that in some limit it may be taken to be very small. We show below that for such transitions, called *compact* in the literature, the boundary conditions are:

1. The volume flow rate is conserved.
2. The pressure is transmitted.

Consider FIgure 3.9.1*a*. For incident waves having frequencies below cutoff, the field will be purely one-dimensional at locations x_1 and x_2, upstream and downstream of the transition, respectively. Then, for the fluid volume limited by normal planes passing through these locations, (3.8.2) gives, when linearized,

$$\frac{1}{\rho_0} \int_{x_1}^{x_2} \frac{\partial \rho'}{\partial t} S(x)\, dx + (uS)_{x_2} - (uS)_{x_1} = 0 \qquad (3.9.4)$$

Now, for monochromatic waves having velocity amplitude u_0, the integral term has a magnitude of the order of $ku_0 \bar{S}(x_2 - x_1)$, where $\bar{S}$ is some mean value of S. On the other hand, the other terms have magnitudes of the order of $u_0 \bar{S}$. Therefore, provided that

$$\frac{x_2 - x_1}{\lambda} \ll 1$$

we may neglect the integral in (3.9.4) so that

$$S_2 u_2 = S_0 u_0 \qquad (3.9.5)$$

This condition simply states that the volume of fluid leaving the first section must be equal to that which enters the second. For a transition having more than two branches, (3.9.5) generalizes to

$$S_0 u_0 = S_1 u_1 + S_2 u_2 + \cdots \qquad (3.9.6)$$

Similarly, use of Euler's equation in linearized form shows that across a two-sided transition, the acoustic pressures are equal; that is,

$$p_0' = p_1' \qquad (3.9.7a)$$

For more than two branches, this condition is

$$p_0' = p_1' = p_2' = \cdots \qquad (3.9.7b)$$

Returning now to the example of Figure 3.9.1a, we place the transition at $x = 0$ and apply (3.9.5) and (3.9.7a) to (3.9.2) and (3.9.3). This gives

$$S_0(A_0 - \tilde{R}_0) = S_1\tilde{T}_1 \tag{3.9.8}$$

$$A_0 + \tilde{R}_0 = \tilde{T}_1 \tag{3.9.9}$$

Therefore,

$$\frac{\tilde{R}_0}{A_0} = \frac{S_1 - S_0}{S_1 + S_2} \tag{3.9.10}$$

$$\frac{\tilde{T}_1}{A_0} = \frac{2S_0}{S_0 + S_1} \tag{3.9.11}$$

Hence, the reflection coefficient is

$$\alpha_r = \left(\frac{S_1 - S_0}{S_0 + S_1} \right)^2 \tag{3.9.12}$$

whereas the transmission coefficient $\alpha_t = 1 - \alpha_r = (S_1/S_0)|\tilde{T}_1|^2/A_0^2$ is

$$\alpha_t = \frac{4S_0 S_1}{(S_0 + S_1)^2} \tag{3.9.13}$$

We observe that these results are symmetric in S_0 and S_1. Therefore they apply whether $S_1 > S_0$ or $S_1 < S_0$.

Tubes with Fluids Having Different Properties

In the example of the two-tube transition, the tubes were assumed to be filled with the same fluid so that the mean densities and the speeds of sound in the two tubes were equal. There are some situations, however, where this is not the case, an obvious instance being that in which different tubes are filled with different fluids. Another example, perhaps more common, is that in which the two tubes have fluids at different temperatures. In either case, the boundary conditions at the transition remain the same; that is, we still have continuity of volume flow and of acoustic pressure. Now, however, the densities and sound speeds are not the same. Therefore, (3.9.8) becomes

$$\frac{S_0}{\rho_0 c_0}(A_0 - \tilde{R}_0) = \frac{S_1}{\rho_1 c_1}\tilde{T}_1 \tag{3.9.14}$$

The second boundary condition, (3.9.9), remains as is.

Equation (3.9.14) shows that the quantities $S_0/\rho_0 c_0$ and $S_1/\rho_1 c_1$ play an important role in this problem. In fact, each is equal to the ratio of the acoustic pressure to the volume flow in the direction of propagation, and may therefore be defined as the acoustic impedances of the two tubes. Thus, if we denote $S_0/\rho_0 c_0$ by $\tilde{Z}_0$, then

$$\tilde{Z}_0 = \frac{\tilde{p}_i'}{\tilde{u}_i S_0} \tag{3.9.15}$$

and similarly for $\tilde{Z}_1$. It should be noted, however, that in this case both impedances are real. Denoting their values by Z_0 and Z_1, the reflection and transmission coefficients can be expressed as

$$\alpha_r = \left(\frac{Z_1 - Z_0}{Z_0 + Z_1}\right)^2 \tag{3.9.16}$$

$$\alpha_t = \frac{4 Z_0 Z_1}{(Z_0 + Z_1)^2} \tag{3.9.17}$$

These are more general than (3.9.12) and (3.9.13) because they show that complete transmission requires only that the groups $\rho c/S$ be equal, and not that the densities, speeds of sound, and areas be separately equal. Thus, if it were possible to modify the properties in the second tube so that the two impedances would match, a maximum amount of energy would be transmitted into the second tube.

Transmission into Several Branches

As a generalization of the problem introduced above, we consider the transmission and reflection of a plane, monochromatic wave at a branching point in a duct, as sketched in Figure 3.9.2. As indicated in the figure, a plane, monochromatic wave, traveling in a tube denoted by the subindex zero, reaches a transition point where the tube is divided into several branches. For the sake of generality, we consider the case of a tube connected to r different branches. A number m of these will be assumed to have finite lengths. Therefore, these m branches will contain both transmitted and reflected waves. The remaining $r - m$ branches are taken to be infinitely long so that they will contain only transmitted waves.

As before, we assume that the transition is compact. Therefore, we may apply the conditions (3.9.6) and (3.9.7b) there. Thus, if the transmitted and reflected waves in the nth branch have pressure amplitudes given by T_n and R_n at the

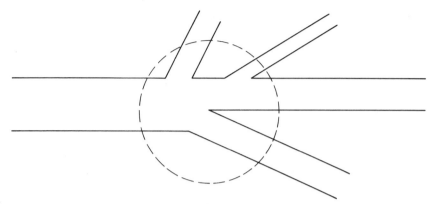

Figure 3.9.2 Transmission into several branches. Diameter of transition circle is assumed small compared to the wavelength of the incident wave.

transition, the boundary condition on the pressure gives

$$A_0 + \tilde{R}_0 = \tilde{T}_n + \tilde{R}_n, \quad n = 1, 2, \dots, m \tag{3.9.18}$$

$$A_0 + \tilde{R}_0 = \tilde{T}_n, \quad n = m+1, m+2, \dots, r \tag{3.9.19}$$

Here, $\tilde{R}_0$ is the amplitude of the reflected wave in the initial tube. Similarly, the continuity of volume flow into the r branches gives

$$\frac{S_0}{\rho_0 c_0}(A_0 - \tilde{R}_0) = \frac{S_1}{\rho_1 c_1}(\tilde{T}_1 - \tilde{R}_1) + \frac{S_2}{\rho_2 c_2}(\tilde{T}_2 - R_2) + \cdots + \frac{S_m}{\rho_m c_m}(\tilde{T}_m - \tilde{R}_m)$$

$$+ \frac{S_{m+1}}{\rho_{m+1} c_{m+1}} \tilde{T}_{m+1} + \cdots + \frac{S_r}{\rho_r c_r} \tilde{T}_r \tag{3.9.20}$$

Introducing the real quantities

$$I_{0n} = \frac{\rho_0 c_0 / S_0}{\rho_n c_n / S_n}, \quad n = 1, 2, \dots, r \tag{3.9.21}$$

(3.9.20) can be written as

$$A_0 - \tilde{R}_0 = \sum_{n=1}^{m} I_{0n}(\tilde{T}_n - \tilde{R}_n) + \sum_{n=m+1}^{r} I_{0n} \tilde{T}_n \tag{3.9.22}$$

Now, the unknown quantities in the problem are R_0; $\tilde{T}_1, \tilde{T}_2, \dots, \tilde{T}_r$; $\tilde{R}_1, \tilde{R}_2, \dots, \tilde{R}_m$; that is, we have $r + m + 1$ unknowns. The number of the equations that has been given so far is $m+(r-m)+1=r+1$, so that m additional equations are needed. These are provided by the conditions that exist at the end of the m branches. Since these conditions may vary from branch to branch, it is convenient to express them in terms of complex impedances, as done in Section 3.1. Thus, the end of the nth finite branch will have a specific impedance given by

$$\tilde{\zeta}_n = \frac{\rho_n c_n}{S_n}(\xi_n + i\eta_n), \quad n = 1, 2, \ldots, m \tag{3.9.23}$$

In other words, the boundary condition there will be

$$\frac{\tilde{p}_n'}{S_n \tilde{u}_n} = \frac{\rho_n c_n}{S_n}(\xi_n + i\eta_n) \tag{3.9.24}$$

Now, since in each branch we have plane, monochromatic, waves, the pressure and volume flow rate in each will be of the form

$$\tilde{p}_n' = e^{-i\omega t}(\tilde{T}_n e^{ik_n x} + \tilde{R}_n e^{-ik_n x}) \tag{3.9.25}$$

$$S_n \tilde{u}_n = e^{-i\omega t}\frac{S_n}{\rho_n c_n}(\tilde{T}_n e^{ik_n x} - \tilde{R}_n e^{-ik_n x}) \tag{3.9.26}$$

Thus, if the nth branch has a length l_n, the boundary condition (3.9.24) gives

$$\frac{\tilde{T}_n e^{ik_n l_n} + \tilde{R}_n e^{-ik_n l_n}}{\tilde{T}_n e^{ik_n l_n} - \tilde{R}_n e^{-ik_n l_n}} = w_n \tag{3.9.27}$$

where

$$w_n = \xi_n + i\eta_n \tag{3.9.28}$$

This gives

$$\tilde{R}_n = \tilde{T}_n e^{2ik_n l_n}\frac{w_n - 1}{w_n + 1} \tag{3.9.29}$$

This may be combined with (3.9.18) to give, for the finite-length branches,

$$\tilde{T}_n = \frac{A_0 + \tilde{R}_0}{1 + \dfrac{w_n - 1}{w_n + 1}e^{2ik_n l_n}} \tag{3.9.30}$$

$$\tilde{R}_n = (A_0 + \tilde{R}_0)\frac{\dfrac{w_n - 1}{w_n + 1}e^{2ik_n l_n}}{1 + \dfrac{w_n - 1}{w_n + 1}e^{2ik_n l_n}} \tag{3.9.31}$$

To simplify the notation, we write these as

$$\tilde{T}_n = \frac{A_0 + \tilde{R}_0}{1+f_n}, \quad n=1,2,\ldots,m \tag{3.9.32}$$

$$\tilde{R}_n = \left(A_0 + \tilde{R}_0\right)\frac{f_n}{1+f_n}, \quad n=1,2,\ldots,m \tag{3.9.33}$$

where

$$f_n = \frac{w_n - 1}{w_n + 1}e^{2ik_n l_n} \tag{3.9.34}$$

These results for $\tilde{T}_n$ and $\tilde{R}_n$ may now be used with (3.9.18), (3.9.19), and (3.9.22) to solve for any desired quantity. A particularly interesting one is $\tilde{R}_0/A_0$ because its squared amplitude defines the reflection coefficient α_r, and this can be used as an indication of the effects of branching. Thus, we eliminate $\tilde{T}_n$ and $\tilde{R}_n$ from (3.9.22) and obtain

$$\frac{\tilde{R}_0}{A_0} = \frac{1 - \displaystyle\sum_{n=1}^{m} I_{0n}\frac{1-f_n}{1+f_n} - \sum_{n=m+1}^{r} I_{0n}}{1 + \displaystyle\sum_{n=1}^{m} I_{0n}\frac{1-f_n}{1+f_n} + \sum_{n=m+1}^{r} I_{0n}} \tag{3.9.35}$$

Other coefficients can be similarly obtained.

EXAMPLE. Tube with one side branch

As an application of (3.9.35), we consider an infinitely long tube that has at some point an opening to which a tube of length l is connected, as indicated in Figure 3.9.3. In effect, we have a case for which $r=2$. Also, since one branch extends to infinity, we have $m=1$. For this case, (3.9.35) gives

$$\frac{\tilde{R}_0}{A_0} = \frac{1 - I_{01}\dfrac{1-f_1}{1+f_1} - I_{02}}{1 + I_{01}\dfrac{1-f_1}{1+f_1} + I_{02}} \tag{3.9.36}$$

If the fluid filling the tube has uniform composition, then $I_{02} = 1$ and $I_{01} = S_1/S_0$

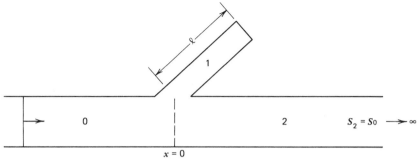

Figure 3.9.3 Side branch.

so that

$$\frac{\tilde{R}_0}{A_0} = \frac{-\dfrac{S_1}{S_0}\dfrac{1-f_1}{1+f_1}}{2+\dfrac{S_1}{S_0}\dfrac{1-f_1}{1+f_1}} \tag{3.9.37}$$

Now, the values of f_1 depend on the type of termination at the end of the side tube. If this is rigid, then $w_1 \to \infty$, so that $f_1 = \exp(2ik_1l_1)$. If it is open to the atmosphere, then $w = 0$ and $f_1 = -\exp(2ik_1l_1)$. Between those two limits, w_1 will have complex values. Let us consider the rigid-termination case. Here, we have

$$\frac{\tilde{R}_0}{A_0} = \frac{S_1}{S_0}\frac{i\sin k_1l_1}{2\cos k_1l_1 - i(S_1/S_0)\sin k_1l_1} \tag{3.9.38}$$

The reflection coefficient is, then,

$$\alpha_r = \frac{(S_1/S_0)^2\sin^2 k_1l_1}{4\cos^2 k_1l_1 + (S_1/S_0)^2\sin^2 k_1l_1} \tag{3.9.39}$$

The transmission coefficient may be computed in terms of $\tilde{T}_2/A_0 = 1 + \tilde{R}_0/A_0$, or from the fact that since no dissipation effects exist, energy is conserved so that $\alpha_t = 1 - \alpha_r$. In either case, we obtain

$$\alpha_t = \frac{1}{1 + (S_1/2S_0)^2\tan^2 k_1l_1} \tag{3.9.40}$$

We can see from this equation that the side tube acts as an acoustic filter in the sense that it filters out certain frequencies, namely, those for which $k_1 l_1 = (2n-1)\pi/2$. At these frequencies, the reflection coefficient is unity, implying, then, that all of the incident energy is returned to the source. This occurs because at those frequencies, the field in the side branch is resonating in response to the incident wave. Thus, during the first part of any given cycle, the incident field pumps all of its energy into the side tube and receives it back during the second part of the cycle.

The success of the procedure outlined above to solve branching problems depends on our ability to specify the boundary conditions at the ends of each branch. If these are known, then the solution, as given by (3.9.35), will contain no unspecified quantities and is therefore preferable to one that does. There are cases, however, when devices other than simple branches are connected to the main tube. For example, a Helmholtz resonator may be mounted on the walls of a tube as indicated in Figure 3.9.4. Such devices are usually represented in terms of their impedances on the assumption that their real and imaginary parts are known.

To derive the result corresponding to (3.9.35) we simply note that the impedance of a device at a transition located at $x=0$ may be represented as

$$\tilde{Z}_n(0) = \frac{\rho_n c_n}{S_n} \frac{\tilde{T}_n + \tilde{R}_n}{\tilde{T}_n - \tilde{R}_n} = X_n + iY_n \tag{3.9.41}$$

Following the same procedure as used before, we find that $\tilde{R}_0/A_0$ still satisfies (3.9.35) provided that we take for f_n the value

$$f_n = \frac{\dfrac{S_n}{\rho_n c_n} \tilde{Z}_n(0) - 1}{\dfrac{S_n}{\rho_n c_n} \tilde{Z}_n(0) + 1} \tag{3.9.42}$$

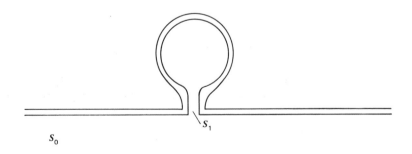

Figure 3.9.4 Tube with Helmholtz resonator as a side branch.

instead of that given by 3.9.34. Thus

$$\frac{\tilde{R}_0}{A_0} = \frac{1 - \sum_{n=1}^{m} \frac{\rho_0 c_0}{S_0} \tilde{Z}_n^{-1}(0) - \sum_{n=m+1}^{r} \frac{\rho_0 c_0}{\rho_n c_n} \frac{S_n}{S_0}}{1 + \sum_{n=1}^{m} \frac{\rho_0 c_0}{S_0} \tilde{Z}_n^{-1}(0) + \sum_{n=m+1}^{r} \frac{\rho_n c_n}{\rho_0 c_0} \frac{S_n}{S_0}} \qquad (3.9.43)$$

where we have used (3.2.21) for I_{0n}. For the important case when the fluid filling the tube and branches is the same, (3.9.43) reduces to

$$\frac{\tilde{R}_0}{A_0} = \frac{1 - \sum_{n=1}^{m} (\rho_0 c_0 / S_0) \tilde{Z}_n^{-1} - \sum_{n=m+1}^{r} (S_n / S_0)}{1 + \sum_{n=1}^{m} (\rho_0 c_0 / S_0) \tilde{Z}_n^{-1} + \sum_{n=m+1}^{r} (S_n / S_0)} \qquad (3.9.44)$$

Finally, if we have a single lateral device attached to a constant-cross-section tube, such as in Figures 3.9.3 and 3.9.4, then

$$\frac{\tilde{R}_0}{A_0} = \frac{-1}{1 + \frac{2 S_0}{\rho_0 c_0} \tilde{Z}_1(0)} \qquad (3.9.45)$$

where $\tilde{Z}_1$ is the acoustical impedance of the device. Therefore, the transmission coefficient $\alpha_t = 1 - |\tilde{R}_0|^2 / A_0^2$ is

$$\alpha_t = \frac{4 S_0}{\rho_0 c_0} \frac{X_1 + \frac{S_0}{\rho_0 c_0} (X_1^2 + Y_1^2)}{1 + \frac{4 S_0}{\rho_0 c_0} \left[X_1 + \frac{S_0}{\rho_0 c_0} (X_1^2 + Y_1^2) \right]} \qquad (3.9.46)$$

where X_1 and Y_1 are the real and imaginary parts of $\tilde{Z}_1(0)$.

EXAMPLE. Tube with a Helmholtz Resonator

The specific acoustic impedance of a Helmholtz resonator was calculated in Section 3.1, and is given by (3.1.61). Remembering that $\tilde{Z} = (1/S)\tilde{Z}_a$, we can write that result as

$$\tilde{Z}_1(0) = \frac{R'}{S_1} + i \frac{\rho_0 c_0^2}{V \omega} \left(1 - \frac{\omega}{\omega_H} \right) \qquad (3.9.47)$$

where $\omega_H = c_0 \sqrt{S_1/lV}$ is the natural frequency of the resonator. Consider for simplicity the case when R' is very small so that we may approximately put $X_1 = 0$. Then α_t becomes, after some rearrangement,

$$\alpha_t = \frac{1}{1 + \dfrac{1}{4} \dfrac{(S_1/S_0)^2 (V/S_1 l)}{(\omega/\omega_H - 1)^2}} \qquad (3.9.48)$$

The variations of α_t with frequency predicted by this equation are displayed in Figure 3.9.5 for three different resonators, all having the same value of the area ratio (S_1/S_0). It is seen that at a frequency equal to the natural frequency of each resonator, there is no transmitted energy. Thus, the Helmholtz resonator, like the side branch treated earlier, may be used to filter out unwanted sounds traveling in tubes.

The above procedure may be used to obtain transmission and reflection coefficients in more complicated situations than that of the example given here. In fact, the whole subject of acoustic-filter theory is based on such a procedure. It should be remembered, however, that the procedure rests on the assumption that (a) the waves are plane, (b) the wavelength is much larger than the transition length, (c) all the tubes are sufficiently narrow so that propagation is well below cutoff conditions, and (d) branch impedances can be evaluated. These assumptions impose nontrivial limitations to the applicability of the method.

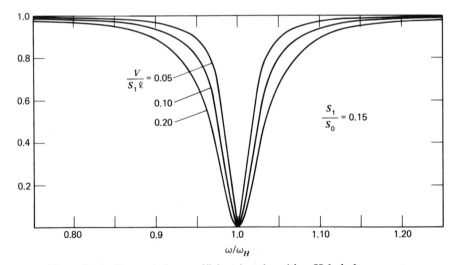

Figure 3.9.5 Transmission coefficient in tube with a Helmholtz resonator.

PROBLEMS

3.9.1 Determine the transmission coefficient for the "muffler" shown below.

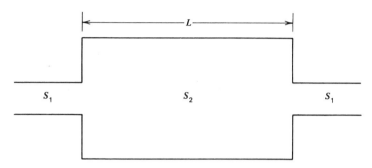

The first and third tubes extend to infinity.

3.9.2 Show that the impedance of a finite-length branch at the transition $\tilde{Z}_n(0) = \tilde{p}'_n(0)/S_n \tilde{u}_n(0)$ may be expressed in terms of the impedance at the end of the branch by means of

$$\tilde{Z}_n(0) = \frac{\rho_n c_n}{S_n} \frac{e^{-ik_n l_n} + \dfrac{w_n - 1}{w_n + 1} e^{ik_n l_n}}{e^{-ik_n l_n} - \dfrac{w_n - 1}{w_n + 1} e^{ik_n l_n}}$$

3.9.3 Show that the side branch, treated as an example at the end of this section, may be replaced by impedance $\tilde{Z}_1(0)$ given by

$$\tilde{Z}_1(0) = \frac{\rho_1 c_1}{S_1} \frac{\cos k_1 l_1}{i \sin k_1 l_1}$$

3.9.4 Consider Equation (3.9.46) for the transmission ocefficient in a tube having a side device. Assume that the imaginary part of the device's impedance has a fixed value Y_1. Determine the value of the real part, X_2, which will optimize the filtering effects of the device.

3.9.5 In Problem 3.9.3, let $k_1 = k + i\alpha$. Show that for $\alpha l_1 \ll 1$,

$$\tilde{Z}_1(0) = -\frac{\rho_1 c_1}{S_1} \frac{\alpha l_1 + i \sin k l_1 \cos k l_1}{\sin^2 k l_1 + (\alpha l_1)^2 \cos^2 k l_1}$$

3.9.6 Consider a three-branch problem. Branches 1 and 3 extend to infinity. The second branch has an impedance given by the result of the previous problem. Determine the transmission coefficient into branch 3 if its cross-section area is $S_3 = S_1$.

3.9.7 Consider a plane wave traveling in a long tube, the walls of which have a small perforation of cross-sectional area S at some point. This perforation would therefore be equivalent to a side branch of very small length. Assuming that the fluids inside and outside the pipe have the same properties, and that the fluid outside extends to infinity so that the boundary condition is $p' = 0$, show that the impedance of the orifice is

$$\tilde{Z}(0) \approx - \frac{\rho c}{S} ikl$$

where l is the thickness of the tube and is such that $kl \ll 1$. Determine α_t.

3.9.8 Determine the transmission coefficient for a tube having two Helmholtz resonators mounted diametrically opposite to each other at plane $x = 0$. The natural frequencies of the resonators are ω_1 and $\omega_2 = 2\omega_1$. Their necks have equal cross-sectional area. Sketch α_t versus ω/ω_1.

3.10   TUBE WITH TEMPERATURE GRADIENT

In the previous section, we considered sound-wave transmission through tubes having sudden changes in their cross-sectional areas or in their mean densities and temperatures. To analyze such problems, we assumed that the transitions between different regions were compact, that is, the regions of transition were assumed to be short relative to the wavelength of the incoming wave. It is therefore of some interest to study a transmission problem involving a noncompact transition. In general, however, treatment of such problems requires advanced mathematical techniques. In what follows, we use a simple type of transition that is not difficult to handle.

Consider the situation depicted in Figure 3.10.1. A wave of frequency ω and pressure amplitude A_1 is traveling in a tube along which the absolute temperature changes smoothly from value T_1 to value T_3. We approximate the smooth variation with

$$T(x) = T_1(1 + mx), \quad 0 \le x \le L \tag{3.10.1}$$

where L is the length of the transition and where

$$m = (T_3 - T_1)/T_1 L \tag{3.10.2}$$

Because of these temperature variations, the acoustic field in the tube must be determined from three wave equations. For monochromatic time dependence,

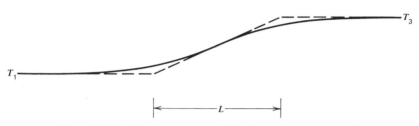

Figure 3.10.1 Temperature transition in a constant-area tube.

these are

$$\frac{d^2\tilde{p}_1}{dx^2} + k_1^2\tilde{p}_1 = 0, \quad x \leqslant 0 \tag{3.10.3}$$

$$\frac{d^2\tilde{p}_2}{dx^2} + \frac{\omega^2}{c^2(x)}\tilde{p}_2 = 0, \quad 0 \leqslant x \leqslant L \tag{3.10.4}$$

$$\frac{d^2\tilde{p}_3}{dx^2} + k_3^2\tilde{p}_3 = 0, \quad x \geqslant L \tag{3.10.5}$$

where $k_1 = \omega/c_1$ and $k_3 = \omega/c_3$.

The solutions of (3.10.3) and (3.10.5) are

$$\tilde{p}_1 = A_1 e^{ik_1 x} + \tilde{B}_1 e^{-ik_1 x}, \quad x \leqslant 0 \tag{3.10.6}$$

$$\tilde{p}_3 = \tilde{A}_3 e^{ik_3(x-L)}, \quad x \geqslant L \tag{3.10.7}$$

where we have assumed that medium 3 extends to infinity. The velocities corresponding to $\tilde{p}_1$ and $\tilde{p}_3$ are

$$\tilde{u}_1 = \frac{1}{\rho_1 c_1}\left(A_1 e^{ik_1 x} - \tilde{B}_1 e^{-ik_1 x}\right) \tag{3.10.8a}$$

$$\tilde{u}_3 = \frac{\tilde{A}_3}{\rho_3 c_3} e^{ik_3(x-L)} \tag{3.10.8b}$$

The pressure and velocity in the transition region are more difficult to obtain because the speed of sound in that region is a function of position. Thus, if the tube is filled with a perfect gas, we have

$$c^2(x) = c_1^2(1+mx)^2, \quad 0 \leqslant x \leqslant L \tag{3.10.9}$$

This gives

$$\frac{d^2\tilde{p}_2}{dx^2} + \frac{k_1^2}{(1+mx)^2}\tilde{p}_2 = 0 \tag{3.10.10}$$

The form of the coefficient of $\tilde{p}_2$ makes it difficult to obtain exact solutions of (3.10.10). However, if mL is small (i.e., if $(T_3/T_1 - 1)$ is small), we may obtain a solution in terms of Airy functions. These are related to Bessel functions of order 1/3, and are tabulated in Chapter 10 of Abramowitz and Stegun. To obtain the equation satisfied by these functions—Airy's equation—we first put $(1+mx)^{-2} \simeq 1 - 2mx$. With $z = k_1x$, this gives

$$\frac{d^2\tilde{p}_2}{dz^2} + \left(1 - \frac{2m}{k_1}z\right)\tilde{p}_2 = 0 \tag{3.10.11}$$

We now let

$$y = \left(\frac{k_1}{2m}\right)^{2/3}\left(1 - \frac{2m}{k_1}z\right) \tag{3.10.12}$$

and obtain

$$\frac{d^2\tilde{p}_2}{dy^2} + y\tilde{p}_2 = 0 \tag{3.10.13}$$

This is Airy's equation. Its solutions are the Airy functions of negative argument $Ai(-y)$, $Bi(-y)$. Thus,

$$\tilde{p}_2 = \tilde{A}_2 Ai(-y) + \tilde{B}_2 Bi(-y) \tag{3.10.14}$$

The acoustic velocity corresponding to this pressure is approximately $\tilde{u}_2 = (1/i\omega\rho_1)(d\tilde{p}_2/dx)$, or

$$\tilde{u}_2 = \frac{(2m/k_1)^{1/3}}{i\omega\rho_1 c_1}\left[\tilde{A}_2 Ai'(-y) + \tilde{B}_2 Bi'(-y)\right] \tag{3.10.15}$$

The coefficients of $\tilde{B}_1$, $\tilde{A}_2$, $\tilde{B}_2$, and $\tilde{A}_3$ may be obtained by applying the boundary conditions at $x=0$ and $x=L$. These conditions are

$$x=0: \quad \tilde{p}_1=\tilde{p}_2, \quad \tilde{u}_1=\tilde{u}_2 \qquad (3.10.16)$$

$$x=L: \quad \tilde{p}_2=\tilde{p}_3, \quad \tilde{u}_2=\tilde{u}_3 \qquad (3.10.17)$$

Using (3.10.16), we obtain

$$\tilde{A}_2 Ai(-y_0)+\tilde{B}_2 Bi(-y_0)=A+\tilde{B}_1 \qquad (3.10.18)$$

$$\tilde{A}_2 Ai'(-y_0)+\tilde{B}_2 Bi'(-y_0)=iN_1(A+\tilde{B}_1) \qquad (3.10.19)$$

In these equations, y_0 represents the value of y corresponding to x_0 (i.e., $y_0=(k_1/2m)^{2/3}$), and $N_1=(k_1/2m)^{1/3}$. The boundary conditions at $x=L$ yield

$$\tilde{A}_2 Ai(-y_L)+\tilde{B}_2 Bi(-y_L)=\tilde{A}_3 \qquad (3.10.20)$$

$$\tilde{A}_2 Ai'(-y_L)+\tilde{B}_2 Bi'(-y_L)=iN_3\tilde{A}_3 \qquad (3.10.21)$$

where $y_L=(k_1/2m)^{2/3}(1-2mL)$ and $N_3=(\rho_1 c_1/\rho_3 c_3)N_1$.
Equations (3.10.18)–(3.10.21) may be expressed as

$$\begin{bmatrix} -1 & Ai(-y_0) & Bi(-y_0) & 0 \\ iN_1 & Ai'(-y_0) & Bi'(-y_0) & 0 \\ 0 & Ai(-y_L) & Bi(-y_L) & -1 \\ 0 & Ai'(-y_L) & Bi'(-y_L) & -iN_3 \end{bmatrix} \begin{bmatrix} \tilde{B}_1 \\ \tilde{A}_2 \\ \tilde{B}_2 \\ \tilde{A}_3 \end{bmatrix} = A_1 \begin{bmatrix} 1 \\ iN_1 \\ 0 \\ 0 \end{bmatrix} \qquad (3.10.22)$$

This system of equations differs from that for transmission through a wall only in that the trigonometric functions have been replaced by Airy functions. Proceeding as we did in Section 3.3, we obtain an equation for the transmitted pressure amplitude $\tilde{A}_3$:

$$\tilde{A}_3 = \frac{A_1}{\det \overline{\overline{A}}}(C_{41}+iN_1C_{42}) \qquad (3.10.23)$$

where $\det \overline{\overline{A}}$ is the determinant of the coefficient matrix and C_{41} and C_{42} are the cofactors of the elements A_{14} and A_{24}, respectively, of that matrix. That is,

$$C_{41} = -\begin{vmatrix} iN_1 & Ai'(-y_0) & Bi'(-y_0) \\ 0 & Ai(-y_L) & Bi(-y_L) \\ 0 & Ai'(-y_L) & Bi'(-y_L) \end{vmatrix} \qquad (3.10.24)$$

and

$$C_{42} = \begin{vmatrix} -1 & Ai(-y_0) & Bi(-y_0) \\ 0 & Ai(-y_L) & Bi(-y_L) \\ 0 & Ai'(-y_L) & Bi'(-y_L) \end{vmatrix} \qquad (3.10.25)$$

Expanding (3.10.24), we obtain

$$C_{41} = -iN_1 \left[Ai(-y_L)Bi'(-y_L) - Ai'(-y_L)Bi(-y_L) \right] \qquad (3.10.26)$$

The quantity in square brackets is the Wronskian of the solutions Ai and Bi, and is equal to $1/\pi$. Therefore,

$$C_{41} = -\frac{i}{\pi}N_1 \qquad (3.10.27)$$

Similarly, we obtain $C_{42} = -1/\pi$ so that

$$\frac{\tilde{A}_3}{A_1} = -\frac{2i}{\pi}\frac{N_1}{\det \overline{\overline{A}}} \qquad (3.10.28)$$

The determinant of the coefficient matrix is

$$\det \overline{\overline{A}} = -\left\{ Ai'(-y_0)Bi'(-y_L) - Ai'(-y_L)Bi'(-y_0) \right.$$

$$-iN_3 \left[Ai'(-y_0)Bi(-y_L) - Ai(-y_L)Bi'(-y_0) \right] \right\}$$

$$-iN_1 \left\{ A_i(-y_0)Bi'(-y_L) - Ai'(-y_L)Bi(-y_0) \right.$$

$$-iN_3 \left[Ai(-y_0)Bi(-y_L) - Ai(-y_L)Bi(-y_0) \right] \right\} \qquad (3.10.29)$$

Separating this expression into its real and imaginary parts, we can write

$$\det \overline{\overline{A}} = [a] + N_1 N_3 [b] + iN_1 [c] + N_3 [d] \qquad (3.10.30)$$

where we have introduced the shorthand notation

$$[a] = Ai'(-y_L)Bi'(-y_0) - Ai'(-y_0)Bi'(-y_L) \qquad (3.10.31)$$

$$[b] = Ai(-y_L)Bi(-y_0) - Ai(-y_0)Bi(-y_0) \qquad (3.10.32)$$

$$[c] = Ai'(-y_L)Bi(-y_0) - Ai(-y_0)Bi'(-y_L) \qquad (3.10.33)$$

$$[d] = Ai'(-y_0)Bi(-y_L) - Ai(-y_L)Bi'(-y_0) \qquad (3.10.34)$$

These results may be used to compute the energy transmitted into medium 3. As before, we measure this in terms of a transmission coefficient defined as the ratio of transmitted to incident intensities. Thus, $\alpha_t = (\rho_1 c_1 / \rho_3 c_3)(|\tilde{A}_3|^2 / A^2)$. Substitution from (3.10.28) yields

$$\alpha_t = \frac{4}{\pi^2} \frac{(\rho_1 c_1 / \rho_3 c_3) N_1^2}{\det \overline{\overline{A}}^2} \tag{3.10.35}$$

Because of the complexity of $\det \overline{\overline{A}}$, it is not possible to obtain a closed-form expression for α_t. However, it is possible to obtain a limiting form of (3.10.35) applicable for low frequencies. First we note that the quantity y_L may be written as

$$y_L = (k_1 L)^{2/3} \frac{1 - 2mL}{(2mL)^{2/3}} \tag{3.10.36}$$

Table 3.10.1 Values of Functions Defined by Equations (3.10.31)–(3.10.34)

y	$[a]$	$[b]$	$[c]$	$[d]$
0.0	0.0	0.0	−0.3183	−0.3183
0.2000	0.0064	0.0636	−0.3175	−0.3179
0.4000	0.0254	0.1266	−0.3115	−0.3149
0.6000	0.0565	0.1876	−0.2956	−0.3069
0.8000	0.0961	0.2439	−0.2619	−0.2916
1.0000	0.1488	0.2924	−0.2166	−0.2670
1.5000	0.2830	0.3536	−0.0079	−0.1585
2.0000	0.3498	0.2862	0.2812	0.0048
2.5000	0.2029	0.0340	0.5108	0.1848
3.0000	−0.0338	−0.1625	0.4333	0.2211
3.5000	−0.3334	−0.2909	0.0349	0.1246
4.0000	−0.3846	−0.1825	−0.4448	−0.0700
4.5000	−0.0703	0.0895	−0.5472	−0.1967
5.0000	0.3481	0.2648	−0.0752	−0.1214
5.5000	0.3939	0.1415	0.5225	0.0872
6.0000	−0.0553	−0.1503	0.5013	0.1855
6.5000	−0.4571	−0.2390	−0.2030	0.0391
7.0000	−0.2167	0.0090	−0.6510	−0.1586
7.5000	0.3701	0.2378	−0.1156	−0.1151
8.0000	0.3781	0.0852	0.6319	0.1094
8.5000	−0.2637	−0.2059	0.3220	0.1461
9.0000	−0.4522	−0.1290	−0.5796	−0.0742
9.5000	0.2064	0.1828	−0.4161	−0.1528
10.0000	0.4775	0.1365	0.5702	0.0634

For a given value of mL, and therefore of $T_3/T_1 - 1$, we may use our results to obtain α_t as a function of frequency. The low-frequency limit $k_1 L \to 0$ is of special interest because it specifies the conditions under which the transition may be considered compact. Stated differently, in the limit $k_1 L \to 0$, the tube is sharply divided into regions of uniform temperatures T_1 and T_3. These regions have characteristic impedances $\rho_1 c_1$ and $\rho_3 c_3$, respectively, so that we should recover the result for the transmission coefficient that was derived in Section 3.1. When $k_1 L \to 0$, $y_L \to y_0$ so that $[a]=[b]=0$ and $[c]=[d]= -1/\pi$. Therefore,

$$\alpha_t \to \frac{4(\rho_1 c_1/\rho_3 c_3)N_1^2}{(N_1 + N_3)^2} \tag{3.10.37}$$

Using the definitions of N_1 and N_3, this gives

$$\alpha_t(k_1 L \to 0) = \frac{4(\rho_1 c_1/\rho_3 c_3)}{(1+\rho_1 c_1/\rho_3 c_3)^2} \tag{3.10.38}$$

As expected, this agrees with (3.1.23).

For finite values of $k_1 L$, we use the tabulated values of the Airy functions to compute the quantities $[a]$, $[b]$, $[c]$, and $[d]$ defined earlier. Table 3.10.1 gives

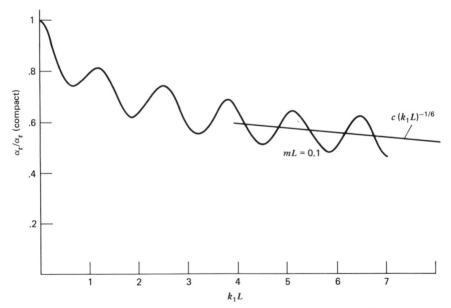

Figure 3.10.2 Variations with frequency of the ratio of transmission coefficient through a smooth transition to the transmission coefficient for a compact transition.

the values of these quantities for some selected values of y and for $y_0 = 0$. Figure 3.10.2 shows the variations with frequency of the actual transmission coefficient relative to its compact counterpart. Thus, as the frequency increases, the "actual" transition transmits less and less energy than the compact-transition results would predict. The figure also shows that the transmission coefficient oscillates with $k_1 L$. This oscillation is due to changes of amplitude of the field in the transition, and is analogous to the oscillations displayed by the transmission coefficient obtained for transmission through a wall (see Figure 3.3.2).

The problem of sound transmission in nonuniform media is of central importance in ocean acoustics due to the vertical temperature and salinity gradients that exist in the oceans. These gradients produce many interesting effects, such as alteration of the direction of propagation and the trapping of acoustic energy within well-defined regions. These effects are of considerable importance because it is through acoustic waves that the oceans are being studied. The interested reader should consult the texts on the subject listed in the bibliography.

PROBLEMS

3.10.1 Using the asymptotic expressions for $Ai(-x)$ and $Bi(-x)$ given by Abramowitz and Stegun, show that when $k_1 L \to \infty$, α_t approaches zero as $(k_1 L)^{-1/6}$.

3.10.2 At a compact interface, the transmission coefficient is given by (3.1.23). This is applicable to waves moving from medium 1 to medium 2 and vice versa. Consider propagation through a linear transition with the temperature decreasing from T_1 at $x = 0$ to T_3 at $x = L$. Show that the transmission coefficient has the same form as before, but that the argument of the Airy functions is now $y = (k_1 L)^{2/3}[(1 + 2mL)/(2mL)^{2/3}]$. Compare the two transmission coefficients.

3.10.3 An acoustic ray may be defined as a line that is tangent to the wave-front normal everywhere. Consider a plane wave approaching an interface at an angle of incidence θ_i. At the interface, the speed of sound changes from c to $c + \delta c$, where $\delta c \ll c$. Use Snell's law to show that the direction of the incident ray will change by an amount given by

$$\delta\theta = \tan\theta_i \frac{\delta c}{c}$$

Use this result to sketch sound rays for a wave moving into regions of increasing or decreasing sound speed.

3.10.4   A wave propagates in a medium of infinite extent. The speed of sound varies linearly with the distance along the vertical. Using the result of Problem 3.10.3, trace the path of a sound ray. Show that this is an arc of a circle.

SUGGESTED REFERENCES

Beyer, R. T. *Nonlinear Acoustics*. This is a modern text on the subject. Chapters 3 and 4 deal with distortion and with short waves, respectively. The introduction is a very readable overview of the field.

Eisner, E. "Complete Solutions of the 'Webster' Horn Equation." This article contains a very extensive annotated bibliography.

Lesser, M. B. and Crighton, D. B. "Physical Acoustics and Method of Matched Asymptotic Expansions." Section 3 of this work relates to acoustic wave guides.

Stewart, G. W. and Lindsay, R. B. *Acoustics: A Text on Theory and Applications*. This book contains material on acoustic filtration.

Whitham, G. B., *Linear and Nonlinear Waves*. Chapter 2 of this advanced text deals with linear dispersive waves.

CHAPTER FOUR
SPHERICAL AND CYLINDRICAL WAVES

Except for an occasional reference to waves of arbitrary geometry, the waves studied so far have been assumed to have plane fronts. This assumption has simplified our calculations greatly, as it has reduced the number of independent variables to two. Unfortunately, plane sound waves are the exception rather than the rule. More common waves have fronts that are not plane, and not much can be said about them unless their geometry is so simple that it can be fitted somehow with one of the few coordinate systems in which the wave equation is separable. In this chapter, we consider two such systems of coordinates: the polar spherical and the polar cylindrical. These systems are very useful in studying some problems of fundamental importance. For example, the spherical polar coordinate system leads to the fundamental point-source concept; this, in turn, plays a dominant role in the theory of sound emission. Similarly, the polar cylindrical system provides a means of studying the practically important problem of sound-wave propagation in circular tubes. The chapter also includes a brief outline of techniques used to handle less restrictive types of waves.

4.1 CENTRALLY SYMMETRIC WAVES

The simplest nonplane wave is the centrally symmetric spherical wave. Here, all the properties of the wave at any point in space depend only on time and on the distance to the origin of a system of coordinates. To study this important type of wave, we first write the wave equation in the system of coordinates depicted in Figure 4.1.1. The angles θ and φ in the figure are the polar and the azimuthal angles, respectively. The latter angle is measured around the polar

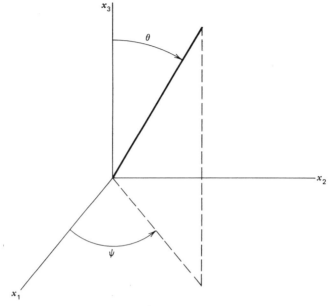

Figure 4.1.1 Spherical-polar coordinates.

axis from any reference line lying on the x_1, x_2 plane. In these coordinates, the wave equation can be written as (see Appendix B)

$$\frac{1}{c_0^2} \frac{\partial^2 \phi}{\partial t^2} = \frac{1}{r^2} \frac{\partial}{\partial r}\left(r^2 \frac{\partial \phi}{\partial r}\right) + \frac{1}{r^2 \sin\theta} \frac{\partial}{\partial \theta}\left(\sin\theta \frac{\partial \phi}{\partial \theta}\right) + \frac{1}{r^2 \sin^2\theta} \frac{\partial^2 \phi}{\partial \varphi^2} \quad (4.1.1)$$

The acoustic pressure and velocity can still be obtained from ϕ by the usual relations. However, $\mathbf{u} = \nabla\phi$ has, in general, three components, that is, $\mathbf{u} = (u_r, u_\theta, u_\varphi)$. These components are given by

$$u_r = \frac{\partial \phi}{\partial r}$$

$$u_\theta = \frac{1}{r} \frac{\partial \phi}{\partial \theta} \qquad\qquad (4.1.2)$$

$$u_\varphi = \frac{1}{r\sin\theta} \frac{\partial \phi}{\partial \varphi}$$

For the centrally symmetric case, the velocity potential is only a function of r and t, so that (4.1.1) reduces to

$$\frac{\partial^2 \phi}{\partial t^2} = c_0^2 \left(\frac{\partial^2 \phi}{\partial r^2} + \frac{2}{r} \frac{\partial \phi}{\partial r} \right) \tag{4.1.3}$$

However, since

$$\frac{\partial^2 (r\phi)}{\partial r^2} = r \left(\frac{\partial^2 \phi}{\partial r^2} + \frac{2}{r} \frac{\partial \phi}{\partial r} \right)$$

we can write (4.1.3) as

$$\frac{\partial^2 (\phi r)}{\partial t^2} = c_0^2 \frac{\partial^2 (\phi r)}{\partial r^2} \tag{4.1.4}$$

This is a one-dimensional wave equation for the quantity $r\phi$. Hence, the general solution of the wave equation for the centrally symmetric case is

$$\phi(r,t) = \frac{f(c_0 t - r) + g(c_0 t + r)}{r} \tag{4.1.5}$$

Thus, $f(c_0 t - r)$ represents a centrally symmetric wave moving away from the origin with uniform speed c_0, whereas $g(c_0 t + r)$ represents a similar wave converging onto the origin, as sketched in Figure 4.1.2. The first type of wave is an *outgoing* wave, whereas the second is an *incoming* wave. We note that the origin is the center of curvature for both wave fronts, so that if $g(c_0 t + r)$ arises owing to reflection from some boundary, then the boundary also has to be a

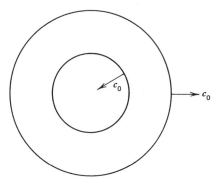

Figure 4.1.2 Outgoing and incoming spherically symmetric waves.

sphere centered at $r=0$. It should be noted that the amplitude of each wave depends on position. This is due mainly to changes of the wave-front areas, which change with r as r^2, and to the requirement that the total energy be conserved. Also, the amplitude of each of these waves approaches infinity as $r \to 0$; that is, the solution is singular at the origin. This singularity will be studied later in some detail. Here, we note that if we compute the volume flow rate through a sphere of radius r enclosing the origin, we obtain

$$Q = 4\pi r^2 u_r$$

Since $u_r = \partial \phi / \partial r$, this yields

$$Q = 4\pi r \left[-f'(c_0 t - r) + g'(c_0 t + r) \right] - 4\pi \left[f(c_0 t - r) + g(c_0 t + r) \right]$$

Thus, if we now let $r \to 0$, we find that the general solution predicts that a finite amount of fluid is emerging from the origin. That is, at $r=0$ we must have some kind of a source. However, if the origin is not a source of fluid, the functions f and g must be such that at $r=0$, $Q=0$. This can be so only if

$$g(c_0 t) = -f(c_0 t) \tag{4.1.6}$$

in which case the solution becomes

$$\phi(r, t) = \frac{f(c_0 t - r) - f(c_0 t + r)}{r} \tag{4.1.7}$$

This solution will be used in Section 4.2 to study a simple reflection problem.

Consider now a situation where only the outgoing wave exists. The acoustic pressure in the wave is

$$p' = -\frac{\rho_0 c_0}{r} f'(c_0 t - r) \tag{4.1.8}$$

Similarly, the radial velocity component (which is the only velocity component) becomes

$$u_r = -\frac{1}{r} f'(c_0 t - r) - \frac{1}{r^2} f(c_0 t - r) \tag{4.1.9}$$

Thus, contrary to the plane-wave case, the pressure and velocity are not proportional to one another. In fact, combination of (4.1.8) and (4.1.9) yields

$$\frac{p'}{u_r} = \frac{\rho_0 c_0}{1 + f/rf'}$$

This shows, however, that as $r \to \infty$,

$$p'/u_r = \rho_0 c_0 \qquad (4.1.10)$$

This result is identical to that applicable in the plane-wave case and was to be expected because for r large, any portion of the wave front can be considered plane.

To compute the intensity for r large, we use (4.1.10) and obtain

$$I(r) = \frac{\rho_0 c_0}{r^2} \langle f'^2 \rangle \qquad (4.1.11)$$

This intensity decays as $1/r^2$. As pointed out earlier, the decay is due to spreading of the wave-front area, which increases as r^2.

Monochromatic Case

The velocity potential for spherically symmetric monochromatic waves may be derived from (2.5.3). Thus, when $\tilde{\Phi} = \tilde{\Phi}(r)$, that equation reduces to

$$\frac{d^2 \tilde{\Phi}}{dr^2} + \frac{2}{r} \frac{d\tilde{\Phi}}{dr} + k^2 \tilde{\Phi} = 0 \qquad (4.1.12)$$

From our previous discussion, it follows that

$$\tilde{\Phi} = \frac{\tilde{A}}{r} e^{\pm ikr} \qquad (4.1.13)$$

However, it is instructive to rederive this result in a different way. In doing so, we introduce some functions needed later. Equation 4.1.12 for $\tilde{\Phi}$ can also be written as

$$\tilde{\Phi}''(z) + \frac{2}{z} \tilde{\Phi}'(z) + \tilde{\Phi}(z) = 0 \qquad (4.1.14)$$

where $z = kr$. This equation is a special case of

$$\tilde{\Phi}''(z) + \frac{2}{z} \tilde{\Phi}'(z) + \left[1 - \frac{n(n+1)}{z^2} \right] \tilde{\Phi}(z) = 0, \quad n = 0, \pm 1, \pm 2, \dots \qquad (4.1.15)$$

Particular solutions are the *spherical Bessel functions*:

First kind

$$j_n(z) = \sqrt{\pi/2z} \, J_{n+1/2}(z) \qquad (4.1.16)$$

Second kind

$$y_n(z) = \sqrt{\pi/2z}\ Y_{n+1/2}(z) \qquad (4.1.17)$$

Third kind

$$h_n^{(1)}(z) = j_n(z) + i y_n(z) = \sqrt{\pi/2z}\ H_{n+1/2}^{(1)}(z) \qquad (4.1.18)$$

$$h_n^{(2)}(z) = j_n(z) - i y_n(z) = \sqrt{\pi/2z}\ H_{n+1/2}^{(2)}(z) \qquad (4.1.19)$$

In these definitions, $J(z)$, $Y(z)$, and $H(z)$ are the Bessel functions of the first, second, and third kinds, respectively.

Some of the properties of these functions are listed in Appendix D. Additional information can be found in Chapter 10 of Abramowitz and Stegun. Here, we note that the pairs (j_n, y_n) and $(h_n^{(1)}, h_n^{(2)})$ are linearly independent for every n, and that for low values of n they can be expressed in terms of elementary functions. For example, for $n = 0$ and $n = 1$ we have

$$j_0(z) = \frac{\sin z}{z}, \qquad y_0(z) = -\frac{\cos z}{z} \qquad (4.1.20)$$

$$j_1(z) = \frac{1}{z^2}(\sin z - z \cos z), \qquad y_1(z) = -\frac{1}{z^2}(\cos z + z \sin z) \quad (4.1.21)$$

These give

$$h_0^{(1)}(z) = -\frac{i}{z} e^{iz}, \qquad h_0^{(2)}(z) = \frac{i}{z} e^{-iz} \qquad (4.1.22)$$

$$h_1^{(1)}(z) = -\frac{1}{z}\left(1 + \frac{i}{z}\right) e^{iz}, \qquad h_1^{(2)} = -\frac{1}{z}\left(1 - \frac{i}{z}\right) e^{-iz} \qquad (4.1.23)$$

For other values of n, these functions may be obtained from the following recurrence relation, applicable to j_n, y_n, $h_n^{(1)}$, and $h_n^{(2)}$:

$$f_{n+1}(z) = \frac{1}{z}(2n+1) f_n(z) - f_{n-1}(z) \qquad (4.1.24)$$

Each of the pairs given by (4.1.20) to (4.1.23) gives the solution of (4.1.15). The pair (4.1.22) is more useful when dealing with propagation problems, $h_0^{(1)}$ corresponding to the outgoing wave and $h_0^{(2)}$ to the incoming wave.

Standing Waves in a Spherical Cavity

As an application of our monochromatic, spherically symmetric solution, we consider wave motion inside a rigid-walled, spherical cavity of radius a. The velocity potential is given by

$$\tilde{\phi} = e^{-i\omega t} \left[A j_0(kr) + \tilde{B} y_0(kr) \right] \tag{4.1.25}$$

In order that $\tilde{\phi}$ be finite at $r = 0$, we require that B be zero [see (4.2.20)]. Thus,

$$\tilde{\phi} = A e^{-i\omega t} j_0(kr) \tag{4.1.26}$$

Now, if the walls of the cavity are assumed rigid, then the normal velocity must vanish there; that is,

$$\left(\frac{\partial \phi}{\partial r} \right)_{r=a} = 0$$

Therefore, we can have centrally symmetric, standing waves inside the sphere for values of k that satisfy

$$j_0'(ka) = 0 \tag{4.1.27}$$

Tabulated values of zeros of j_n' can be found in the literature.[1] However, for $n = 0$, they can be found directly from $j_0(z) = \sin z / z$. Thus, taking the derivative and evaluating it at $z = ka$ yields

$$\tan ka = ka \tag{4.1.28}$$

The values of ka that satisfy this equation specify the allowable wavelengths, and therefore the frequencies of oscillation. These values may be obtained from (4.1.28) by trial and error or by graphical methods (see Problem 4.1.1). The first two roots are

$$k_1 a = 4.493$$

$$k_2 a = 7.725$$

The lowest frequency of oscillation is therefore given by

$$\omega_1 = 4.493(c_0 / a) \tag{4.1.29}$$

---

[1] *Royal Mathematical Society Tables*, Vol. 7: *Bessel Functions*, Part III, Zeros and Associated Values (Cambridge University Press, 1960).

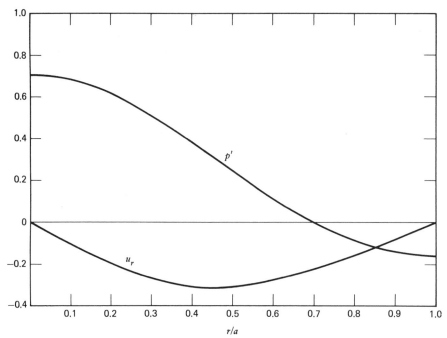

Figure 4.1.3 Instantaneous spatial variations of acoustic pressure and velocity in a spherical cavity for the lowest-frequency, symmetric-oscillation case.

This, however, is not the absolute lowest frequency, for as we will see in Section 4.3, a lower one appears when the potential is allowed to depend on θ.

Figure 4.1.3 displays the instantaneous variations with r/a of the pressure and velocity for the case having the root $k_1 = 4.493/a$. It should be noted that the pressure has one nodal line at a distance from the wall approximately equal to $0.3a$. Since for $k_1 = 4.493/a$ the wavelength is $\lambda_1 = 1.40a$, we find that the nodal line is at a distance from the wall equal to 0.214λ, a distance not too different from the corresponding one in front of a plane reflector.

PROBLEMS

4.1.1 By plotting the functions $f(ka) = ka$ and $g(ka) = \tan ka$ versus ka, show that the third root of (4.1.27) is $k_3 a \approx 10.90$.

4.1.2 Determine the locations of the velocity nodal lines for the root $k_2 = 7.725/a$ in the spherical-cavity example.

4.2 PROBLEMS WITH SPHERICAL SYMMETRY

The general solution for the centrally symmetric geometry can be used to study a variety of problems. In this section, we consider a few of these in order to display some of the salient features of spherical waves.

Radially Pulsating Sphere

We first study the sound field produced by a sphere whose surface pulsates radially. The problem is important because it provides the basis for the study of sound emission.

Consider, then, a sphere of initial radius a, whose center of mass coincides with the origin of a system of coordinates, as shown in Figure 4.2.1. The sphere's center remains fixed at that position, but at time t_0 the points on its surface begin to move radially with a velocity U given by

$$U = U(t) \tag{4.2.1}$$

Provided the amplitude of the motion is small, the time dependence of U can be arbitrary. If the sphere is imbedded in an infinite medium, there will be no reflected waves so that the velocity potential in the fluid around the sphere may be expressed as

$$\phi(r, t) = \frac{1}{r} f\left(t - \frac{r}{c_0}\right) \tag{4.2.2}$$

The form of f can be obtained from the boundary condition on the surface of the sphere. This condition is that the fluid velocity should be equal to the velocity of the surface. However, because of our assumption regarding the

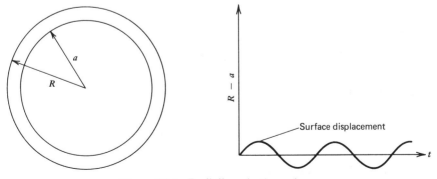

Figure 4.2.1 Radially pulsating sphere.

amplitude of the motion, we can apply this condition at $r=a$ rather than at the instantaneous position of the sphere's surface. This is entirely analogous to the flat-piston problem treated in Section 3.6.

Now, the fluid velocity has only a radial component, and this is given by

$$u_r(r,t) = -\frac{1}{rc_0}f'\left(t-\frac{r}{c_0}\right) - \frac{1}{r^2}f\left(t-\frac{r}{c_0}\right) \tag{4.2.3}$$

Applying the boundary condition, we obtain

$$f'\left(t-\frac{a}{c_0}\right) + \frac{c_0}{a}f\left(t-\frac{a}{c_0}\right) = -ac_0U(t) \tag{4.2.4}$$

This is a first-order differential equation that can be used to obtain f. This can be done for any value of a, but it is instructive to consider, first, the case when a is small or, more precisely, when $a \ll \lambda$, where λ is some typical wavelength. Let T be the time over which f changes considerably; that is, a typical period if the motion is oscillatory. Then,

$$|f'(t-a/c_0)| \sim |f|T^{-1} \sim |f|\omega$$

Therefore, the ratio of the magnitude of the first term in (4.2.4) relative to that of the second is

$$\frac{|f'|}{|(c_0/a)f|} \sim \frac{a}{c_0/\omega} \sim \frac{a}{\lambda} \ll 1$$

In those conditions, our differential equation reduces to

$$f(t-a/c_0) = -a^2U(t) \tag{4.2.5}$$

from which we obtain

$$r\phi(r,t) = f(t-r/c_0) = -a^2U\left[t-(r-a)/c_0\right] \tag{4.2.6}$$

Further, for points r such that $r \gg a$, this reduces to

$$\phi(r,t) = -\frac{a^2}{r}U\left(t-\frac{r}{c_0}\right) \tag{4.2.7}$$

In particular, if $U(t)$ is periodic, that is, if

$$U(t) = U_0e^{-i\omega t}$$

then

$$\tilde{\phi}(r,t) = -\frac{a^2 U_0}{r} e^{i(kr-\omega t)} \tag{4.2.8}$$

Therefore, in the limit $a/\lambda \ll 1$, the sphere radiates spherical monochromatic waves with an amplitude proportional to the rate at which the sphere changes its volume.

Let us return to the full differential equation for f. Writing $\xi = t - a/c_0$, we have

$$f'(\xi) + \frac{c_0}{a} f(\xi) = -ac_0 U\left(\xi + \frac{a}{c_0}\right) \tag{4.2.9}$$

This can also be written as

$$\frac{d}{d\xi}\left[e^{c_0\xi/a}f(\xi)\right] = -ac_0 e^{c_0\xi/a} U\left(\xi + \frac{a}{c_0}\right) \tag{4.2.10}$$

Thus, integrating between $-\infty$ (when $U=0$) and ξ, we obtain

$$f(\xi) = -ac_0 e^{-c_0\xi/a} \int_{-\infty}^{\xi} e^{c_0\xi'/a} U\left(\xi' + \frac{a}{c_0}\right) d\xi' \tag{4.2.11}$$

We now put $\tau = \xi' + a/c_0$ so that

$$f(\xi) = -ac_0 \exp\left[-\frac{c_0}{a}\left(\xi + \frac{a}{c_0}\right)\right] \int_{-\infty}^{\xi + a/c_0} e^{c_0\tau/a} U(\tau) \, d\tau \tag{4.2.12}$$

Hence,

$$\phi(r,t) = -\frac{ac_0}{r} \exp\left[-\frac{c_0}{a}\left(t - \frac{r-a}{c_0}\right)\right] \int_{-\infty}^{t-(r-a)/c_0} U(\tau) e^{c_0\tau/a} \, d\tau \tag{4.2.13}$$

Thus, the signal arriving at point r at time t depends on all the signals sent by the sphere from time $t = -\infty$ to time $t - (r-a)/c_0$. Also, a signal produced by the sphere at time t will be heard at position r at a time $t + (r-a)/c_0$. The quantity

$$t' = t - \frac{r-a}{c_0} \tag{4.2.14}$$

is therefore referred to as the *retarded time*. In terms of t', we have

$$\phi(r, t') = -\frac{ac_0}{r} e^{-c_0 t'/a} \int_{-\infty}^{t'} e^{c_0 \tau/a} U(\tau)\, d\tau \tag{4.2.15}$$

We can illustrate this solution by considering some simple motions.

Harmonic Pulsations. Here the sphere begins to pulsate, at $t=0$, about $r=a$ according to

$$U(t) = \mathrm{Re}\left(U_0 e^{-i\omega t}\right) \tag{4.2.16}$$

Hence,

$$\tilde{\phi}(r, t) = -\frac{ac_0 U_0}{r} \exp\left[-\frac{c_0}{a}\left(t - \frac{r-a}{c_0}\right)\right]$$

$$\times \int_0^{t-(r-a)/c_0} \exp\left[\frac{c_0}{a}\left(1 - i\frac{\omega a}{c_0}\right)\tau\right] d\tau \tag{4.2.17}$$

Now, $\omega/c_0 = k$, so that upon integration we obtain

$$\tilde{\phi}(r, t) = -\frac{a^2 U_0}{r} \frac{\exp\left[-ikc_0\left(t - \frac{r-a}{c_0}\right)\right] - \exp\left[-\frac{c_0}{a}\left(t - \frac{r-a}{c_0}\right)\right]}{1 - ika}$$

$$\tag{4.2.18}$$

This result contains both the transient and steady-state solutions. The latter is usually of more interest, and is obtained from (4.2.18) by letting $t - (r-a)/c_0$ become large. Thus,

$$\tilde{\phi}(r, t) = -\frac{a^2 U_0}{r} \frac{\exp\left[-ikc_0\left(t - \frac{r-a}{c_0}\right)\right]}{1 - ika} \tag{4.2.19}$$

or

$$\tilde{\phi}(r, t) = -\frac{a^2 U_0}{r} \frac{e^{-i[\omega t - k(r-a)]}}{1 - ika} \tag{4.2.20}$$

For $ka \to 0$, this reduces to the simplified solution obtained earlier.

Although not needed in the present section, we give the acoustic pressure, velocity, and intensity. Thus, since $\tilde{p}' = -\rho_0 \partial \tilde{\phi}/\partial t$, $\tilde{u}_r = \partial \tilde{\phi}/\partial r$, we have

$$\tilde{p}' = i\rho_0 \omega \tilde{\phi} \tag{4.2.21}$$

$$\tilde{u}_r = ik(1 + i/kr)\tilde{\phi} \tag{4.2.22}$$

Therefore,

$$I = \tfrac{1}{2}k^2 \rho_0 c_0 |\tilde{\phi}|^2 \tag{4.2.23}$$

Substitution from (4.2.20) yields

$$I = \tfrac{1}{2}\rho_0 c_0 k^2 \left(\frac{a^2 U_0}{r}\right)^2 \frac{1}{1 + (ka)^2} \tag{4.2.24}$$

The above solution shows that the radial velocity is given by the sum of two terms:

$$\tilde{u}_{r_1} = ik\tilde{\phi}, \qquad \tilde{u}_{r_2} = -\tilde{\phi}/r$$

The first depends on distance as $1/r$, whereas the second varies as $1/r^2$. Hence, when r is large, only the first contributes to the acoustic field and is therefore referred to as the far-field velocity. Similarly, as $r \to 0$, the second term is dominant. Therefore, it is referred to as the near-field velocity. Reference to our intensity computations show that the near-field velocity does not contribute to the acoustic intensity. In fact, when $kr \to 0$, the intensity $\langle p'u \rangle$ vanishes. This means that in the vicinity of the sphere, the fluid behaves as though it were incompressible.

Impulsive Motion of Short Duration. As a second example we consider a simple, short-duration surface motion. Suppose that at time $t = -\varepsilon/2$ the sphere's surface starts moving outwards with constant velocity U_0, and that at $t = \varepsilon/2$ the motion of the sphere stops, as depicted in Figure 4.2.2. The motion then consists simply of an outward displacement of the sphere's surface so that the fluid surrounding the sphere is compressed by a small amount. One might expect the solution to be a compressive wave traveling into the fluid, but as we will see, this is not the case.

Now, at time $t = \varepsilon/2$, the radius of the sphere will be equal to $a + \varepsilon U_0$. In order for our solution to be applicable, we require that

$$\varepsilon U_0 \ll a \tag{4.2.25}$$

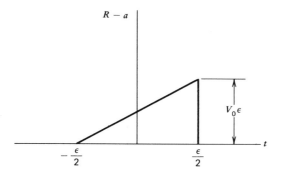

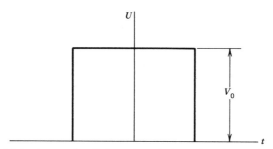

Figure 4.2.2 Outward radial motion of short duration and amplitude.

The surface velocity is given by

$$U = \begin{cases} 0, & t < -\varepsilon/2 \\ U_0, & -\varepsilon/2 \leqslant t \leqslant \varepsilon/2 \\ 0, & t > \varepsilon/2 \end{cases} \qquad (4.2.26)$$

For such a velocity, our solution for the potential is

$$\phi(r,t') = \begin{cases} 0, & t' < -\varepsilon/2 \\ -\dfrac{ac_0 U_0}{r} e^{-c_0 t'/a} \displaystyle\int_{-\varepsilon/2}^{t'} e^{c_0 \tau/a}\,d\tau, & -\dfrac{\varepsilon}{2} \leqslant t' \leqslant \dfrac{\varepsilon}{2} \\ -\dfrac{ac_0 U_0}{r} e^{-c_0 t'/a} \displaystyle\int_{-\varepsilon/2}^{\varepsilon/2} e^{c_0 \tau/a}\,d\tau, & t' > \dfrac{\varepsilon}{2} \end{cases} \qquad (4.2.27)$$

where $t' = t - (r-a)/c_0$. Integrating, we obtain

$$\phi(r, t' < -\varepsilon/2) = 0$$

$$\phi\left(r, -\frac{\varepsilon}{2} \leqslant t' \leqslant \frac{\varepsilon}{2}\right) = -\frac{a^2 U_0}{r}\left[1 - \exp\left[-\frac{c_0}{a}\left(\frac{\varepsilon}{2} + t'\right)\right]\right] \qquad (4.2.28)$$

$$\phi\left(r, t' \geqslant \frac{\varepsilon}{2}\right) = -\frac{2a^2 U_0}{r} e^{-c_0 t'/a} \sinh\left(\frac{c_0 \varepsilon}{2a}\right)$$

In the limit $\varepsilon \to 0$, these expressions reduce to

$$\phi(r, t') = \begin{cases} 0, & t' < \varepsilon/2 \\[2mm] -\dfrac{ac_0 U_0}{r}\left(\dfrac{\varepsilon}{2} + t'\right), & -\dfrac{\varepsilon}{2} \leqslant t' \leqslant \dfrac{\varepsilon}{2} \\[2mm] -\dfrac{ac_0 U_0}{r}\varepsilon e^{-c_0 t'/a}, & t' > \dfrac{\varepsilon}{2} \end{cases} \qquad (4.2.29)$$

Figure 4.2.3 shows these results schematically in terms of t'.

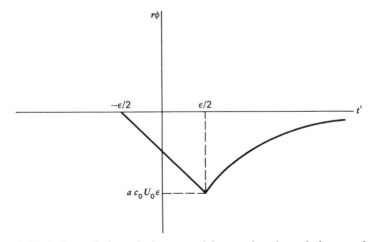

Figure 4.2.3 Variations of the velocity potential as a function of the retarded time $t' = t - r/c_0$.

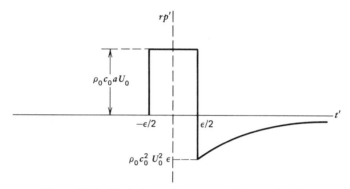

Figure 4.2.4   Variations of pressure with retarded time.

The acoustic pressure is given by

$$p'(r, t' < -\varepsilon/2) = 0$$

$$p'\left(r, -\frac{\varepsilon}{2} \leqslant t' \leqslant \frac{\varepsilon}{2}\right) = \frac{\rho_0 c_0 a U_0}{r} \exp\left[-\frac{c_0}{a}\left(\frac{\varepsilon}{2} + t'\right)\right] \qquad (4.2.30)$$

$$p'\left(r, t' > \frac{\varepsilon}{2}\right) = \frac{2\rho_0 c_0 a U_0}{r} \sinh\left(\frac{c_0 \varepsilon}{2a}\right) e^{-c_0 t'/a}$$

These results are shown schematically (for ε small) in Figure 4.2.4.

Since $t' = t - (r - a)/c_0$, the above results give the variations of $r\phi$ and rp' with time at some fixed location r.

The variations of rp' with distance (for a fixed time $t \gg \varepsilon$, say) can be obtained directly from the above figures, since for a given t', r is given by $r = a + c_0(t - t')$. Thus, for fixed t and with the variations of t' shown graphically, we obtain the variations for rp' with distance displayed in Figure 4.2.5. As time increases, the profile moves to the right without changing its amplitude so that p' has a similar profile but its amplitude decays with r as $1/r$.

These results show clearly that the disturbance, although created by a local compression, contains both compressions and expansions. This is a particular example of a general result for disturbances that are initially confined to a finite region. It applies to both spherical and cylindrical waves, but not to plane waves. To derive this result, we consider the spherically symmetric case. Here,

$$\phi(r, t) = \frac{1}{r} f(c_0 t - r) + \frac{1}{r} g(c_0 t + r) \qquad (4.2.31)$$

Consider again only outgoing waves. Here the pressure and velocity at any point

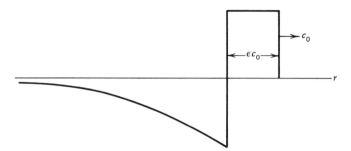

Figure 4.2.5 Variations of rp' with distance.

are

$$p'(r,t)= -\frac{\rho_0 c_0}{r} f'(c_0 t - r) \tag{4.2.32}$$

$$u_r(r,t)= -\frac{1}{r^2} f(c_0 t - r) - \frac{1}{r} f'(c_0 t - r) \tag{4.2.33}$$

Combining these two equations, we obtain

$$\left(\frac{p'}{\rho_0 c_0} - u_r\right) r^2 = f(c_0 t - r) \tag{4.2.34}$$

Now, suppose that at $t = t_0$ the wave is confined to the region inside a spherical shell whose limiting radii are r_1 and r_2, respectively, as shown in Figure 4.2.6. Then (4.2.34) shows that

$$f(c_0 t_0 - r) = 0, \quad \begin{cases} r > r_2 \\ r < r_1 \end{cases} \tag{4.2.35}$$

For $t > t_0$, we have [since $f(c_0 t - r)$ represents an outgoing wave]

$$f(c_0 t - r) = 0, \quad \begin{cases} r > r_2 + c_0(t - t_0) \\ r < r_1 + c_0(t - t_0) \end{cases} \tag{4.2.36}$$

In other words, $f(c_0 t - r) \neq 0$ only inside a shell of thickness $r_2 - r_1$. Consider now the acoustic pressure multiplied by r; it is given by

$$rp' = -\rho_0 c_0 \frac{df}{d(c_0 t - r)} \tag{4.2.37}$$

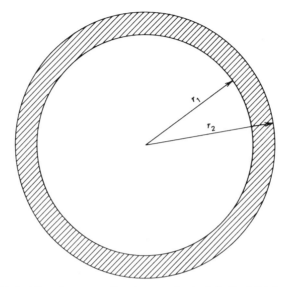

Figure 4.2.6 Disturbance confined to a spherical shell of thickness $r_2 - r_1$.

Holding t constant, we can write

$$rp'(r; t_0) \, dr = \frac{df}{dr} \, dr \cdot \rho_0 c_0 \tag{4.2.38}$$

so that upon integration between any two locations outside the shell $r_1 \leqslant r_2$, we obtain

$$\int_{r<r_1}^{r>r_2} rp' \, dr = 0 \tag{4.2.39}$$

which shows that p' must contain both rarefactions and expansions. Similarly, if we hold r constant in (4.2.37) and integrate with respect to t, between $t_1 < t_0 + (r_0 - r_2)/c_0$ and $t_2 > t_0 + (r_0 - r_1)/c_0$, we obtain

$$\int_{t_1}^{t_2} p' \, dt = 0 \tag{4.2.40}$$

that is, the mean pressure fluctuation is zero.

Initial-Value Problem

Problems in which the initial distributions of pressure and velocity are prescribed can be solved by means of the general solution in terms of some integrals containing the initial values of p' and u_r. We begin with the general solution for the case when both incoming and outgoing waves are present. Thus,

$$-\frac{1}{\rho_0 c_0} p'(r,t) = \frac{1}{r} f'(c_0 t - r) + \frac{1}{r} g'(c_0 t + r) \qquad (4.2.41)$$

and

$$u_r(r,t) = \frac{\partial}{\partial r} \frac{1}{r} \left[f(c_0 t - r) + g(c_0 t + r) \right] \qquad (4.2.42)$$

Now, for $t=0$, $f'(c_0 t - r) = -df/dr$. Therefore, if $P_0(r)$ denotes the initial pressure disturbance distribution, (4.2.41) can be integrated with respect to r to give

$$-f(-r) + g(r) = -\frac{1}{\rho_0 c_0} \int r P_0(r)\, dr \qquad (4.2.43)$$

The integration is over the complete region where $P_0 \neq 0$. Similarly, if $U_0(r)$ represents the initial velocity, we have, from (4.2.42)

$$f(-r) + g(r) = r \int U_0(r)\, dr \qquad (4.2.44)$$

By elimination, we obtain from (4.2.43) and (4.2.44)

$$f(-r) = \frac{1}{2} \left[r \int U_0(r)\, dr + \frac{1}{\rho_0 c_0} \int r P_0(r)\, dr \right] \qquad (4.2.45)$$

and

$$g(r) = \frac{1}{2} \left[r \int U_0(r)\, dr - \frac{1}{\rho_0 c_0} \int r P_0(r)\, dr \right] \qquad (4.2.46)$$

By means of these equations, we can determine the function $g(c_0 t + r)$ for positive values of its argument, and the function $f(c_0 t - r)$ for negative arguments. Now, the argument of g is always positive. Therefore, (4.2.46) is sufficient to determine that function. However, the argument of $f(c_0 t - r)$ may become positive so that an additional equation is needed. Usually such an equation is provided by physical considerations. For example, in the important case when the origin is not a source, the potential there should be finite so that we would impose the additional requirement given by (4.1.6). Thus, if ξ denotes a positive variable, this requirement can be written as

$$f(\xi) = -g(\xi), \quad \xi \geqslant 0 \tag{4.2.47}$$

Since $g(\xi)$ is completely determined by the initial conditions, this equation determines f for positive values of its argument. Therefore, (4.2.45) to (4.2.47) provide the solution to the initial-value problem in a spherically symmetric region that includes the origin. If the origin is not included, other boundary conditions can be used instead of (4.2.47), depending on the geometry of the included region.

It should be pointed out that the outlined solution can also be used for the initial-value problem involving plane, one-dimensional waves. All that is required is deletion of the multiplicative factor r in (4.2.45) and (4.2.46), and to specify suitable boundary conditions. In the plane-wave case, however, a disturbance may consist entirely of compressions or expansions.

Returning to the centrally symmetric case, we point out several results that follow from the solution. Most important, perhaps, is the division of the initial disturbance into two parts—an outgoing wave and an incoming wave. The precise profile of these waves depends on the details of the initial disturbance. Further, each of these waves may contain compressions and expansions, as shown by the following example.

Bursting Balloon.   Consider a spherical mass of gas inside a spherical balloon of radius a. Initially, the gas inside the balloon is at rest, but its pressure exceeds that of the surrounding gas (which has otherwise the same properties) by a small amount Δ. Thus, the initial pressure disturbance is

$$P_0 = p'(r,0) = \begin{cases} \Delta, & r \leqslant a \\ 0, & r > a \end{cases}$$

and the initial velocity $U_0(r)$ is zero. At $t = 0$, the gas is allowed to expand freely, say, by sudden removal of the balloon's membrane. The resulting motion is then prescribed by (4.2.45) and (4.2.46), which give

$$f(-r) = \frac{1}{2\rho_0 c_0} \int r P_0(r)\,dr \tag{4.2.48}$$

$$g(r) = -\frac{1}{2\rho_0 c_0} \int r P_0(r)\,dr \tag{4.2.49}$$

Substituting the value of P_0 and integrating, we obtain

$$f(-r) = \begin{cases} r^2\Delta/4\rho_0 c_0, & 0 \leqslant r \leqslant a \\ a^2\Delta/4\rho_0 c_0, & r > a \end{cases} \tag{4.2.50}$$

$$g(r) = \begin{cases} -r^2\Delta/4\rho_0 c_0, & 0 \leqslant r \leqslant a \\ -a^2\Delta/4\rho_0 c_0, & r > a \end{cases} \tag{4.2.51}$$

Therefore, the function $g(c_0 t + r)$ is given by

$$g(c_0 t + r) = \begin{cases} -\dfrac{\Delta}{4\rho_0 c_0}(c_0 t + r)^2, & 0 \leqslant c_0 t + r \leqslant a \\[2mm] -\dfrac{a^2\Delta}{4\rho_0 c_0}, & a < c_0 t + r \end{cases} \tag{4.2.52}$$

Since we do not have a source at $r = 0$, we must have $f(\xi) = -g(\xi)$ for $\xi \geqslant 0$. Therefore, the above value of $g(c_0 t + r)$ prescribes $f(c_0 t - r)$ when $c_0 t - r \geqslant 0$:

$$f(c_0 t - r) = \begin{cases} \dfrac{\Delta}{4\rho_0 c_0}(c_0 t - r)^2, & 0 \leqslant c_0 t - r \leqslant a \\[2mm] a^2\Delta/4\rho_0 c_0, & a < c_0 t - r \end{cases} \tag{4.2.53}$$

For negative values of its argument, f is given by (4.2.50) so that when $c_0 t - r$ is negative, we have

$$f(c_0 t - r) = \begin{cases} \dfrac{\Delta}{4\rho_0 c_0}(r - c_0 t)^2, & -a \leqslant c_0 t - r \leqslant 0 \\[2mm] a^2\Delta/4\rho_0 c_0, & c_0 t - r \leqslant -a \end{cases}$$

Combining this with (4.2.53), we may write

$$f(c_0 t - r) = \begin{cases} \dfrac{\Delta}{4\rho_0 c_0}(r - c_0 t)^2, & -a \leqslant r - c_0 t \leqslant a \\[2mm] a^2\Delta/4\rho_0 c_0, & a < r - c_0 t < -a \end{cases} \tag{4.2.54}$$

These results show that the function f is variable only within a shell of thickness $2a$. To see corresponding pressure and velocity variations, we first consider field points outside the sphere $r = a$. In this case, $c_0 t + r > a$ for all t so that the

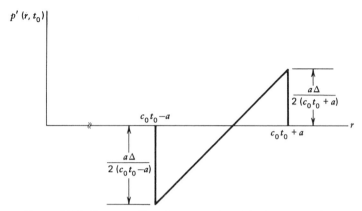

Figure 4.2.7   Pressure disturbance produced by bursting balloon.

function $g(c_0t+r)$ is a constant. The acoustic pressure, given by (4.2.41), is therefore

$$p'(r,t) = \begin{cases} \dfrac{1}{2r}(r-c_0t)\Delta, & -a \leqslant r - c_0t \leqslant a \\ 0, & a < r - c_0t < -a \end{cases} \qquad (4.2.55)$$

This is shown schematically in Figure 4.2.7, where p' is shown as a function of r for a fixed time. For $c_0t_0 - a < r < c_0t_0 + a$, we have an outgoing pulse consisting of a sudden compression, followed by a linear expansion, and terminated by a second sudden compression that brings the pressure back to its original value. An interesting feature of the profile is that the maximum negative pressure is larger than the maximum compression. This difference vanishes for very small initial radius a. In any event, the amplitude of the wave diminishes with distance as $1/r$ as the wave moves out, as is required of spherical waves.

Similarly, Figure 4.2.8 shows the time dependence of the pressure variations at a given distance $r = r_0$ from the origin. Thus, at a given location one would first experience a sudden pressure increase, followed by a linear decrease, and then by a sudden compression of the same magnitude as the initial one, which brings the pressure back to its mean value. This is a typical "N" wave.[2] To understand how this profile was formed, it is necessary to consider points $r \leqslant a$, where g' may have nonzero values. Thus, when c_0t+r has values between 0 and a,

---

[2] Actual N waves have features that are not displayed by our profile. For example, in the present case (ideal and linear), the width of the pulse is constant; its decay depends only on geometrical spreading of its wave-front area; and its recompressive tail is sharp. None of these results is applicable when nonlinearities and dissipation are included.

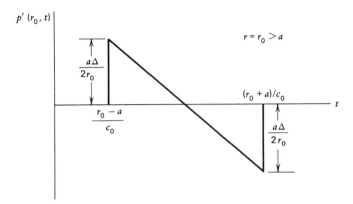

Figure 4.2.8 Variations with time of the acoustic pulse produced by a bursting balloon at a given location.

$g'(c_0t+r)$ is, from (4.2.51),

$$g'(c_0t+r) = -\frac{\Delta}{2\rho_0c_0}(c_0t+r) \tag{4.2.56}$$

Similarly, the function $f'(c_0t-r)$ is

$$f'(c_0t-r) = -\frac{\Delta}{2\rho_0c_0}(r-c_0t), \quad -a \leqslant r-c_0t \leqslant a \tag{4.2.57}$$

Now, the acoustic pressure is given by the sum of terms proportional to f' and g'. We write the total pressure as

$$p'(r,t) = \tfrac{1}{2}\Delta(p'_f + p'_g) \tag{4.2.58}$$

where

$$p'_f = \frac{1}{r}(r-c_0t), \quad -a \leqslant r-c_0t \leqslant a$$

$$p'_g = \frac{1}{r}(r+c_0t), \quad 0 \leqslant r+c_0t \leqslant a$$

Outside the specified ranges, these quantities are zero. Therefore, when $t=0_-$, p'_f and p'_g are equal to unity for values of r between 0 and a, and equal to zero otherwise. This is, of course, the prescribed initial overpressure. At $t=0_+$, the high-pressure mass is suddenly released. On physical grounds, we should expect

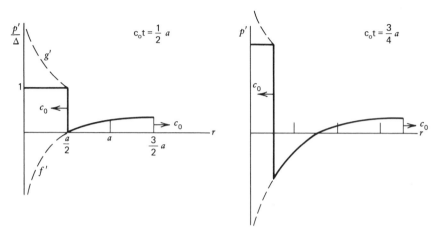

Figure 4.2.9 Incoming and outgoing waves immediately after bursting.

a compression wave moving toward the region $r > a$, where the pressure is lower. Similarly, we should also expect an expansion wave moving toward the region $r < a$, where the pressure is higher. Both waves travel with speed c_0 so that at time t the regions outside the spherical shell limited by the radii $r_0 = c_0 t + a$ and $r_1 = a - c_0 t$ have not yet experienced any changes. This is displayed in Figure 4.2.9, where the nondimensional pressure p'/Δ, as given by the above equations, is plotted versus r for several values of t. We see that the incoming wave is a purely expansive wave, whereas the outgoing wave has a compressive front followed by an expansive tail. These waves interact to form a pressure profile that meets all the physical requirements mentioned above.

The acoustic pressure near the origin is quite interesting, for as Figure 4.2.10 shows, when the incoming wave approaches it, its amplitude increases markedly. Similarly, immediately after the incoming wave has reached the origin, the outgoing wave is seen to have an almost equally large amplitude (but of opposite sign). Obviously, one could say that the incoming expansive wave is being reflected as a compressive wave. However, it remains to explain the origin of the large amplitudes shown as the incoming wave approaches the origin, and as it is reflected. The answer lies simply in the fact that when the wave is converging onto $r = 0$, its wave-front area is decreasing as r^2 so that its amplitude has to increase proportionately in order that the energy of the wave remain constant. This is, of course, merely the opposite effect to that which occurs when a spherical wave spreads outwardly. One has to remember, however, that large amplitudes are inadmissible within the linear model used to derive the results given by (4.2.56) and (4.2.57) above. Also, as our discussion in Section 3.7 shows, discontinuous waves of expansion (i.e., expansive shock waves), such as

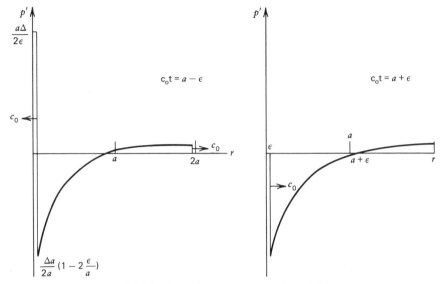

Figure 4.2.10 Acoustic pressure near the origin.

those in Figures 4.2.9 and 4.2.10, do not exist. Such waves necessarily have smooth leading fronts.

PROBLEMS

4.2.1 The radial fluid velocity produced by a harmonically pulsating sphere may be written as

$$\tilde{u}_r = -\frac{A}{r^2}(1 - ikr)e^{i(kr - \omega t)}$$

From this result, determine the distance, in terms of kr, beyond which the amplitude of the second term is equal to at least 99 percent of that given by the equation above. Also, determine the distance below which the first term is at least 99 percent of that given by the full equation.

4.2.2 A sphere of radius a begins to pulsate sinusoidally at $t=0$ with uniform surface velocity $U = U_0 \sin \omega t$. The pulsations end at $t = t_0$. Determine the acoustic field due to this motion.

4.2.3 A small pulsating sphere is placed inside an elastic material of density ρ_1, speed of sound c_1, and radius R. The two spheres $r = a$ and $r = R$ are concentric. The arrangement is immersed in an infinite body of fluid of density ρ_2 and speed of sound c_2. Determine the transmission coefficient for medium 2 in the limit $a \ll R$.

4.2.4 Consider an irrotational flow of an inviscid, incompressible fluid. A velocity potential exists that satisfies Laplace's equation

$$\nabla^2 \phi = 0$$

Show that in the near-field, the velocity potential for the pulsating sphere satisfies this equation.

4.3 AXIALLY SYMMETRIC SPHERICAL WAVES

In the last two sections, we studied some examples of centrally symmetric waves. These were considered first as they provide the simplest examples of waves that do not have plane fronts. However, it is clear that in order to have this type of wave, one must also have centrally symmetric initial and boundary conditions. These conditions impose a severe restriction on the use of our centrally symmetric solution.

In this section, we consider slightly less restrictive types of waves, namely, spherical waves whose properties are symmetric only with respect to a line. Later in the section we will also include a brief discussion of spherical waves that lack even this symmetry.

The axially symmetric case is of interest because it will enable us to treat several important problems. One problem is the interaction of a plane wave with a sphere in an unbounded fluid. The interaction is symmetric with respect to an axis that passes through the center of the sphere and is parallel to the direction of propagation. If we align the axis of symmetry with the polar axis of a spherical-polar system of coordinates, it is clear that such an interaction would be independent of the azimuthal angle φ. On the other hand, it will depend on the polar angle θ, as well as on the distance from the origin. Thus, the potential of a spherical wave that is symmetric with respect to the polar axis satisfies

$$\frac{\partial^2 \phi}{\partial t^2} = c_0^2 \left[\frac{\partial^2 \phi}{\partial r^2} + \frac{2}{r} \frac{\partial \phi}{\partial r} + \frac{1}{r^2 \sin\theta} \frac{\partial}{\partial \theta} \left(\sin\theta \frac{\partial \phi}{\partial \theta} \right) \right] \qquad (4.3.1)$$

For monochromatic waves of frequency ω, this becomes

$$\frac{\partial^2 \tilde{\phi}}{\partial r^2} + \frac{2}{r} \frac{\partial \tilde{\phi}}{\partial r} + \frac{1}{r^2 \sin\theta} \frac{\partial}{\partial \theta} \left(\sin\theta \frac{\partial \tilde{\phi}}{\partial \theta} \right) + k^2 \tilde{\phi} = 0 \qquad (4.3.2)$$

where $\tilde{\phi} = g(r, \theta)e^{-i\omega t}$. If we assume that $g(r, \theta)$ can be written as $g(r, \theta) = R(r)\Theta(\theta)$, then substitution into (4.3.2) gives

$$\frac{1}{R}\left(r^2\frac{d^2R}{dr^2} + 2r\frac{dR}{dr} + r^2k^2R\right) = -\frac{1}{\Theta\sin\theta}\frac{d}{d\theta}\left(\sin\theta\frac{d\Theta}{d\theta}\right) \qquad (4.3.3)$$

By assumption, the left-hand side of this equation depends only on r, whereas the right-hand side depends only on θ. This can be so only if both sides are equal to some constant. Denoting this constant by M^2, we obtain two equations, namely,

$$\frac{d^2R}{dr^2} + \frac{2}{r}\frac{dR}{dr} + \left(k^2 - \frac{M^2}{r^2}\right)R = 0 \qquad (4.3.4)$$

$$\frac{1}{\sin\theta}\frac{d}{d\theta}\left(\sin\theta\frac{d\Theta}{d\theta}\right) + M^2\Theta = 0 \qquad (4.3.5)$$

It may be shown that in order that the last equation have finite solutions for $\theta = 0$ and π, M^2 must be given by

$$M^2 = n(n+1), \quad n = 0, 1, 2, \dots$$

Therefore, the radial part of the solution becomes

$$\frac{d^2R}{dr^2} + \frac{2}{r}\frac{dR}{dr} + \left[k^2 - \frac{n(n+1)}{r^2}\right]R = 0 \qquad (4.3.6)$$

whereas (4.3.5) becomes

$$\frac{1}{\sin\theta}\frac{d}{d\theta}\left(\sin\theta\frac{d\Theta}{d\theta}\right) + n(n+1)\Theta = 0 \qquad (4.3.7)$$

The radial equation is identical to (4.1.15) so that its solutions are spherical Bessel functions. The general solution of (4.3.7) may be expressed as

$$\Theta = A_n P_n(\cos\theta) + B_n Q_n(\cos\theta) \qquad (4.3.8)$$

where $P_n(\cos\theta)$ is the Legendre polynomial of degree n, and $Q_n(\cos\theta)$ is the Legendre function of the second kind, of degree n. However, because $Q_n(\cos\theta)$ is not finite at $\theta = 0$ and $\theta = \pi$, we require only the Legendre polynomials

$P_n(\cos\theta)$. The first few of these polynomials are given by

$$P_0(\cos\theta) = 1$$

$$P_1(\cos\theta) = \cos\theta$$

$$P_2(\cos\theta) = \tfrac{1}{2}(3\cos^2\theta - 1)$$

For higher values of n, $P_n(z)$ may be obtained from

$$(n+1)P_{n+1} - (2n+1)zP_n + nP_{n-1} = 0$$

Several properties of these functions are listed in Appendix E. Here we note that they are orthogonal in the range $-1 \leqslant z \leqslant 1$, and satisfy the normalization condition

$$\int_{-1}^{1} P_n(z)P_m(z)\,dz = \begin{cases} 0, & n \neq m \\ \dfrac{2}{2n+1}, & n = m \end{cases} \qquad (4.3.9)$$

In terms of these functions, and of the solutions to the radial-factor equation given by (4.1.18) and (4.1.19), the monochromatic solution to the wave equation in polar coordinates with azimuthal symmetry can be written as

$$\tilde{\phi}(r,\theta,t) = \sum_{n=0}^{\infty} \left[A_n h_n^{(1)}(kr) + B_n h_n^{(2)}(kr) \right] P_n(\cos\theta)e^{-i\omega t} \qquad (4.3.10)$$

Also, in view of the definitions of $h_n^{(1)}$ and $h_n^{(2)}$, this can be written as

$$\tilde{\phi}(r,\theta,t) = \sum_{n=0}^{\infty} \left[A_n' j_n(kr) + B_n' y_n(kr) \right] P_n(\cos\theta)e^{-i\omega t} \qquad (4.3.11)$$

Both forms are equivalent. The form given by (4.3.10) is convenient to use when a description in terms of propagating waves is desired. Thus, the function $h_n^{(1)}(kr)\exp(-i\omega t)$ represents an outgoing wave, whereas $h_n^{(2)}(kr)\exp(-i\omega t)$ represents an incoming one.

Standing Waves in a Spherical Cavity

We reconsider the problem of standing waves in a spherical cavity. The problem was discussed in Section 4.1 by means of the spherically symmetric solution. At present, we would like to determine the lowest-frequency axisymmetric mode of oscillation in the cavity; in order to do this, we make use of the solution obtained above for the velocity potential. In this case, (4.3.11) is more convenient than (4.3.10). Thus, since y_n is not finite at $r=0$, we set $B'_n=0$ and obtain

$$\tilde{\phi}(r,\theta,t)= \sum_{n=0}^{\infty} A'_n j_n(kr) P_n(\cos\theta) e^{-i\omega t} \qquad (4.3.12)$$

The term with $n=0$ gives the spherically symmetric case, for which it was found that the lowest allowable value of ka was given by $ka=4.49$. We now consider the term with $n=1$. This is explicitly given by

$$\tilde{\phi}_1 = A_1 j_1(kr) \cos\theta\, e^{-i\omega t} \qquad (4.3.13)$$

As in the spherically symmetric case, the allowable values of k are to be determined from the condition that the radial velocity vanishes as $r=a$:

$$\left(\frac{\partial\phi}{\partial r}\right)_{r=a} = 0$$

This gives

$$j'_1(ka)=0 \qquad (4.3.14)$$

From (4.1.21), we have

$$j_1(z)= \frac{1}{z^2}(\sin z - z\cos z)$$

so that (4.3.14) becomes

$$\frac{\sin ka}{ka} + \frac{2\cos ka}{(ka)^2} = \frac{2\sin ka}{(ka)^3} \qquad (4.3.15)$$

Rearranging this, we can write

$$\tan ka = \frac{2ka}{2-(ka)^2} \qquad (4.3.16)$$

This is an implicit equation from which the allowed values of ka can be found. Its lowest root is $ka=0$. The next is

$$k_1 a = 2.08\cdots$$

and is lower than the one found for centrally symmetric oscillations.

The pressure and velocity components corresponding to this root are given by

$$p'(r, \theta, t) = -\rho_0 \omega A j_1(k_1 r) \cos\theta \sin\omega t \qquad (4.3.17)$$

$$u_r(r, \theta, t) = A k_1 j_1'(kr) \cos\theta \cos\omega t \qquad (4.3.18)$$

$$u_\theta(r, \theta, t) = -\frac{A}{r} j_1(kr) \sin\theta \cos\omega t \qquad (4.3.19)$$

To get an idea of what the oscillation is like, we plot in Figure 4.3.1 the variations of pressure, and velocity magnitude $u = \sqrt{u_r^2 + u_\theta^2}$, at the instant when these quantities have their largest values. Figures 4.3.1a shows the acoustic pressure on a meridional plane of the spherical cavity. It is seen that the pressure has a nodal line on $\theta = \pi/2$, and that for a given value of θ (other than $\pi/2$), the magnitude of the pressure increases with r.

Figure 4.3.1b shows lines having constant fluid velocity magnitude. These lines are given by $u = $ constant, where

$$u(r, \theta, t) = Ak \left\{ \left[j'(k_1 r) \cos\theta \right]^2 + \left[\frac{1}{k_1 r} j_1(k_1 r) \sin\theta \right]^2 \right\}^{1/2} \cos\omega t \quad (4.3.20)$$

The figure shows u at a time when $\cos\omega t = 1$ (when $p' = 0$). At other times, the motion subsides in magnitude but the constant-magnitude lines remain the same. The numbers on those lines indicate the relative magnitude of the velocity. As one might have anticipated, the locations of maximum velocity magnitude generally correspond to those of minimum pressure.

Rigid Sphere in a Sound Wave

As a second example, we consider a rigid sphere suspended freely in a fluid in which a plane monochromatic wave is being propagated. To simplify the problem, gravity effects are not included. The interaction results in two effects that are of interest. The first effect is the oscillatory motion that the sphere will acquire under the influence of the waves. The second is the spreading of some of the incident energy into all directions around the sphere. Referring to Figure 4.3.2, it is clear that in the vicinity of the polar axis, the sphere will reflect some of the incident energy into a direction opposite to that of the incident. At other points, however, the reflection will take place at other angles. The net result is a complicated pattern of waves emanating from the sphere. These are usually referred to as *scattered waves*, and the process that produces them is called *scattering*. Scattering results in a net loss of energy from the incident wave, a loss that may be of interest in some applications.

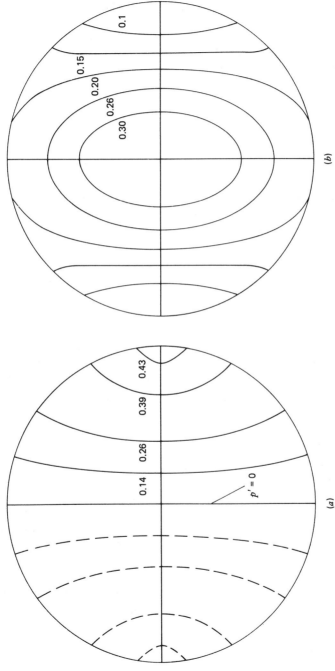

Figure 4.3.1 Variations of pressure and velocity magnitudes in a spherical cavity for the case of lowest frequency of oscillation. (a) Acoustic pressure; (b) fluid velocity.

229

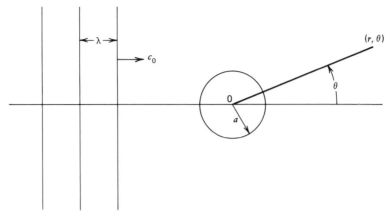

Figure 4.3.2 Plane wave falling on a sphere.

The sphere problem can be studied best by using a system of coordinates fixed on the center of the sphere and moving with it. However, since such a system is accelerating, we must add to the right-hand side of the fluid's equation of motion a fictitious force equal to $-\rho(dU/dt)$, where U is the velocity of the accelerating frame. (This is very much like adding Coriolis and centrifugal forces to the equations of motion, when the frame is rotating.) In the present case, U is the velocity of the sphere and will be denoted by U_p. Thus, with respect to a system of coordinates moving with the sphere, the linearized Euler equation becomes

$$\rho_0 \frac{\partial \mathbf{v}}{\partial t} + \nabla p' = -\rho_0 \frac{d\mathbf{U}_p}{dt} \tag{4.3.21}$$

where $\mathbf{v}$ is the fluid's velocity as seen by an observer moving with the sphere. For such an observer, the boundary condition on the sphere's surface, $r=a$, is

$$\mathbf{v} \cdot \mathbf{n} = 0 \tag{4.3.22}$$

where $\mathbf{n}$ is a unit normal vector pointing into the fluid. If we denote by $\mathbf{u}$ the fluid velocity as seen by a fixed observer whose position coincides instantaneously with the center of the sphere, then

$$\mathbf{u} = \mathbf{v} + \mathbf{U}_p \tag{4.3.23}$$

so that in terms of $\mathbf{u}$, (4.3.21) can be written as

$$\rho_0 \frac{\partial \mathbf{u}}{\partial t} + \nabla p' = 0 \tag{4.3.24}$$

This is our original Euler's equation; when combined with the continuity equation (which is not affected by the above transformation), it yields the wave equation. The only difference is that the boundary condition on **u** is

$$\mathbf{u} \cdot \mathbf{n} = \mathbf{U}_p \cdot \mathbf{n}, \quad r = a \tag{4.3.25}$$

The velocity potential satisfies (4.3.1), which is a linear equation. Therefore, ϕ can be written as the sum of two contributions:

$$\phi = \phi_i + \phi_s \tag{4.3.26}$$

The first, ϕ_i, is the potential of the incident waves. The second, ϕ_s, is the potential of the scattered waves. Since the incident wave is a plane, one-dimensional wave, $\tilde{\phi}_i$ is given simply by

$$\tilde{\phi}_i = A e^{i(kx - \omega t)} \tag{4.3.27}$$

where A is its amplitude and ω its frequency, and its direction of propagation x coincides with the polar axis of a system of coordinates.

The scattered-wave potential is given by either (4.3.10) or (4.3.11), but since the fluid surrounding the sphere is unbounded, it is convenient to use the former equation and set the coefficient of $h_n^{(2)}$ (which represents incoming waves) equal to zero. Thus,

$$\tilde{\phi}_s = \sum_{n=0}^{\infty} C_n' h_n^{(1)}(kr) P_n(\cos\theta) e^{-i\omega t} \tag{4.3.28}$$

where the coordinates r and θ are measured in a plane passing through the center of the sphere, as shown in Figure 4.3.2. The quantities C_n' are to be determined from the condition given by (4.3.25). Since the sphere's motion will be in the direction of propagation, we can write

$$\mathbf{U}_p = \mathbf{i}\tilde{U}_{p0} e^{-i\omega t} \tag{4.3.29}$$

where **i** is a unit vector along x. Since $\mathbf{i} \cdot \mathbf{n} = \cos\theta$, we can write that condition in terms of the velocity potentials as

$$\left(\frac{\partial \tilde{\phi}_i}{\partial r}\right)_{r=a} + \left(\frac{\partial \tilde{\phi}_s}{\partial r}\right)_{r=a} = \tilde{U}_{p0} \cos\theta\, e^{-i\omega t} \tag{4.3.30}$$

The incident-wave potential, as given by (4.3.27), is not suitable for substitution into this equation. It is more convenient to express it in terms of spherical harmonics. The required expansion is (with $x = r\cos\theta$)

$$e^{ikr\cos\theta} = \sum_{n=0}^{\infty} i^n (2n+1) j_n(kr) P_n(\cos\theta) \tag{4.3.31}$$

Denoting, for convenience, the quantities C_n' by

$$C_n' = C_n i^n (2n+1) \tag{4.3.32}$$

and dropping the superscript on $h_n^{(1)}$, the boundary condition yields

$$\sum_{n=0}^{\infty} k i^n (2n+1) \left[A j_n'(ka) + C_n h_n'(ka) \right] P_n(\cos\theta) = \tilde{U}_{p0} \cos\theta \tag{4.3.33}$$

For $n = 1$, the explicit forms for $P_n(\cos\theta)$ show that $P_1(\cos\theta) = \cos\theta$. Hence, C_1 is such that

$$A j_1'(ka) + C_1 h_1'(ka) = \frac{\tilde{U}_{p0}}{3ik} \tag{4.3.34}$$

For $n \neq 1$, each of the terms appearing inside the square brackets in the infinite series given by (4.3.33) has to vanish. Hence,

$$C_n = -A \frac{j_n'(ka)}{h_n'(ka)}, \quad n \neq 1 \tag{4.3.35}$$

However, these are not sufficient to specify the solution to the problem, as U_{p0} also is unknown. Another equation is therefore needed. This is provided by the sphere's equation of motion, which in general form is

$$m_s \frac{d\mathbf{U}_p}{dt} = \mathbf{F} + \mathbf{F}_e \tag{4.3.36}$$

where m_s is the mass of the sphere, $\mathbf{F}$ is the force with which the fluid acts on the sphere, and $\mathbf{F}_e$ is the external force (e.g., gravity). In this case, $\mathbf{F}_e = 0$, and since the fluid is inviscid, $\mathbf{F}$ is given by

$$\mathbf{F} = -\int_{A_s} p\mathbf{n} \, dA \tag{4.3.37}$$

where A_s represents the area of the sphere. The pressure is $p = p_0 + p'$, but p_0 contributes nothing to the force on the sphere. The force is due to p', which for monochromatic waves is given by

$$\tilde{p}' = i\rho_0\omega\tilde{\phi} \qquad (4.3.38)$$

Thus, taking the x component of (4.3.36) and (4.3.37), we obtain, after substitution from (4.3.38),

$$\tilde{U}_{p0}e^{-i\omega t} = \frac{\rho_0}{m_s}\int_{A_s}\tilde{\phi}\cos\theta\,dA \qquad (4.3.39)$$

with $dA = 2\pi a^2\sin\theta\,d\theta$. Now, $\tilde{\phi} = \tilde{\phi}_i + \tilde{\phi}_s$, where, except for the time factor, $\tilde{\phi}_i$ is given by (4.3.31) and where $\tilde{\phi}_s$ is given by (4.3.28) and (4.3.32). Substituting these values into (4.3.39), and observing that because of the orthogonality conditions given by (4.3.9) for $P_n(\cos\theta)$, the only term in ϕ that contributes to the integral is that with $n=1$, we obtain

$$\frac{\tilde{U}_{p0}}{3ik} = \frac{2\pi a^2\rho_0}{km_s}\left[Aj_1(ka) + C_1h_1(ka)\right]\int_0^\pi \cos^2\theta\sin\theta\,d\theta \qquad (4.3.40)$$

The numerical value of the integral is 2/3, so that the quantity multiplying the square brackets can be written as

$$\frac{4\pi a^2\rho_0}{3km_s} = \frac{\delta}{ka} \qquad (4.3.41)$$

where

$$\delta = \rho_0/\rho_p \qquad (4.3.42)$$

and where ρ_p is the density of the sphere. Thus,

$$\frac{\tilde{U}_{p0}}{3ik} = \frac{\delta}{ka}\left[Aj_1(ka) + C_1h_1(ka)\right] \qquad (4.3.43)$$

To find C_1, we equate (4.3.34) and (4.3.43) and obtain

$$C_1 = -A\frac{bj_1'(b) - \delta j_1(b)}{bh_1'(b) - \delta h_1(b)} \qquad (4.3.44)$$

where

$$b = ka \qquad (4.3.45)$$

Since the amplitude A of the incident wave is known, the above result for C_1, together with (4.3.35) for the other values of C_n, formally completes the solution. In particular, C_1 suffices to determine U_{p0}. However, it is given in a form that is not very useful. A more tractable form can be obtained by expressing the spherical Bessel functions appearing in (4.3.44) in terms of elementary functions. To do this, we first express the derivatives of $j_1(b)$ and of $h_1(b)$ by means of

$$f_1'(b) = f_0(b) - \frac{2}{b} f_1(b) \tag{4.3.46}$$

where f represents either j_1 or h_1. With these changes, (4.3.44) becomes

$$C_1 = -A \frac{bj_0(b) - (2+\delta)j_1(b)}{bh_0(b) - (2+\delta)h_1(b)} \tag{4.3.47}$$

All the functions appearing here can be expressed in terms of elementary functions. Thus, from (4.1.20) to (4.1.23), we have

$$\left. \begin{array}{l} bj_0(b) = \sin b \\ b^2 j_1(b) = (\sin b - b\cos b) \end{array} \right\} \tag{4.3.48}$$

$$\left. \begin{array}{l} bh_0(b) = -ie^{ib} \\ b^2 h_1(b) = -(b+i)e^{ib} \end{array} \right\} \tag{4.3.49}$$

With these substitutions, our equation for C_1 becomes

$$C_1 = -A \frac{b^2 \sin b - (2+\delta)(\sin b - b\cos b)}{(2+\delta)b + i(2+\delta - b^2)} e^{-ib} \tag{4.3.50}$$

Returning to (4.3.43) and substituting this value of C_1, we obtain

$$\tilde{U}_{p0} = 3ik \frac{\delta}{b^3} A \left[\sin b - b\cos b + \frac{b^2 \sin b - (2+\delta)(\sin b - b\cos b)}{(2+\delta)b + i(2+\delta - b^2)}(b+i) \right] \tag{4.3.51}$$

After some algebra, this can be simplified to

$$\tilde{U}_{p0} = 3ik\delta A \frac{(\sin b + i\cos b)}{(2+\delta)b + i(2+\delta - b^2)} \tag{4.3.52}$$

or

$$\tilde{U}_{p0} = 3ik\delta A \frac{e^{-ib}}{(2+\delta-b^2)+i(2+\delta)b} \tag{4.3.53}$$

If the incident wave has a fluid velocity amplitude U_0, the quantity A is given by

$$A = U_0/ik \tag{4.3.54}$$

Therefore,

$$\frac{\tilde{U}_{p0}}{U_0} = \frac{3}{2}\frac{\delta}{1+\delta/2} \frac{e^{-i(b+\eta)}}{\sqrt{b^2+(1-b^2/(2+\delta))^2}} \tag{4.3.55}$$

where

$$\eta = \tan^{-1}\frac{b}{1-b^2/(2+\delta)} \tag{4.3.56}$$

The amplitude of the sphere's velocity oscillations is given by the real part of (4.3.55). A quantity related to this, and simpler to plot, is the magnitude of $\tilde{U}_{p0}$. Thus,

$$\left|\frac{\tilde{U}_{p0}}{U_0}\right| = \frac{3}{2}\frac{\delta}{1+\delta/2} \frac{1}{\left[b^2+(1-b^2/(2+\delta))^2\right]^{1/2}} \tag{4.3.57}$$

Let us consider the limits $b \ll 1$ and $b \gg 1$ separately. For $b = ka \ll 1$, we have

$$\left.\begin{array}{c} \left|\dfrac{\tilde{U}_{p0}}{U_0}\right| \approx \dfrac{3}{2}\dfrac{\delta}{1+\delta/2} \\[2mm] \eta \approx (ka) \end{array}\right\} \quad ka \ll 1 \tag{4.3.58}$$

In this limit, the sphere moves almost in phase with the fluid, but with a different amplitude. This amplitude depends only on the density ratio δ. Thus, for heavy spheres in air, δ is very small so that $|\tilde{U}_{p0}/U_0| \ll 1$. In the opposite limit, say gas bubbles in water,[3] $\delta \gg 1$, so that $|\tilde{U}_{p0}| \simeq 3U_0$.

---

[3]Provided the bubbles are sufficiently small so that they may be considered rigid.

The result given by (4.3.58) can be derived move easily using the theory of irrotational, incompressible flows. The reasons for this theory to be applicable in acoustics when ka is very small may be seen as follows. The quantity ka is basically the ratio of the sphere's diameter to the wavelength. Now, in a sound wave, the density, pressure, and so on change appreciably over a distance comparable to λ. Therefore, the changes of these quantities over a distance a, relative to the maximum change, are of the order of ka, and when $ka \ll 1$ they can be neglected. That is, the flow near the sphere is then nearly incompressible. The situation is similar to that discussed in Problem 4.2.4.

Returning to our result for $|U_{p0}/U_0|$, we consider the opposite limit, namely, $ka \gg 1$. Here we obtain

$$\left. \begin{array}{c} \left| \dfrac{\tilde{U}_{p0}}{U_0} \right| \approx \dfrac{3\delta}{(ka)^2} \\[2mm] \eta \approx \pi \end{array} \right\} \quad ka \gg 1 \qquad (4.3.59)$$

Thus, for ka large, the sphere's velocity amplitude (and also its displacement amplitude) is very small compared with the fluid's velocity amplitude. This was to be expected, as in this limit the sphere is too large for the wave to move it appreciably.

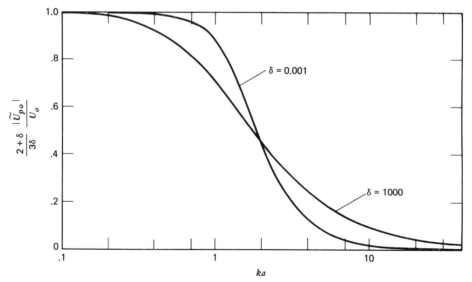

Figure 4.3.3 Changes with frequency of the ratio of a sphere's velocity amplitude to fluid velocity amplitude.

For a fixed value of a, (4.3.58) and (4.3.59) give the low- and high-frequency limits. For intermediate frequencies, (4.3.57) has to be used. This equation is displayed graphically in Figure 4.3.3.

It should be added that these results for $|\tilde{U}_{p0}/U_0|$ are of limited validity because they were obtained under the assumption that viscous effects can be ignored. As we will see in Chapter 6, this is generally not possible.

Scattering by a Sphere

We now turn the problem around and ask: What are the effects of the sphere on the incoming waves? As remarked earlier, the sphere will spread some of the incident energy in all directions. In addition, since the sphere moves in response to the incident waves, it will produce, due to its motion, waves whose direction of propagation does not, in general, coincide with the direction of the incident wave.

The intensity and power of the scattered waves are of interest, and may be obtained from

$$\tilde{\phi}_s = \sum_{n=0}^{\infty} i^n(2n+1)C_n h_n(kr)P_n(\cos\theta)e^{-i\omega t} \tag{4.3.60}$$

This gives, for the scattered-wave pressure,

$$\tilde{p}_s' = i\omega\rho_0 \sum_{n=0}^{\infty} i^n(2n+1)C_n h_n(kr)P_n(\cos\theta)e^{-i\omega t} \tag{4.3.61}$$

The two velocity components u_r and u_θ are given by (4.1.2). Thus, the tangential velocity due to the scattered waves is

$$\tilde{u}_{\theta s} = \sum_{n=0}^{\infty} i^n(2n+1)C_n h_n(kr)(dP_n/d\theta)e^{-i\omega t} \tag{4.3.62}$$

By definition,

$$\frac{dP_n(\cos\theta)}{d\theta} = P_n^1(\cos\theta) \tag{4.3.63}$$

where $P_n^1(\cos\theta)$ is the associated Legendre polynomial of order 1 and degree n. Hence,

$$\tilde{u}_{\theta s} = \sum_{n=0}^{\infty} i^n(2n+1)C_n h_n(kr)P_n^1(\cos\theta)e^{-i\omega t} \tag{4.3.64}$$

The radial velocity component in the scattered wave is

$$\tilde{u}_{rs} = k \sum_{n=0}^{\infty} i^n (2n+1) C_n h'_n(kr) P_n(\cos\theta) e^{-i\omega t} \tag{4.3.65}$$

Consider the radial scattered intensity. This is given by $\langle u_{rs} p'_s \rangle$ or, since the waves are monochromatic, by $\frac{1}{2} \text{Re}(\tilde{u}_{rs} \tilde{p}'^*_s)$. Therefore,

$$I_s = \frac{1}{2}\rho_0 \omega k \, \text{Re} \sum_n \sum_m (2n+1)(2m+1) i^{m-n} P_n(\cos\theta)$$

$$\times P_m(\cos\theta) C_m C_n^* \left[-i h'_m(kr) h_n^*(kr) \right] \tag{4.3.66}$$

The associated power may be obtained by integrating I_s over the area of a sphere of radius r:

$$\Pi_s = 2\pi r^2 \int_0^\pi I_s \sin\theta \, d\theta \tag{4.3.67}$$

but in view of (4.3.9),

$$\int_0^\pi P_n(\cos\theta) P_m(\cos\theta) \sin\theta \, d\theta = \begin{cases} 0, & m \neq n \\ \dfrac{2}{2n+1}, & m = n \end{cases} \tag{4.3.68}$$

Hence,

$$\Pi_s = 2\pi\rho_0 \omega k \sum_{n=0}^{\infty} (2n+1)|C_n|^2 r^2 \text{Re}\left[-i h'_n(kr) h_n^*(kr) \right] \tag{4.3.69}$$

The real part of the quantity in square brackets may be evaluated as follows. First, eliminate h'_n by means of

$$h'_n(z) = h_{n-1}(z) - \frac{n+1}{z} h_n(z) \tag{4.3.70}$$

Then, since by definition $h_n = j_n + i y_n$, we have

$$h'_n h_n^* = (j_n - i y_n)(j_{n-1} + i y_{n-1}) - \frac{n+1}{z}(j_n + i y_n)(j_n - i y_n)$$

$$= j_n j_{n-1} + y_n y_{n-1} - \frac{n+1}{z}(j_n^2 + y_n^2) + i(j_n y_{n-1} - y_n j_{n-1}) \tag{4.3.71}$$

For real values of their arguments, the functions $j_n(z)$ and $y_n(z)$ are real. Therefore,

$$\text{Re}\left[-ih'_n(kr)h_n^* \right] = j_n y_{n-1} - y_n j_{n-1} \qquad (4.3.72)$$

The value of the quantity on the right-hand side is simply $1/(kr)^2$. Consequently, the scattered power is

$$\Pi_s = 2\pi\rho_0 c_0 \sum_{n=0}^{\infty} (2n+1)|C_n|^2 \qquad (4.3.73)$$

A related quantity is the *scattering cross section* σ_s. This is defined as that area of the incident beam that transmits a power equal to the scattered power:

$$\sigma_s I_{\text{inc}} = \Pi_s \qquad (4.3.74)$$

Since the incident wave is plane, $I_{\text{inc}} = \frac{1}{2}\rho_0 k\omega|A|^2$ so that

$$\sigma_s = 4\pi a^2 \sum_{n=0}^{\infty} \frac{(2n+1)}{b^2} \frac{|C_n|^2}{|A|^2} \qquad (4.3.75)$$

This completes the solution. However, before it can be used, the infinite series must be evaluated, and this is generally quite involved, owing primarily to the complexity of the coefficients C_n. These depend strongly on $b=ka$, as (4.2.5) and (4.3.44) show. For example, from (4.2.35), (4.3.48), and (4.3.49), we have

$$C_0 = A\frac{b\cos b - \sin b}{b-i} e^{-ib} \qquad (4.3.76)$$

so that

$$|C_0|^2 = |A|^2 \frac{(b\cos b - \sin b)^2}{1+b^2} \qquad (4.3.77)$$

Similarly, C_1 [given by (4.3.50)] may, after some algebra, be written as

$$C_1 = -\frac{A}{2}\left[1 + \frac{(2+\delta)b - i(2+\delta-b^2)}{(2+\delta)b + i(2+\delta-b^2)} \right] \qquad (4.3.78)$$

If we use the angle η defined in (4.3.56), this can also be written as

$$C_1 = -\tfrac{1}{2}A\left[1 - e^{-2i(b-\eta)} \right] \qquad (4.3.79)$$

Therefore,

$$|C_1|^2 = |A|^2 \sin^2(b-\eta) \tag{4.3.80}$$

For n greater than one, the algebraic complexity of the C_n's increases considerably, and it is not possible to write them out as compactly as C_0 and C_1. For general values of $b(=ka)$, one must therefore evaluate (4.3.75) numerically. However, when ka is very small or very large, it is possible to find suitable approximations. Thus, when ka is very small, the leading terms in (4.3.75) are those with $n=0$ and $n=1$. In that limit, our results above give

$$C_0 = -(i/3)Ab^3 \tag{4.3.81}$$

For C_1, we note that as $b \to 0$, $b - \eta \simeq \frac{1}{6}\{(\delta-1)/[\delta/2+1]\}b^3$. Therefore,

$$C_1 = -\frac{i}{6}A\frac{\delta-1}{\delta/2+1}b^3 \tag{4.3.82}$$

The largest contribution to (4.3.75) arising from the term $n=2$ can be shown to be proportional to b^{10}, and is therefore negligible when compared to $|C_0|^2$ and $|C_1|^2$ when $b \to 0$. Therefore, in the limit as $b \to 0$, the scattering cross section becomes

$$\frac{\sigma_s}{\pi a^2} = \frac{4}{9}\left[1 + \frac{3}{4}\left(\frac{\delta-1}{\delta/2+1}\right)^2\right](ka)^4 \tag{4.3.83}$$

Thus, when $ka \ll 1$, σ_s is a very small fraction of the sphere's cross-sectional area. It should be noticed that although the contributions of the $n=0$ and $n=1$ terms are of the same order of magnitude, the contribution of the $n=1$ term vanishes when the density of the sphere is equal to the density of the fluid. This is as it should be because the $n=1$ term has its origin in the motion of the sphere, relative to the fluid, and this relative motion vanishes when the density of the sphere is equal to that of the fluid [see Equation (4.3.58)].

We now return to Equation (4.3.66), which gives the scattered intensity as a function of position. When $ka \to 0$, the sums in that equation may also be evaluated without difficulty. Thus, when $ka \to 0$, the leading terms in each sum are those proportional to C_0 and C_1 so that

$$I_s = \frac{1}{2}\rho_0\omega k \, \mathrm{Re}\{-i[C_0^* h_0^* - 3i\cos\theta C_1^* h_1^*(kr) + \cdots]$$

$$\times [C_0 h_0'(kr) + 3i\cos\theta C_1 h_1'(kr) + \cdots]\} \tag{4.3.84}$$

where we have used the explicit values of $P_0(\cos\theta)$ and of $P_1(\cos\theta)$. Introducing the shorthand notation

$$\gamma_{mn} = -ih'_m(kr)h_n^*(kr) \tag{4.3.85}$$

we can write (4.3.84) as

$$I_s = \tfrac{1}{2}\rho_0\omega k\,\mathrm{Re}\big\{|C_0|^2\gamma_{00} + 9|C_1|^2\gamma_{11}\cos^2\theta$$

$$+ 3i\cos\theta\big[C_1C_0^*\gamma_{10} - C_0C_1^*\gamma_{01}\big]\big\} \tag{4.3.86}$$

The real parts of γ_{00} and γ_{11} have already been evaluated. They are

$$\mathrm{Re}(\gamma_{00}) = \mathrm{Re}(\gamma_{11}) = 1/(kr)^2$$

The remaining quantities γ_{mn} may be evaluated easily in the far-field. Thus, for $kr\to\infty$, the explicit expressions for $h_0, h'_0; h_1, h'_1$ given in Appendix D yield

$$i\gamma_{10} = 1/(kr)^2, \qquad i\gamma_{01} = -1/(kr)^2$$

Finally, the approximations for C_0 and C_1, valid for $ka\to 0$, are given by (4.3.81) and (4.3.82). When these are combined with the above expressions for γ_{10} and γ_{01}, they give

$$C_1C_0^*i\gamma_{10} - C_0C_1^*i\gamma_{01} = \frac{|A|^2b^6}{9(kr)^2}\frac{\delta-1}{\delta/2+1} \tag{4.3.87}$$

Substitution of these results in (4.3.84) yields

$$I_s = \frac{\rho_0 c_0}{2r^2}|A|^2b^6\left[\frac{1}{9} + \frac{1}{4}\left(\frac{\delta-1}{\delta/2+1}\right)^2\cos^2\theta + \frac{1}{3}\left(\frac{\delta-1}{\delta/2+1}\right)\cos\theta\right] \tag{4.3.88}$$

If this is normalized with the intensity of the incident wave, $I_{inc} = \tfrac{1}{2}\rho_0\omega k|A|^2$, we obtain

$$\frac{I_s}{I_{inc}} = \frac{1}{9}\left(\frac{a}{r}\right)^2(ka)^4\left[1 + \frac{3}{2}\left(\frac{\delta-1}{\delta/2+1}\right)\cos\theta\right]^2 \tag{4.3.89}$$

The angular dependence of this scattered intensity displays (through the density ratio δ) the effects of the sphere motion. Thus, when $\delta = 1$, the sphere moves with the fluid, and (4.3.89) shows that the amplitude of the scattered intensity is constant on any spherical surface of large radius around the sphere. When δ decreases, the sphere's velocity also decreases, and vanishes when $\delta=0$. In that limit, the scattered intensity has a maximum value in the $\theta=\pi$ direction. This is opposite to the direction of the incident wave. In the opposite case, when $\delta\gg1$, the maximum occurs in the forward direction. These limiting values of I_s/I_{inc} are displayed in Figure 4.3.4. It should be remembered, however, that these curves apply only to the case $ka\ll1$. As ka increases, the complexity of the intensity pattern increases, and as pointed out earlier, it is necessary to resort to numerical techniques to compute the scattered intensity and power. However, in the limit $ka\to\infty$, it is again possible to obtain limiting expressions for these quantities. Here we consider only the limiting form of the scattering cross section σ_s. When ka is very large, the sphere does not move so that all of the coefficients C_n may be obtained from (4.3.35):

$$C_n = -A\frac{j_n'(b)}{h_n'(b)}, \quad n=0,1,2,\dots \tag{4.3.90}$$

For large values of $b(=ka)$, the spherical Bessel functions oscillate rapidly, and it is difficult to see from (4.3.90) the limiting form of C_n. It is therefore convenient to introduce some functions $\delta_n'(b)$, called the *phase angles*, that may be used to evaluate numerically the coefficients C_n. First, we note that j_n' may be written as

$$j_n' = \tfrac{1}{2}(h_n' + h_n'^*)$$

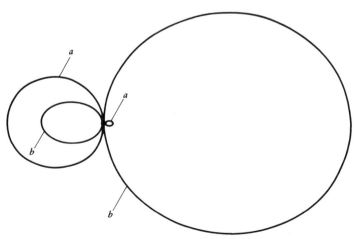

Figure 4.3.4 Distribution of scattered-wave intensity around a sphere. (a) $\delta\ll1$; (b) $\delta\gg1$.

where h_n^* is the complex conjugate of h_n. This gives

$$C_n = -\tfrac{1}{2}A\left(1+\frac{h_n'^*}{h_n'}\right)$$

(4.3.91)

The ratio $h_n'^*/h_n'$ is a ratio of complex-conjugate quantities. Therefore, its amplitude is unity so that we can write

$$\frac{h_n'^*}{h_n'} = -e^{-2i\delta_n'}$$

(4.3.92)

where the phase angle δ_n' depends on n and on ka. This gives

$$|C_n|^2 = |A|^2 \sin^2\delta_n'$$

(4.3.93)

Substituting this into (4.3.75), we obtain

$$\sigma_s = 4\pi a^2 \frac{1}{b^2} \sum_{n=0}^{\infty} (2n+1)\sin^2\delta_n'$$

(4.3.94)

If the sphere is at rest, this result applies whatever ka is. If the sphere is not held fixed, (4.3.94) still applies for all ka provided that for $n=1$ we use $\delta_1' = b - \eta$ [see Equation (4.3.80)].

Values of δ_n' may be computed from information on the spherical Bessel functions given in Chapter 10 of Abramowitz and Stegun. Also, values of $\delta_n(x)$ for $n=2$ to $n=9$ and for values of x up to 5 are given in Table XV of Morse and Feshbach. For $ka\to\infty$, we may use the asymptotic approximations for δ_n' given in Appendix D. These show that for $x>n$, the values of $\delta_n'(x)$ are relatively large, and that for $n>x$, their magnitude is very small. We may therefore approximate the sum (4.3.94) by

$$\sigma_s \simeq 4\pi a^2 \frac{1}{b^2} \sum_{n=0}^{b} (2n+1)\sin^2\delta_n'(b)$$

(4.3.95)

Further, since b is large, $\sin^2\delta_n'$ oscillates rapidly between 0 and 1 so that is may be replaced by its average value of $\tfrac{1}{2}$. Thus,

$$\sigma_s = 2\pi a^2 \frac{1}{b^2} \sum_{n=0}^{b} (2n+1) = 2\pi a^2 \frac{(b+1)^2}{b^2}$$

(4.3.96)

But since b is large, we can write

$$\sigma_s \approx 2\pi a^2, \qquad b\to\infty$$

(4.3.97)

This is an interesting result, for it shows that the scattering cross section is *twice* the cross-sectional area of the sphere. Stated differently, a sphere of cross-sectional area S will remove from the incident wave all the energy contained in a tube of cross-sectional area $2S$. This somewhat puzzling result, known as Babinet's principle in the theory of optics, arises because when $ka \gg 1$, the sphere is reflecting the intercepted energy back to the source, and is producing a scattered intensity field in the forward direction of exactly the same magnitude as that which is reflected.

These effects are easier to observe with light than with sound because of the much smaller wavelengths of the visible part of the electromagnetic wave spectrum.

Arbitrary Spherical Waves

We conclude this section with a brief discussion of monochromatic spherical waves having no simplifying symmetries: spherical waves that depend on all three spatial coordinated r, θ, and φ. The potential for such waves satisfies (4.1.1), which for monochromatic waves becomes

$$\frac{1}{r^2} \frac{\partial}{\partial r}\left(r^2 \frac{\partial \tilde{\phi}}{\partial r} \right) + \frac{1}{r^2 \sin\theta} \frac{\partial}{\partial \theta}\left(\sin\theta \frac{\partial \tilde{\phi}}{\partial \theta} \right) + \frac{1}{r^2 \sin^2\theta} \frac{\partial^2 \tilde{\phi}}{\partial \varphi^2} + k^2 \tilde{\phi} = 0 \quad (4.3.98)$$

The equation is separable so that if we put $\tilde{\phi} = f(r)g(\theta)h(\varphi)$, we obtain

$$r^2 \frac{d^2 f}{dr^2} + 2r \frac{df}{dr} + (k^2 r^2 - M^2)f = 0 \qquad (4.3.99)$$

$$\frac{1}{\sin\theta} \frac{d}{d\theta}\left(\sin\theta \frac{dg}{d\theta} \right) + \left(M^2 - \frac{q^2}{\sin^2\theta} \right)g = 0 \qquad (4.3.100)$$

$$\frac{d^2 h}{d\varphi^2} + q^2 h = 0 \qquad (4.3.101)$$

where M^2 and q^2 are the separation constants. The solutions of (4.3.101) are $\sin(q\varphi)$ and $\cos(q\varphi)$, but if the velocity potential ϕ is to be single-valued, it is necessary that q be an integer, for otherwise $(r, \theta, \varphi + 2\pi) \neq (r, \theta, \varphi)$. Thus, $q = 0, \pm 1, \pm 2, \ldots$ Denoting this integer by m (4.3.100) becomes

$$\frac{1}{\sin\theta} \frac{d}{d\theta}\left(\sin\theta \frac{dg}{d\theta} \right) + \left(M^2 - \frac{m^2}{\sin^2\theta} \right)g = 0 \qquad (4.3.102)$$

This equation, like (4.3.5), has solutions that are finite at $\theta = 0$ and $\theta = \pi$ for values of $M^2 = n(n+1)$, where n is an integer. Thus,

$$\frac{1}{\sin\theta}\frac{d}{d\theta}\left(\sin\theta\frac{dg}{d\theta}\right) + \left[n(n+1) - \frac{m^2}{\sin^2\theta}\right]g = 0 \qquad (4.3.103)$$

When $m=0$, the solutions of this equation are the Legendre polynomials $P_n(\cos\theta)$. For nonzero m, the solutions of (4.3.103), which are finite at $\theta=0$ and $\theta=\pi$, are the associated Legendre polynomials $P_n^m(\cos\theta)$, where, for positive n and m,

$$P_n^m(\cos\theta) = (\sin\theta)^{m/2}\frac{d^m P_n(\cos\theta)}{d(\cos\theta)^m} \qquad (4.3.104)$$

For negative values of m and n, the associated Legendre polynomials can be expressed in terms of those with positive m and n, but here we will require only the latter and these are given by (4.3.104).

Equation (4.3.103) has another independent solution, written as $Q_n^m(\cos\theta)$, but this becomes infinite at $\theta=0$ and $\theta=\pi$ and therefore does not apply to physical situations where those points are part of the region in which the potential is required.

The products $\sin(m\varphi)P_n^m(\cos\theta)$ and $\cos(m\varphi)P_n^m(\cos\theta)$ are periodic functions of φ and θ so that a nodal pattern, prescribed by the values of m and n, will be defined on the surface of a given spherical surface. Thus, the number of nodal lines that cross the equatorial plane $\theta=\pi/2$ is equal to the integer m. On the other hand, the number of nodal lines that cross a constant φ plane is $n-2$. (The associated Legendre polynomials also vanish at the poles $\theta=0$ and $\theta=\pi$.) The surface of a sphere can therefore be thought of as being subdivided into spherical rectangles, or *tesserae*, by the nodal lines. Therefore, the above products are sometimes called *tesseral harmonics*. Similarly, the functions

$$Y_n^m(\theta,\varphi) = P_n^m(\cos\theta)e^{im\varphi} \qquad (4.3.105)$$

are called spherical harmonics of the first kind.

Returning to the radial-factor equation, (4.3.99), we see that with $M^2 = n(n+1)$ it becomes identical to (4.1.15) so that its solutions are spherical Bessel functions. Therefore, for monochromatic waves, the solution to the wave equation in spherical coordinates may be written as

$$\tilde{\phi}(r,\theta,\varphi,t) = \sum_n\sum_m\left[A_n h_n^{(1)}(kr) + B_n h_n^{(2)}(kr)\right]Y_n^m(\theta,\varphi)e^{-i\omega t} \quad (4.3.106)$$

or as

$$\tilde{\phi}(r,\theta,\varphi,t)=\sum_n \sum_m \left[A'_n j_n(kr) + B'_n y_n(kr) \right] Y_n^m(\theta,\varphi) e^{-i\omega t} \quad (4.3.107)$$

PROBLEMS

4.3.1   The first nonzero root of (4.3.16) was found to be $k_1 a = 2.08$. Determine the next higher root. Show that the acoustic mode corresponding to that root has two nodal lines. Sketch a few lines having constant pressure.

4.3.2   Obtain explicit values for $P_1^1, P_2^1, P_2^2, P_3^1, P_3^2,$ and P_3^3. Verify that P_3^2 and P_3^3 have only one nodal line.

4.3.3   Let **x** and **x′** be the position vectors of two points whose coordinates in a spherical polar system of coordinates are r, θ, φ and r', θ', φ', respectively. Be expressing **x** and **x′** in terms of their cartesian components $x_1, x_2,$ and x_3, show that the distance between the two points may be expressed as

$$|\mathbf{x}-\mathbf{x'}|^2 = r^2 + r'^2 - 2rr'\left[\cos\theta\cos\theta' + \sin\theta\sin\theta'\cos(\varphi-\varphi')\right]$$

4.3.4   Show that a unit vector along the radial direction in a spherical polar system of coordinates may be expressed as

$$\mathbf{e}_r = \mathbf{i}\sin\theta\cos\varphi + \mathbf{j}\sin\theta\sin\varphi + \mathbf{k}\cos\varphi$$

where **i**, **j**, and **k** are unit vectors along the axis $x, y,$ and z of a cartesian system of coordinates, with z corresponding to the polar axis.

4.3.5   Using the results of the previous problem, show, by actual integration, that the integral of a uniform pressure over the surface of a sphere vanishes. What would be the result if instead of a spherical surface we had some other surface?

4.3.6   When a plane sound wave falls on a *fixed* sphere, the scattered intensity may be computed from the results obtained in this section by setting $\delta = 0$. Another procedure is to derive it from the basic solution for the velocity potential, with C_n given by (4.3.35) for all n. Use this approach to show that in the low-frequency limit, $ka \rightarrow 0$,

$$I_s/I_{\text{inc}} = \tfrac{1}{9}(ka)^4\left(1 - \tfrac{3}{2}\cos\theta\right)^2 (a/r)^2$$

4.4 CIRCULARLY CYLINDRICAL WAVES

We now consider circularly cylindrical waves. As their name implies, these are waves whose wave fronts are circular cylinders. However, as in the spherical-wave case treated at the end of the last section, the field properties are, in general, not constant on the wave-front surface. The system of coordinates used to treat this type of wave is the cylindrical, and is depicted in Figure 4.4.1. In the figure, x, σ, and φ are the coordinates of some point P. The coordinate x gives the distance to some origin 0 along the cylinder axis, and the quantity σ is the distance from P to the x axis measured along the normal to the axis. Different values of σ generate coaxial cylinders of circular cross section.

In this system of coordinates, the wave equation is

$$\frac{1}{c_0^2} \frac{\partial^2 \phi}{\partial t^2} = \frac{1}{\sigma} \frac{\partial}{\partial \sigma} \left(\sigma \frac{\partial \phi}{\partial \sigma} \right) + \frac{1}{\sigma^2} \frac{\partial^2 \phi}{\partial \varphi^2} + \frac{\partial^2 \phi}{\partial x^2} \tag{4.4.1}$$

The pressure is given, as before, by $-\rho_0(\partial\phi/\partial t)$. The three components of the fluid velocity are

Axial

$$u = \frac{\partial \phi}{\partial x} \tag{4.4.2}$$

Transverse

$$v = \frac{\partial \phi}{\partial \sigma} \tag{4.4.3}$$

Azimuthal

$$w = \frac{1}{\sigma} \frac{\partial \phi}{\partial \varphi} \tag{4.4.4}$$

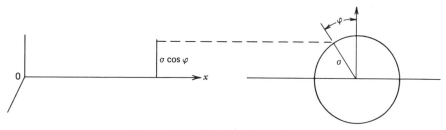

Figure 4.4.1 Cylindrical coordinates.

When the field does not depend on the axial distance x, the origin may be placed anywhere on that axis (the axis of symmetry). In that case, the transverse coordinate σ becomes identical to the radial distance to the origin. The transverse velocity is then along the radius and is therefore referred to as radial, although that name is also frequently given to the transverse velocity defined by (4.4.3).

Monochromatic Waves

For harmonic time dependence, (4.4.1) becomes

$$\frac{\partial^2 \tilde{\phi}}{\partial \sigma^2} + \frac{1}{\sigma} \frac{\partial \tilde{\phi}}{\partial \sigma} + \frac{1}{\sigma^2} \frac{\partial^2 \tilde{\phi}}{\partial \varphi^2} + \frac{\partial^2 \tilde{\phi}}{\partial x^2} + k^2 \tilde{\phi} = 0 \tag{4.4.5}$$

where $k = \omega/c_0$. Assuming a solution of the form $\tilde{\phi} = f(x)g(\sigma)h(\varphi)\exp(-i\omega t)$, we obtain

$$\frac{d^2 f}{dx^2} + N^2 f = 0 \tag{4.4.6}$$

$$\frac{d^2 g}{d\sigma^2} + \frac{1}{\sigma} \frac{dg}{d\sigma} + \left[(k^2 - N^2) - \frac{M^2}{\sigma^2} \right] g = 0 \tag{4.4.7}$$

$$\frac{d^2 h}{d\varphi^2} + M^2 h = 0 \tag{4.4.8}$$

where N^2 and M^2 are the separation constants. The solution for h is

$$h(\varphi) = a \cos M\varphi + b \sin M\varphi$$

If the fluid occupies an angular sector around the x axis smaller than 2π, then the values of M are determined by the boundary conditions at the ends of the sector, and may therefore be nonintegral. On the other hand, if the fluid occupies the complete range $0 \leqslant \varphi \leqslant 2\pi$, it is clear that for ϕ to be single valued, the quantity M must be zero or an integer. Thus,

$$h(\varphi) = a_m \cos m\varphi + b_m \sin m\varphi, \quad m = 0, 1, 2, \dots \tag{4.4.9}$$

The equation for $f(x)$ has solutions of the form $f(x) = \exp(\pm iNx)$. Since N plays the role of the wave number along x, we denote N by k_x. Therefore, (4.4.7) can be written as

$$g''(\sigma) + \frac{1}{\sigma} g'(\sigma) + \left(k_\sigma^2 - \frac{m^2}{\sigma^2} \right) g(\sigma) = 0 \tag{4.4.10}$$

where

$$k_\sigma^2 = k^2 - k_x^2 \tag{4.4.11}$$

The equation for $g(\sigma)$ will be recognized as Bessel's differential equation of integral order m (see Section 3.8). Its general solution is

$$g(\sigma) = A J_m(k_\sigma \sigma) + B Y_m(k_\sigma \sigma) \tag{4.4.12}$$

For small values of $k_\sigma \sigma$, $Y_m \sim (k_\sigma \sigma)^{-m}$. Therefore, if the axis $\sigma = 0$ is part of the region of interest, and if ϕ is to be finite there, we must put $B = 0$. In that case, the velocity potential is given by

$$\tilde{\phi}(x, \sigma, \varphi, t) = \sum_{m=0}^{\infty} (A_m \cos m\varphi + B_m \sin m\varphi)(C_m e^{ik_x x} + D_m e^{-ik_x x}) J_m(k_\sigma \sigma) e^{-i\omega t}$$

$$\tag{4.4.13}$$

When the axis $\sigma = 0$ is not part of the region of interest, as in the region exterior to a cylinder, the function Y_m should be retained in the solution.

The meaning of each term of (4.4.13) will be made clearer later when we consider waves inside tubes. It should be noted that for large values of σ, the equation for $g(\sigma)$ becomes $g'' + k_\sigma^2 g = 0$. Therefore, we should expect that, far from the axis $\sigma = 0$, the velocity potential should depend on σ in a nearly sinusoidal manner. This also follows from the expansions for J_m for large values of its argument. Thus, as $z \to \infty$, we have

$$J_m(z) \sim \sqrt{\frac{2}{\pi z}} \cos\left(z - \frac{m\pi}{2} - \frac{\pi}{4}\right)$$

This expansion shows that the intensity in a cylindrical wave decays with distance from the axis as $1/\sigma$. This is as it should be because the area of the wave front increases with σ.

Another useful representation for the transverse factor $g(\sigma)$ is in terms of the Hankel functions $H_m^{(1)}$ and $H_m^{(2)}$, defined by

$$H_m^{(1)}(z) = J_m(z) + i Y_m(z) \tag{4.4.14}$$

$$H_m^{(2)}(z) = J_m(z) - i Y_m(z) \tag{4.4.15}$$

These functions can be used to represent outgoing and ingoing waves, respectively. This can be seen by considering their asymptotic expansions for large values of their arguments. Thus,

$$H_m^{(1)}(z) \sim \sqrt{\frac{2}{\pi z}} \, \exp\left[i\left(z - \frac{m\pi}{2} - \frac{\pi}{4} \right) \right] \tag{4.4.16}$$

$$H_m^{(2)}(z) \sim \sqrt{\frac{2}{\pi z}} \, \exp\left[-i\left(z - \frac{m\pi}{2} - \frac{\pi}{4} \right) \right] \tag{4.4.17}$$

Waves inside Circular Tubes

One important application of our monochromatic-wave solution is to the study of wave propagation in circular ducts. Here, we will consider only cylinders with rigid walls. Other kinds may be treated in a similar fashion, using impedance-type boundary conditions.

The velocity potential for positive-going waves in the tube is given by

$$\tilde{\phi}(x, \sigma, \varphi, t) = \sum_{m=0}^{\infty} \left[A_m \cos(m\varphi) + B_m \sin(m\varphi) \right] J_m(k_\sigma \sigma) e^{i(k_x x - \omega t)}$$

$$\tag{4.4.18}$$

For a given frequency ω, the value of k is fixed ($= \omega / c_0$) so that (4.4.11) sets a limit on the values that k_σ and k_x can have. Also, just as with waves in a two-dimensional duct, not all values of the transverse wave number, k_σ in this case, are allowed. The allowed values must be such that the transverse velocity component $v = \partial \phi / \partial \sigma$ vanishes on the wall of the cylinder. Thus, if the cylinder has an inner radius a, the condition is

$$J_m'(k_\sigma a) = 0 \tag{4.4.19}$$

For a given value of m, there will be infinitely many values of $k_\sigma a$ that satisfy this equation. If we use the index s to order them, those values of $k_\sigma a$ may be expressed as

$$k_{\sigma ms} a = \pi \alpha_{ms}, \quad m, s = 0, 1, 2, \ldots \tag{4.4.20}$$

Some values of $\pi \alpha_{ms}$ are given in Table 4.4.1. Since there is a double infinite set of them, the velocity potential is given by the double sum

$$\tilde{\phi} = \sum_{m=0}^{\infty} \sum_{s=0}^{\infty} \tilde{\phi}_{ms} \tag{4.4.21}$$

Table 4.4.1 Some Roots $\pi\alpha_{ms}$ of $J'_m(k_\sigma a)=0$

		s	
m	0	1	2
0	0	3.832	7.016
1	1.841	5.331	8.536
2	3.054	6.706	9.969

where $\tilde{\phi}_{ms}$ is the velocity potential that corresponds to a given set of values of m and s. The solution for $m=0$, $s=0$ should be obtained separately because the Bessel function $J_0(\pi\alpha_{00}\sigma/a)=0$ does not represent properly the radial dependence of $\tilde{\phi}$. Thus, for $m=0$, $s=0$, (4.4.11) gives

$$k^2-k_x^2=0$$

so that (4.4.10) becomes

$$g''+\frac{1}{\sigma}g'=0 \tag{4.4.22}$$

The solution to this equation, which is finite at $\sigma=0$, is $g=$ constant so that the corresponding potential is

$$\tilde{\phi}_{00}=A_{00}e^{i(kx-\omega t)} \tag{4.4.23}$$

This is simply a plane wave. To see what the other terms in (4.4.21) represent, we write that equation as

$$\tilde{\phi}=\tilde{\phi}_{00}+\sum_{s=1}^{\infty}\tilde{\phi}_{0s}+\sum_{m=1}^{\infty}\tilde{\phi}_{m0}+\sum_{m=1}^{\infty}\sum_{s=1}^{\infty}\tilde{\phi}_{ms} \tag{4.4.24}$$

where, except for $\tilde{\phi}_{00}$, $\tilde{\phi}_{ms}$ is given by

$$\tilde{\phi}_{ms}=[A_{ms}\cos m\varphi+B_{ms}\sin m\varphi]J_m(\pi\alpha_{ms}\sigma/a)e^{i(k_{xms}x-\omega t)} \tag{4.4.25}$$

Dropping the time factor, we have

$$\tilde{\phi}=A_{00}e^{ikx}+\sum_{s=1}^{\infty}A_{0s}J_0(\pi\alpha_{0s}\sigma/a)e^{ik_{0s}x}$$

$$+\sum_{m=1}^{\infty}[A_{m0}\cos m\varphi+B_{m0}\sin m\varphi]e^{ik_{xm0}x}J_m(\pi\alpha_{m0}\sigma/a)$$

$$+\sum_{s=1}^{\infty}\sum_{m=1}^{\infty}[A_{ms}\cos m\varphi+B_{ms}\sin m\varphi]J_m(\pi\alpha_{ms}\sigma/a)e^{ik_{xms}} \tag{4.4.26}$$

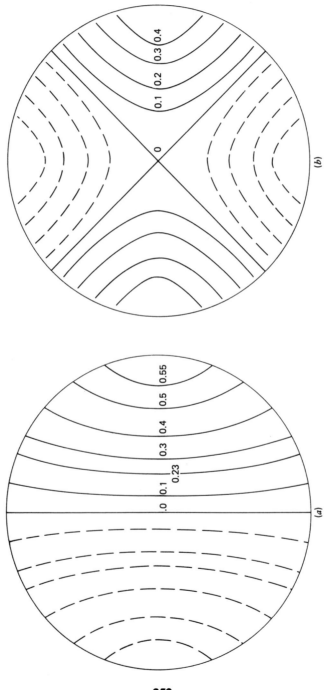

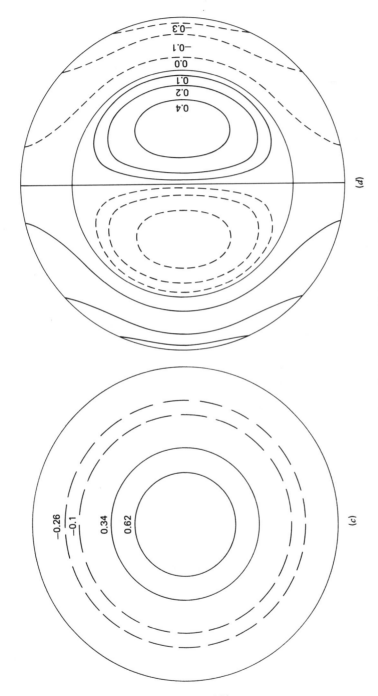

Figure 4.4.2 Acoustic pressure on a plane normal to the tube axis. (a) Lowest nonplane mode; (b) the 2,0 mode; (c) first radial mode (0,1); (d) lowest mode having radial and tangential dependence (the 1,1 mode).

Each term in this series represents a mode of wave propagation in the tube. As stated above, the $0, 0$ mode prepresents the purely plane mode. Similarly, the $0, s$ modes, with $s > 0$, represent modes that have transverse but no azimuthal dependence. Further, depending on the value of s, they may have nodal circles on any cross section of the tube. Such modes are usually referred to as radial modes, although it is clear that if they are propagated (i.e., if $k_{x0s} > 0$), they also depend on x. The $m, 0$ modes depend on all three spatial coordinates, and have nodal lines for values of φ that depend on m. Their transverse dependence is such that they have no nodal circles. Because of their azimuthal structure, these modes are sometimes referred to as "tangential."

The remaining modes are combination modes having both radial and azimuthal nodal lines. To obtain the locations of some of these lines, we consider a few modes in ascending order of the corresponding values of $\pi\alpha_{ms}$. The lowest value is $\pi\alpha_{00} = 0$ and corresponds to the plane mode. On any cross section in the tube, this has uniform pressure and velocity. The next root is $\pi\alpha_{10} = 1.841$ and corresponds to an acoustic mode of oscillation having one nodal line. That is, the acoustic pressure in the tube vanishes for some angle φ, as sketched in Figure 4.4.2a. It should be added that for a given driving frequency, the $1, 0$ mode is the nonplane mode (i.e., a mode having σ or φ dependence) most likely to be propagated. This follows from the definition of k_{xms}, that is, from

$$k_{xms} = k\sqrt{1 - (\pi\alpha_{ms}/ka)^2} \tag{4.4.27}$$

because the quantity inside the radical then has (for a given ka) the largest possible positive values.

The next mode, the $2, 0$ mode, is similar to the $1, 0$ mode but has two nodal lines, as shown in Figure 4.4.2b.

The first nontangential oscillation that appears is the $0, 1$ mode, given by $\pi\alpha_{01} = 3.8317$. Here, on a cross section of the duct, the fluid merely performs a transverse oscillation, as shown in Figure 4.4.2c. The next root is $\pi\alpha_{11}$. It involves both tangential and radial oscillations, as displayed in Figure 4.4.2d.

It is clear from these examples that the numbers m, s give the number of nodal lines along φ and σ, respectively, for each allowed mode of oscillation. Further, the characteristic frequencies corresponding to these oscillations, that is, the cutoff frequencies corresponding to each mode, are given by the quantities $k_{\sigma ms}c_0 = \pi\alpha_{ms}c_0/a$. We denote these by Ω_{ms}. Thus,

$$\Omega_{ms} = \pi\alpha_{ms}c_0/a \tag{4.4.28}$$

In terms of Ω_{ms}, the x component of the wave number can be written as

$$k_{xms} = k\sqrt{1 - (\Omega_{ms}/\omega)^2} \tag{4.4.29}$$

showing that propagation along the duct axis is dispersive. Therefore, just as in the two-dimensional duct, a given mode will or will not be propagated, depending on whether the excitation frequency ω is larger or smaller than the corresponding Ω_{ms}. Thus, if

$$\omega > \Omega_{ms} \tag{4.4.30}$$

the given mode (m, s) will be propagated, if excited, and the opposite will be true if $\omega < \Omega_{ms}$. The lowest cutoff frequency is

$$\omega = \Omega_{10} = 0.5861\pi c_0/a \tag{4.4.31}$$

so that, as stated in Chapter 3, if ω is smaller than this value, only the plane-wave mode will be propagated.

Excitation of Acoustic Modes.   In addition to having a cutoff frequency smaller than the excitation frequency, a given mode must be excited before it can be propagated. The situation is similar to that studied in Section 3.5 for the two-dimensional channel. The main difference between the two is that in this case the transverse dependence is described by a Bessel function instead of a sinusoidal function. However, because the orthogonality conditions for the Bessel functions have not been introduced so far, we study excitation in a very long cylindrical tube of radius a. The excitation may be provided by several means; for example, by a rotating fan, or by a membrane executing some kind of small-amplitude oscillation that is not uniform over its entire surface. For our purposes, it will be sufficient to assume that at $x=0$ some mechanism exists that produces there an excitation velocity in the x direction given by

$$\tilde{U}_e = U_{e0}(\sigma, \varphi)e^{-i\omega t} \tag{4.4.32}$$

The boundary condition at $x=0$ is $u = U_e$ for all t. Therefore, using (4.4.26), we obtain

$$U_{e0}(\sigma, \varphi) = ikA_{00} + \sum_{s=1}^{\infty} A_{0s}J_0(\pi\alpha_{0s}\sigma/a)ik_{x0s}$$

$$+ \sum_{m=1}^{\infty}\sum_{s=0}^{\infty} ik_{xms}[A_{ms}\cos m\varphi + B_{ms}\sin m\varphi]J_m(\pi\alpha_{ms}\sigma/a) \tag{4.4.33}$$

This is a Fourier–Bessel series expansion for $U_{e0}(\sigma, \varphi)$. The constant term of this series is simply the average of $U_{e0}(\sigma, \varphi)$ over the area of the tube. To see this, we integrate both sides of (4.3.33) over the area of the tube:

$$\frac{1}{\pi a^2} \int_0^{2\pi} d\varphi \int_0^a U_{e0} \sigma \, d\sigma$$

$$= ik A_{00} + \frac{1}{\pi a^2} \sum_{s=1}^{\infty} ik_{x0s} A_{0s} \int_0^a \sigma J_0(\pi \alpha_{0s} \sigma/a) \, d\sigma$$

$$(4.4.34)$$

The integral of the Bessel function on the right-hand side gives

$$\int_0^a \sigma J_0 \left(\frac{\pi \alpha_{0s} \sigma}{a} \right) d\sigma = \frac{a}{\pi \alpha_{0s}} J_1(\pi \alpha_{0s}) \tag{4.4.35}$$

But $J_1(\pi \alpha_{0s}) = -J_0'(\pi \alpha_{0s}) = 0$ for all s. Therefore,

$$ik A_{00} = \overline{U}_e \tag{4.4.36}$$

where $\overline{U}_e$ is the area average of the excitation velocity. It then follows that if the excitation velocity in the x direction has zero average, the plane-wave mode will not be excited.

Returning to the evaluation of the other coefficients, we write (4.4.33) as

$$U_{e0}(\sigma, \varphi) - \overline{U}_e = \sum_{s=1} ik_{x0s} A_{0s} J_0(\pi \alpha_{0s} \sigma/a)$$

$$+ \sum_{s=0} \sum_{m=1} \left[A_{ms} \cos m\varphi + B_{ms} \sin m\varphi \right] J_m(\pi \alpha_{ms} \sigma/a) ik_{xms}$$

$$(4.4.37)$$

To evaluate A_{ms} and B_{ms}, we make use of the orthogonality conditions for $\sin m\varphi, \cos m\varphi,$ and $J_m(\pi \alpha_{ms} \sigma/a)$. The conditions for the sinusoidal functions are those used earlier:

$$\int_0^{2\pi} \sin m\varphi \cos m\varphi \, d\varphi = 0 \tag{4.4.38}$$

$$\int_0^{2\pi} \cos m\varphi \cos m'\varphi \, d\varphi = \pi \delta_{mm'}, \quad m', m > 0 \tag{4.4.39}$$

The orthogonality conditions for the Bessel functions are stated in Appendix C. In the present case, where $J'(\pi\alpha_{ms})=0$ and $s+m>0$ (so that $\pi\alpha_{ms}$ does not vanish), we have

$$\int_0^a \sigma J_m\left(\frac{\pi\alpha_{ms}\sigma}{a}\right) J_{m'}\left(\frac{\pi\alpha_{m's'}\sigma}{a}\right) d\sigma$$

$$= \tfrac{1}{2}a^2\left[1 - \frac{m^2}{(\pi\alpha_{ms})}\right][J_m(\pi\alpha_{ms})]^2\delta_{ss'}\delta_{mm'} \qquad (4.4.40)$$

Multiplying both sides of (4.4.37) by

$$\sigma\cos m'\varphi J_m(\pi\alpha_{m's'}\sigma/a)$$

and integrating over the area of the tube, we obtain (since $\int_0^{2\pi}\cos m'\varphi\, d\varphi\equiv 0$)

$$\int_0^{2\pi} d\varphi \int_0^a \left(U_{e0}-\overline{U}_e\right)\sigma\cos m'\varphi J_m(\pi\alpha_{m's'}\sigma/a)\, d\sigma$$

$$= \sum_{m=1}\sum_{s=0} ik_{xms}\int_0^{2\pi} d\varphi\left[A_{ms}\cos m\varphi\cos m'\varphi + B_{ms}\sin m\varphi\cos m'\varphi\right]$$

$$\times \int_0^a \sigma J_{m'}(\pi\alpha_{m's'}\sigma/a) J_m(\pi\alpha_{ms}\sigma/a)\, d\sigma \qquad (4.4.41)$$

In view of (4.4.38) and (4.4.39), the integral over φ vanishes unless $m=m'$, in which case the right-hand side of (4.4.41) reduces to

$$\sum_{s=0} ik_{xm's}\pi A_{m's}\int_0^a \sigma J_{m'}(\pi\alpha_{m's'}\sigma/a) J_{m'}(\pi\alpha_{m's}\sigma/a)\, d\sigma$$

But in view of (4.4.40), this integral vanishes unless $s=s'$ so that

$$\int_0^{2\pi} d\varphi \int_0^a \left(U_{e0}-\overline{U}_e\right)\sigma\cos m'\varphi J_{m'}(\pi\alpha_{m's'}\sigma/a)\, d\sigma$$

$$= ik_{xm's'}\frac{\pi a^2}{2}A_{m's'}\left[1 - \frac{m'^2}{(\pi\alpha_{m's'})^2}\right]J_{m'}^2(\pi\alpha_{m's'}) \qquad (4.4.42)$$

Therefore, when $m > 0$, we have

$$A_{ms} = \frac{(1/\pi)\int_0^{2\pi} V_{ms}(\varphi)\cos m\varphi \, d\varphi}{ik_{xms}\left[1 - \dfrac{m^2}{(\pi\alpha_{ms})^2}\right]J_m^2(\pi\alpha_{ms})} \tag{4.4.43}$$

where

$$V_{ms}(\varphi) = \frac{2}{a^2}\int_0^a (U_{e0} - \overline{U}_e)\sigma J_m(\pi\alpha_{ms}\sigma/a)\,d\sigma \tag{4.4.44}$$

Similarly,

$$B_{ms} = \frac{(1/\pi)\int_0^{2\pi} V_{ms}(\varphi)\sin m\varphi \, d\varphi}{ik_{xms}\left[1 - \dfrac{m^2}{(\pi\alpha_{ms})^2}\right]J_m^2(\pi\alpha_{ms})} \tag{4.4.45}$$

Finally, for $m = 0$, $s > 0$, we multiply (4.4.37) by $\sigma J_0(\pi\alpha_{0s'}\sigma/a)$ and integrate with respect to σ and φ as before, to obtain

$$A_{0s} = \frac{(1/\pi)\int_0^{2\pi} V_{0s}\,d\varphi}{2ik_{x0s}J_0^2(\pi\alpha_{0s})} \tag{4.4.46}$$

The integrals on the numerators of (4.4.43), (4.4.45), and (4.4.46) represent the coefficients of the Fourier-series expansion of $V_{ms}(\varphi)$:

$$V_{ms} = \tfrac{1}{2}a_{0s} + \sum_{m=1}^{\infty} a_{ms}\cos m\varphi + \sum_{m=1} b_{ms}\sin m\varphi \tag{4.4.47}$$

If we use this notation for the integrals appearing in the equations for A_{ms}, B_{ms}, and A_{0s}, we obtain

$$\tilde{\phi} = \overline{U}_e e^{i(kx - \omega t)} + \sum_{s=1} \frac{a_{0s}J_0(\pi\alpha_{0s}\sigma/a)}{2ik_{x0s}J_0^2(\pi\alpha_{0s})}e^{i(k_{x0s}x - \omega t)}$$

$$+ \sum_{m=1}\sum_{s=0} \frac{[a_{ms}\cos m\varphi + b_{ms}\sin m\varphi]J_m(\pi\alpha_{ms}\sigma/a)}{ik_{xms}\left[1 - \dfrac{m^2}{(\pi\alpha_{ms})^2}\right]J_m^2(\pi\alpha_{ms})}e^{i(k_{xms}x - \omega t)} \tag{4.4.48}$$

It should be noted that if $U_{e0}(\sigma, \varphi) = \bar{U}_e$, then $a_{0s} = a_{ms} = b_{ms} = 0$ so that the only mode that is excited is the plane-wave mode.

EXAMPLE.

We consider a tube of radius a fitted, at $x = 0$, with a disc, also of radius a. The disc is rotating with angular velocity in such a manner that its axis of rotation coincides with the tube axis. We will assume that during rotation, the disc displaces the fluid in front of it in the direction of the tube axis. Clearly, if the disc is flat, no such displacement will be produced because the fluid, assumed inviscid, will not be affected by the disc's rotation. However if the disc surface has a structure that depends on σ or on φ, then as it rotates, it will displace some fluid in front of it in the direction of its normal (i.e., along the x axis). The amount of fluid displaced by a given element of the disc depends on the distance of that element from the mean location of all elements, which we take as $x = 0$. Since we are using linear equations, these displacements should be very small. Owing to the disc's rotation, every element of fluid in front of the disc is displaced by a small amount, which we may express as

$$\tilde{x}_f(0, \sigma, \varphi, t) = x_{f0}(\sigma, \varphi) e^{-i\Omega t} \tag{4.4.49}$$

Therefore, the excitation velocity is

$$\tilde{U}_e = -i\Omega x_{f0}(\sigma, \varphi) e^{-i\Omega t}$$

$$= U_{e0}(\sigma, \varphi) e^{-i\Omega t} \tag{4.4.50}$$

To obtain the modal coefficients, we need to know $x_{f0}(\sigma, \varphi)$. For convenience, we take the dependence of this quantity on σ and φ to be very simple. Suppose that the disc profile depends sinusoidally on φ and linearly on σ:

$$x_{f0}(\sigma, \varphi) = \varepsilon(\sigma/a) \cos n\varphi, \quad n = 1, 2, \ldots \tag{4.4.51}$$

where ε is the amplitude of the disc "bumpiness." Therefore,

$$U_{e0} = -i\Omega(\varepsilon/a)\sigma \cos n\varphi \tag{4.4.52}$$

Several results follow directly from this equation. First, since $\bar{U}_e$ is zero for this surface, the plane-wave mode will not be excited. Second, since we used $\cos n\varphi$ to describe the φ dependence, all the coefficients b_{ms} are zero. Further, the only coefficient a_{ms} that is not zero is that for which $m = n$. Hence,

$$\tilde{\phi} = \sum_{s=0}^{\infty} \frac{a_{ns} \cos n\varphi \, e^{i(k_{xns}x_x - \Omega t)}}{ik_{xns}\left[1 - \dfrac{n^2}{(\pi\alpha_{ns})^2}\right] J_n^2(\pi\alpha_{ns})} \tag{4.4.53}$$

where

$$a_{ns} = \frac{2}{\pi a^2} \int_0^{2\pi} \cos n\varphi \, d\varphi \int_0^a \left(U_{e0} - \overline{U}_e\right)\sigma J_n(\pi\alpha_{ns}\sigma/a) \, d\sigma \tag{4.4.54}$$

With (4.4.50), this becomes

$$a_{ns} = -\frac{2}{a^2} i\Omega\left(\frac{\varepsilon}{a}\right)\int_0^a \sigma^2 J_n(\pi\alpha_{ns}\sigma/a) \, d\sigma \tag{4.4.55}$$

The integral on the right-hand side is (see Gradshteyn and Ryzhik, page 683)

$$\frac{a^3}{\pi\alpha_{ns}} J_{n+1}(\pi\alpha_{ns})$$

Further, the Bessel functions satisfy the following recurrence relation

$$J_{n+1}(z) = \frac{n}{z} J_n(z) - J_n'(z) \tag{4.4.56}$$

but since $J_n'(\pi\alpha_{ns}) = 0$, the integral can be written as

$$\int_0^a \sigma^2 J_n(\pi\alpha_{ns}\sigma/a) \, d\sigma = \frac{a^3 n}{(\pi\alpha_{ns})^2} J_n(\pi\alpha_{ns}) \tag{4.4.57}$$

Therefore,

$$a_{ns} = -2in\Omega\varepsilon \frac{J_n(\pi\alpha_{ns})}{(\pi\alpha_{ns})^2} \tag{4.4.58}$$

Substituting this into (4.4.53) and using (4.4.29) for k_{xns} yields

$$\tilde{\phi} = -2\varepsilon c_0 \sum_{s=0}^{\infty} \frac{\cos n\varphi \, e^{i(k_{xns} - \Omega t)}}{\sqrt{1 - (\Omega_{ns}/\Omega)^2} \left[(\pi\alpha_{ns}/n)^2 - 1\right] J_n(\pi\alpha_{ns})} \tag{4.4.59}$$

where Ω_{ns} is the cutoff frequency for the n, s mode and is given by (4.4.28). Thus, for a given value of $n > 0$, all n, s modes having cutoff frequencies smaller than the disc's rotational frequency Ω will be excited and propagated in the tube. Since the number of roots $\pi\alpha_{ns}$ increases with both s and n, it is clear that as Ω increases, more and more modes will be excited. However, for a fixed rotational frequency, only a finite number of them will be propagated. In fact, if Ω is sufficiently low, it may happen that $\Omega < \pi\alpha_{10}c_0/a$ so that no waves will be propagated in the tube.

PROBLEMS

4.4.1 Determine the acoustic intensity of the propagated waves in the circular tube of the example given in this section. Specialize your results for the following case: $n = 1, \Omega = 360 \sec^{-1}, c_0/a = 100 \sec^{-1}$.

4.4.2 In the above problem, determine the number N of modes that are propagated in the tube as a function of the disc's frequency Ω.

4.4.3 Consider a closed but hollow rigid cylinder of radius a and length L. Allowing transverse modes, determine the lowest three frequencies of oscillation for the fluid filling the cylinder.

4.4.4 Determine the velocity potential in a circular tube of radius a and length L. One end of the tube is filled with a membrane oscillating along the axial direction of the tube with velocity $\overline{U}_e = U_{e0}(\sigma, \varphi)\cos \omega t$. The other end is closed by a rigid reflector.

4.5 NONMONOCHROMATIC CYLINDRICAL WAVES

We return to the purely axisymmetric case and consider nonmonochromatic cylindrical waves, which have some distinctive features worthy of study.

For the axially symmetric case, the wave equation in cylindrical coordinates reduces to

$$\frac{\partial^2 \phi}{\partial t^2} = c_0^2 \left(\frac{\partial^2 \phi}{\partial r^2} + \frac{1}{r} \frac{\partial \phi}{\partial r} \right) \tag{4.5.1}$$

where r is the distance from the axis of symmetry. Except for the numerical factor of the second term on the right-hand side, this equation is the same as (4.1.3), which describes centrally symmetric waves. However, this small difference is sufficient to prevent us from obtaining a general solution of (4.5.1) following the simple procedure that was used to obtain the general solution of (4.1.3).

One procedure that may be used to obtain the general solution for the arbitrary time-dependence case is to construct it from the monochromatic time-dependence solution using the Fourier-transform method. Thus, if we take the Fourier transform of (4.5.1), we obtain

$$\frac{d^2\tilde{\phi}}{dr^2} + \frac{1}{r}\frac{d\tilde{\phi}}{dr} + \left(\frac{\omega}{c_0}\right)^2\tilde{\phi} = 0 \qquad (4.5.2)$$

where, as before, the transformation is defined by

$$\tilde{\phi} = \frac{1}{\sqrt{2\pi}}\int_{-\infty}^{\infty}\phi e^{i\omega t}\,dt \qquad (4.5.3)$$

The solution to (4.5.2) is given by a combination of Bessel functions of order zero:

$$\tilde{\phi} = \tilde{F}(\omega)J_0\left(\frac{\omega r}{c_0}\right) + \tilde{G}(\omega)Y_0\left(\frac{\omega r}{c_0}\right) \qquad (4.5.4)$$

where the quantities $\tilde{F}(\omega)$ and $\tilde{G}(\omega)$ represent the Fourier amplitudes and generally depend on the frequency.

Equation (4.5.4) can also be expressed in terms of traveling waves by means of the Hankel function $H_0^{(1)}$ and $H_0^{(2)}$:

$$\tilde{\phi} = \tilde{f}(\omega)H_0^{(1)}\left(\frac{\omega r}{c_0}\right) + \tilde{g}(\omega)H_0^{(2)}\left(\frac{\omega r}{c_0}\right) \qquad (4.5.5)$$

Let us derive the potential corresponding to the first term on the right-hand side of (4.5.5). It represents an outgoing wave, and is given by

$$\phi = \frac{1}{\sqrt{2\pi}}\int_{-\infty}^{\infty}\tilde{f}(\omega)H_0^{(1)}\left(\frac{\omega r}{c_0}\right)e^{-i\omega t}\,d\omega \qquad (4.5.6)$$

To integrate this, we must use an explicit representation of $H_0^{(1)}$ as a function of $(\omega r/c_0)$. A useful one is

$$H_0^{(1)}\left(\frac{\omega r}{c_0}\right) = \frac{2}{\pi i}\int_0^{\infty}\exp\left[i\left(\frac{\omega r}{c_0}\right)\cosh w\right]dw \qquad (4.5.7)$$

Substituting this into (4.5.6), we obtain

$$\phi(r,t) = \frac{1}{\sqrt{2\pi}}\frac{2}{\pi i}\int_{-\infty}^{\infty}\tilde{f}(\omega)e^{-i\omega t}\left[\int_0^{\infty}\exp\left[i\left(\frac{\omega r}{c_0}\right)\cosh w\right]dw\right]d\omega \qquad (4.5.8)$$

Interchanging the order of integration, this can be written as

$$\phi(r,t) = \frac{2}{\pi i} \int_0^\infty dw \left\{ \frac{1}{\sqrt{2\pi}} \int_{-\infty}^\infty \tilde{f}(\omega) \exp\{ -i\omega[t - (r/c_0)\cosh w]\} \, d\omega \right\}$$

$$(4.5.9)$$

The quantity in braces can be recognized as the inverse of $\tilde{f}(\omega)$, with time t replaced by $t - (r/c_0)\cosh w$. Thus,

$$\phi(r,t) = \frac{2}{\pi i} \int_0^\infty f\left(t - \frac{r}{c_0}\cosh w\right) dw \qquad (4.5.10)$$

For the cylindrical case, this function is the equivalent of the function $f(t-r/c_0)$ appearing in the general solution of the wave equation for the centrally symmetric case. Before studying some of the implications of this result, let us show that it is a solution of (4.5.1). In doing so, we will obtain the time and space derivatives needed to compute the acoustic pressure and velocity. Thus,

$$\frac{\partial \phi}{\partial t} = \int_0^\infty f'\left(t - \frac{r}{c_0}\cosh w\right) dw \qquad (4.5.11)$$

$$\frac{\partial \phi}{\partial r} = -\frac{1}{c_0} \int_0^\infty \cosh w \, f'\left(t - \frac{r}{c_0}\cosh w\right) dw \qquad (4.5.12)$$

Hence, with $\cosh^2 w - \sinh^2 w = 1$, we obtain

$$\frac{\partial^2 \phi}{\partial t^2} - c_0^2\left(\frac{\partial^2 \phi}{\partial r^2} + \frac{1}{r}\frac{\partial \phi}{\partial r}\right) = -\int_0^\infty \sinh^2 w \, f''\left(t - \frac{r}{c_0}\cosh w\right) dw$$

$$+ \frac{c_0}{r} \int_0^\infty \cosh w \, f'\left(t - \frac{r}{c_0}\cosh w\right) dw \quad (4.5.13)$$

The second integral on the right-hand side can be integrated by parts to yield

$$\int_0^\infty \cosh w \, f'\left(t - \frac{r}{c_0}\cosh w\right) dw = \sinh w \, f'\left(t - \frac{r}{c_0}\cosh w\right)\Bigg|_0^\infty$$

$$+ \frac{r}{c_0} \int_0^\infty \sinh^2 w \, f''\left(t - \frac{r}{c_0}\cosh w\right) dw$$

$$(4.5.14)$$

Substituting into (4.5.13), we obtain

$$\frac{\partial^2\phi}{\partial t^2} - c_0^2\left(\frac{\partial^2\phi}{\partial r^2} + \frac{1}{r}\frac{\partial\phi}{\partial r}\right) = \frac{c_0}{r}\lim_{w\to\infty}\sinh w f'\left(t - \frac{r}{c_0}\cosh w\right) \quad (4.5.15)$$

Thus, in order that (4.5.10) be a solution of (4.5.1), we require that the function f be such that its derivative vanishes rapidly as its argument goes to $-\infty$. The requirement is not as restrictive as it appears because it can be satisfied if $f'(t)\to 0$ as $t\to -\infty$. For example, if $f(t)$ is zero for $-\infty < t_0 < t$, the requirement is satisfied.

Similarly, the second term in (4.5.5) leads to a potential of the term

$$\phi = \frac{2}{\pi i}\int_0^\infty g\left(t + \frac{r}{c_0}\cosh w\right) dw \quad (4.5.16)$$

where the function g is such that $g'(t)\to 0$ rapidly as $t\to\infty$. Hence, the general solution in the axisymmetric case becomes

$$\phi(r,t) = \int_0^\infty f\left(t - \frac{r}{c_0}\cosh w\right) dw + \int_0^\infty g\left(t + \frac{r}{c_0}\cosh w\right) dw \quad (4.5.17)$$

where the numerical factor has been absorbed in f. This result can also we written as

$$\phi(r,t) = \int_{-\infty}^{t-r/c_0}\frac{f(\eta)\,d\eta}{\sqrt{(t-\eta)^2 - r^2/c_0^2}} + \int_{t+r/c_0}^\infty\frac{g(\eta)\,d\eta}{\sqrt{(\eta-t)^2 - r^2/c_0^2}} \quad (4.5.18)$$

This representation in terms of outgoing and incoming waves can be used to point out some special features of the solution. Consider the outgoing part:

$$\phi(r,t) = \int_{-\infty}^{t-r/c_0}\frac{f(\eta)\,d\eta}{\sqrt{(t-\eta)^2 - r^2/c_0^2}} \quad (4.5.19)$$

The pressure and velocity corresponding to this potential are from (4.5.11) and (4.5.12), given by

$$p' = -\rho_0\int_{-\infty}^{t-r/c_0}\frac{f'(\eta)\,d\eta}{\sqrt{(t-\eta)^2 - r^2/c_0^2}} \quad (4.5.20)$$

$$u_r = -\frac{1}{r}\int_{-\infty}^{t-r/c_0}(t-\eta)\frac{f'(\eta)\,d\eta}{\sqrt{(t-\eta)^2 - r^2/c_0^2}} \quad (4.5.21)$$

These equations show that the field at location r and time t is determined by the values of the function f between $-\infty$ and the retarded time $t - r/c_0$. That is, it depends on almost all the past history of f.

The appearance of the retarded time $t - r/c_0$ is not surprising. It implies, for example, that if the motion begins at $t=0$, the field at r is zero for times $t < r/c_0$. Thus, a cylindrical wave, like any other wave, has a well-defined front. However, contrary to the plane or to the spherical case, a traveling cylindrical wave has no well-defined tail. That is, after the wave has reached a given location, the acoustic field decreases slowly with time and vanishes at $t \to \infty$. To see this, we consider the case when the function $f(t)$ differs from zero only in a finite interval, that is, $f(t) \neq 0$ for $t_1 < t < t_2$. This may be produced by a sound source that acts only during a finite length of time. Here, the lower limit of the integrals in (4.5.19) to (4.5.21) is t_1. The upper limit becomes t_2 provided we consider the field only at times $t > r/c_0 + t_2$. For such times, the result for the pressure, for example gives

$$p' = -\rho_0 \int_{t_1}^{t_2} \frac{f'(\eta)\,d\eta}{\sqrt{(t-\eta)^2 - (r/c_0)^2}}, \quad t > \frac{r}{c_0} + t_2$$

$$= -\rho_0 \int_{t_1}^{t_2} \frac{f'(\eta)\,d\eta}{\sqrt{(t-\eta+r/c_0)(t-\eta-r/c_0)}} \tag{4.5.22}$$

To obtain a suitable approximation for p' applicable for long times, we write this as

$$p' = -\frac{\rho_0}{t} \int_{t_1}^{t_2} \frac{f'(\eta)\,d\eta}{\left(1 - \dfrac{r/c_0+\eta}{t}\right)^{1/2}\left(1 + \dfrac{r/c_0-\eta}{t}\right)^{1/2}} \tag{4.5.23}$$

If we now let $t \to \infty$ and expand the quantities in the denominator of the integrand, we obtain

$$p' \approx -\frac{\rho_0}{t} \int_{t_1}^{t_2} \left(1 + \frac{r/c_0+\eta}{2t}\right)\left(1 - \frac{r/c_0-\eta}{2t}\right) f'(\eta)\,d\eta$$

$$\approx -\frac{\rho_0}{t} \int_{t_1}^{t_2} \left[1 - \frac{r/c_0^2}{2t} + \frac{\eta}{t} + \cdots\right] f'(\eta)\,d\eta \tag{4.5.24}$$

Since $f(t_2) = f(t_1) = 0$, the first two terms in the integrand contribute nothing.

The third term can be integrated by parts, and gives

$$p' \approx \frac{\rho_0}{t^2} \int_{t_1}^{t_2} f(\eta) \, d\eta, \quad t \to \infty \tag{4.5.25}$$

This result shows that the pressure decays slowly with time.

Consider, now, the velocity as given by (4.5.21). Proceeding as we did for p', we obtain, for large t,

$$u_r \approx -\frac{1}{r} \int_{t_1}^{t_2} \left(1 - \frac{\eta}{t}\right) \left[1 - \frac{r/c_0^2}{2t} + \frac{\eta}{t} + \cdots \right] f'(\eta) \, d\eta \tag{4.5.26}$$

or

$$u_r \approx \frac{r}{4c_0^2 t^3} \int_{t_1}^{t_2} f(\eta) \, d\eta, \quad t \to \infty \tag{4.5.27}$$

Although the velocity dies out faster than the pressure, both quantities decay slowly with time. The reasons for this slow decay originate from the fact that a purely cylindrical wave can be produced only by some mechanism that acts along an infinite line coinciding with the axis of symmetry. We may therefore say that the field at some point r is due to contributions arising from all points along that line. The relative amplitude and phase of each contribution depend on their source in the line. Thus, for large times, the field at r will be made of contributions originating at farther and farther points in the line. Since the line is of infinite extent, these contributions continue to arrive at $t \to \infty$, but their amplitudes are smaller and smaller, as indicated by our asymptotic approximations.

Another interesting result from the approximations derived above is that in the tail of a cylindrical wave, the pressure and velocity are not in phase. Thus, using (4.5.25) and (4.5.27), we have

$$\frac{p'}{\rho_0 c_0 u_r} \approx \frac{c_0 t}{r} \tag{4.5.28}$$

This applies provided $t - r/c_0 \gg t_2 > t_1$. Now, at a fixed time t, the front portion of the wave is at a distance $c_0(t - t_1)$, whereas the signal corresponding to time t_2 is at a distance $c_0(t - t_2)$ from the line of symmetry. Therefore, the result given by (4.5.28) [which is limited to distances r such that $r \ll c_0(t - t_2)$] applies only to the tail end of the wave. In other parts of the wave, the ratio of pressure to velocity does not differ significantly from the characteristic impedance $\rho_0 c_0$ of the fluid. To show this, we again use the full results for the pressure and the velocity, but now we consider the case when $t_1 < t - r/c_0 < t_2$. Here, we have for the pressure

$$p' = -\rho_0 \int_{t_1}^{t-r/c_0} \frac{f'(\eta)\, d\eta}{\sqrt{(t-\eta+r/c_0)(t-\eta-r/c_0)}}, \quad t_1 \leqslant \eta \leqslant t - \frac{r}{c_0} < t_2$$

(4.5.29)

Because of the range of integration, the following inequality results:

$$t_1 + \frac{r}{c_0} - \eta \leqslant \frac{r}{c_0} \leqslant t - \eta \leqslant \frac{r}{c_0} + t_2 - \eta \qquad (4.5.30)$$

Hence, when t_2 is small compared to r/c_0 the quantity $t - \eta$ differs little from r/c_0 for all values of η within the range of integration. Thus,

$$p' \approx -\rho_0 \sqrt{\frac{c_0}{2r}} \int_{t_1}^{t-r/c_0} \frac{f'(\eta)\, d\eta}{\sqrt{t-\eta-r/c_0}}, \qquad \frac{c_0 t_2}{r} \ll 1 \qquad (4.5.31)$$

Following the same procedure, we obtain for u_r

$$u_r \approx -\frac{1}{c_0} \sqrt{\frac{c_0}{2r}} \int_{t_1}^{t-r/c_0} \frac{f'(\eta)\, d\eta}{\sqrt{t-\eta-r/c_0}}, \qquad \frac{c_0 t_2}{r} \ll 1 \qquad (4.5.32)$$

so that

$$p' = \rho_0 c_0 u_r, \qquad r \gg c_0 t_2 > c_0 t_1 \qquad (4.5.33)$$

This result is not surprising, as the condition $r \gg c_0 t_2$ implies that is applies only in the far-field. However, the result is applicable only in the range $t_1 \leqslant t - r/c_0 \leqslant t_2$. As discussed earlier, this represents the main part of the wave.

PROBLEMS

4.5.1 Show that a cylindrical wave, like a spherical wave, must contain compressions and expansions.

4.5.2 Consider a situation in which both outgoing and incoming cylindrical waves exist. Show that if the axis $r = 0$ is not a source of fluid, the velocity potential is given by

$$\phi = \int_0^\infty \left[f\left(t - \frac{r}{c_0}\cosh w\right) - f\left(t + \frac{r}{c_0}\cosh w\right) \right] dw$$

CHAPTER FIVE
SOUND EMISSION

In this chapter we study some of the basic mechanisms that are responsible for the production of sound waves. Some examples of sound emission have already been seen; for instance, the emission due to a piston in a tube, or due to the release of some initial disturbance. Here, we consider the problem of sound emission in a more systematic manner, with the objective of better understanding the various types of sound sources that exist. To do this, we consider emission into unbounded media. The presence of boundaries gives rise to reflected waves that, in principle, can be handled by the methods introduced in the last section of this chapter.

We begin by reconsidering the pulsating-sphere problem treated in the last chapter, as it is a good example of the most fundamental type of sound source.

5.1 RADIATION FROM A PULSATING SPHERE

We consider the sphere of Section 4.2. The sphere is executing pulsations of small amplitude about some mean radius a such that all points on its surface move with the same velocity $U(t) \ll c_0$. When the pulsations are harmonic with a single frequency ω, we found that at any point $r \geqslant a$ in the fluid, the velocity potential is given by

$$\tilde{\phi}(r, t) = -\frac{U_0 a^2 e^{i[k(r-a)-\omega t]}}{r[1-ika]} \tag{5.1.1}$$

The quantity $U_0 a^2$ can be expressed in terms of the instantaneous volume flow rate crossing some closed control surface enclosing the sphere, which is due to the sphere's change of volume. Denoting this quantity by $Q(t)$, we have, in general

$$Q = \int_A u_r \, dA \qquad (5.1.2)$$

Because of the boundary condition, $u_r(a, t) = U(t)$. Therefore, if A is a sphere of radius a, (5.1.2) becomes

$$Q(t) = U(t) \int_0^{2\pi} d\varphi \int_0^{\pi} a^2 \sin\theta \, d\theta = 4\pi a^2 U(t) \qquad (5.1.3)$$

In the particular case when $U = \mathrm{Re}\{U_0 e^{-i\omega t}\}$, we can also write $Q = \mathrm{Re}\{Q_0 e^{-i\omega t}\}$, so that $U_0 = (1/4\pi a^2)Q_0$. In terms of Q_0, our monochromatic potential can be written as

$$\tilde{\phi}(r, t) = -\frac{Q_0}{4\pi r} \frac{e^{i[k(r-a)-\omega t]}}{1 - ika} \qquad (5.1.4)$$

This clearly shows that the pressure disturbances at some point in the fluid are proportional to the amount of fluid being added or removed; for example, the larger the volume flow rate Q_0, the louder the resulting sound waves.

The acoustic pressure and velocity are, respectively,

$$\tilde{p}' = i\rho_0 \omega \tilde{\phi} \qquad (5.1.5)$$

and

$$\tilde{u}_r = ik\tilde{\phi}\left(1 + \frac{i}{kr}\right) \qquad (5.1.6)$$

As in Section 4.2, we note that $\tilde{u}_r$ contains a term that is significant only in the near-field $kr \ll 1$ and that does not contribute to the emitted intensity. The equation for $\tilde{u}_r$ can be written as

$$\tilde{u}_r = U_0 e^{-i\omega t + ik(r-a)} \frac{1 - ikr}{1 - ika} \left(\frac{a}{r}\right)^2 \qquad (5.1.7)$$

Therefore, the intensity at some point r is, with $I = \frac{1}{2}\mathrm{Re}(\tilde{p}'\tilde{u}^*)$,

$$I = \frac{1}{2}\rho_0 k\omega |\tilde{\phi}|^2 = \frac{1}{2}\rho_0 k\omega \left(\frac{Q_0}{4\pi r}\right)^2 \frac{1}{1 + (ka)^2} \qquad (5.1.8)$$

The total emitted power is obtained by integrating the intensity over a closed area enclosing the sphere:

$$\Pi = \tfrac{1}{2}\rho_0 c_0 \frac{(kQ_0)^2/4\pi}{1+(ka)^2} \tag{5.1.9}$$

For a given Q_0, this is proportional to ω^2. The power emitted per unit surface area of the sphere is

$$\Pi' = \rho_0 c_0 \frac{(ka)^2}{1+(ka)^2} \tfrac{1}{2} U_0^2 \tag{5.1.10}$$

For reasons that will become clear later, the coefficient of $\tfrac{1}{2} U_0^2$ in this expression is called the *radiation resistance R_r*.

Forces on the Pulsating Sphere

Also of interest are the forces that the fluid exerts on the sphere. These forces are important because they determine how much work the pulsating sphere can do on the surrounding fluid and, therefore, how much acoustic power is emitted. Now, the fluid forces on the sphere are radial, and have a magnitude given by

$$F = -4\pi a^2 p'(a, t) \tag{5.1.11}$$

or, in terms of the complex potential $\tilde{\phi}$,

$$\tilde{F} = -4\pi a^2 \rho_0 i\omega \tilde{\phi}(a, t) \tag{5.1.12}$$

Using $\tilde{\phi}$ as given by (5.1.1), we obtain

$$\tilde{F} = 4\pi\rho_0 a^3 \frac{1}{1-ika} i\omega U_0 e^{-i\omega t} \tag{5.1.13}$$

Since $U_0 e^{-i\omega t} = \tilde{U}$, this can also be written as

$$\tilde{F} = 4\pi\rho_0 a^3 \frac{1+ika}{1+(ka)^2} \tilde{U} i\omega$$

Further, since $-i\omega\tilde{U} = \dot{\tilde{U}}$ is the acceleration of the sphere's surface, we can write this as

$$\tilde{F} = -4\pi\rho_0 a^3\omega \frac{ka}{1+(ka)^2} \tilde{U} - 4\pi\rho_0 a^3 \frac{1}{1+(ka)^2} \dot{\tilde{U}} \tag{5.1.14}$$

The force that the body exerts on the fluid is just the negative of this, and is

$$\tilde{F}_f = 4\pi a^2 \rho_0 c_0 \frac{(ka)^2}{1+(ka)^2} \tilde{U} + 4\pi\rho_0 a^3 \frac{1}{1+(ka)^2} \dot{\tilde{U}} \tag{5.1.15}$$

The first part of the force on the sphere is called the *dissipative part* because it results in a net amount of dissipation. This follows from the fact that this part is in phase with the velocity so that by necessity the time average of the rate at which the sphere does work is positive. To show this, we compute the work done on the fluid. This is given by

$$\langle F_f U \rangle = \tfrac{1}{2} \operatorname{Re}(\tilde{F}_f \tilde{U}^*) \tag{5.1.16}$$

Therefore, the average rate at which each unit surface area does work is

$$\Pi' = \frac{\langle F_f U \rangle}{4\pi a^2} = \rho_0 c_0 \frac{(ka)^2}{1+(ka)^2} \tfrac{1}{2} U_0^2 \tag{5.1.17}$$

Comparison with (5.1.10) shows that this energy input is radiated in the form of acoustic waves. Returning to (5.1.15), it is clear that we may think of the dissipative part of the force as a "resistance" due to radiation. Therefore, if we write the resistive force per unit area as $R_r \tilde{U}$, then

$$\Pi' = R_r \tfrac{1}{2} U_0^2 \tag{5.1.18}$$

where R_r, the radiation resistance is given by

$$R_r = \rho_0 c_0 \frac{(ka)^2}{1+(ka)^2} \tag{5.1.19}$$

The maximum value of R_r (and therefore the maximum radiated power per unit area) is $\rho_0 c_0$, the characteristic resistance of the fluid. This value is approached for large ka, as sketched in Figure 5.1.1.

The second part of the force in (5.1.15) may be written as

$$-G = \tfrac{4}{3}\pi a^3 \rho_0 \frac{3}{1+(ka)^2} \frac{dU}{dt} \tag{5.1.20}$$

This is called the *inertial part* of the force because it is proportional to the instantaneous acceleration of the sphere's surface. Since it is 90 deg. out of phase with respect to the surface velocity, it does not lead to any energy expenditure

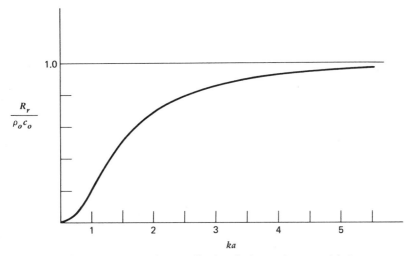

Figure 5.1.1 Variations of normalized radiation resistance with frequency.

(in particular, it produces no acoustic waves). The force G is also said to be *reactive* because it always acts in a direction opposite to the motion of the sphere's surface. This reaction is seen from (5.1.20) to be proportional to a mass of fluid. In fact, for some purposes, it is convenient to think of the coefficient of dU/dt in (5.1.20) as representing a mass of fluid that oscillates with the sphere and therefore increases the sphere's inertia. To clarify these statements, let us consider the near-field fluid velocity. This is given by

$$\tilde{U}_{nf} = \left(\frac{a}{r}\right)^2 \frac{\tilde{U}}{1-ika} \tag{5.1.21}$$

where $\tilde{U} = U_0 e^{-i\omega t}$ is the sphere's surface velocity. The average kinetic energy associated with U_{nf} is

$$\langle T_{nf} \rangle = \tfrac{1}{2}\rho_0 \int_V \langle U_{nf}^2 \rangle \, dV \tag{5.1.22}$$

where the volume of integration is that occupied by the fluid. By assumption, this extends to infinity. Thus, since $dV = r^2 \sin\theta \, dr \, d\theta \, d\varphi$, we obtain

$$\langle T_{nf} \rangle = \tfrac{1}{2}\rho_0 \int_0^{2\pi} \int_0^\pi \int_a^\infty \tfrac{1}{2}\tilde{U}_{nf}\tilde{U}_{nf}^* r^2 \sin\theta \, d\theta \, d\varphi \, dr \tag{5.1.23}$$

Substitution from (5.1.21) yields, upon integration.

$$\langle T_{nf} \rangle = \tfrac{1}{2}\left(\tfrac{4}{3}\pi a^3 \rho_0\right)\frac{3}{1+(ka)^2}U_{rms}^2 \qquad (5.1.24)$$

where $U_{rms} = (1/\sqrt{2})U_0$ is the root-mean-squared surface velocity. If we write (5.1.24) in the usual form for the average kinetic energy for an oscillating body of mass M (i.e., $\langle T_{nf} \rangle = \tfrac{1}{2}MU_{rms}^2$), we see that the mass M_0 of fluid related to this near-field is given by

$$M_0 = \rho_0 V_s \frac{3}{1+(ka)^2} \qquad (5.1.25)$$

where V_s is the volume of the sphere. If we remember that the near-field is significant only in the vicinity of the sphere,[1] we may say that the near-field occupies a volume V_{nf} around the sphere given by

$$V_{nf} = V_s \frac{3}{1+(ka)^2} \qquad (5.1.26)$$

The situation is illustrated schematically in Figure 5.1.2. It should be noted that as ka increases, the near-field region and the inertial force decrease, while the radiated intensity increases. Opposite effects occur for decreasing ka.

The mass M_0 given by (5.1.25) represents, in effect, an increase of the mass of the vibrating surface, which is due to the fluid's reaction. The situation is somewhat analogous to that which occurs when a *translating* rigid body accelerates in an incompressible fluid. There, it is found that the fluid exerts a force on the body that may be symbolically[2] expressed as

$$G = -\bar{\alpha}\rho_0 V_b \frac{du}{dt} \qquad (5.1.27)$$

where V_b is the volume of the body and $\bar{\alpha}$ is a second-order tensor that depends on the shape and size of the body. Thus, in the case of the translating body, we can write

$$M_b \frac{du}{dt} = G + F \qquad (5.1.28)$$

[1] The "near-field" extends to infinity, as (5.1.21) shows. However, as we move far from the sphere, the contributions of this field to $\langle T_{nf} \rangle$ are negligible, as they decrease as $1/r^4$.

[2] The actual expression gives the ith component of G as a linear combination of the components of (du/dt). Thus, $G_i = -\rho_0 V_0 \alpha_{ij}(du_j/dt)$.

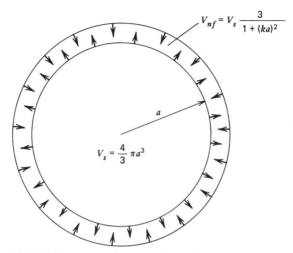

$$V_{nf} = V_s \frac{3}{1 + (ka)^2}$$

$$V_s = \frac{4}{3}\pi a^3$$

Figure 5.1.2 Volume around a pulsating sphere due to added mass M_0.

where F represents all other forces applied on the sphere. Using (5.1.27), this may be written as

$$\left(M_b + \bar{\alpha}\rho_0 V_b\right)\frac{du}{dt} = F \tag{5.1.29}$$

This shows that the net effect of G on the translating body can be taken into account by adding to its mass a mass $\bar{\alpha}\rho_0 V_b$. This quantity is therefore called the *added mass*, and $\bar{\alpha}$ is called the coefficient of virtual inertia.

If this analogy is extended to the pulsating sphere, (5.1.25) shows that the coefficient of virtual inertia for that case is

$$\bar{\alpha} = \frac{3}{1 + (ka)^2} \tag{5.1.30}$$

Pulsating Bubble

The significance of the added-mass concept may be easier to grasp by considering a gas bubble immersed in a liquid host and performing radial pulsations at its natural frequency of oscillation. The value of this frequency is of interest, as it determines the frequency of the waves emitted by the bubble. As in other free-oscillation situations, we require an oscillating mass and a restoring, springlike force. For the bubble, there are two contributions to the restoring force: (a) surface tension and (b) compressibility of the gas inside the bubble.

For simplicity, we ignore surface-tension effects, although they could be taken into account easily (see Problem 5.1.2). Thus, "springiness" will be due to the increase and decrease in pressure that take place when the bubble pulsates. The question remains: What is the oscillating mass? Clearly, the surface of the bubble is massless, as it merely separates the gas from its liquid host. However, the liquid surrounding the bubble is oscillating radially, and our argument above indicates that its effect on the pulsating sphere is equivalent to an increase of the surface inertia by an amount $M_0 = \alpha \rho_0 V_s$. This, then, is the oscillating mass, and according to (5.1.25) it is given by

$$M_0 = \frac{4\pi a^3 \rho_0}{1 + (ka)^2}$$

where a is the mean radius of the bubble. Now, if the instantaneous radial displacement of the bubble surface, measured from the equilibrium position, is $\varepsilon(t)$, then the equation of motion becomes

$$M_0 \frac{d^2\varepsilon}{dt^2} = 4\pi a^2 (p_b - p_0) \tag{5.1.31}$$

where p_0 is the pressure outside the bubble and p_b is the pressure inside. In equilibrium, we have $p_b = p_0$, and the volume of the bubble is $V_0 = \frac{4}{3}\pi a^3$. Thus, if we assume that the gas in the bubble behaves as a perfect gas, and if we take the oscillations to be adiabatic (as we should, since we are dealing with ideal fluids), then

$$p_b = p_0 (V_0/V)^\gamma \tag{5.1.32}$$

where γ is the ratio of specific heats of the gas in the bubble. Hence, with $V = \frac{4}{3}\pi(a+\varepsilon)^3$ and $\varepsilon \ll a$, we obtain

$$p_b - p_0 \approx -3\gamma p_0 \varepsilon / a \tag{5.1.33}$$

and our equation of motion becomes

$$\frac{d^2\varepsilon}{dt^2} + \frac{3\gamma p_0}{\rho_0 a^2} \left[1 + (ka)^2\right] \varepsilon = 0 \tag{5.1.34}$$

Thus, the natural frequency of oscillation is given by

$$\omega^2 = \frac{3\gamma p_0}{\rho_0 a^2} \left[1 + \left(\frac{\omega a}{c_0}\right)^2\right] \tag{5.1.35}$$

where c_0 is the speed of sound in the host fluid. Since the gas in the bubble is perfect, its speed of sound is $c_g^2 = \gamma p_0 / \rho_b$, so that

$$\omega^2 = \frac{3\rho_b}{\rho_0 a^2} c_g^2 \left[1 + \left(\frac{\omega a}{c_0} \right)^2 \right] \tag{5.1.36}$$

Solving for ω, we obtain

$$\omega = \frac{c_g}{a} \sqrt{\frac{3\delta^{-1}}{1 - 3\delta^{-1}(c_g/c_0)^2}} \tag{5.1.37}$$

where $\delta = \rho_0 / \rho_b$ is the ratio of liquid density to gas density. Since this ratio is very large, and since c_g is usually smaller than c_0, we may write

$$\omega = \frac{c_g}{a} \sqrt{3 \frac{\rho_b}{\rho_0}} \tag{5.1.38}$$

a result that could have been obtained earlier by assuming a priori that ka would be very small.

Simple Source

Before we study more general sound-emission problems, we return to the monochromatic potential for the pulsating sphere. This is given by (5.1.4), or

$$\tilde{\phi}(r, t) = -\frac{Q_0}{4\pi r} \frac{e^{i[k(r-a)-\omega t]}}{1 - ika}$$

We notice that in the far-field (i.e., when $kr \gg 1$), when $a \ll \lambda$, we obtain a potential in which the radius of the sphere does not appear. Therefore, if we could, somehow, introduce at $r = 0$ an amount of fluid at the rate Q, we would obtain at a distance r the same effects produced by a finite source. This leads, of course, to the point-source concept, so useful in incompressible fluid dynamics. The only difference is that here the fluid must be introduced at a fluctuating rate rather than at a constant rate.

The potential due to a point source is obtained from $\tilde{\phi}$ above by taking the limit $a \to 0$ but keeping Q_0 constant. This yields

$$\tilde{\phi}(r, t) = -\frac{Q_0}{4\pi r} e^{i(kr - \omega t)} \tag{5.1.39}$$

This is the so-called *simple-source potential*. The corresponding intensity and power are

$$I = \frac{\rho_0 \omega^2 Q_0^2}{32 \pi^2 r^2 c_0} \tag{5.1.40}$$

and

$$\Pi = \frac{\rho_0 \omega^2 Q_0^2}{8 \pi c_0} \tag{5.1.41}$$

To obtain the corresponding result for nonmonochromatic time dependence, we return to Section 4.2. We notice from (4.2.7) that as the pulsating-sphere radius decreases to zero, the potential becomes

$$\phi = - \frac{a^2}{r} U\left(t - \frac{r}{c_0}\right) \tag{5.1.42}$$

where $U(t)$ is the velocity of the sphere's surface. Two important forms of this result can be written that do not show explicitly the radius a. Thus, since the rate at which fluid volume is generated is $Q(t) = 4\pi a^2 U(t)$, it follows that

$$\phi = - \frac{1}{4\pi r} Q\left(t - \frac{r}{c_0}\right) \tag{5.1.43}$$

This result is the nonmonochromatic equivalent of (5.1.39).

The second important form follows from the fact that $Q(t)$ is equal to the rate at which the sphere volume V_s changes, that is, $Q(t) = \dot{V}_s(t)$. Thus,

$$\phi = - \frac{1}{4\pi r} \dot{V}_s\left(t - \frac{r}{c_0}\right) \tag{5.1.44}$$

Either (5.1.43) or (5.1.44) may be used as the basis for the study of sound emission by vibrating bodies: (5.1.44) in terms of the body changes of volume and (5.1.43) in terms of the sources of fluid representing the volume changes.

PROBLEMS

5.1.1   When surface tension is included, the pressure inside a small gas bubble is larger than that outside by an amount $4\sigma/a$, where σ is the coefficient of surface tension and a is the radius of the bubble. Determine the natural frequency of pulsation when σ is finite.

5.1.2 Compute the kinetic energy associated with the far-field velocity. Explain your results in physical terms.

5.1.3 An air bubble of radius $a = 0.1$ cm executes harmonic pulsations, at its natural frequency, of amplitude $\varepsilon = 0.01a$. If the bubble is immersed in water at 20°C, determine the emitted acoustic power.

5.1.4 Consider a body of volume V immersed in a fluid of infinite extent. The body is pulsating in such a manner that its volume changes at the rate $\dot{V}$. Far from the body, the sound waves produced by the pulsations are spherical. Show that for a body of arbitrary shape, the far-field potential is

$$\phi = -\frac{1}{4\pi r}\dot{V}\left(t - \frac{r}{c_0}\right)$$

[*Hint*: Near the body, ϕ satisfies Laplace's equation (see Section 4.2). The leading term in the solution of $\nabla^2\phi = 0$ is $\phi = A/r$, where A is a constant.]

5.1.5 With reference to Figure 5.1.2, show that when $ka \ll 1$, the added mass effects may be accounted for by assuming that the pulsating sphere has an effective radius a_e equal to $4^{1/3}a$.

5.2 INHOMOGENEOUS WAVE EQUATION

In this section we show that the simple-source potential is a solution of the wave equation everywhere except at the origin. This exception arises because when we derived the basic equations of motion, we did not allow for the possibility of having sources within the fluid. We now include that possibility. Thus, if $Q(\mathbf{x}, t)$ represents the rate at which fluid volume is added at time t and position $\mathbf{x}$, per unit volume of fluid, the mass-conservation balance should be

$$\frac{d}{dt}\int_V \rho\, dV = -\int_A \rho\mathbf{u}\cdot\mathbf{n}\, dA + \int \rho Q(\mathbf{x}, t)\, dV \qquad (5.2.1)$$

Proceeding as in Section 1.4, we obtain

$$\frac{\partial\rho}{\partial t} + \nabla\cdot(\rho\mathbf{u}) = \rho Q(\mathbf{x}, t) \qquad (5.2.2)$$

In linearized form, with $p' = c_0^2\rho'$, this reduces to

$$\frac{\partial p'}{\partial t} + \rho_0 c_0^2\nabla\cdot\mathbf{u} = \rho_0 c_0^2 Q(\mathbf{x}, t) \qquad (5.2.3)$$

The momentum equation also needs modification because the sources of fluid increase the momentum entering a fluid element. This momentum addition is $\mathbf{u}(\mathbf{x}, t)Q(\mathbf{x}, t)$ per unit volume. Therefore, the equation describing the conservation of momentum is

$$\int_{\tau}\left[\frac{\partial(\rho u_i)}{\partial t}+\frac{\partial}{\partial x_j}(\rho u_i u_j)\right]d\tau=-\int_{\tau}\nabla p\,d\tau+\int_{\tau}u_i Q\rho\,d\tau \qquad (5.2.4)$$

Thus,

$$\frac{\partial(\rho u_i)}{\partial t}+\frac{\partial}{\partial x_j}(\rho u_i u_j)=-\frac{\partial p}{\partial x_i}+u_i Q\rho \qquad (5.2.5)$$

However, when this equation is linearized, we recover the usual Euler equation

$$\rho_0\frac{\partial \mathbf{u}}{\partial t}+\nabla p'=0 \qquad (5.2.6)$$

From (5.2.2) and (5.2.6), we can obtain an equation for the velocity potential:

$$-\frac{\partial^2\phi}{\partial t^2}+c_0^2\nabla^2\phi=c_0^2 Q(\mathbf{x}, t) \qquad (5.2.7)$$

or for the pressure disturbance:

$$\frac{\partial^2 p'}{\partial t^2}-c_0^2\nabla^2 p'=\rho_0 c_0^2\left(\frac{\partial Q}{\partial t}\right) \qquad (5.2.8)$$

These are examples of the inhomogeneous scalar wave equation, about which much is known. In particular, if the source distribution is prescribed, it is possible to obtain a solution in terms of an integral over a volume enclosing the sources. To obtain the solution, we first consider the case where the distribution of sources is limited to the origin. That is, $Q(\mathbf{x}, t)=0$ unless $\mathbf{x}=0$. This type of distribution can be described in terms of the Dirac-delta function $\delta(\mathbf{x})=\delta(x)\delta(y)\delta(z)$ defined by

$$\int_V f(\mathbf{x})\delta(\mathbf{x})\,dV(x)=\begin{cases} f(0) & \text{if } V \text{ includes the origin} \\ 0 & \text{otherwise} \end{cases} \qquad (5.2.9)$$

Thus, a point source at $\mathbf{x}=0$, of strength $Q(t)$, can be represented by

$$Q(\mathbf{x}, t)=\delta(\mathbf{x})Q(t) \qquad (5.2.10)$$

Suppose that $Q(t)$ is harmonic; that is, $\tilde{Q}(t) = Q_0 e^{-i\omega t}$. Then we can assume that ϕ also will depend harmonically on time so that the equation for ϕ becomes

$$\nabla^2\tilde{\phi} + k^2\tilde{\phi} = \delta(\mathbf{x})Q \qquad (5.2.11)$$

We want to show that a solution to this equation is

$$\tilde{\phi} = -\frac{e^{ikr}}{4\pi r}Q \qquad (5.2.12)$$

First, we show that outside $r = 0$ this is a solution to the time-independent wave equation

$$\nabla^2\tilde{\phi} + k^2\tilde{\phi} = 0 \qquad (5.2.13)$$

To accomplish this, we must show that for $r \neq 0$

$$\nabla^2\frac{e^{ikr}}{r} = -k^2\frac{e^{ikr}}{r} \qquad (5.2.14)$$

This easily can be shown to be the case by expressing the Laplacian operator in spherical polar coordinates and by taking derivatives. However, for later applications we will use the indicial notation introduced in Chapter 1. Thus, since $r = \sqrt{x_j x_j}$ and $\delta x_j/\delta x_i = \delta_{ij}$, we obtain the following results:

$$\frac{\partial r}{\partial x_i} = \frac{x_i}{r} \qquad \text{or} \qquad \nabla r = \frac{\mathbf{x}}{r}$$

$$\frac{\partial r^{-1}}{\partial x_i} = -\frac{x_i}{r^3} \qquad \text{or} \qquad \nabla\frac{1}{r} = -\frac{\mathbf{x}}{r^3} \qquad (5.2.15)$$

These can be combined with $(\partial/\partial x_i)\exp(ikr) = ik\exp(ikr)(\partial r/\partial x_i)$ to yield

$$\nabla\frac{e^{ikr}}{r} = ik\left(\frac{\mathbf{x}}{r^2} - \frac{\mathbf{x}}{r^3}\right)e^{ikr}$$

$$= \frac{\mathbf{x}}{r^3}(ikr - 1)e^{ikr} \qquad (5.2.16)$$

Similarly, since $\partial x_i/\partial x_i = 3$, $\partial r^{-2}/\partial x_i = -2x_i/r^4$, and $\partial r^{-3}/\partial x_i = -3x_i/r^5$, we obtain

$$\frac{\partial^2}{\partial x_i \partial x_i}\frac{e^{ikr}}{r} = -k^2\frac{e^{ikr}}{r}$$

as required. Consider, now, a region that includes the origin. Here we must show that (5.2.12) satisfies (5.2.11), or

$$-Q\left[\nabla^2\frac{e^{ikr}}{r}+k^2\frac{e^{ikr}}{r}\right]=4\pi\delta(\mathbf{x})Q \tag{5.2.17}$$

That is, we want to show that the function $\exp(ikr)/r$ satisfies

$$\nabla^2\frac{e^{ikr}}{r}+k^2\frac{e^{ikr}}{r}=-4\pi\delta(\mathbf{x}) \tag{5.2.18}$$

Integrating (5.2.18) over a small volume that includes the origin, we obtain -4π on the right-hand side. Therefore,

$$\int_V\nabla^2\frac{e^{ikr}}{r}dV+k^2\int_V\frac{e^{ikr}}{r}dV=-4\pi \tag{5.2.19}$$

Let us denote the first volume integral by I_1 and use the divergence theorem to transform it into an area integral. Thus, if A_ε is a sphere of small-radius ε with outside normal $\mathbf{n}$, then

$$I_1=\int_{A_\varepsilon}\mathbf{n}\cdot\nabla\frac{e^{ikr}}{r}dA=\int_0^{2\pi}d\varphi\int_0^\pi\left[\frac{d}{dr}\frac{e^{ikr}}{r}\right]_{r=\varepsilon}\varepsilon^2\sin\theta\,d\theta \tag{5.2.20}$$

The quantity inside the square brackets, evaluated at $r=\varepsilon$, is simply $(ik\varepsilon-1)/\varepsilon^2$. Hence, in the limit as $\varepsilon\to0$, we obtain $I_1=-4\pi$.

The second integral in (5.2.19) gives $4\pi\int_0^\varepsilon r\exp(ikr)\,dr$, and this vanishes as $\varepsilon\to0$. Thus, if $\mathbf{x}=0$ is included in the volume of integration,

$$\int\left[\nabla^2\frac{e^{ikr}}{r}+k^2\frac{e^{ikr}}{r}\right]dV=-4\pi \tag{5.2.21}$$

From these results, we conclude that

$$\tilde{\phi}(\mathbf{x},t)=-\frac{Q_0}{4\pi r}e^{ikr-i\omega t} \tag{5.2.22}$$

is a solution of (5.2.7) for the case when there is an isolated point source at the origin. If the singularity is not at the origin, but at some other point $\mathbf{x}'$, say, then

$$\tilde{\phi}(\mathbf{x},t)=-\frac{e^{ik|\mathbf{x}-\mathbf{x}'|}}{4\pi|\mathbf{x}-\mathbf{x}'|}Q_0e^{-i\omega t} \tag{5.2.23}$$

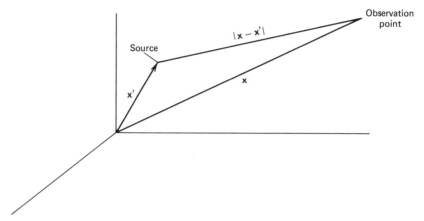

Figure 5.2.1 Locations of source and field points.

The quantity $|\mathbf{x} - \mathbf{x}'|$ replaces r as the distance between source and field point, as shown in Figure 5.2.1. The quantity

$$g(\mathbf{x}|\mathbf{x}') = -\frac{e^{ik|\mathbf{x}-\mathbf{x}'|}}{4\pi|\mathbf{x}-\mathbf{x}'|} \qquad (5.2.24)$$

is an example of a class of functions called the Green's functions, which are solutions of the inhomogeneous wave equation. The function g given above is the Green's function for unbounded media. It plays an important role in emission problems, and it is the basis to study a variety of other important problems. Here we merely point out that $g(\mathbf{x}|\mathbf{x}')$ is the time-independent potential produced at $\mathbf{x}$ by a simple source at $\mathbf{x}'$. We also note that

$$g(\mathbf{x}|\mathbf{x}') = g(\mathbf{x}'|\mathbf{x}) \qquad (5.2.25)$$

that is, source and field points are interchangeable. This is a particular example of an important principle in acoustics known as the reciprocity principle, which states that under some conditions, the field produced at a point $\mathbf{x}$, say, due to a source at $\mathbf{x}'$, is the same as that produced at $\mathbf{x}'$ by a source at $\mathbf{x}$. This principle is discussed in Sec. 5.11.

Similarly, if we have several sources, located at $\mathbf{x}'_0, \mathbf{x}'_1, \mathbf{x}'_2, \ldots$ and of strengths $Q_0, Q_1, Q_2, \ldots$, respectively, then, since the wave equation is linear, we can add their contributions to the field at $\mathbf{x}$:

$$\tilde{\phi}(\mathbf{x}, t) = -\frac{1}{4\pi} \sum_\alpha \frac{Q_\alpha(\mathbf{x}'_\alpha)e^{ik|\mathbf{x}-\mathbf{x}'_\alpha|-i\omega t}}{|\mathbf{x}-\mathbf{x}'_\alpha|} \qquad (5.2.26)$$

Linear Array

While (5.2.26) may be used to obtain the field due to a variety of arrays of simple sources, we consider here only an array of n equal sources, each of strength Q_0, evenly spaced along a straight line as shown in Figure 5.2.2. Let us take this line to coincide with the x axis. Then $x'_\alpha = (x_\alpha, 0, 0)$. Now, since the sources are evenly spaced, and the first one is at the origin, we have $x_\alpha = (\alpha - 1) d$, where d is the spacing. Hence

$$\tilde{\phi} = -\frac{1}{4\pi} Q_0 e^{-i\omega t} \sum_{\alpha=1}^{n} \frac{\exp\left\{ ik\sqrt{[x-(\alpha-1)d]^2 + y^2} \right\}}{\sqrt{[x-(\alpha-1)d]^2 + y^2}} \qquad (5.2.27)$$

The sum in this equation may be evaluated explicitly for points in the far-field of the array. For these we have

$$|\mathbf{x} - \mathbf{x}'_\alpha| = \sqrt{x^2 + y^2 - 2(\alpha-1)xd + (\alpha-1)^2 d^2} \approx r\left[1 - \frac{(\alpha-1)xd}{r^2} \right] \qquad (5.2.28)$$

where $r^2 = x^2 + y^2$. We note that if only the leading term of this expansion is used in (5.2.27), we obtain

$$\tilde{\phi} = -\frac{1}{4\pi r} n Q_0 e^{i(kr-\omega t)} \qquad (5.2.29)$$

This represents a point source at the origin, of strength $Q_T = nQ_0$, which is

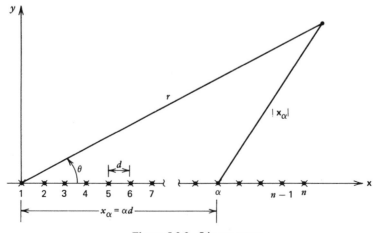

Figure 5.2.2 Linear array.

simply the total strength of the array. Retaining the two terms in (5.2.28) yields

$$\tilde{\phi} = -\frac{1}{4\pi r} Q_T e^{i(kr-\omega t)} \frac{1}{n} \sum_{\alpha=1}^{n} e^{-ikd(\alpha-1)\cos\theta} \tag{5.2.30}$$

where θ is the angle between **x** and the axis of the array. Equation (5.2.30) also represents a point source at the origin, but this time the field is not uniform around the origin. Rather, it is modified by a *directivity* function

$$D(\theta) = \frac{1}{n} \sum_{\alpha=1}^{n} e^{-idk(\alpha-1)\cos\theta} \tag{5.2.31}$$

The finite sum in this equation may be evaluated by using the well-known finite sum of the geometric progression:

$$\sum_{\alpha=1}^{n} q^{\alpha-1} = \frac{q^n-1}{q-1}$$

Thus

$$D(\theta) = \frac{1}{n} \frac{e^{-ikdn\cos\theta}-1}{e^{-ikd\cos\theta}-1} \tag{5.2.32}$$

This may be simplified by introducing an angle Θ defined by

$$\Theta = \tfrac{1}{2} kd\cos\theta \tag{5.2.33}$$

This yields

$$|D| = \left|\frac{\sin n\Theta}{n\sin\Theta}\right| \tag{5.2.34}$$

where $|D|$ is the amplitude of the directivity. Equation (5.2.34) may be used to plot the field pattern around the origin for various values of n and dk (see Problem 5.2.4).

Continuous Distributions

If instead of a discrete distribution of sources we have a continuous one, the potential becomes

$$\tilde{\phi}(\mathbf{x}, t) = -\frac{1}{4\pi} \int \frac{Q(\mathbf{x}')e^{ik|\mathbf{x}-\mathbf{x}'|-i\omega t}}{|\mathbf{x}-\mathbf{x}'|} dV(\mathbf{x}') \tag{5.2.35}$$

In deriving this equation, we have assumed that the sources had a strength that depended on position but vibrated in phase with the same frequency ω. Different frequencies can be included in terms of Fourier transforms. We can, however, also write the solution for ϕ by using the point-source result derived in Section 5.1. That result showed that if we had a point source at the origin whose volume flow rate was given by $Q(t)$, then the potential at $\mathbf{x}$ was given by $\phi(\mathbf{x}, t) = -Q(t - r/c_0)/4\pi r$. Therefore, if the point source is at $\mathbf{x}'$ and its strength is a function of position [i.e., $Q = Q(\mathbf{x}, t)$],

$$\tilde{\phi}(\mathbf{x}, t) = -\frac{1}{4\pi|\mathbf{x} - \mathbf{x}'|} Q\left(\mathbf{x}', t - \frac{|\mathbf{x} - \mathbf{x}'|}{c_0}\right) \tag{5.2.36}$$

where $t - |\mathbf{x} - \mathbf{x}'|/c_0$ is the retarded time and the $\mathbf{x}'$ in the argument of Q implies that Q is a function of position. Thus, for a continuous source distribution, the potential is given by an integral over the distribution, or

$$\tilde{\phi}(\mathbf{x}, t) = -\frac{1}{4\pi} \int_V \frac{Q\left(\mathbf{x}', t - \frac{|\mathbf{x} - \mathbf{x}'|}{c_0}\right)}{|\mathbf{x} - \mathbf{x}'|} \, dV(\mathbf{x}') \tag{5.2.37}$$

In terms of this representation, the pressure fluctuation is given by

$$p'(\mathbf{x}, t) = \frac{1}{4\pi} \frac{\partial}{\partial t} \int_V \frac{\rho_0 Q\left(\mathbf{x}', t - \frac{|\mathbf{x} - \mathbf{x}'|}{c_0}\right)}{|\mathbf{x} - \mathbf{x}'|} \, dV(\mathbf{x}') \tag{5.2.38}$$

Since $\rho_0 Q$ is the mass addition rate, this equation shows that unless mass is added at a time-dependent rate, there will be no emission of sound waves from the distribution.

Equation (5.2.35) gives the potential in terms of a volume distribution of sources. However, there are cases when the distributions are limited to certain lines or surfaces. We consider these separately.

Line Distributions.   Consider, first, the special case where Q vanishes everywhere except along a line whose cross-sectional area is $\delta A(\mathbf{x}')$, as shown in Figure 5.2.3. Then,

$$Q\delta V = Q\delta A(\mathbf{x}')\delta l \tag{5.2.39}$$

Let $\delta A \to 0$ and $Q \to \infty$ such that $Q\delta A \to Q_l$, where Q_l represents the volume flow rate per unit length. Then, for a simple harmonic source distribution along this

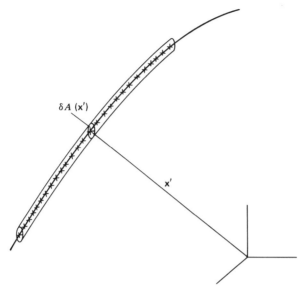

Figure 5.2.3 Line distribution of sources.

line,

$$\tilde{\phi}(\mathbf{x}, t) = -\frac{e^{-i\omega t}}{4\pi} \int Q_l(\mathbf{x}') \frac{e^{ik|\mathbf{x}-\mathbf{x}'|}}{|\mathbf{x}-\mathbf{x}'|} \, dl(\mathbf{x}') \tag{5.2.40}$$

or if the distribution is not monochromatic,

$$\phi(\mathbf{x}, t) = -\frac{1}{4\pi} \int \frac{Q_l\left(\mathbf{x}', t - \dfrac{|\mathbf{x}-\mathbf{x}'|}{c_0}\right)}{|\mathbf{x}-\mathbf{x}'|} \, dl(\mathbf{x}') \tag{5.2.41}$$

where $\mathbf{x}'$ is the position on the line where the source strength per unit length is $Q_l(\mathbf{x}')$.

EXAMPLES

(1) Monochromatic Time Dependence. As an example, we consider the case where the distribution is rectilinear and uniform and extends to infinity in both directions, as shown in Figure 5.2.4. Furthermore, we limit this calculation to the harmonic dependence on time. Thus, with

$$Q_l = Q_0 e^{-i\omega t} \tag{5.2.42}$$

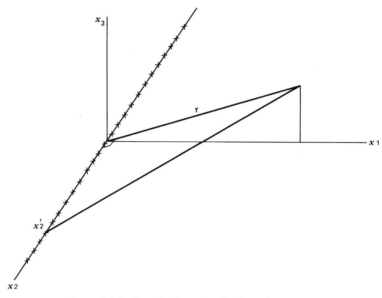

Figure 5.2.4   Straight-line distribution of sources.

and with the uniform distribution coinciding with the x_2 axis, we obtain

$$\tilde{\phi} = -\frac{Q_0 e^{-i\omega t}}{4\pi} \int_{-\infty}^{\infty} \frac{e^{ik|\mathbf{x}-\mathbf{e}_2 x_2'|}}{|\mathbf{x}-\mathbf{e}_2 x_2'|}\, dx_2' \tag{5.2.43}$$

Now, $|\mathbf{x}-\mathbf{e}_2 x_2'| = \sqrt{r^2 + x_2'^2} = r\sqrt{1+z^2}$, so that

$$\tilde{\phi} = -\frac{Q_0 e^{-i\omega t}}{4\pi} \int_{-\infty}^{\infty} \frac{e^{ikr\sqrt{1+z^2}}}{\sqrt{1+z^2}}\, dz \tag{5.2.44}$$

When divided by πi, the integral is one of the definitions of $H_0(kr)$, the Hankel function of order zero. Hence,

$$\tilde{\phi} = -\frac{i}{4} Q_0 e^{-i\omega t} H_0(kr) \tag{5.2.45}$$

This is equivalent to the solution presented in Section 4.4 for the case of waves with axial symmetry.

(2) General Time Dependence. When the source strength does not vary harmonically in time, we have, for an infinite line source along the x_2 axis,

$$\phi = -\frac{1}{4\pi} \int_{-\infty}^{\infty} \frac{Q_l\left[x_2', t - \dfrac{|\mathbf{x} - \mathbf{e}_2 x_2'|}{c_0}\right]}{|\mathbf{x} - \mathbf{e}_2 x_2'|} dx_2' \tag{5.2.46}$$

Here $|\mathbf{x} - \mathbf{e}_2 x_2'|$ is the distance between observation and source points, and $\mathbf{e}_2$ is a unit vector along the x_2 axis. Thus,

$$|\mathbf{x} - \mathbf{e}_2 x_2'| = \sqrt{x_1^2 + x_3^2 + x_2'^2} \tag{5.2.47}$$

If we introduce the notation $r = \sqrt{x_1^2 + x_3^2}$, (5.2.46) becomes

$$\phi = -\frac{1}{4\pi} \int_{-\infty}^{\infty} \frac{Q_l\left[x_2', t - \dfrac{\sqrt{r^2 + x_2'^2}}{c_0}\right]}{\sqrt{r^2 + x_2'^2}} dx_2' \tag{5.2.48}$$

If $Q_l(x_2', t)$ is specified, then (5.2.48) can be used to obtain the potential at $\mathbf{x}$. This equation describes cylindrical waves produced by a line source whose strength varies along its length. When the strength is constant, we should recover a result equivalent to that presented in Section 4.5. Thus, if $Q_l = Q_l(t)$ only, the integrand in (5.2.48) is symmetric with respect to x_2', and we can write

$$\phi = -\frac{1}{2\pi} \int_{0}^{\infty} \frac{Q_l\left[t - \dfrac{\sqrt{r^2 + x_2'^2}}{c_0}\right]}{\sqrt{r^2 + x_2'^2}} dx_2' \tag{5.2.49}$$

If we now let $\eta = t - \sqrt{r^2 + x_2'^2}\,/c_0$,

$$\phi = -\frac{1}{2\pi} \int_{-\infty}^{t-r/c_0} \frac{Q_l(\eta)\,d\eta}{\sqrt{(t-\eta)^2 - r^2/c_0^2}} \tag{5.2.50}$$

This may be recognized as the outgoing part of the general solution of the wave equation with axial symmetry that was presented in Section 4.5. The corresponding pressure and velocity are given by (4.5.20) and (4.5.21), respectively.

Surface Distribution.   This is an important type of distribution, because vibrating surfaces can be simulated in terms of suitable source distributions.

Suppose that sources are distributed on some surface of thickness $\delta\tau$. Then,

$$Q\delta V = Q\delta\tau\delta A$$

so that if we let $\delta\tau \to 0$ and $Q \to \infty$ in such a way that $Q\delta\tau = Q_a$, where Q_a represents the volume flow rate per unit surface area, we obtain

$$\tilde{\phi}(\mathbf{x}, t) = -\frac{1}{4\pi}\int_A Q_a(\mathbf{x}')\frac{e^{ik|\mathbf{x}-\mathbf{x}'|-i\omega t}}{|\mathbf{x}-\mathbf{x}'|}dA(\mathbf{x}') \qquad (5.2.51)$$

Here, $Q_a(\mathbf{x}')\,dA(\mathbf{x}')$ is the volume flow rate emitted by a surface element located at $\mathbf{x}'$. We would like to express Q_a in terms of the velocity of the surface we are attempting to represent with sources. For this purpose, consider a volume element in the form of a pillbox enclosing part of the surface distribution, as shown in Figure 5.2.5. The continuity equation, integrated over a volume V

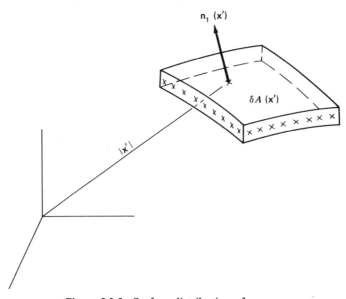

Figure 5.2.5   Surface distribution of sources.

bounded by surface area A, is

$$\int_V \frac{\partial \rho'}{\partial t} dV = - \int_A \rho_0 \mathbf{n} \cdot \mathbf{u} \, dA + \int_A \rho_0 Q_a \, dA \tag{5.2.52}$$

Let us apply this equation to the volume shown in the figure, assuming that the thickness of the pillbox is very small compared to its other dimensions. As the thickness is decreased to zero, the contributions to the area integrals arising from the thin sides can be neglected. Similarly, the volume integral can also be neglected, so that if the area of the wide face of the pillbox is δA, then

$$Q_a = \mathbf{n}_1 \cdot \mathbf{u}_1 + \mathbf{n}_2 \cdot \mathbf{u}_2 \tag{5.2.53}$$

In terms of the velocity potential, this can be written as

$$Q_a = \mathbf{n}_1 \cdot \nabla \phi + \mathbf{n}_2 \cdot \nabla \phi \tag{5.2.54}$$

This shows, incidentally, that while the velocity potential is continuous across the surface [see Equation (5.2.51)], its normal gradient is not.

In some cases, simplifications can be made in the boundary condition given by (5.2.54). Thus, when the surface is plane:

$$\left(\frac{\partial \phi}{\partial n} \right)_1 = \left(\frac{\partial \phi}{\partial n} \right)_2 = \frac{\partial \phi}{\partial n} \tag{5.2.55}$$

and so

$$Q_a = 2 \frac{\partial \phi}{\partial n} \tag{5.2.56}$$

This gives

$$\phi(\mathbf{x}, t) = - \frac{1}{2\pi} \int \frac{\partial \phi}{\partial n} \frac{e^{ik|\mathbf{x}-\mathbf{x}'|-i\omega t}}{|\mathbf{x}-\mathbf{x}'|} dA(\mathbf{x}') \tag{5.2.57}$$

Therefore, if the normal velocity of the surface is known, (5.2.49) can be integrated to yield the velocity potential.

EXAMPLE

A simple example of a planar distribution of sources is provided by a plane wall of infinite extent, oscillating along its normal with a velocity given by $U_0 \cos \omega t$. The velocity potential for this problem may be written directly from the general solution to the wave equation:

$$\tilde{\phi} = \frac{U_0}{ik} e^{i(kx - \omega t)} \qquad (5.2.58)$$

Let us see how this simple result follows from our equations above. Since $\partial \phi / \partial n$ is the normal velocity, (5.2.57) gives

$$\tilde{\phi} = -\frac{1}{2\pi} U_0 e^{-i\omega t} \int \frac{e^{ik|\mathbf{x} - \mathbf{x}'|}}{|\mathbf{x} - \mathbf{x}'|} dA(\mathbf{x}') \qquad (5.2.59)$$

The distance $|\mathbf{x} - \mathbf{x}'|$ between source point and observation point may be expressed conveniently in terms of the polar-cylindrical coordinates, shown in Figure 5.2.6. Since the plane is of infinite extent, we may place the origin of the coordinate system anywhere on it. In particular, we may place it so that the observation point falls on the coordinate axis that is along the normal to the plane. In these conditions, $\mathbf{x} = (0, 0, z)$ and $\mathbf{x}' = (r', \theta', 0)$ so that

$$|\mathbf{x} - \mathbf{x}'| = \sqrt{z^2 + r'^2} \qquad (5.2.60)$$

and $dA(\mathbf{x}') = r' \, d\theta' \, dr'$. Therefore,

$$\tilde{\phi} = -\frac{1}{2\pi} U_0 e^{-i\omega t} \int_0^\infty \int_0^{2\pi} \frac{r' e^{ik\sqrt{z^2 + r'^2}}}{\sqrt{z^2 + r'^2}} dr' \, d\theta' \qquad (5.2.61)$$

Integrating with respect to θ' and putting $\zeta = \sqrt{z^2 + r'^2}$, this becomes

$$\tilde{\phi} = -U_0 e^{-i\omega t} \int_z^\infty e^{ik\zeta} d\zeta \qquad (5.2.62)$$

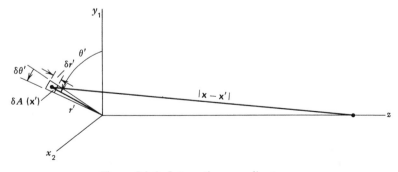

Figure 5.2.6 Integration coordinates.

Because of the upper limit, this integral does not converge. However, if we put $k \to k + i\varepsilon$ (this allows for dissipation, as explained in Section 3.6), the integral is finite. Further, if after integration we let $\varepsilon \to 0$, we obtain (5.2.58).

The procedure of letting k become complex may be found more acceptable if one realizes that no source distribution can be of infinite extent. Thus, the above manipulation is equivalent to having a distribution whose strength decreases very slowly with distance ($\varepsilon \ll 1$), but vanishes at infinity.

PROBLEMS

5.2.1 Consider a spherical distribution of monochromatic sources. The source strength is Q_0 throughout. The radius of the distribution is a. Show that the potential is given by

$$\tilde{\phi} = -\frac{Q_0}{kr} \frac{\sin ka - ka\cos ka}{k^2} e^{i(kr - \omega t)}$$

5.2.2 Consider a distribution of monochromatic sources that consists of a plane circular ring of radius a. Determine the potential on a line perpendicular to the plane of the ring and passing through the ring axis. Show that in the far-field the ring may be replaced with a point source.

5.2.3 Show that the solution to (5.2.37) for the distribution given by (5.2.10) is

$$\phi(r, t) = -\frac{1}{4\pi r} Q\left(t - \frac{r}{c_0}\right)$$

5.2.4 Consider the amplitude $|D|$ of the directivity function for the linear array. Show that the pattern it represents has nodes and antinodes for certain values of Θ. Select a value of n larger than two. Plot $|D|$ as a function of θ for two different values of kd.

5.2.5 Consider two simple sources separated by a distance d. The strength of the sources is the same, but their frequencies are not. Show that the potential at some point equidistant from either source is

$$\tilde{\phi} = \frac{A}{r} \cos\left(\frac{\omega_2 - \omega_1}{2} t\right) \exp i\left(kr - \frac{\omega_2 + \omega_1}{2} t\right)$$

where r is the distance to either source and where ω_1 and ω_2 are the two frequencies. From this result, determine the pressure fluctuation and sketch the variations of this quantity with time.

5.3 EMISSION FROM A PISTON IN AN INFINITE WALL

An important example of the simulation of vibrating bodies in terms of surface distributions is provided by a flat, circular piston of radius a mounted flush in an infinite wall, as sketched in Figure 5.3.1. Let the piston oscillate with a small amplitude and with a velocity given by $\tilde{U} = U_0 \exp(-i\omega t)$. The acoustic field in front of the piston will be symmetric about the x axis so that for the purpose of computing ϕ, we may consider an observation point on the xy plane. The results will, of course, be valid for any other field point with the same r and θ. As discussed earlier, each element of area $dA(\mathbf{x}')$ on the piston surface produces at $\mathbf{x}$ an element of potential

$$\delta\tilde{\phi} = -\frac{\tilde{\mathbf{U}} \cdot \mathbf{n}}{2\pi} \frac{e^{ik|\mathbf{x}-\mathbf{x}'|}}{|\mathbf{x}-\mathbf{x}'|} \delta A(\mathbf{x}') \qquad (5.3.1)$$

where $\mathbf{n}$ is a unit normal to $dA(\mathbf{x}')$. Integrating over the wall surface, we obtain

$$\tilde{\phi}(\mathbf{x}, t) = -\frac{1}{2\pi} \tilde{U} \int_{A_p} \frac{e^{ik|\mathbf{x}-\mathbf{x}'|}}{|\mathbf{x}-\mathbf{x}'|} dA(\mathbf{x}') \qquad (5.3.2)$$

because $\tilde{U}$ is a constant over the piston's surface and is zero otherwise. Of course, if the piston were not mounted flush with the infinite wall we would have to include in our integral the contributions from all other exposed portions of the piston's surface. This would greatly increase the complexity of the problem.

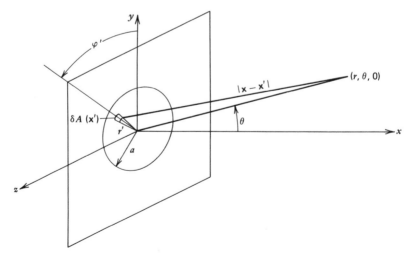

Figure 5.3.1 Flat piston in an infinite, plane wall.

Now, if $\mathbf{i}$, $\mathbf{j}$, and $\mathbf{k}$ are unit vectors along the x, y, and z axes, respectively, then

$$\mathbf{x} = \mathbf{i} r \cos \theta + \mathbf{j} r \sin \theta$$

$$\mathbf{x}' = \mathbf{j} r' \cos \varphi' + \mathbf{k} r' \sin \varphi'$$

Therefore,

$$|\mathbf{x} - \mathbf{x}'| = \sqrt{r^2 + r'^2 - 2rr' \sin \theta \cos \varphi'} \tag{5.3.3}$$

It is clear that exact integration of (5.3.2) with $|\mathbf{x} - \mathbf{x}'|$ as given above is difficult. Fortunately, several approximations can be made to simplify the problem. Most important is the far-field approximation, which is valid when $r \gg a$. To obtain this approximation, we expand $|\mathbf{x} - \mathbf{x}'|$ in terms of r'/r and retain only the first two terms. Thus, since

$$|\mathbf{x} - \mathbf{x}'| = r \sqrt{1 + (r'/r)^2 - 2(r'/r) \sin \theta \cos \varphi'}$$

we obtain

$$|\mathbf{x} - \mathbf{x}'| \simeq r - r' \sin \theta \cos \varphi' \tag{5.3.4}$$

Hence, for $r \gg a$, (5.3.2) becomes

$$\tilde{\phi}(\mathbf{x}, t) = -\frac{\tilde{U}}{2\pi r} \int_0^a \int_0^{2\pi} r' e^{ikr - ikr' \sin \theta \cos \varphi} \, dr' \, d\varphi' \tag{5.3.5}$$

It will be noticed that in the denominator of (5.3.2) we have put $|\mathbf{x} - \mathbf{x}'| = r$. The next term would have given a correction to the integral of the order of $(r'/r)^2$. Rewriting (5.3.5),

$$\tilde{\phi}(\mathbf{x}, t) = -\frac{\tilde{U} e^{ikr}}{2\pi r} \int_0^a r' \, dr' \int_0^{2\pi} e^{-ikr' \sin \theta \cos \varphi'} \, d\varphi' \tag{5.3.6}$$

The integral over φ' is one of the representations of the Bessel function of zero order:

$$J_0(z) = \frac{1}{2\pi} \int_0^{2\pi} e^{-iz \cos \varphi'} \, d\varphi' \tag{5.3.7}$$

The second integral yields

$$\int_0^a r' J_0(kr' \sin \theta) \, dr' = a^2 \frac{J_1(ka \sin \theta)}{ka \sin \theta} \tag{5.3.8}$$

Therefore,

$$\tilde{\phi}(\mathbf{x}, t) = -\frac{a^2 U_0}{2r} e^{i(kr-\omega t)} \frac{2 J_1(ka \sin \theta)}{ka \sin \theta} \tag{5.3.9}$$

This is the far-field potential due to generation, by the piston, of a volume flow rate $Q_p = \pi a^2 U_0$. In terms of Q_p, we have

$$\tilde{\phi}(\mathbf{x}, t) = -\frac{Q_p}{2\pi r} e^{i(kr-\omega t)} \frac{2 J_1(ka \sin \theta)}{ka \sin \theta} \tag{5.3.10}$$

Comparison with (5.2.12) shows that this is a point-source potential modified by a *directivity* term giving the θ dependence. From (5.3.10), we see that the fluid velocity has both radial and tangential components. For large values of r, the radial velocity decays with r as $1/r$, whereas the tangential component decays as $1/r^2$. Hence, for sufficiently large r only the radial component is significant, and is given by $\tilde{u}_r \approx ik\tilde{\phi}$. The pressure is given by (5.1.5) so that $I_r = \frac{1}{2}\rho_0 c_0 k^2 |\tilde{\phi}|^2$.

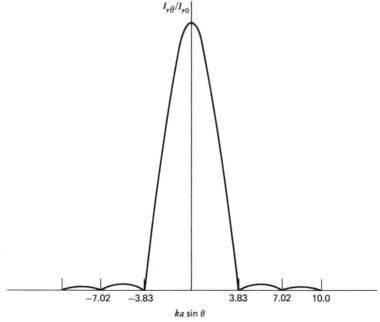

Figure 5.3.2 Normalized intensity pattern.

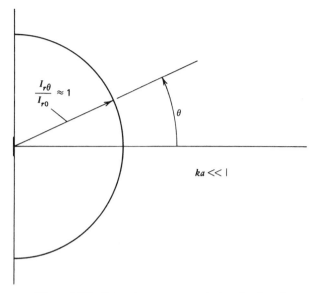

Figure 5.3.3 Intensity pattern variation for $ka \to 0$.

This yields

$$I_r = \frac{\rho_0 \omega^2 a^4 U_0^2}{8 c_0 r^2} \left[\frac{2 J_1(ka \sin \theta)}{ka \sin \theta} \right]^2 \tag{5.3.11}$$

This result displays the typical ω^2 dependence of monopole radiation [see Equation (5.1.9)]. Also, it shows that for any value of ka, the intensity is largest along the line $\theta = 0$, where it has the value

$$I_{r0} = \frac{\rho_0 \omega^2 a^4 U_0^2}{8 c_0 r^2} \tag{5.3.12}$$

This follows from (5.3.11) because near $x = 0$, $J_1(x) \approx \frac{1}{2} x$. Figure 5.3.2 shows the ratio $I_{r\theta}/I_{r0}$ versus $ka \sin \theta$. The zeros of $I_{r\theta}$ correspond to the zeros of J_1. In practical applications it is more important to determine the intensity along a given direction θ. One then uses a polar plot of $I_{r\theta}/I_{r0}$ versus θ, as in Figures 5.3.3 and 5.3.4. We see in Figure 5.3.2 that the main intensity lobe falls between values of $ka \sin \theta$ equal to -3.83 and 3.83. Therefore, as $ka \to 0$, the main lobe will be limited by larger and larger angles, with the limiting uniform distribution shown in Figure 5.3.3. This is as it should be, since in the limit $ka \to 0$ we merely

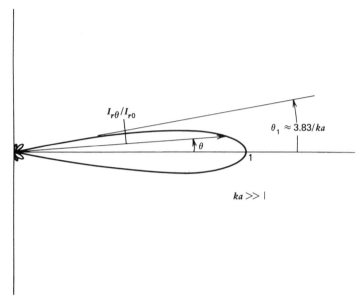

Figure 5.3.4 Intensity pattern variation for *ka* large.

have a point source. In the other extreme, when $ka \rightarrow \infty$, the main lobe is limited by smaller and smaller angles, as sketched in Figure 5.3.4.

Axial Pressure

Also of interest is the acoustic pressure for points near the piston face. We obtain this from $\tilde{\phi}$ by means of the usual relationship. Thus,

$$\tilde{p}'(\mathbf{x}, t) = \frac{i\omega}{2\pi} \rho_0 \tilde{U} \int \frac{e^{ik|\mathbf{x} - \mathbf{x}'|}}{|\mathbf{x} - \mathbf{x}'|} dA(\mathbf{x}') \qquad (5.3.13)$$

where $|\mathbf{x} - \mathbf{x}'| = \sqrt{r^2 + r'^2 - 2rr' \sin\theta \cos\varphi'}$. Integration is fairly simple for observation points that lie along the axis passing through the piston center. For these points, $\theta = 0$ and $r = x$, so that

$$\tilde{p}'(x, t) = \frac{i\omega}{2\pi} \rho_0 \tilde{U} \int_0^{2\pi} d\varphi' \int_0^a \frac{e^{ik\sqrt{x^2 + r'^2}} r' \, dr'}{\sqrt{x^2 + r'^2}} \qquad (5.3.14)$$

Setting $z = \sqrt{x^2 + r'^2}$, the integral over r' is changed to

$$\int_x^{\sqrt{a^2 + x^2}} e^{ikz} \, dz = \frac{1}{ik} \left[e^{ik\sqrt{x^2 + a^2}} - e^{ikx} \right] \qquad (5.3.15)$$

Therefore,

$$\tilde{p}'(x,t) = \rho_0 c_0 \tilde{U}\left[e^{ik\sqrt{x^2+a^2}} - e^{ikx} \right] \qquad (5.3.16)$$

This can be rewritten as

$$\tilde{p}'(x,t) = \rho_0 c_0 \tilde{U} \exp\left[i\frac{k}{2}\left(\sqrt{x^2+a^2}+x\right)\right]\left[\exp\left[i\frac{k}{2}\left(\sqrt{x^2+a^2}-x\right)\right]\right.$$

$$\left. - \exp\left[-i\frac{k}{2}\left(\sqrt{x^2+a^2}-x\right)\right]\right] \qquad (5.3.17)$$

which shows that the axial pressure is

$$\tilde{p}' = 2\rho_0 c_0 U_0 \sin\frac{k}{2}\left[\sqrt{x^2+a^2}-x\right]\exp\left\{i\left[\tfrac{1}{2}k\left(\sqrt{x^2+a^2}+x\right)-\omega t+\pi/2\right]\right\}$$

$$(5.3.18)$$

It is of interest to compare this pressure with the pressure produced by a piston of infinite extent; that is,

$$\tilde{p}'(x,t) = \rho_0 c_0 U_0 e^{i(kx-\omega t)} \qquad (5.3.19)$$

Both (5.3.18) and (5.3.19) represent acoustic waves moving away from the piston, but there are several important differences. For example, in the infinite-wall case, the phase velocity and the amplitude of the waves remain constant, whereas in the finite-piston case, these quantities depend on position. Thus, the rms pressure in the latter case is, from (5.3.18),

$$p_{rms} = \sqrt{2}\,\rho_0 c_0 U_0 \sin\frac{k}{2}\left[\sqrt{x^2+a^2}-x\right] \qquad (5.3.20)$$

whereas in the former case it is $\rho_0 c_0 U_0/\sqrt{2}$. Contrary to the infinite-wall case, the rms pressure in front of the finite piston is not constant. In fact, for some values of ka, it oscillates between positive and negative values. Between these extrema, there are nodal points whose locations are given by the solutions of

$$\sqrt{1+\xi_n^2} - \xi_n = 2\pi n/ka, \quad n=1,2,3,\dots \qquad (5.3.21)$$

where $\xi_n = x_n/a$ is the nondimensional distance to the nth node. Transposing ξ_n to the right-hand side of (5.3.21) and squaring, we obtain

$$\xi_n = \frac{1-(2\pi n/ka)^2}{4\pi n/ka} \qquad (5.3.22)$$

Since only positive values of ξ are possible, this equation shows that the nth node will not occur if $ka \leqslant 2\pi n$. Further, no nodes will exist at all if $ka \leqslant 2\pi$; that is, if the wavelength is equal to or larger than the piston radius. On the other hand, when ka is sufficiently large, several nodes exist. These are crowded together near the piston, but the distance between any two consecutive nodes increases with distance from the piston. In particular, for a given value of a (but with $a > \lambda$), the last node occurs for

$$\xi_1 = (a/2\lambda)\left[1 - (\lambda/a)^2\right] \qquad (5.3.23)$$

This shows that there are no nodes beyond a distance x from the piston equal to $a^2/2\lambda$. Thus, if $\lambda \ll a$, the nodal region is large, but if $\lambda = O(a)$, this region is limited to the near-field of the piston.

Beyond $\xi = \xi_1$, the pressure first increases until the last maximum value is reached at $\xi = (a/\lambda)[1 - (\lambda/2a)^2]$, and then decreases monotonically with distance, as sketched in Figure 5.3.5. Far from the piston (i.e., where $\xi \gg 1$), the pressure amplitude can be approximated by

$$|p'| = 2\rho_0 c_0 U_0 \sin \frac{ka}{4\xi} \qquad (5.3.24)$$

If $\xi \gg ka$, this becomes

$$|p'| = \frac{1}{2x} \rho_0 \omega U_0 a^2 \qquad (5.3.25)$$

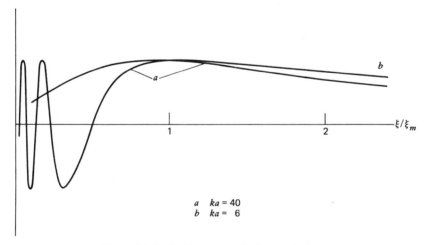

$$
\begin{array}{ll}
a & ka = 40 \\
b & ka = 6
\end{array}
$$

Figure 5.3.5 Axial pressure in front of piston.

Thus, the amplitude of the wave eventually decreases as $1/x$. This is as it should be, since to an observer in the far-field, the piston appears as a point source. In fact, the above expression for the pressure amplitude also applies, in the far-field, to off-axis positions (with r instead of x in the denominator).

Let us return to the axial rms pressure for the case when nodes occur (e.g., curve a in Figure 5.3.5). The oscillatory pattern shown is highly reminiscent of the interference pattern that is formed in front of a plane reflector owing to the interaction of incident and reflected plane waves. In the present case, the interference is between the spherical waves that are emitted simultaneously by each point of the piston surface but arrive at the x axis at different times. For example, a wave that originates at the edge of the piston will have to travel a distance $\sqrt{x^2 + a^2}$ to reach any location x on the axis, whereas the distance traveled by a wave produced at the center of the piston surface will have to travel only a distance x. Because of this difference, there will be interference between the two waves. In fact, (5.3.16) gives the impression that the complete pattern is formed only by waves originating at the edge of the piston and interacting with a plane wave, but in reality, the pattern is due to spherical waves emanating from all the points on the piston surface.

Forces on the Piston

We now come to the important problem of evaluating the forces that the fluid exerts on the piston. These are, as we know, a force in phase with the piston's velocity that accounts for radiation losses, and a force proportional to the piston's acceleration that accounts for an increase of the piston's mass. To compute these forces, we need to evaluate the pressure fluctuations on the piston face. The pressure at an arbitrary point $\mathbf{x}$ is

$$\tilde{p}'(\mathbf{x}, t) = -i\omega\rho_0 \tilde{U} \int_{A_p} \frac{e^{ik|\mathbf{x}-\mathbf{x}'|}}{2\pi|\mathbf{x}-\mathbf{x}'|} \, dA(\mathbf{x}') \qquad (5.3.26)$$

As we saw earlier, this implies that an element of area $\delta A(\mathbf{x}')$, located at $\mathbf{x}'$, produces at $\mathbf{x}$ an excess pressure given by

$$\delta\tilde{p}'(\mathbf{x}, t) = -i\omega\rho_0 \tilde{U} \frac{e^{ik|\mathbf{x}-\mathbf{x}'|}}{2\pi|\mathbf{x}-\mathbf{x}'|} \delta A(\mathbf{x}') \qquad (5.3.27)$$

If $\mathbf{x}_1$ is on the piston, the forces exerted on an element of area $\delta A(\mathbf{x}_1)$ will be

$$\delta F(\mathbf{x}_1, t) = -\delta p'(\mathbf{x}_1, t)\delta A(\mathbf{x}_1) \qquad (5.3.28)$$

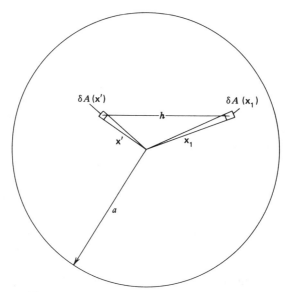

Figure 5.3.6 Surface elements on piston's face.

Substitution from (5.3.27), with $\mathbf{x}=\mathbf{x}_1$, yields

$$\delta\tilde{F}(\mathbf{x}_1,t)=\frac{i\omega\rho_0}{2\pi}\,\tilde{U}\frac{e^{ikh}}{h}\,\delta A(\mathbf{x}')\delta A(\mathbf{x}_1) \tag{5.3.29}$$

where h is the distance between $\mathbf{x}_1$ and $\mathbf{x}'$, as shown in Figure 5.3.6. Therefore, the total force on the piston will be

$$\tilde{F}=\frac{i\omega\rho_0\tilde{U}}{2\pi}\int\int\frac{e^{ikh}}{h}\,dA(\mathbf{x}')\,dA(\mathbf{x}_1) \tag{5.3.30}$$

However, the element of force produced at $\mathbf{x}'$ by an element at $\mathbf{x}_1$ is

$$\delta F(\mathbf{x}',t)=\delta F(\mathbf{x}_1,t) \tag{5.3.31}$$

That is, the force at $\mathbf{x}'$ due to an element at $\mathbf{x}_1$ is the same as the force at $\mathbf{x}_1$ due to an element at $\mathbf{x}'$. Hence,

$$\tilde{F}=\frac{i\omega\rho_0\tilde{U}}{\pi}\int_{A_1}\int_{A'}\frac{e^{ikh}}{h}\,dA(\mathbf{x}')\,dA(\mathbf{x}_1) \tag{5.3.32}$$

where each pair of elements $dA(\mathbf{x}')\,dA(\mathbf{x}_1)$ should be taken only once. Consider

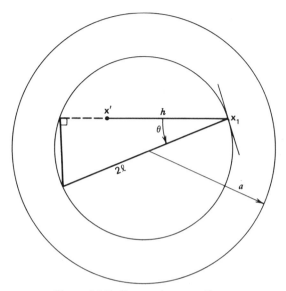

Figure 5.3.7 Integration coordinates.

first the integral over $dA(\mathbf{x}')$ in terms of the coordinates shown in Figure 5.3.7. If the element $\mathbf{x}_1$ is at the edge of a circle of radius l, then

$$\int_{A(\mathbf{x}')} \frac{e^{ikh}}{h} dA(\mathbf{x}') = \int_{-\pi/2}^{\pi/2} d\theta \int_0^{2l\cos\theta} \frac{e^{ikh}}{h} h\, dh$$

$$= \frac{2}{ik} \int_0^{\pi/2} \left[e^{2ilk\cos\theta} - 1 \right] d\theta$$

$$= \frac{\pi}{ik} \frac{2}{\pi} \int_0^{\pi/2} \left[\cos(2lk\cos\theta) - 1 \right] d\theta$$

$$+ \frac{\pi}{k} \frac{2}{\pi} \int_0^{\pi/2} \sin(2lk\cos\theta)\, d\theta \qquad (5.3.33)$$

The two integrals appearing within the braces represent tabulated functions. Thus,

$$\frac{2}{\pi} \int_0^{\pi/2} \cos(z\cos\theta)\, d\theta = J_0(z) \qquad (5.3.34)$$

and

$$\frac{2}{\pi} \int_0^{\pi/2} \sin(z\cos\theta)\, d\theta = \mathbf{H}_0(z) \qquad (5.3.35)$$

The function $H_0(z)$ is known as Struve's function of zero order. Some of its properties are given in Chapter 12 of Abramowitz and Stegun. In terms of J_0 and H_0, we have

$$\int_{A(\mathbf{x}')} \frac{e^{ikh}}{h} dA(\mathbf{x}') = \frac{\pi}{ik} \left[J_0(2lk) - 1 \right] + \frac{\pi}{k} H_0(2lk) \qquad (5.3.36)$$

The second integration in (5.3.32) can now be performed, and gives (since l varies between 0 and a)

$$\int \int \frac{e^{ikh}}{h} dA(\mathbf{x}') \, dA(\mathbf{x}_1) = \frac{2\pi^2}{ik} \int_0^a l \left[J_0(2lk) - 1 \right] dl$$

$$+ \frac{2\pi^2}{k} \int_0^a l H_0(2lk) \, dl \qquad (5.3.37)$$

If we now put $z = 2lk$ in these integrals, we obtain

$$\int \int \frac{e^{ikh}}{h} dA(\mathbf{x}') \, dA(\mathbf{x}_1) = \frac{\pi^2}{2ik^3} \int_0^{2ka} z \left[J_0(z) - 1 \right] dz + \frac{\pi^2}{2k^3} \int_0^{2ka} z H_0(z) \, dz$$

$$(5.3.38)$$

Since

$$\int_0^{2ka} z J_0(z) \, dz = 2ka J_1(2ka)$$

and

$$\int_0^{2ka} z H_0(z) \, dz = 2ka H_1(2ka)$$

we have for the force

$$\tilde{F} = \frac{\rho_0 \omega \tilde{U} \pi}{2k^3} \left[2ka J_1(2ka) - 2(ka)^2 \right] + \frac{i\omega \tilde{U} \rho_0 a}{k^2} H_1(2ka) \qquad (5.3.39)$$

Rearranging,

$$\tilde{F} = \rho_0 c_0 \pi a^2 \left[1 - \frac{J_1(2ka)}{ka} \right] \tilde{U} - \rho_0 4\pi a^3 \frac{H_1(2ka)}{(2ka)^2} \dot{\tilde{U}} \qquad (5.3.40)$$

Thus, the radiation resistance R_r is given by

$$R_r = \rho_0 c_0 \left[1 - \frac{J_1(2ka)}{ka} \right]$$ (5.3.41)

and is sketched in Figure 5.3.8. For small ka, this gives $R_r \approx \frac{1}{2}\rho_0 c_0 (ka)^2$. Similarly, when z is small, the series expansion of $\mathbf{H}_1(z)$ is

$$\mathbf{H}_1(z) \simeq \frac{2}{\pi} \left[\frac{z^2}{1^2 \cdot 3} - \frac{z^4}{1^2 \cdot 3^2 \cdot 5} + \cdots \right]$$

The leading term gives

$$\frac{\mathbf{H}_1(2ka)}{(2ka)^2} \approx \frac{2}{3\pi}$$ (5.3.42)

Therefore, for $2ka$ small, one effect of the fluid on the piston is to increase its effective mass by an amount

$$4\pi a^3 \rho_0 \frac{\mathbf{H}_1(2ka)}{(2ka)^2} \approx \pi a^2 \rho_0 \frac{8a}{3\pi}$$ (5.3.43)

When $ka \to 0$, the second term in (5.3.40) may be taken into account by supposing that the vibrating piston carries with it a mass of fluid equal to that contained in a cylinder whose base is the piston and whose height is $8a/3\pi$.

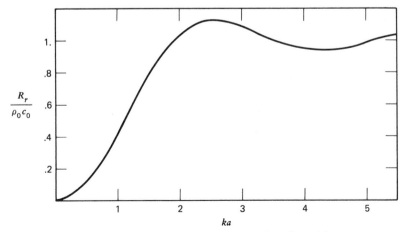

Figure 5.3.8 Radiation resistance as a function of ka.

Applications to Helmholtz Resonators

The above results may be used to obtain an effective length for the neck of a Helmholtz resonator, and also to obtain one contribution to the real part of the resonator's specific impedance. Let us begin with the effective length l_e. This quantity refers to the length of the oscillatory lumped mass in the neck of the resonator. Suppose that the resonator is mounted in a wall such that its open end is flush with the wall. Then it is clear that the lumped mass in the neck will radiate sound in the same manner as the piston considered in this section. Therefore, if the only boundary limiting the region in which emission takes place is the wall in which the resonator is mounted, then the lumped mass must be increased by the "added mass" given by (5.3.43) for $ka \ll 1$, this addition is equivalent to increasing the length of the oscillating mass by an amount equal to $(8a/3\pi)$. Thus,

$$l_e = l(1 + 8a/3\pi l), \quad ka \ll 1 \tag{5.3.44}$$

Using this length, the natural frequency of a Helmholtz resonator becomes

$$\omega_H = c_0 \sqrt{\frac{S}{lV(1 + 8a/3\pi l)}} \tag{5.3.45}$$

Experimental studies performed by various investigators using different resonators shown that the predicted frequencies differ from those obtained experimentally by amounts which vary from a few percent up to as much as 25 percent. This is not bad, especially if one considers the many fundamental assumptions that were used in the derivation of (3.1.45). Nevertheless, because of its importance in many practical applications, efforts have often been made to improve the theory. The interested reader should consult the recent article on the subject by Panton and Miller.

The second quantity of interest is the resistive part, R', of the resonator's specific impedance. In an ideal fluid, this is due to radiation only, and will be therefore denoted by R'_r. For a resonator mounted on a wall as described above, R'_r is clearly given by the radiation resistance R_r divided by πa^2. Thus, when $ka \ll 1$ we obtain from (5.3.41)

$$R'_r = \frac{1}{2\pi} \rho_0 c_0 k^2 \tag{5.3.46}$$

so that the specific impedance of the resonator becomes (see Eq. 3.1.61)

$$\tilde{z}_a = \frac{1}{2\pi} \rho_0 c_0 k^2 + i\rho_0 \left(\frac{c_0^2 S}{\omega V} - \omega l_e \right) \tag{5.3.47}$$

When real-fluid effects are taken into account, the real part of $\overset{\star}{z}_a$ must be modified to include the dissipation that takes place in the neck of the resonator. This is done in Problem 6.5.3, where it is shown that when $ka \ll 1$ the dominant resistive part is that due to radiation so that (5.3.47) may be applied to real fluids also.

PROBLEMS

5.3.1 Consider emission from circular loudspeakers mounted on an infinite baffle. Determine the maximum loudspeaker diameters that will produce nearly uniform fields in front of the loudspeakers at 50, 500 and 5000 Hz.

5.3.2 From the tabulated values of the Struve functions, sketch $H_1(2ka)/(2ka)^2$. At what frequency is the mass loading on the piston a maximum?

5.3.3 Consider a rectangular piston of sides $2l_x$ and $2l_y$, mounted in a wall of infinite extent. The wall coincides with the xy plane, and the coordinate origin is placed on the center of the piston's surface. The piston oscillates along its normal direction with a velocity given by $U_0 \cos \omega t$. Show that in the far-field of the piston, the velocity potential may be written as

$$\tilde{\phi} = -\frac{Q_p}{2\pi r} D e^{i(kr - \omega t)}$$

where $Q_p = 4l_x l_y U_0$ and where the directivity factor D is given by

$$D = \frac{\sin(kl_x x/r)}{(kl_x x/r)} \frac{\sin(kl_y y/r)}{(kl_y y/r)}$$

5.3.4 In the above problem, let θ represent the angle between the z axis and the radius vector r. Then, $x = r \sin \theta \cos \varphi$, $y = r \sin \theta \sin \varphi$, where φ is measured from the x axis. Sketch the directivity pattern for $\varphi = 0$ and $\varphi = \pi/2$ when $kl_x \gg 1$ and $kl_y \ll 1$.

5.3.5 Determine the resonance frequency of a Helmholtz resonator whose specific impedance is given by (5.3.48). Compare your results with (5.3.45).

5.3.6 Consider a tube mounted inside a large wall. One end of the tube is open and is flush with the wall. The other end is inside the wall and contains a membrane moving with velocity $U_0 \cos \omega t$. At the open end, the acoustic pressure is approximately zero, and the fluid velocity is u_L. As far as emission is concerned, this fluid velocity is equivalent to a moving piston mounted in the wall. Determine the radiated power as a function of the membrane's velocity and frequency.

5.4 COMPACT DISTRIBUTIONS OF SOURCES

Another important type of source distribution is that which is confined to a given three-dimensional region; that is, to a limited volume, as shown schematically in Figure 5.4.1. The quantity ℓ appearing in the figure represents a typical dimension of the extent of the distribution. The distribution can be used to illustrate the importance of the retarded time. Consider the result for the pressure:

$$p'(\mathbf{x}, t) = \frac{1}{4\pi} \frac{\partial}{\partial t} \int_V \frac{\rho_0 Q\left(\mathbf{x}', t - \dfrac{|\mathbf{x} - \mathbf{x}'|}{c_0}\right)}{|\mathbf{x} - \mathbf{x}'|} \, dV(\mathbf{x}') \qquad (5.4.1)$$

This gives the pressure at $\mathbf{x}$ in terms of the contributions of the sources located at $\mathbf{x}'$ inside V. Consider now a point in the far-field where $|\mathbf{x}| \gg |\mathbf{x}'|$. Since

$$|\mathbf{x} - \mathbf{x}'| = \sqrt{(\mathbf{x} - \mathbf{x}') \cdot (\mathbf{x} - \mathbf{x}')}$$

$$= |\mathbf{x}| \sqrt{1 - 2\frac{\mathbf{x} \cdot \mathbf{x}'}{|\mathbf{x}|^2} + \frac{|\mathbf{x}'|^2}{|\mathbf{x}|^2}}$$

the first approximation to $|\mathbf{x} - \mathbf{x}'|$ is

$$|\mathbf{x} - \mathbf{x}'| = |\mathbf{x}| = r$$

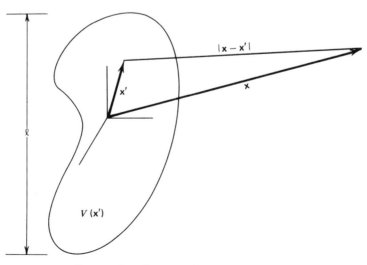

Figure 5.4.1 Compact distribution of sources.

Now, $|\mathbf{x} - \mathbf{x}'|$ appears both in the retarded time and in the denominator of (5.4.1). In the denominator, we may use the first approximation directly. Thus,

$$p' = \frac{1}{4\pi r}\frac{\partial}{\partial t}\int \rho_0 Q\left(\mathbf{x}', t - \frac{|\mathbf{x} - \mathbf{x}'|}{c_0}\right) dV(\mathbf{x}') \qquad (5.4.2)$$

As $\mathbf{x}'$ is varied within the distribution, the retarded time $t - |\mathbf{x} - \mathbf{x}'|/c_0$ changes, and we may expect the signals produced by the various sources to arrive at $\mathbf{x}$ at different times. Of course, we also get time-varying signals from a given source, as Q also depends on time. It is therefore convenient to estimate the relative magnitudes of the various contributions to the changes in retarded time. Since the source has a typical dimension of length l, the variation of the retarded time due to location of the sources has a magnitude of the order of l/c_0. If, on the other hand, a typical period of the fluctuations of Q with time is $2\pi/\omega = \lambda/c_0$, where λ is the corresponding wavelength, then the variations of retarded time due to location, relative to those due to time changes, are given by

$$l/\lambda$$

Therefore, if $l \ll \lambda$, we can neglect the variations of retarded time due to source location and obtain

$$p'(\mathbf{x}, t) \approx \frac{1}{4\pi r}\frac{\partial}{\partial t}\int_V \rho_0 Q\left(\mathbf{x}', t - \frac{r}{c_0}\right) dV(\mathbf{x}') \qquad (5.4.3)$$

These approximations simplify, considerably, the problem of sound emission from this type of distribution. In fact, comparison with (5.1.43) shows that we have replaced the whole volume distribution by a single-point source of strength

$$\int_V Q\left(\mathbf{x}', t - \frac{r}{c_0}\right) dV(\mathbf{x}')$$

The simplications are obviously useful, and are generally correct when the above restrictions on l are satisfied. However, in the special case when the total strength vanishes instantaneously, that is, when

$$\int Q \, dV(\mathbf{x}') = 0 \qquad (5.4.4)$$

then the approximate result obtained above for the far-field predicts that $p' = 0$. This is incorrect, as can be seen by considering a rigid body oscillating about some position with small amplitude. The body is clearly equivalent to a distribution of sources. That such a motion produces some sound waves is clear on physical grounds. However, if we compute the volume flow rate through a surface coinciding instantaneously with the surface of the body, we obtain

$Q(t) = \int \mathbf{n} \cdot \mathbf{u} \, dA$. But since the body is rigid, this gives $Q = 0$. It may be thought that this is so because the local source strength at every point on the surface of the body is identically zero, but in general this is not the case, as can be seen by computing the volume flow across a *fixed* control area in front of a small-area element of the oscillating body. Owing to the relative motion, the volume flow rate is

$$\delta Q(\mathbf{x}', t) = \mathbf{n} \cdot \mathbf{U}_b \, \delta A(\mathbf{x}') \tag{5.4.5}$$

where $\mathbf{U}_b$ is the local velocity of the body. Thus, the local volume flow rate does not vanish. Our argument above simply shows that the total distribution for a rigid oscillating body is such that the individual contributions of the various sources cancel out completely.

This example points to the origin of our incorrect and paradoxical result. By neglecting, entirely, the variations of retarded time, we have erroneously allowed source cancellation to an extent that does not take place when a rigid body oscillates in a fluid. This problem was discussed by Stokes more than a hundred years ago in his work on sound emission from oscillating bodies. A particularly enlightening portion of Stokes' discussion may be found in Rayleigh's book (*Theory of Sound*, Volume II, Article 324).

The same argument holds for our compact distribution. By neglecting the variations of retarded time due to source location, we have allowed the sources to cancel each other. To correct this situation, we must retain these variations in the original formulation or, as we will see below, we must replace the source distribution with another type that *does* radiate sound when retarded-time effects are neglected.

These arguments point to an emission mechanism of a kind different from the monopole or point-source radiation. In Section 5.5 we will introduce the new type by considering the emission of sound from a small, rigid sphere oscillating in a fluid.

PROBLEMS

5.4.1 In Problem 5.2.1, obtain the leading approximation to the far-field potential by setting $|\mathbf{x} - \mathbf{x}'| = r$ in (5.2.35). Compare your result to that given in Problem 5.2.1.

5.5 OSCILLATING SPHERE

To illustrate some of the remarks made in Section 5.4 and to introduce a second kind of sound-generation mechanism, we consider the simplest example of a body oscillating about a mean position without changing its volume: a rigid sphere oscillating rectilinearly in an unbounded fluid. Clearly, the instantaneous volume-flow rate is zero so that the approximation given by (5.4.3) is not applicable.

The problem can be solved by using a coordinate system fixed in space and applying approximate boundary conditions. However, it is simpler to use a coordinate system fixed on the center of the sphere. Since this system is accelerating with respect to a fixed system, we must add to the right-hand side of the equation of motion a fictitious force $-\rho_0 \, dU/dt$, where U is the velocity of the body and ρ_0 is the density of the fluid. Thus, with respect to a system moving with the body, the *linearized* Euler equation is

$$\rho_0 \frac{\partial \mathbf{v}}{\partial t} + \nabla p' = -\rho_0 \frac{d\mathbf{U}}{dt} \tag{5.5.1}$$

where $\mathbf{v}$ is the fluid velocity as seen by an observer moving with the body. The fluid velocity as seen by an observer in a fixed frame is

$$\mathbf{u} = \mathbf{v} + \mathbf{U} \tag{5.5.2}$$

so that in terms of $\mathbf{u}$, we have

$$\rho_0 \frac{\partial \mathbf{u}}{\partial t} + \nabla p' = 0 \tag{5.5.3}$$

This is the usual Euler equation, and when combined with the continuity equation, results in the wave equation. The only difference is that the boundary conditions are now applied exactly on $r = a$, where a is the sphere radius. For the inviscid case, we have only one boundary condition on the body:

$$\mathbf{u} \cdot \mathbf{n} = \mathbf{U} \cdot \mathbf{n} \quad \text{on} \quad r = a \tag{5.5.4}$$

where $\mathbf{n}$ is a unit normal vector. At infinity, the fluid should be at rest, or $\mathbf{u} = 0$.

For simplicity, we consider only the case where the sphere's motion is harmonic, that is, $\tilde{\mathbf{U}} = U_0 \exp(-i\omega t)$. The velocity potential will also be harmonic, and therefore satisfies the Helmholtz equation

$$\nabla^2 \tilde{\phi} + k^2 \tilde{\phi} = 0 \tag{5.5.5}$$

Using polar spherical coordinates, with the polar axis along the direction of motion as shown in Figure 5.5.1, (5.5.5) can be written as

$$\frac{1}{r^2} \frac{\partial}{\partial r} \left(r^2 \frac{\partial \tilde{\phi}}{\partial r} \right) + \frac{1}{r^2 \sin\theta} \frac{\partial}{\partial \theta} \left(\sin\theta \frac{\partial \tilde{\phi}}{\partial \theta} \right) + k^2 \tilde{\phi} = 0 \tag{5.5.6}$$

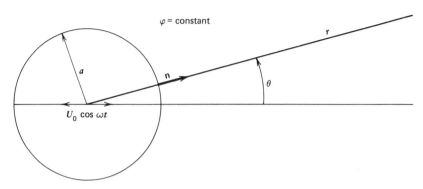

Figure 5.5.1 Oscillating sphere.

This does not contain azimuthal derivatives, because the field has symmetry about the direction of motion.

For outgoing waves, the solution to (5.5.6) is obtained from (4.3.10) by setting B_n [the coefficient of $h_n^{(2)}(kr)$] equal to zero. Thus,

$$\tilde{\phi}(r,\theta,t) = \sum_{n=0}^{\infty} A_n h_n^{(1)}(kr) P_n(\cos\theta) e^{-i\omega t} \tag{5.5.7}$$

Since, on the sphere's surface, the unit normal vector is along the radius, we have $\mathbf{n}\cdot\mathbf{u} = u_r$ and $\mathbf{n}\cdot\mathbf{U} = U_0\cos\theta$, where θ is the angle made by the position vector with the direction of motion. Therefore, the boundary condition can be written as

$$u_r = \frac{\partial\phi}{\partial r} = U_0\cos\theta \quad \text{on} \quad r = a \tag{5.5.8}$$

Taking the derivative of (5.5.7), evaluating it on $r = a$, and dropping the superscript on $h_n^{(1)}$, we have

$$U_0\cos\theta = \sum_{n=0}^{\infty} kA_n h_n'(ka) P_n(\cos\theta)$$

Since $P_n(\cos\theta)$ is equal to $\cos\theta$ only when $n = 1$, it is clear that, except for $n = 1$, all of the A_n's are identically zero. Therefore,

$$A_1 = \frac{U_0}{kh_1'(ka)} \tag{5.5.9}$$

so that the velocity potential is given by

$$\tilde{\phi}(r,\theta,t) = U_0 \frac{h_1(kr)}{kh_1'(ka)} \cos\theta \, e^{-i\omega t} \tag{5.5.10}$$

We note that because ϕ depends on r and θ, we have both u_r and a tangential component u_θ. To compute these and the pressure, we use (4.3.49) for $h_1(z)$:

$$h_1(z) = -\left(\frac{1}{z} + \frac{i}{z^2}\right)e^{iz} \tag{5.5.11}$$

so that

$$h_1'(z) = \frac{e^{iz}}{z^3}\left[2z + i(2-z^2)\right] \tag{5.5.12}$$

Making use of these results, we can write

$$\tilde{\phi}(r,\theta,t) = U_0 \cos\theta \frac{a^3}{r^2} \frac{ikr-1}{2-2ika-(ka)^2} e^{ik(r-a)-i\omega t} \tag{5.5.13}$$

From this, we obtain the acoustic pressure and the velocity components. The pressure is

$$\tilde{p}' = i\omega\rho_0 U_0 \cos\theta \frac{a^3}{r^2} \frac{ikr-1}{2-2ika-(ka)^2} e^{ik(r-a)-i\omega t} \tag{5.5.14}$$

The radial velocity component is

$$\tilde{u}_r = U_0\left(\frac{a}{r}\right)^3 \cos\theta \frac{2-2ikr-(kr)^2}{2-2ika-(ka)^2} e^{ik(r-a)-i\omega t} \tag{5.5.15}$$

Similarly, the tangential fluid velocity $u_\theta = (1/r)(\partial\phi/\partial\theta)$ is

$$\tilde{u}_\theta = -U_0\left(\frac{a}{r}\right)^3 \sin\theta \frac{ikr-1}{2-2ika-(ka)^2} e^{ik(r-a)-i\omega t} \tag{5.5.16}$$

Since $i = \exp[i(\pi/2)]$, these equations show that p' and u_θ are 90 deg. out of phase so that the time average of their product vanishes for all r (i.e., $I_\theta = 0$).

It should be noticed that the tangential velocity decays more rapidly with r than the radial velocity so that for sufficiently large values of r, the field is purely radial. On the other hand, near the sphere, both radial and tangential velocities are equally significant. The near field of a small sphere is of particular interest. Thus, when $ka \ll l$ and $r \to a$ we obtain

$$u_r = U_0 \left(\frac{a}{r}\right)^3 \cos\theta \cos\omega t \qquad (5.5.17)$$

$$u_\theta = \tfrac{1}{2} U_0 \left(\frac{a}{r}\right)^3 \sin\theta \cos\omega t \qquad (5.5.18)$$

Since neither of these equations contains k, they represent an incompressible flow field. This also follows from our discussion in Section 4.2.

It is of interest to see graphically how this field depends on position around the sphere. Since the flow is symmetric about the polar axis, we may obtain a graphical representation of the flow by plotting instantaneous streamlines on any plane containing that axis. Streamlines are lines that are everywhere tangent to the local fluid velocity. The equations for the streamlines may be obtained by using the general techniques explained in texts on fluid mechanics. At present, however, we derive equations for the streamlines corresponding to (5.5.17) and (5.5.18) by noting that if $d\mathbf{s}$ is an element of length on a streamline, their defining property may be expressed as

$$d\mathbf{s} \times \mathbf{u} = 0$$

Since the streamlines are planar, we need consider only an element $d\mathbf{s}$ on the r, θ plane. This may be expressed as $d\mathbf{s} = \mathbf{e}_r dr + \mathbf{e}_\theta r d\theta$, where $\mathbf{e}_r$ and $\mathbf{e}_\theta$ are unit vectors along the r and θ directions, respectively. Since $\mathbf{u} = u_r \mathbf{e}_r + u_\theta \mathbf{e}_\theta$, we obtain, from $d\mathbf{s} \times \mathbf{u} = 0$,

$$u_\theta \, dr = r u_r \, d\theta$$

The streamlines are given by those values of r and θ that satisfy this differential equation. To obtain these, we substitute u_r and u_θ from (5.5.17) and (5.5.18). This gives

$$\tfrac{1}{2} \sin\theta \, dr - r \cos\theta \, d\theta = 0$$

which yields, upon integration,

$$r / \sin^2\theta = C$$

where the constant C is different for different streamlines. For a given value of C, this equation describes the coordinates of a streamline. Figure 5.5.2 shows a few streamlines at an instant when the sphere is moving to the right. When the

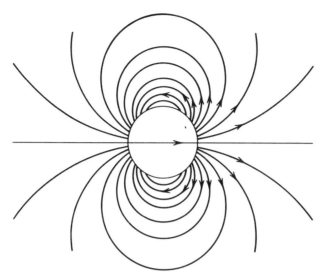

Figure 5.5.2 Streamlines around an oscillating sphere.

sphere moves in the opposite direction, the streamlines remain the same but the arrows in the figure, which indicate the direction of the flow, are reversed. The figure clearly shows that the field near the sphere represents a reciprocating fluid motion between the front and the rear of the sphere. Thus, a volume flow produced by a surface element in one side of the sphere is compensated by a flow toward an element symmetrically located in the rear of the sphere. This simple example explains why a rigid body oscillating in a fluid produces no net volume flow rate, as was pointed out by Stokes in his study of sound emission from oscillating bodies.

Let us return to the acoustic field and compute the radial intensity $I_r = \frac{1}{2}\mathrm{Re}\langle \tilde{p}'\tilde{u}_r^* \rangle$, or

$$I_r = \frac{1}{2}\frac{\rho_0 \omega U_0^2 \cos^2\theta}{4+(ka)^4}\frac{a^6}{r^5}\mathrm{Re}\left[i(ikr-1)(2+2ikr-k^2r^2)\right] \qquad (5.5.19)$$

The real part of the quantity in square brackets is just $(kr)^3$. Consequently,

$$I_r = \rho_0 c_0 \left(\frac{a}{r}\right)^2 \frac{(ka)^4}{4+(ka)^4}\frac{1}{2}U_0^2\cos^2\theta \qquad (5.5.20)$$

This varies with frequency as ω^4 instead of the typical ω^2 dependence for monopole radiation. We therefore expect this new mechanism to be relatively

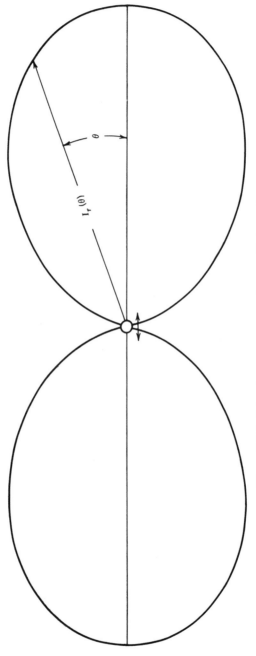

Figure 5.5.3 Intensity pattern around an oscillating sphere.

more efficient at higher frequencies than the simple source. The intensity depends strongly on the polar angle, being maximum along the polar axis $\theta=0$. If the axial value is denoted by I_0, then $I_r/I_0 = \cos^2 \theta$, independent of ka. Thus, for a fixed r, the intensity variations with θ are as sketched in Figure 5.5.3. The total emitted power is obtained by integrating I_r over a large sphere of radius r:

$$\Pi = \int_0^{2\pi} d\varphi \int_0^\pi I_r r^2 \sin \theta \, d\theta$$

Substitution from (5.5.20), together with

$$\int_0^\pi \cos^2 \theta \sin \theta \, d\theta = \tfrac{2}{3}$$

yields a power emitted per unit area of the sphere:

$$\frac{\Pi}{4\pi a^2} = \tfrac{1}{3}\rho_0 c_0 \frac{(ka)^4}{4+(ka)^4} \tfrac{1}{2} U_0^2 \tag{5.5.21}$$

Force on the Sphere

We now compute the force acting on the oscillating sphere. It is given by

$$\mathbf{F} = -\int_A p' \mathbf{n} \, dA \tag{5.5.22}$$

Clearly, this force does not have an azimuth component. Also, it has no component perpendicular to the direction of motion; that is, no lift. The only nonvanishing component is along the polar axis. Since a unit vector along this axis makes an angle θ with the unit normal vector $\mathbf{n}$, we have

$$F = -\int_0^{2\pi} d\varphi \int_0^\pi p'(a,t) a^2 \sin \theta \cos \theta \, d\theta \tag{5.5.23}$$

Using the result given in (5.5.14), we obtain, using complex notation,

$$\tilde{F} = -2\pi i \omega \rho_0 U_0 a^3 \frac{(ika-1)e^{-i\omega t}}{2-2ika-(ka)^2} \int_0^\pi \sin \theta \cos^2 \theta \, d\theta \tag{5.5.24}$$

The numerical value of the integral is $\tfrac{2}{3}$ so that

$$\tilde{F} = -i\tfrac{4}{3}\pi a^3 \omega \rho_0 U_0 \frac{(ika-1)e^{-i\omega t}}{2-2ika-(ka)^2} \tag{5.5.25}$$

This can also be written as

$$\tilde{F} = \tfrac{4}{3}\pi a^3 \rho_0 \omega U_0 e^{-i\omega t} \frac{-(ka)^3 + i\left[2 + (ka)^2\right]}{4 + (ka)^4} \tag{5.5.26}$$

We write this as the sum of two terms, one proportional to the velocity and the other proportional to the acceleration $\dot{\tilde{U}}$:

$$\tilde{F} = -\tfrac{4}{3}\pi a^3 \rho_0 \frac{(ka)^3}{4 + (ka)^4} \omega \tilde{U} - \tfrac{4}{3}\pi a^3 \rho_0 \frac{2 + (ka)^2}{4 + (ka)^4} \dot{\tilde{U}} \tag{5.5.27}$$

The first term, being in phase with velocity, leads to dissipation. Comparison with (5.5.21) shows that the dissipation is due to radiation. As in the cases of the pulsating sphere and the circular piston, this part of the force can be represented by means of the radiation resistance R_r. Thus, dividing this part of the force by $4\pi a^2$, we obtain

$$R_r = \rho_0 c_0 \cdot \frac{1}{3} \frac{(ka)^4}{4 + (ka)^4} \tag{5.5.28}$$

In view of its definition, this result could also have been obtained from (5.5.21). For small ka, R_r is proportional to $(ka)^4$, and we should expect a correspondingly small radiation. This is as it should be, since $ka \ll 1$ implies low frequencies and, consequently, nearly complete cancellation.

   The other part of the force represents an increase of the mass of the sphere equal to

$$M_0 \bar{\alpha} = \tfrac{4}{3}\pi a^3 \rho_0 \frac{2 + (ka)^2}{4 + (ka)^4} \tag{5.5.29}$$

As in Section 5.2, $\bar{\alpha}$ is called the coefficient of virtual inertia and is given by

$$\bar{\alpha} = \frac{2 + (ka)^2}{4 + (ka)^4} \tag{5.5.30}$$

In the limit, as $ka \to 0$, we obtain $\bar{\alpha} = \tfrac{1}{2}$. This gives the incompressible limit for the added mass of a rigid sphere accelerating in an ideal fluid. The maximum value of $\bar{\alpha}$, however, occurs at $ka \approx 0.91$ and is $\bar{\alpha} = 0.603$.

PROBLEMS

5.5.1 Compare the intensity emitted by an oscillating sphere with the scattered-wave intensity produced by a movable sphere in a sound wave [i.e., term $n=1$ in Equation (4.3.69)]. For simplicity, consider only the case $ka \ll 1$.

5.5.2 Consider a circular cylinder of radius a oscillating vertically with velocity $U_0 \cos \omega t$ in a fluid of infinite extent. Show that the acoustic pressure produced by the motion is

$$\tilde{p}'(r, t) = \rho_0 c_0 U_0 \cos \theta \frac{H_1(kr)}{H_1'(ka)} e^{-i\omega t}$$

where θ is the angle between the plane of oscillation and the observation point. Reduce this result for the case $ka \to 0$, $kr \to \infty$.

5.5.3 From the result of Problem 5.5.2, determine the fluid forces acting on the cylinder. What is the radiation resistance? Show that in the limit $ka \to 0$, the added mass per unit length of cylinder is $\pi \rho_0 a^2$.

5.5.4 Explain why loudspeakers are usually placed in air-tight cabinets.

5.6 RADIATION FROM FLUCTUATING FORCES

In the same way that a pulsating sphere was used to introduce the point source, we use the oscillating sphere to introduce the fundamental mechanism responsible for sound generation by bodies oscillating in a fluid without a change of volume: the fluctuating point force. To do this, we consider the result derived in Section 5.5 for the acoustic pressure due to an oscillating sphere, and try to determine what the motion corresponds to as the diameter of the sphere is made smaller and smaller. First, we write (5.5.14) as

$$2\pi r^2 \frac{e^{-ik(r-a)}}{ikr-1} \tilde{p}' = i\omega \rho_0 U_0 \cos \theta \frac{2\pi a^3}{2 - 2ika - (ka)^2} e^{-i\omega t} \tag{5.6.1}$$

From (5.5.25) it follows that the right-hand side of this equation is equal to

$$-\frac{3}{2} \frac{\tilde{F} \cos \theta}{ika - 1}$$

Therefore, we can write $\tilde{p}'$ as

$$\tilde{p}' = -\frac{3\tilde{F}}{4\pi r^2} \cos \theta \frac{ikr-1}{ika-1} e^{ik(r-a)} \tag{5.6.2}$$

where $\tilde{F} = \tilde{F}_0 \exp(-i\omega t)$. We now take the limit $ka \to 0$ but keep $\tilde{F}$ finite, and obtain

$$\tilde{p}' = -\frac{f_0 \cos\theta}{4\pi r^2}(ikr-1)e^{ikr-i\omega t} \tag{5.6.3}$$

where we have put $f_0 = 3F_0$. The right-hand side of this equation may be written conveniently as the dot product between two vectors. Thus, from (5.2.16), we see that

$$\frac{\mathbf{x}}{r}(ikr-1)\frac{e^{ikr}}{r^2} = \nabla\frac{e^{ikr}}{r}$$

If we introduce a vector $\mathbf{f}_0$ such that $\mathbf{f}_0 \cdot \mathbf{x}/r = f_0 \cos\theta$, we may write (5.6.3) as

$$\tilde{p}' = -\mathbf{f}_0 \cdot \nabla \frac{e^{ikr-i\omega t}}{4\pi r} \tag{5.6.4}$$

This form shows that as $a \to 0$, we can replace the oscillating sphere by a point force, acting along the direction of motion.

Point-Force Distribution

Now that we have made the distinction between emission by fluid addition and emission by force fluctuations, we reconsider the equations of motion, assuming there are no sources in the fluid but assuming there are point forces acting on it. The equation of continuity is (2.3.1), and the linearized momentum equation is now

$$\rho_0 \frac{\partial \mathbf{u}}{\partial t} + \nabla p' = \mathbf{F}(\mathbf{x}, t) \tag{5.6.5}$$

where $\mathbf{F}(\mathbf{x}, t)$ is the force acting on a volume element at $\mathbf{x}$, per unit volume of fluid. Since in some cases $\mathbf{F}$ may be nonconservative (i.e., $\nabla \times \mathbf{F} \neq 0$), it is not generally possible to introduce a velocity potential. However, we can eliminate $\mathbf{u}$ from the above equation using the linearized equation of continuity. Taking cross derivatives of (2.3.1) and (5.6.5), we obtain

$$\frac{\partial^2 p'}{\partial t^2} - c_0^2 \nabla^2 p' = -c_0^2 \nabla \cdot \mathbf{F} \tag{5.6.6}$$

As with any other vector function, we may express $\mathbf{F}$ as the gradient of a scalar plus the curl of a vector; that is, $\mathbf{F} = \nabla U + \nabla \times \mathbf{V}$. Since $\nabla \cdot (\nabla \times \mathbf{V}) = 0$, and since the quantity $\nabla \cdot \mathbf{F}$ appears as a source term, it is clear that only the conservative part of $\mathbf{F}$, ∇U will produce sound waves by this mechanism. This means, for example, that a dissipative force will not directly produce sound waves (although the heat it produces will, as discussed in Section 5.10).

Equation (5.6.6) is another inhomogeneous wave equation for the acoustic pressure, this time with $c_0^2 \nabla \cdot \mathbf{F}$ instead of $c_0^2 Q$ on the right-hand side. Therefore, if $\mathbf{F}$ is given, the solution of (5.6.6) is

$$\tilde{p}'(\mathbf{x}, t) = -\frac{1}{4\pi} \int_V \frac{\dfrac{\partial F_i}{\partial x_i'}\left(\mathbf{x}', t - \dfrac{|\mathbf{x} - \mathbf{x}'|}{c_0}\right)}{|\mathbf{x} - \mathbf{x}'|} \, dV(\mathbf{x}') \tag{5.6.7}$$

The notation on the argument of the spatial derivative of $\mathbf{F}$ means that after the derivative with respect to x_i is taken, it should be evaluated at $\mathbf{x} = \mathbf{x}'$ and at the retarded time $t - |\mathbf{x} - \mathbf{x}'|/c_0$. For example, if we have a distribution of monochromatic point forces, all acting in phase, then $\mathbf{F}(\mathbf{x}, t) = \mathbf{F}_0(\mathbf{x}) \exp(-i\omega t)$ and

$$\nabla \cdot \mathbf{F}(\mathbf{x}, t) = e^{-i\omega t} \nabla \cdot \mathbf{F}_0(\mathbf{x}) \tag{5.6.8}$$

Therefore, the spatial derivative of $\mathbf{F}$ evaluated as prescribed in the integrand of (5.6.7) is

$$\frac{\partial F_i}{\partial x_i'} = \exp\left[-i\omega\left(t - \frac{|\mathbf{x} - \mathbf{x}'|}{c_0}\right)\right] \nabla_{\mathbf{x}'} \cdot F_0(\mathbf{x}') \tag{5.6.9}$$

where $\nabla_{\mathbf{x}'}$ operates on the primed variables. Equation (5.6.9) gives

$$\tilde{p}'(\mathbf{x}, t) = -\frac{e^{-i\omega t}}{4\pi} \int_V \frac{e^{ik|\mathbf{x} - \mathbf{x}'|} \nabla_{\mathbf{x}'} \cdot \mathbf{F}_0}{|\mathbf{x} - \mathbf{x}'|} \, dV(\mathbf{x}') \tag{5.6.10}$$

Simple-Point Forces

We continue the analogy between emission by fluctuating forces and emission by volume addition and consider the simple-point force: a force concentrated at one point and fluctuating harmonically in time. This is, of course, equivalent to a small oscillating sphere, but it is instructive to rederive the result from the general result given by (5.6.10) above. We therefore consider a point force at the origin of a cartesian system of coordinates x_1, x_2, x_3 acting harmonically in a direction along the x_1 axis. To represent $\mathbf{F}(\mathbf{x}, t)$, we require again the delta function

$$\delta(\mathbf{x}) = \delta(x_1)\delta(x_2)\delta(x_3) \tag{5.6.11}$$

In terms of $\delta(\mathbf{x})$, we have

$$\mathbf{F}(\mathbf{x}, t) = \mathbf{e}_1 \delta(\mathbf{x}) F_0 e^{-i\omega t} \tag{5.6.12}$$

where $\mathbf{e}_1$ is a unit vector along x_1 and F_0 is the "strength" of the point force. Taking the divergence of $\mathbf{F}$, we obtain

$$\frac{\partial F_i}{\partial x_i} = F_0 e^{-i\omega t} \delta(x_2) \delta(x_3) \delta'(x_1) \tag{5.6.13}$$

where $\delta'(x_1)$ is defined in such a way that

$$\int_{-\infty}^{\infty} g(x_1) \delta'(x_1)\, dx_1 = -g'(0) = -\left(\frac{\partial g}{\partial x_1}\right)_{x_1 = 0} \tag{5.6.14}$$

The derivative $\partial F_i / \partial x_i$ evaluated at the retarded time is then

$$\frac{\partial F_i}{\partial x_i'}\left(\mathbf{x}', t - \frac{|\mathbf{x} - \mathbf{x}'|}{c_0}\right) = -F_0 e^{ik|\mathbf{x} - \mathbf{x}'| - i\omega t} \delta'(x_1) \delta(x_2) \delta(x_3) \tag{5.6.15}$$

We substitute this into our working equation for $\tilde{p}'$:

$$\tilde{p}'(\mathbf{x}, t) = -\frac{F_0 e^{-i\omega t}}{4\pi} \int \int \int \frac{e^{ik|\mathbf{x} - \mathbf{x}'|}}{|\mathbf{x} - \mathbf{x}'|} \delta'(x_1') \delta(x_2') \delta(x_3')\, dx_1'\, dx_2'\, dx_3' \tag{5.6.16}$$

In view of the properties of $\delta'(x)$ and of $\delta(x)$, this gives

$$\tilde{p}'(\mathbf{x}, t) = \frac{F_0 e^{-i\omega t}}{4\pi} \left[\frac{\partial}{\partial x_1'} \frac{e^{ik|\mathbf{x} - \mathbf{x}'|}}{|\mathbf{x} - \mathbf{x}'|}\right]_{\mathbf{x}' = 0} \tag{5.6.17}$$

Since $|\mathbf{x} - \mathbf{x}'|$ can be written as

$$|\mathbf{x} - \mathbf{x}'| = \sqrt{(x_j - x_j')(x_j - x_j')} \tag{5.6.18}$$

we have

$$\frac{\partial}{\partial x_1'} \frac{e^{ik|\mathbf{x} - \mathbf{x}'|}}{|\mathbf{x} - \mathbf{x}'|} = \frac{1}{|\mathbf{x} - \mathbf{x}'|} \frac{-ik(x_1 - x_1')}{|\mathbf{x} - \mathbf{x}'|} e^{ik|\mathbf{x} - \mathbf{x}'|}$$

$$+ e^{ik|\mathbf{x} - \mathbf{x}'|} \frac{(x_1 - x_1')}{|\mathbf{x} - \mathbf{x}'|^3} \tag{5.6.19}$$

Evaluating this at $x'=0$ as required and denoting $|\mathbf{x}|$ by r, we obtain

$$\left[\frac{\partial}{\partial x_1'}\frac{e^{ik|\mathbf{x}-\mathbf{x}'|}}{|\mathbf{x}-\mathbf{x}'|}\right]_{x'=0}=\frac{x_1}{r^3}(1-ikr)e^{ikr} \qquad (5.6.20)$$

Now, x_1, the component of r along the x_1 axis, is just $x_1=r\cos\theta$, where θ is the angle between $\mathbf{F}$ and $\mathbf{x}$. Therefore,

$$\tilde{p}'(\mathbf{x},t)=-\frac{F_0\cos\theta}{4\pi r^2}(ikr-1)e^{ikr-i\omega t} \qquad (5.6.21)$$

which corresponds to (5.6.3). Another useful form for this result can be obtained by writing $F_0\cos\theta$ as the vector product between $\mathbf{F}_0$ and a unit vector along $\mathbf{x}$; that is, with $\mathbf{x}/r$. Further, since $(\mathbf{x}/r)(ikr-1)[\exp(ikr)/r^2]=\nabla[\exp(ikr)/r]$, then

$$\tilde{p}'(\mathbf{x},t)=-\mathbf{F}_0\cdot\nabla\frac{e^{i(kr-\omega t)}}{4\pi r} \qquad (5.6.22)$$

The result is identical to that obtained from the oscillating sphere potential, as it should be. We notice that the pressure is given by the first spatial derivative of a simple-source potential, a fact that gives us a clue as to how to represent fluctuating forces in terms of simple sources. Finally, for future reference, we note that if the point force is located at $\mathbf{x}'$, then

$$\tilde{p}'(\mathbf{x},t)=-\mathbf{F}_0\cdot\nabla\frac{e^{ik|\mathbf{x}-\mathbf{x}'|-i\omega t}}{4\pi|\mathbf{x}-\mathbf{x}'|} \qquad (5.6.23)$$

Arbitrary Time Dependence

When the point force is not monochromatic, we write, instead of (5.6.12),

$$\mathbf{F}(\mathbf{x},t)=\mathbf{e}_1F(t)\delta(\mathbf{x}) \qquad (5.6.24)$$

Proceeding as before, we substitute this into (5.6.7) and obtain

$$p'(\mathbf{x},t)=\frac{1}{4\pi}\left[\frac{\partial}{\partial x_1'}\frac{F\left(t-\dfrac{|\mathbf{x}-\mathbf{x}'|}{c_0}\right)}{|\mathbf{x}-\mathbf{x}'|}\right]_{x_1'=0} \qquad (5.6.25)$$

Using the chain rule for differentiation, we can write for the derivative of the numerator

$$\frac{\partial}{\partial x_1'} F\left(t - \frac{|\mathbf{x} - \mathbf{x}'|}{c_0}\right) = \frac{\partial F\left(t - \dfrac{|\mathbf{x} - \mathbf{x}'|}{c_0}\right)}{\partial\left(t - \dfrac{|\mathbf{x} - \mathbf{x}'|}{c_0}\right)} \frac{-1}{c_0} \frac{\partial |\mathbf{x} - \mathbf{x}'|}{\partial x_1'}$$

$$= \frac{x_1 - x_1'}{c_0 |\mathbf{x} - \mathbf{x}'|} \frac{\partial}{\partial t} F\left(t - \frac{|\mathbf{x} - \mathbf{x}'|}{c_0}\right) \tag{5.6.26}$$

Therefore,

$$p' = \frac{x_1}{4\pi r}\left[\frac{1}{rc_0} \frac{\partial F(t - r/c_0)}{\partial t} + \frac{1}{r^2} F\left(t - \frac{r}{c_0}\right)\right] \tag{5.6.27}$$

where $r = |\mathbf{x}|$. This can also be written as

$$p'(x, t) = -\frac{x_1}{r} \frac{\partial}{\partial r} \frac{1}{4\pi r} F\left(t - \frac{r}{c_0}\right) \tag{5.6.28}$$

a result that was to be expected.

PROBLEMS

5.6.1   Consider a line distribution of point forces. The line falls on the x_2 axis and the forces act along the x_1 axis. Show that the acoustic pressure due to the distribution is

$$p' = \frac{x_1}{4\pi} \int \frac{1}{|\mathbf{x} - \mathbf{e}_1 x_1'|^2}\left[\frac{1}{c_0} \frac{\partial}{\partial t} F\left(t - \frac{|\mathbf{x} - \mathbf{e}_1 x_1'|}{c_0}\right) + \frac{F\left(t - \dfrac{|\mathbf{x} - \mathbf{e}_1 x_1'|}{c_0}\right)}{|\mathbf{x} - \mathbf{e}_1 x_1'|}\right] dx_1'$$

5.6.2   Consider the far-field of a distribution of point forces limited to some volume V. The leading term in the expansion of $|\mathbf{x} - \mathbf{x}'|$ is $|\mathbf{x} - \mathbf{x}_0'|$, where $\mathbf{x}_0'$ is some representative point inside V. Show that the leading term of (5.6.10) gives

$$\tilde{p}' = -\frac{e^{ik|\mathbf{x} - \mathbf{x}_0'| - i\omega t}}{4\pi r} \int_A \mathbf{F}_0 \cdot \mathbf{n} \, dA$$

where A is a surface enclosing V. Since $\mathbf{F}_0 \equiv 0$ on A, this gives $\tilde{p}' = 0$. Is this correct? Explain.

5.6.3 For monochromatic waves, show that the acoustic velocity $\mathbf{u}$ at a point where $\mathbf{F}$ vanishes is

$$\mathbf{u} = \frac{i}{4\pi\rho_0\omega} \int_V [1 - ik|\mathbf{x}-\mathbf{x}'|] \frac{(\mathbf{x}-\mathbf{x}')\nabla_{\mathbf{x}'}\cdot\mathbf{F}_0(\mathbf{x}')}{|\mathbf{x}-\mathbf{x}'|^3} e^{ik|\mathbf{x}-\mathbf{x}'|-i\omega t} \, dV(\mathbf{x}')$$

5.6.4 By direct differentiation, show that

$$\nabla_{\mathbf{x}'}\cdot\frac{\mathbf{F}_0(\mathbf{x}')e^{ik|\mathbf{x}-\mathbf{x}'|}}{|\mathbf{x}-\mathbf{x}'|} = \frac{e^{ik|\mathbf{x}-\mathbf{x}'|}}{|\mathbf{x}-\mathbf{x}'|}\nabla_{\mathbf{x}'}\cdot\mathbf{F}_0(\mathbf{x}')$$

$$+ \frac{e^{ik|\mathbf{x}-\mathbf{x}'|}}{|\mathbf{x}-\mathbf{x}'|^3}[1-ik|\mathbf{x}-\mathbf{x}'|]\mathbf{F}_0(\mathbf{x}')\cdot(\mathbf{x}-\mathbf{x}')$$

5.6.5 Consider a distribution of point forces that vanishes outside a region V. Using the result given above (Problem 5.6.4), show that the pressure due to the distribution is

$$\tilde{p}'(\mathbf{x}, t) = \frac{e^{-i\omega t}}{4\pi} \int_V \frac{e^{ik|\mathbf{x}-\mathbf{x}'|}[1-ik|\mathbf{x}-\mathbf{x}'|]\mathbf{F}_0(\mathbf{x}')\cdot(\mathbf{x}-\mathbf{x}')}{|\mathbf{x}-\mathbf{x}'|^3} \, dV(\mathbf{x}')$$

5.6.6 From the result of Problem 5.6.5, show that the leading term in the far-field is

$$\tilde{p}' = -\frac{ik}{4\pi} e^{ik|\mathbf{x}-\mathbf{x}'_0|-i\omega t} \frac{(\mathbf{x}-\mathbf{x}'_0)\cdot\mathbf{G}}{|\mathbf{x}-\mathbf{x}'_0|^2}$$

where $\mathbf{G} = \int_V \mathbf{F}_0(\mathbf{x}') \, dV(\mathbf{x}')$ is the total strength of the distribution, and $\mathbf{x}_0$ is some representative point inside V (compare with Problem 5.6.2). Show that the far-field velocity and intensity are, respectively,

$$\tilde{u}_r = -\frac{ik}{4\pi\rho_0 c_0} \frac{(\mathbf{x}-\mathbf{x}'_0)\cdot\mathbf{G}}{|\mathbf{x}-\mathbf{x}'_0|^2} e^{ik|\mathbf{x}-\mathbf{x}'_0|-i\omega t}$$

$$I_r = \frac{k^2}{32\pi^2\rho_0 c_0} \frac{|\mathbf{G}|^2\cos^2\theta}{|\mathbf{x}-\mathbf{x}'_0|^2}$$

where θ is the angle between $\mathbf{x}-\mathbf{x}'_0$ and $\mathbf{G}$.

5.7 ACOUSTIC DIPOLES

Since the acoustic pressure produced by a fluctuating force is given by the product of a vector with a spatial derivative of a point-source potential, we may expect that similar results can be produced by placing two point sources at a small distance from each other and letting the distance approach zero. We already know that the combined strength of these two sources must vanish so that we conclude they have, at any given time, opposite strengths. Since this must hold at all times, the sources must be identical and 180 deg. out of phase. Now, each of the sources produces at **x** a velocity potential of the form given by (5.2.23). Therefore, the two sources, at $\mathbf{x}'_1$ and $\mathbf{x}'_2$, respectively, result in a potential given by

$$\tilde{\phi}(\mathbf{x}, t) = Q_0 e^{-i\omega t} \left(\frac{e^{ik|\mathbf{x} - \mathbf{x}'_1|}}{4\pi |\mathbf{x} - \mathbf{x}'_1|} - \frac{e^{ik|\mathbf{x} - \mathbf{x}'_2|}}{4\pi |\mathbf{x} - \mathbf{x}'_2|} \right) \tag{5.7.1}$$

This applies for any two sources at $\mathbf{x}_1$ and $\mathbf{x}_2$, respectively, giving zero volume flow rate across some surface enclosing both. We are currently interested in the special case when the sources are near each other. We therefore put $\mathbf{x}'_1 = \mathbf{x}' - \mathbf{d}/2$ and $\mathbf{x}'_2 = \mathbf{x}' + \mathbf{d}/2$, where **d** is the distance between the sources as shown in Figure 5.7.1, and obtain the desired potential in the limit as $\mathbf{d} \to 0$. Thus,

$$\tilde{\phi}(\mathbf{x}, t) = \lim_{\mathbf{d} \to 0} \frac{Q_0 e^{-i\omega t}}{4\pi} \left(\frac{e^{ik|\mathbf{x} - \mathbf{x}' + \mathbf{d}/2|}}{|\mathbf{x} - \mathbf{x}' + \mathbf{d}/2|} - \frac{e^{ik|\mathbf{x} - \mathbf{x}' - \mathbf{d}/2|}}{|\mathbf{x} - \mathbf{x}' - \mathbf{d}/2|} \right)$$

$$= - \lim_{\mathbf{d} \to 0} Q_0 \mathbf{d} e^{-i\omega t} \cdot \nabla_{\mathbf{x}'} \frac{e^{ik|\mathbf{x} - \mathbf{x}'|}}{4\pi |\mathbf{x} - \mathbf{x}'|} \tag{5.7.2}$$

If we take the limit $\mathbf{d} \to 0$ while allowing $Q_0 \to \infty$ in such a way that $Q_0 \mathbf{d} = \mu$, and

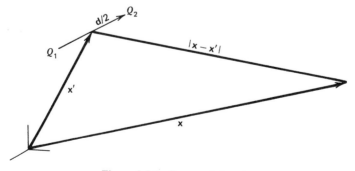

Figure 5.7.1 Source-sink pair.

we use the fact that for any function of $x - x'$, $\nabla_{x'} f = -\nabla_x f = -\nabla f$, then

$$\tilde{\phi}(x, t) = \mu \cdot \nabla \frac{e^{ik|x-x'|-i\omega t}}{4\pi|x-x'|} \tag{5.7.3}$$

where μ is evaluated at x'. This result is the so-called *dipole potential*. If θ is the angle between μ and x, it can also be written as

$$\tilde{\phi}(r, \theta, t) = \frac{1}{4\pi r^2} \mu \cos \theta (ikr - 1) e^{ikr - i\omega t} \tag{5.7.4}$$

where $r = |x - x'|$. It is clear that this is equivalent to the point-force results. The main differences are that a fluctuating force F has been replaced by a dipole of strength μ. We use this potential to compute the intensity of the waves emitted by the dipole. Thus, the acoustic pressure and radial velocity are, respectively,

$$\tilde{p}' = i\rho_0 \omega \mu \cos \theta (ikr - 1) \frac{e^{ikr - i\omega t}}{4\pi r^2} \tag{5.7.5}$$

$$\tilde{u}_r = \mu \cos \theta \frac{2 - 2ikr - (kr)^2}{4\pi r^3} e^{ikr - i\omega t} \tag{5.7.6}$$

The radial intensity is given by

$$I_r = \frac{1}{2} \frac{\mu^2 \cos^2 \theta \rho_0 \omega}{16\pi^2 r^5} \text{Re}\{i(ikr - 1)[2 + 2ikr - (kr)^2]\} \tag{5.7.7}$$

or

$$I_r = \frac{\mu^2 \cos^2 \theta \rho_0 \omega^4}{32\pi^2 r^2 c_0^3} \tag{5.7.8}$$

The total power emitted by the dipole is

$$\Pi_d = \frac{\mu^2 \omega^4 \rho_0}{32\pi^2 c_0^3} \int_0^{2\pi} d\varphi \int_0^\pi \sin \theta \cos^2 \theta \, d\theta \tag{5.7.9}$$

The area integral is $4\pi/3$ so that

$$\Pi_d = \frac{\mu^2 \omega^4 \rho_0}{24\pi c_0^3} \tag{5.7.10}$$

Again, the fourth-power dependence on the frequency should be noticed and compared with the simple-source power that varied as ω^2.

Also of interest is the relative magnitude of power emitted by the dipole to that emitted by the source. Using (5.7.10) and (5.2.4) and with $\mu = Qd$, we obtain

$$\frac{\Pi_d}{\Pi_s} = \frac{d^2\omega^2}{3c_0^2} = \tfrac{1}{3}(kd)^2 \tag{5.7.11}$$

Since $kd = 2\pi(d/\lambda) \ll 1$, we see that indeed the self-cancellation between two sources produces a sound field with an associated power whose magnitude is much smaller than that of a single source.

Line Distribution

We now study the special case when the dipole distribution or, equivalently, the point-force distribution is such that it vanishes everywhere except along a line. Consider an element of volume enclosing a point $\mathbf{x}'$ on this line, as shown schematically in Figure 5.7.2. The total strength in δV is $\mu\delta V = \mu\delta A\delta l$. If we let $\delta A \rightarrow 0$ and $\mu \rightarrow \infty$ in such a way that the limit of their product is finite and equal to μ_l say, the potential induced at $\mathbf{x}$ by a line element δl at $\mathbf{x}'$ is

$$\delta\tilde{\phi}(\mathbf{x}, t) = \mu_l(\mathbf{x}') \cdot \nabla \frac{e^{ik|\mathbf{x}-\mathbf{x}'|-i\omega t}}{4\pi|\mathbf{x}-\mathbf{x}'|}\delta l(\mathbf{x}') \tag{5.7.12}$$

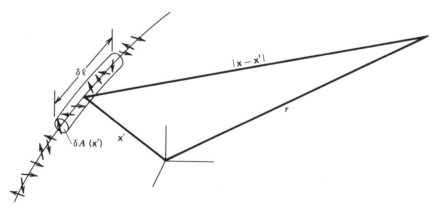

Figure 5.7.2 Line distribution of dipoles.

Hence, the potential due to the line distribution is

$$\tilde{\phi}(\mathbf{x}, t) = \nabla \cdot \int \boldsymbol{\mu}_l(\mathbf{x}') \frac{e^{ik|\mathbf{x}-\mathbf{x}'|-i\omega t}}{4\pi|\mathbf{x}-\mathbf{x}'|} \, dl(\mathbf{x}') \tag{5.7.13}$$

where the integration is to be carried over the total extent of the distribution.

Oscillating String.   As an example of the type of emission problem that can be analyzed in terms of a line distribution of dipoles, we consider a string oscillating in a plane with an amplitude that is very small everywhere, but which can vary from point to point along the string. To simulate this variation, we require a distribution of dipoles of varying strength. Since the amplitude of the motion is small everywhere, we can place the dipoles on some straight line that coincides with the mean position of the string. We take this line to be the x_1 axis, and assume that the motion is limited to the x_2, x_3 plane so that $\boldsymbol{\mu}_l = \mathbf{e}_2 \mu(x_1)$. Therefore, the velocity potential is

$$\tilde{\phi}(\mathbf{x}, t) = \frac{1}{4\pi} \frac{\partial}{\partial x_2} \int \mu(x_1') \frac{e^{ik\sqrt{r^2+x_1'^2}}}{\sqrt{r^2+x_1'^2}} e^{-i\omega t} \, dx_1' \tag{5.7.14}$$

where r represents the distance to the x_1 axis, as shown in Figure 5.7.3.

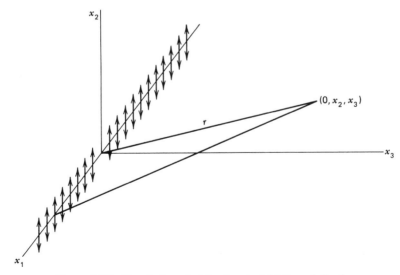

Figure 5.7.3   Simulation of string by straight line of dipoles.

It is clear that unless $\mu(x_1)$ has a very special form, it will be quite difficult to evaluate this integral exactly. Perhaps the simplest case occurs when the dipole strength is constant and the line has an infinite length. This would correspond to an infinitely long string oscillating as a whole, up and down, in a fluid. In fact, when the string has a circular cross section, this problem can be solved easily from the wave equation in cylindrical coordinates by using the method of separation of variables. Here, however, we use the above integral representation to obtain ϕ. Thus, when $\mu(x_1)$ is constant and $-\infty \leqslant x_1' \leqslant \infty$,

$$\tilde{\phi}(\mathbf{x}, t) = \frac{\mu}{4\pi} \frac{\partial}{\partial x_2} \int_{-\infty}^{\infty} \frac{e^{ikr\sqrt{1+z^2}}}{\sqrt{1+z^2}} e^{-i\omega t} \, dz \qquad (5.7.15)$$

The integral is $\pi i H_0(kr)$. Since $\partial r/\partial x_2 = x_2/r$ and $x_2 = r\cos\theta$, this yields

$$\tilde{\phi}(\mathbf{x}, t) = i\frac{\mu}{4} kH_0'(kr)\cos\theta e^{-i\omega t} \qquad (5.7.16)$$

Further the derivative of H_0 with respect to its argument is just $-H_1$. Therefore,

$$\tilde{\phi}(\mathbf{x}, t) = -i\frac{\mu}{4} H_1(kr)k\cos\theta e^{-i\omega t} \qquad (5.7.17)$$

The acoustic pressure and radial velocity are

$$\tilde{p}' = \frac{\rho_0\mu}{4} \omega k H_1(kr)\cos^2\theta e^{-i\omega t} \qquad (5.7.18)$$

$$\tilde{u}_r = -\frac{i\mu k^2 \cos\theta}{4} H_1'(kr)e^{-i\omega t} \qquad (5.7.19)$$

These equations can be used to obtain the acoustic intensity radiated per unit length of cylinder. Substituting in the usual equation for I_r, we have

$$I_r = \tfrac{1}{32}\rho_0 c_0 |\mu|^2 k^4 \cos\theta \, \text{Re}\big[-iH_1^*(kr)H_1'(kr) \big] \qquad (5.7.20)$$

The real part of the quantity inside the square brackets can be evaluated using some of the properties of H_1. Since $H_1 = J_1 + iY_1$,

$$\text{Re}\big[-iH_1^* H_1' \big] = J_1 Y_1' - Y_1 J_1' \qquad (5.7.21)$$

Using the fact that both J_1' and Y_1' obey relations of the form

$$Z_1'(z) = Z_0(z) - \frac{1}{z}Z_1(z) \qquad (5.7.22)$$

we obtain

$$\mathrm{Re}\left[-iH_1^*H_1'\right] = Y_0 J_1 - J_0 Y_1 \qquad (5.7.23)$$

The right-hand side of this equation is a particular case of the Wronskian W of J_n and Y_n:

$$W\left[J_n(z), Y_n(z)\right] = Y_n J_{n+1} - J_n Y_{n+1} = 2/\pi z \qquad (5.7.24)$$

Thus, with $z = kr$, we have

$$\mathrm{Re}\left[-iH_1^*H_1'\right] = 2/\pi kr \qquad (5.7.25)$$

so that the radiated intensity per unit length of string is

$$I_r = \frac{|\mu|^2 k^3 \rho_0 c_0}{16\pi r}\cos^2\theta \qquad (5.7.26)$$

Finally, the power emitted per unit length is $\int_0^{2\pi} I_r r\,d\theta$, so that

$$\Pi = \frac{|\mu|^2 k^3 \rho_0 c_0}{16} \qquad (5.7.27)$$

This completes the solution of the problem. However, before the solution can be used to predict absolute pressure amplitude, intensity, and so on, we must evaluate μ in terms of the physical parameters specifying the string. If the cross section of the string is circular, μ can be evaluated easily by noting that there can be no net fluid flow through the surface of the string. Stated differently, the normal velocity component of the fluid must be equal to the corresponding component of the string velocity when evaluated on the surface of the string. Thus,

$$U_0 \mathbf{e}_2 \cdot \mathbf{n} = \tilde{u}_r, \quad \text{on} \quad r = a \qquad (5.7.28)$$

where $\mathbf{n}$ is a unit normal vector, as shown in Figure 5.7.4. Using (5.7.19) evaluated on $r = a$, we obtain

$$\mu = \frac{4iU_0}{k^2 H_1'(ka)} = \frac{4ia^2 U_0}{(ka)^2 H_1'(ka)} \qquad (5.7.29)$$

This is sufficient to evaluate μ. However, if $ka \ll 1$, we can obtain an explicit

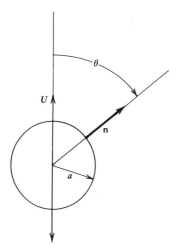

Figure 5.7.4 Circular cross-section string in fluid.

dependence of μ on a. Thus, for $z \ll 1$,

$$z^2 H_1'(z) \approx z^2 + \frac{2i}{\pi} z^2 (\ln z + \cdots) + \frac{2i}{\pi} \qquad (5.7.30)$$

Since $z^2 \ln z \to 0$ as $z \to 0$,

$$(ka)^2 H_1'(ka) \approx 2i/\pi \qquad (5.7.31)$$

so that in the limit $ka \ll 1$,

$$\mu = 2\pi a^2 U_0 \qquad (5.7.32)$$

and the power emitted per unit length of cylinder is

$$\Pi = \frac{\pi}{4c_0^2} a^4 \rho_0 \omega^3 \qquad (5.7.33)$$

It should be noted that the power emitted is proportional to the third power of the frequency and not to the fourth as it is supposed to be for dipole radiation. The reason for the difference is that (5.7.33) represents the power per unit length of string.

PROBLEMS

5.7.1 From the asymptotic expansions for $H_1(z)$ and $H_1'(z)$, show directly that $\text{Re}(-iH_1^* H_1') = 2/\pi z$.

5.7.2 Show that for $ka \to 0$, the radiation resistance per unit length of oscillating cylinder is

$$R_r = \tfrac{1}{2}\pi\rho_0 c_0 (ka)^3$$

5.7.3 Show that the acoustic potential for a surface distribution of dipoles is

$$\tilde{\phi} = \frac{1}{4\pi}\nabla_x \cdot \int_A \mu_A(x') \frac{e^{ik|x-x'|-i\omega t}}{|x-x'|}\, dA(x')$$

where μ_A is the dipole strength per unit area.

5.7.4 Consider a plane and circular distribution of dipoles. In the far-field of the distribution, the acoustic field is equivalent to that of a dipole at the origin. Determine the strength of this dipole if the radius of the circle is a.

5.7.5 Reconsider Problem 5.7.4 and assume that the dipole strength is constant over the circle's surface. Using the far-field approximation used for the field due to a circular piston mounted in a wall, show that

$$\tilde{\phi} = ik\pi a^2 (\mu_A \cdot \nabla r) \frac{2 J_1(ka\sin\theta)}{ka\sin\theta}\left[\frac{1}{4\pi r}e^{i(kr-\omega t)}\right]$$

Note that the quantity in square brackets is the potential due to a simple source at the origin. However, the emission is typical of dipole emission, because the intensity is proportional to ω^4.

5.7.6 If in the last problem the dipoles are aligned with the x_1 axis, $\mu_A \cdot \nabla_r = \mu_A \cos\theta$ so that we may think of the far-field result as being "monopole" emission with a directivity D given by

$$D = \frac{2 J_1(ka\sin\theta)}{ka\sin\theta}\cos\theta$$

The radial intensity is proportional to D^2. With reference to Figures 5.3.3 and 5.3.4, compare the radial-intensity pattern in the far-field of the circular dipole with that of the piston in a wall.

5.8 FAR-FIELD OF COMPACT FORCE DISTRIBUTION

To conclude our discussion of point-force distributions, we consider a distribution limited to a relatively small volume V. Outside this volume there are no sources of sound, and except for the sound waves, the fluid is at rest. An example of such a distribution would be a finite-length string oscillating in a fluid at rest. The acoustic pressure due to an arbitrary distribution of point forces is

$$p'(\mathbf{x}, t) = -\frac{1}{4\pi} \int_V \frac{\dfrac{\partial F_i}{\partial x_i'}\left(\mathbf{x}', t - \dfrac{|\mathbf{x} - \mathbf{x}'|}{c_0}\right)}{|\mathbf{x} - \mathbf{x}'|} \, dV(\mathbf{x}') \qquad (5.8.1)$$

We would like to obtain from this an approximate equation, applicable in the far-field of the distribution. We proceed as we did in Section 5.4, where the compact distribution of point sources was discussed. For simplicity, we consider a system of coordinates whose origin is within the distribution, as shown in Figure 5.8.1. For points $\mathbf{x}$ in the far-field of the distribution, $|\mathbf{x}'| \ll |\mathbf{x}|$, and we may be tempted to drop $\mathbf{x}'$ compared to $\mathbf{x}$, both in the denominator of (5.8.1) and in the retarded time. If this is done, we would obtain (with $|\mathbf{x}| = r$)

$$p'(\mathbf{x}, t) = -\frac{1}{4\pi r} \int_V \nabla_{\mathbf{x}'} \cdot F(\mathbf{x}', t - r/c_0) \, dV(\mathbf{x}') \qquad (5.8.2)$$

This is incorrect, as can be seen from the following considerations. The integrand appearing in this equation is just the divergence of $\mathbf{F}$ evaluated at $\mathbf{x}'$ and at a retarded time $t - r/c_0$. Therefore, at any given instant, this integral can be transformed into an area integral over any closed surface enclosing the compact distribution. However, outside V, F is identically zero so that our approximate equation states that a nonvanishing, compact force distribution emits no sound. If this were correct, a compact distribution made, for example, of a single-point force would not be capable of radiating acoustic energy.

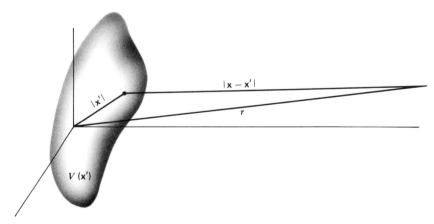

Figure 5.8.1 System of coordinates for compact distribution.

The reason for this physically incorrect result is that (5.8.2) is too crude an approximation of the original, exact integral. What is needed is a form of that equation that predicts no emission only if the force cancellation is indeed complete, and that predicts a correct radiation-field behavior (i.e., decaying as $1/r$) when the cancellation is not total. This form can be obtained by extracting (before neglecting variations of retarded time) those terms that account for the small but nonzero emitted intensity. We first note that

$$\frac{1}{|x-x'|} \nabla_{x'} \cdot F(x', t) = \nabla_{x'} \cdot \frac{F(x', t)}{|x-x'|} - F(x', t) \cdot \nabla_{x'} \frac{1}{|x-x'|} \qquad (5.8.3)$$

but since $\nabla_{x'} |x-x'|^{-1} = -\nabla_x |x-x'|^{-1}$ and $F(x', t)$ is not a function of x, the right-hand side of this relationship can be written as

$$\nabla_{x'} \cdot \frac{F(x', t)}{|x-x'|} + \nabla_x \cdot \frac{F(x', t)}{|x-x'|}$$

where $\nabla = \nabla_x$. Evaluating these quantities at the retarded time, as required, and substituting the result into (5.8.1), we obtain

$$p'(x, t) = -\frac{1}{4\pi} \int \nabla_{x'} \cdot \frac{F\left(x', t - \dfrac{|x-x'|}{c_0}\right)}{|x-x'|} \, dV(x')$$

$$-\frac{1}{4\pi} \nabla \cdot \int \frac{F\left(x', t - \dfrac{|x-x'|}{c_0}\right)}{|x-x'|} \, dV(x') \qquad (5.8.4)$$

Without making any approximations, the first integral can be transformed into an area integral over a surface enclosing V so that its value is zero. Therefore the acoustic pressure due to the distribution is

$$p'(x, t) = -\frac{1}{4\pi} \nabla \cdot \int_V \frac{F\left(x', t - \dfrac{|x-x'|}{c_0}\right)}{|x-x'|} \, dV(x') \qquad (5.8.5)$$

Remembering that the del operator acts only on the unprimed coordinates, we can write

$$p'(\mathbf{x}, t) = -\frac{1}{4\pi} \int \left[F_i\left(\mathbf{x}', t - \frac{|\mathbf{x} - \mathbf{x}'|}{c_0}\right) \frac{\partial}{\partial x_i} \frac{1}{|\mathbf{x} - \mathbf{x}'|} \right.$$

$$\left. + \frac{1}{|\mathbf{x} - \mathbf{x}'|} \frac{\partial}{\partial x_i} F_i\left(\mathbf{x}', t - \frac{|\mathbf{x} - \mathbf{x}'|}{c_0}\right) \right] dV(\mathbf{x}') \qquad (5.8.6)$$

The spatial derivatives appearing in this equation are

$$\frac{\partial}{\partial x_i} \frac{1}{|\mathbf{x} - \mathbf{x}'|} = -\frac{x_i - x_i'}{|\mathbf{x} - \mathbf{x}'|^3}$$

$$\frac{\partial}{\partial x_i} F_i\left(\mathbf{x}', t - \frac{|\mathbf{x} - \mathbf{x}'|}{c_0}\right) = -\frac{x_i - x_i'}{c_0|\mathbf{x} - \mathbf{x}'|} \frac{\partial F_i\left(\mathbf{x}', t - \dfrac{|\mathbf{x} - \mathbf{x}'|}{c_0}\right)}{\partial\left(t - \dfrac{|\mathbf{x} - \mathbf{x}'|}{c_0}\right)}$$

$$= -\frac{x_i - x_i'}{c_0|\mathbf{x} - \mathbf{x}'|} \frac{\partial}{\partial t} F_i\left(\mathbf{x}', t - \frac{|\mathbf{x} - \mathbf{x}'|}{c_0}\right) \qquad (5.8.7)$$

Substituting these above,

$$p'(\mathbf{x}, t) = \frac{1}{4\pi} \int_V \frac{x_i - x_i'}{|\mathbf{x} - \mathbf{x}'|^2} \left[\frac{1}{c_0} \frac{\partial}{\partial t} F_i\left(\mathbf{x}', t - \frac{|\mathbf{x} - \mathbf{x}'|}{c_0}\right) \right.$$

$$\left. + \frac{F_i\left(\mathbf{x}', t - \dfrac{|\mathbf{x} - \mathbf{x}'|}{c_0}\right)}{|\mathbf{x} - \mathbf{x}'|} \right] dV(\mathbf{x}') \qquad (5.8.8)$$

The result is still exact. To simplify it, we consider the far-field of a compact distribution, as shown schematically in Figure 5.8.2. Here $\mathbf{x}'$ represents, as before, the position vector of an arbitrary point within the distribution, and $\mathbf{x}_0$ is a fixed point in the distribution, say its geometrical center. We are interested in

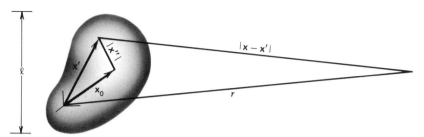

Figure 5.8.2 Far-field of compact distribution of forces.

distributions having values of $\mathbf{x}'' = \mathbf{x}' - \mathbf{x}_0$ such that $|\mathbf{x} - \mathbf{x}'| \approx |\mathbf{x} - \mathbf{x}_0|$. Now, let ω be a typical frequency of the emitted waves. Then, the magnitude of the first term within the square brackets in (5.8.8) is of the order of $|\mathbf{F}|\omega/c_0$, whereas that of the second term is $|\mathbf{F}|/|\mathbf{x} - \mathbf{x}'|$. Hence, the magnitude of the first term relative to that of the second is of the order of

$$\frac{|\mathbf{x} - \mathbf{x}'|}{c_0/\omega} \sim \frac{|\mathbf{x} - \mathbf{x}'|}{\lambda}$$

where λ is a typical wavelength. Therefore, for values of $|\mathbf{x} - \mathbf{x}'| \gg \lambda$, the second term can be neglected:

$$p'(x, t) = \frac{1}{4\pi c_0} \int \frac{x_i - x_i'}{|\mathbf{x} - \mathbf{x}'|^2} \frac{\partial}{\partial t} F_i\left(\mathbf{x}', t - \frac{|\mathbf{x} - \mathbf{x}'|}{c_0}\right) dV(\mathbf{x}') \qquad (5.8.9)$$

Now, the distance $|\mathbf{x} - \mathbf{x}'|$ is given by

$$|\mathbf{x} - \mathbf{x}'| = |\mathbf{x} - \mathbf{x}_0 - \mathbf{x}''| = \sqrt{|\mathbf{x} - \mathbf{x}_0|^2 - 2\mathbf{x}'' \cdot (\mathbf{x} - \mathbf{x}_0) + |\mathbf{x}''|^2} \qquad (5.8.10)$$

Hence, provided that $|\mathbf{x}''| \sim l$ is sufficiently small compared to $|\mathbf{x} - \mathbf{x}_0|$, we can approximate (5.8.9) by

$$p'(\dot{\mathbf{x}}, t) = \frac{1}{4\pi c_0} \frac{x_i - x_{0i}}{|\mathbf{x} - \mathbf{x}'|^2} \int \frac{\partial}{\partial t} F_i\left(\mathbf{x}', t - \frac{|\mathbf{x} - \mathbf{x}'|}{c_0}\right) dV(\mathbf{x}') \qquad (5.8.11)$$

where we have retained the variation of retarded time due to changes in $\mathbf{x}'$. If these variations *can be* neglected, then

$$p'(\mathbf{x}, t) = \frac{1}{4\pi c_0} \frac{x_i - x_{0i}}{|\mathbf{x} - \mathbf{x}_0|^2} \int \frac{\partial}{\partial t} F_i \left(\mathbf{x}', t - \frac{|\mathbf{x} - \mathbf{x}_0|}{c_0} \right) dV(\mathbf{x}') \qquad (5.8.12)$$

In the far-field, the intensity is given by $I = \langle p'^2 \rangle / \rho_0 c_0$. Therefore, using different subindices on $\mathbf{x}$ and $\mathbf{F}$ to avoid confusion when evaluating p'^2, we obtain

$$I = \frac{(x_i - x_{0i})}{16\pi^2 \rho_0 c_0^2} \frac{(y_j - y_{0j})}{|\mathbf{x} - \mathbf{x}_0|^4} \int \int \left\langle \frac{\partial}{\partial t} F_i \left(\mathbf{x}', t - \frac{|\mathbf{x} - \mathbf{x}_0'|}{c_0} \right) \right.$$

$$\times \left. \frac{\partial}{\partial t} F_j \left(\mathbf{y}', t - \frac{|\mathbf{x} - \mathbf{y}_0'|}{c_0} \right) \right\rangle dV(\mathbf{x}') \, dV(\mathbf{y}') \qquad (5.8.13)$$

There are the desired results. They predict a nonvanishing sound field, provided the force distribution is fluctuating. Also, the amplitude of the pressure fluctuation decays, as it should, inversely proportional to the distance from the distribution.

In arriving at these results, we have made two approximations: (a) We took the fraction $(x_i - x_i') / |\mathbf{x} - \mathbf{x}'|^2$ outside the integral. (b) We neglected $|\mathbf{x}''|$ compared to $|\mathbf{x} - \mathbf{x}_0|$ in evaluating the retarded time. An idea of the effects of the first approximation can be obtained by considering the far-field of a simple-point force located at $\mathbf{x}' = 0$. For this case, we have $\mathbf{F} = \mathbf{F}_0 \exp(-i\omega t) \delta(\mathbf{x})$, and so this yields

$$\mathbf{F} \left(\mathbf{x}', t - \frac{|\mathbf{x} - \mathbf{x}'|}{c_0} \right) = \mathbf{F}_0 e^{-i\omega t + ikr} \delta(\mathbf{x})$$

By direct substitution into (5.8.12),

$$\tilde{p}'(\mathbf{x}, t) = \frac{-i\omega}{4\pi c_0} \frac{F_0 \cos \theta}{r^2} e^{-i\omega t + ikr} \qquad (5.8.14)$$

where θ is the angle between F and x. Comparison with the results of Section 5.6 shows that (5.8.14) is indeed the far-field approximation to the result obtained there.

Neglecting variations of retarded time due to changes of $\mathbf{x}'$ within the distribution generally requires that the changes of the retarded time due to $\mathbf{x}'$ be small compared with a typical period of the force fluctuations. These fluctuations produce waves of frequency $\omega = 2\pi(c_0 / \lambda)$ so that the period is of the order of λ/c_0. Since the extent of the distribution is l, the quantity $|\mathbf{x}'|/c_0$ can change by an amount of the order of l/c_0. Therefore, in order that we may neglect changes of $\mathbf{x}'$ when computing the retarded time, we require that $l \ll \lambda$. However, when the instantaneous-force distribution vanishes, that is, when $\int \mathbf{F}(\mathbf{x}', t) \, dV = 0$, the results [see Equation (5.8.12)] predict that no sound is emitted. A similar situation was found in Section 5.4 for the case of a compact distribution of point sources. There it was pointed out that if we wanted to neglect the variations of retarded times, it was necessary to replace the vanishing-source distribution by a nonvanishing distribution of dipoles.

To illustrate the problem for a compact distribution of point forces, we consider two different motions of a short string held rigidly at both ends. In the first, the string is sustaining a transverse standing wave in its lowest characteristic frequency. Figure 5.8.3 shows schematically how the string looks at some fixed time. Now, the displacement of the string takes place on the vertical plane, and is given by

$$y_s = A \cos(\pi x_1/l) \cos \omega t \qquad (5.8.15)$$

The motion can be simulated with point forces, on the x_1 axis, of strength

$$\mathbf{F}(\mathbf{x}, t) = \mathbf{e}_2 F_0 \cos(\pi x_1/l) \cos \omega t, \quad -l/2 \le x_1 \le l/2 \qquad (5.8.16)$$

It is clear that the total force in this compact (line) distribution does not vanish (except when $\cos \omega t = 0$). Therefore we can use our approximate equation for the

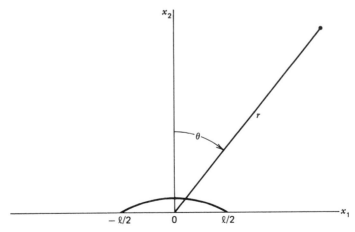

Figure 5.8.3 String oscillating in its lowest characteristic mode.

far-field. Substitution of $\mathbf{F}$ into (5.8.12) yields

$$p'(\mathbf{x}, t) = \frac{F_0 \omega}{4\pi c_0} \frac{x_2}{r^2} \sin(\omega t - kr) \int_{-l/2}^{l/2} \cos\frac{\pi x_1'}{l} \, dx' \qquad (5.8.17)$$

where $x_2 = r\cos\theta$. Thus,

$$p' = -\frac{1}{2} \frac{F_0 \omega l}{\pi^2 c_0 r} \sin(\omega t - kr)\cos\theta \qquad (5.8.18)$$

The intensity is

$$I = \frac{F_0^2 \cos^2\theta}{4\rho_0 c_0 \pi^4 r^2} \left(\frac{\omega l}{c_0}\right)^2 \qquad (5.8.19)$$

Since the force on the string is proportional to its acceleration, (5.8.19) shows that for a given velocity amplitude the emitted intensity is proportional to the fourth power of the frequency and to the square of the cosine of the angle between the direction of $\mathbf{F}$ and the observation point. This is just dipole emission, and we have simply replaced the whole distribution by an equivalent point dipole at the origin. This simple example shows how our approximate equation can lead to meaningful predictions when the total strength does not vanish.

Let us now consider the same string but this time assuming it is executing a vibration corresponding to the second characteristic mode, as sketched in Figure 5.8.4. In this case, the frequency of the string is 2ω, and the displacement is given by

$$y_s = A\sin(2\pi x_1/l)\cos(2\omega t), \quad -l/2 \le x_1 \le l/2 \qquad (5.8.20)$$

Therefore, as seen in Figure 5.8.4 or from direct integration, the instantaneous total force vanishes so that if we neglect the variations of retarded time in our equations, we will find that p' is identically zero. Physically, this is incorrect. One procedure that can remedy the situation is to replace the above distribution of forces with a nonvanishing distribution, and then neglect the variations of retarded time in that distribution. Another is to revert to (5.8.11), and evaluate the sound field without neglecting variations of retarded times in the force distribution. Thus, for harmonic time dependence and for $\mathbf{x}_0 = 0$, that equation yields

$$\tilde{p}' = -\frac{2i\omega F_0 x_2}{4\pi c_0 r^2} e^{-2i\omega t} \int_{-l/2}^{l/2} \sin\frac{2\pi x_1'}{l} e^{i2k|\mathbf{x}-\mathbf{x}'|} \, dx_1' \qquad (5.8.21)$$

where we have assumed that the amplitude of the motion is so small that the

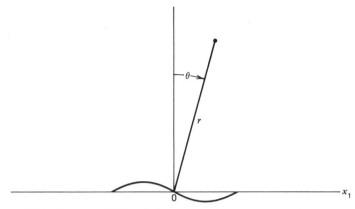

Figure 5.8.4   String oscillating in its second characteristic mode of vibration.

forces can be thought of as being placed on the x_1 axis. If we denote by θ the angle between the x_2 axis and the position vector $\mathbf{x}$, then in the far-field the quantity $|\mathbf{x}-\mathbf{x}'| = \sqrt{r^2 - 2x_1 x_1' + x_1'^2}$ can be written as

$$|\mathbf{x}-\mathbf{x}'| \approx r - x_1' \sin\theta \tag{5.8.22}$$

Therefore,

$$\tilde{p}' = -\frac{2i\omega F_0 \cos\theta}{4\pi c_0 r} e^{2i(kr-\omega t)} \int_{-l/2}^{l/2} \sin\frac{\pi x_1'}{l} e^{-2ikx_1' \sin\theta} \, dx_1'$$

Consider now, for simplicity, the case $kl \ll 1$; then

$$e^{-2ikx_1' \sin\theta} \approx 1 - 2ikx_1' \sin\theta \tag{5.8.23}$$

so that

$$\tilde{p}' = -\frac{\omega F_0 k \sin 2\theta}{2\pi c_0 r} \cos 2(kr-\omega t) \int_{-l/2}^{l/2} x_1' \sin\frac{2\pi x_1'}{l} \, dx_1' \tag{5.8.24}$$

The integral is equal to $l^2/2\pi$ so that the acoustic pressure is

$$p' = -\frac{F_0 \sin 2\theta}{4\pi^2 r} \left(\frac{\omega l}{c_0}\right)^2 \cos 2(kr-\omega t) \tag{5.8.25}$$

Similarly, the far-field radial velocity and the acoustic intensity are, respectively,

$$u_r = \frac{F_0 \sin 2\theta}{4\pi^2 rc_0} \left(\frac{\omega l}{c_0}\right)^2 \sin 2(kr - \omega t) \tag{5.8.26}$$

and

$$I = \frac{F_0^2 \sin^2 2\theta}{32\pi^4 r^2 \rho_0 c_0} \left(\frac{\omega l}{c_0}\right)^4 \tag{5.8.27}$$

The last result is of interest for several reasons. First, the acoustic energy radiated by the string in this example is a small fraction of the energy emitted in the previous case, when the string was oscillating in its first mode. In fact, the ratio of intensities for the same value of F_0 is

$$\frac{I_2}{I_1} = \frac{1}{8}\left(\frac{\omega l}{c_0}\right)^2 \sin^2 \theta \tag{5.8.28}$$

which shows that $I_2 \ll I_1$ for all values of θ.

Second, the intensity now varies with θ as $\sin^2 2\theta$, instead of the typical $\cos^2 \theta$ dependence of dipole emission. The present directivity pattern is sketched in Figure 5.8.5, and should be compared to that for dipole emission (Figure 5.5.3).

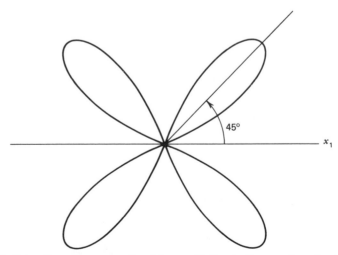

Figure 5.8.5 Intensity pattern around a string oscillating in its second mode of vibration.

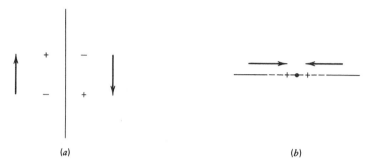

(a)   (b)

Figure 5.8.6 Quadrupoles. (a) Lateral quadrupole; (b) longitudinal quadrupole.

Thirdly, since for a given velocity amplitude of the string F_0 is proportional to ω^2, the intensity is proportional to ω^6, instead of to the typical ω^4 dependence for dipole radiation.

These results point to another type of mechanism for sound emission, different from both source and dipole radiation. This new mechanism can be visualized easily by considering the string of the above example directly in terms of the forces with which it acts on the fluid. Since, at every instant, one-half of the string is moving up, say, while the other is moving down, the forces exerted by each half are always of equal magnitude but act along opposite directions. In fact, this is why some cancellation takes place. Therefore, as far as the sound field is concerned, we can simulate the string by a *fluctuating-force pair*. If we now let the distance between the two forces decrease and let their magnitudes increase proportionately, we obtain another point singularity. The nature of this singularity is determined easily, for as we know, each force can be modeled by two point sources. Therefore, the fluctuating-force pair can be simulated by a quadrupole. We refer to this type of sound emission as *quadrupole* emission. For example, in the case of the string, the source arrangement would be as shown in Figure 5.8.6a. This is known as a lateral quadrupole. It is clear that other arrangements are possible. In fact, any combination of four sources, having each a strength of equal magnitude but having a total zero strength (such as the longitudinal arrangement shown in Figure 5.8.6b), will also form a quadrupole.

PROBLEMS

5.8.1 The far-field pressure due to a line distribution of point forces aligned with the x_1 axis is

$$p'(\mathbf{x}, t) = \frac{1}{4\pi c_0} \frac{x_i - x_{0i}}{|\mathbf{x} - \mathbf{x}_0|^2} \int \frac{\partial}{\partial t} F_i\left(\mathbf{x}', t - \frac{|\mathbf{x} - \mathbf{x}_0|}{c_0}\right) dx_1'$$

Suppose that the line has a length l, and that the fluctuating forces have a spatial and temporal distribution described by

$$\mathbf{F}(\mathbf{x}, t) = \mathbf{e}_2 F_0(x_1)\delta(x_1 - V_0 t)$$

where V_0 is a constant having the dimensions of a velocity. Thus, the force at some point x_1 on the line acts only at time $t = x_1/V_0$. Show that the far-field pressure is

$$p' = \frac{1}{4\pi r} M \cos\theta \left(\frac{dF_0}{dx_1}\right)_{x_1 = M(c_0 t - r)}, \quad -\frac{l}{2} \leqslant V_0 t - r \leqslant \frac{l}{2}$$

where $M = V_0/c_0$ and θ is the angle between the x_2 axis and the observation point.

5.8.2 For the force distribution of Problem 5.8.1, consider the following cases:

$$F_0(x_1) = f_0 \cos\frac{\pi x_1}{l}, \qquad F_0(x_1) = f_0 \sin\frac{2\pi x_1}{l}$$

For each of these, determine the far-field pressure variations and sketch these variations in time and space.

5.8.3 Consider the string of Figure 5.8.4. Show that if the variations of retarded time cannot be neglected, then the pressure is

$$p'(x, t) = -\frac{F_0 \omega x_2}{4\pi c_0} \int_{-l/2}^{l/2} \frac{\cos\frac{\pi x_1'}{l} \sin\omega\left[t - \frac{\sqrt{\sigma^2 + (x_1 - x_1')^2}}{c_0}\right]}{\sigma^2 + (x_1 - x_1')^2} dx_1'$$

where $\sigma^2 = x_2^2 + x_3^2$. For $x_1' \ll r$, approximate $\sqrt{\sigma^2 + (x_1 - x_1')^2}$ in the numerator by $r - x_1' \sin\theta$ and derive a first-order correction to the far-field results given by (5.8.18).

5.9 ACOUSTIC QUADRUPOLES

In this section, we derive equations for the acoustic pressure produced by quadrupole distributions in a fluid, and use them to give a brief description of the physical origins of quadrupole emission. First, we form a quadrupole by means of a distribution of more elementary sources. Instead of forming the quadrupole by arranging four sources of suitable magnitude, we form it by

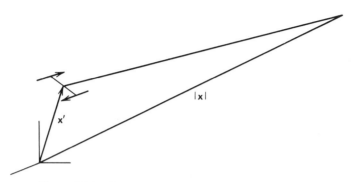

Figure 5.9.1 Two point forces producing a quadrupole.

considering the limiting case of two point forces $\mathbf{F}_1$ and $\mathbf{F}_2$ located at a small distance from each other, as indicated in Figure 5.9.1. We already know that the combined strength should vanish instantaneously (i.e., $\mathbf{F}_1 + \mathbf{F}_2 = 0$). Therefore, the forces must be parallel and of equal but opposite magnitude. Hence, if the forces are located at $\mathbf{x}_1$ and $\mathbf{x}_2$, respectively, the acoustic pressure at $\mathbf{x}$ will be given by the sum of the fields produced separately by each:

$$\tilde{p}'(\mathbf{x}, t) = -\mathbf{F}(\mathbf{x}_1) \cdot \nabla_{\mathbf{x}} \frac{e^{ik|\mathbf{x} - \mathbf{x}_1'| - i\omega t}}{4\pi|\mathbf{x} - \mathbf{x}_1|} + \mathbf{F}(\mathbf{x}_2) \cdot \nabla_{\mathbf{x}} \frac{e^{ik|\mathbf{x} - \mathbf{x}_2'| - i\omega t}}{4\pi|\mathbf{x} - \mathbf{x}_2|}$$

$$= \nabla_{\mathbf{x}} \cdot \left[\mathbf{F}(\mathbf{x}_2) \frac{e^{ik|\mathbf{x} - \mathbf{x}_2|}}{4\pi|\mathbf{x} - \mathbf{x}_2|} - \mathbf{F}(\mathbf{x}_1) \frac{e^{ik|\mathbf{x} - \mathbf{x}_1|}}{4\pi|\mathbf{x} - \mathbf{x}_1|} \right] e^{-i\omega t} \qquad (5.9.1)$$

since $\mathbf{x}_1$ and $\mathbf{x}_2$ are not functions of $\mathbf{x}$. Now, $\mathbf{x}_2 = \mathbf{x}' + \delta\mathbf{x}'/2$ and $\mathbf{x}_1 = \mathbf{x}' - \delta\mathbf{x}'/2$. Hence,

$$\tilde{p}'(\mathbf{x}, t) = \nabla_{\mathbf{x}} \cdot \left[\mathbf{F}\left(\mathbf{x}' + \frac{\delta\mathbf{x}'}{2}\right) \frac{e^{ik|\mathbf{x} - \mathbf{x}' + \delta\mathbf{x}'/2|}}{|\mathbf{x} - \mathbf{x}' + \delta\mathbf{x}'/2|} \right.$$

$$\left. - \mathbf{F}\left(\mathbf{x}' - \frac{\delta\mathbf{x}'}{2}\right) \frac{e^{ik|\mathbf{x} - \mathbf{x}' - \delta\mathbf{x}'/2|}}{|\mathbf{x} - \mathbf{x}' - \delta\mathbf{x}'/2|} \right] \frac{e^{-i\omega t}}{4\pi} \qquad (5.9.2)$$

Further, since $\mathbf{F}(\mathbf{x}_1) = \mathbf{F}(\mathbf{x}_2) = \mathbf{F}(\mathbf{x}')$, this can be written as

$$\tilde{p}'(\mathbf{x}, t) \approx -\frac{1}{4\pi} \frac{\partial}{\partial x_i} \left[F_i(\mathbf{x}') \delta x_j' \frac{\partial}{\partial x_j'} \frac{e^{ik|\mathbf{x} - \mathbf{x}'| - i\omega t}}{|\mathbf{x} - \mathbf{x}'|} \right] \qquad (5.9.3)$$

The acoustic pressure due to the quadrupole is obtained by taking the limits

$\delta x'_j \to 0$ and $F_i \to \infty$. Denoting the limiting value of the product between these two quantities by $T_{ij}(\mathbf{x}')$, we obtain

$$\tilde{p}'(\mathbf{x}, t) = T_{ij}(\mathbf{x}') \frac{\partial^2}{\partial x_i \, \partial x_j} \frac{e^{ik|\mathbf{x}-\mathbf{x}'|-i\omega t}}{4\pi|\mathbf{x}-\mathbf{x}'|} \tag{5.9.4}$$

where we have used $\partial|\mathbf{x}-\mathbf{x}'|/\partial x_j = -\partial|\mathbf{x}-\mathbf{x}'|/\partial x'_j$. Since $T_{ij}(\mathbf{x}')$ does not depend on $\mathbf{x}$, we can write

$$\tilde{p}'(\mathbf{x}, t) = \frac{1}{4\pi} \frac{\partial^2}{\partial x_i \, \partial x_j} T_{ij}(\mathbf{x}') \frac{e^{ik|\mathbf{x}-\mathbf{x}'|-i\omega t}}{|\mathbf{x}-\mathbf{x}'|} \tag{5.9.5}$$

The equivalent result for one nonmonochromatic quadrupole is

$$p'(\mathbf{x}, t) = \frac{1}{4\pi} \frac{\partial^2}{\partial x_i \, \partial x_j} \frac{T_{ij}\left(\mathbf{x}', t - \dfrac{|\mathbf{x}-\mathbf{x}'|}{c_0}\right)}{|\mathbf{x}-\mathbf{x}'|} \tag{5.9.6}$$

whereas for a distribution of such quadrupoles, we have

$$p'(\mathbf{x}, t) = \frac{1}{4\pi} \frac{\partial^2}{\partial x_i \, \partial x_j} \int_V \frac{T_{ij}\left(\mathbf{x}', t - \dfrac{|\mathbf{x}-\mathbf{x}'|}{c_0}\right)}{|\mathbf{x}-\mathbf{x}'|} \, dV(\mathbf{x}') \tag{5.9.7}$$

By means of suitable transformations, this equation can be written as

$$p'(\mathbf{x}, t) = \frac{1}{4\pi} \int \frac{\dfrac{\partial^2}{\partial x_i \, \partial x_j} T_{ij}\left(\mathbf{x}', t - \dfrac{|\mathbf{x}-\mathbf{x}'|}{c_0}\right)}{|\mathbf{x}-\mathbf{x}'|} \, dV(\mathbf{x}') \tag{5.9.8}$$

This shows that p' is the solution of

$$\frac{\partial^2 p'}{\partial t^2} - c_0^2 \nabla^2 p' = c_0^2 \frac{\partial T_{ij}}{\partial x_i \, \partial x_j} \tag{5.9.9}$$

Now, if we attempt to derive this inhomogeneous equation directly from basic balance principles, as we did in the monopole and dipole cases, we would find it necessary to add to the right-hand side of the momentum equation a term equal to $\partial T_{ij}/\partial x_j$. The linearized form of that equation would be

$$\rho_0 \frac{\partial u_i}{\partial t} = -\frac{\partial p'}{\partial x_i} - \frac{\partial T_{ij}}{\partial x_j} \qquad (5.9.10)$$

We would like to determine the physical nature of the quantity T_{ij}. To do this, we integrate (5.9.10) over a fixed volume V and use the divergence theorem. This yields

$$\frac{d}{dt} \int \rho_0 u_i \, dV = -\int (p'\delta_{ij} + T_{ij})n_j \, dA \qquad (5.9.11)$$

where A is the surface area bounding V. The left-hand side is the time rate of change of the linearized ith component of the momentum inside V, so that the area integral on the right-hand side is the net flow of that component out of V in a unit time. Hence, we can think of T_{ij} as being a kind of momentum flux density (momentum flux per unit area). In an ideal fluid, the momentum flux density tensor is given by [see Equation (1.5.6)]

$$\pi_{ij} = p\delta_{ij} + \rho u_i u_j \qquad (5.9.12)$$

so that if we introduce a perturbation to the state of rest by letting $p = p_0 + p'$, and so on, as before, the leading contribution to the momentum flux density fluctuation, $\pi_{ij} - p_0 \delta_{ij}$, is

$$\pi'_{ij} = p'\delta_{ij} + \rho_0 u_i u_j \qquad (5.9.13)$$

Hence, upon comparison with (5.9.11), we obtain

$$T_{ij} = \rho_0 u_i u_j \qquad (5.9.14)$$

The nine quantities $\rho_0 u_i u_j$ are contributions to the momentum flux and therefore represent stresses acting on the fluid. They are usually referred to as *Reynolds stresses*.

Whereas in the cases of monopole and dipole emission the basic mechanisms responsible for sound emission were found to be fluctuating mass addition and fluctuation forces, respectively, we find that the mechanism responsible for quadrupole radiation is a *fluctuating stress*. This is correct, but poses a question as to how we can have a fluctuating stress in a fluid that is at rest except for the sound wave, and in which all the changes of pressure, density, and so on are due to the acoustic waves and are therefore very small. The answer is that aside from very special motions of solid bodies, such as the motion used above to introduce quadrupoles, it is not possible to have a fluctuating stress under the conditions listed above. Thus, the quantity $\rho_0 u_i u_j$ is a second-order quantity, and was, in fact, eliminated from the main equations through linearization in Chapter 2. Hence, we conclude that quadrupole radiation is of no importance in the problem of sound radiation in an ideal fluid at rest, except for some very special solid-body motions.

The situation changes radically when there are, in an unbounded fluid otherwise at rest, regions where the momentum flux fluctuations are not negligible. This occurs, for example, in the turbulent exhausts of jet airplanes. Far from the airplane there are no solid bodies to produce monopole or dipole sound. However, the fluctuating turbulent stresses in the exhausts act as quadrupoles producing, at large distances from jet, the roar that one associates with passing jet airplanes.

Equation (5.9.7) is the basis of the modern theory of aerodynamic sound, first proposed by Lighthill in an article that appeared in 1952. In that article, he derived an exact wave equation (Lighthill's wave equation) in which the nonlinear terms were cast in the role of sources. By emphasizing the importance of retarded-time variations when computing sound fields radiated by compact sources, he was able to explain, for the first time, the mechanisms responsible for jet noise.

Lighthill's theory has also been used to explain other types of acoustical phenomena, such as the tones emitted by wires in a cross-wind—the Aeolian tones. At present, aerodynamic sound continues to be one of the most active areas of research, and the reader is encouraged to consult some of the works in that area listed at the end of this chapter.

5.10 SOUND EMISSION BY HEAT RELEASE

We return to our discussion of sound-emission mechanisms in linear flow fields. So far, we have introduced two such mechanisms: fluctuating mass addition and fluctuating forces. As we have seen, the properties of the sound waves emitted by these fluctuations can be studied by adding source terms to the linearized continuity and momentum equations, respectively. In addition to these two conservation equations, we also have a conservation equation for the internal energy of the fluid [e.g., Equation (1.6.6)]. By analogy with the other two conservation equations, we may expect that time-dependent sources of internal energy will produce sound waves. However, the question remains as to what we mean by sources of internal energy. The answer is provided by the first law of thermodynamics, which indicates that the internal energy of a fluid element can be changed by the performance of work or by the addition of heat. Therefore, one type of internal-energy source could be a small mechanical device (e.g., a balloon) expanding and contracting in a fluid, and therefore doing work on it. As we say in Section 5.1, this motion is equivalent to a source of mass in the effects it produces.

The other possible source of internal energy, namely heat addition, is also similar to mass addition because generally one of the effects of adding heat to a fluid element is to reduce the element's density. This reduction produces an expansion of the heated element, which (like the expansion of a balloon) produces sound waves. This analogy, first pointed out by Chu, is important, as it enables us to study sound waves produced by unsteady heat release (from an unsteady flame, say) in terms of pistonlike motions. However, it should be remembered that in this chapter we are dealing with sound emission in *ideal* fluids: fluids for which the viscosity and the heat conductivity vanish. Clearly, no heat transfer can take place in such fluids (unless it is through radiation, which, however, is not accounted for in our basic equations). Therefore, no sound waves could be emitted by heat addition. On the other hand, heat addition in real fluids does generate sound waves in several situations of practical interest, and we would like to retain the possibility of studying such generation without bringing in the complete set of equations. To accomplish this, we adopt a model in which the fluid has "heat spots" where heat can be added and removed. Outside these spots, the fluid is assumed to behave as ideal so that no heat is transferred between the hot spot and the main bulk of the fluid. Although seemingly crude, the assumption is, in fact, not bad in some real situations where, as we will see in the next chapter, heat transfer between a hot region and the fluid next to it is significant only in the vicinity of the region.

Using such a model, the equations describing the fluid motion are those given in Chapter 1, except that a heat-source term now appears in the entropy equation. In linearized form, the equations are

$$\frac{\partial \rho'}{\partial t} + \rho_0 \nabla \cdot \mathbf{u} = 0 \tag{5.10.1}$$

$$\rho_0 \frac{\partial \mathbf{u}}{\partial t} + \nabla p' = 0 \tag{5.10.2}$$

$$\rho_0 c_{p0} \frac{\partial T'}{\partial t} - \beta_0 T_0 \frac{\partial p'}{\partial t} = \dot{q}(\mathbf{x}', t) \tag{5.10.3}$$

Here, $\dot{q}(\mathbf{x}', t)$ is the amount of heat released per unit volume of fluid and per unit time at location $\mathbf{x}'$ and at time t. We take this quantity as known, although there are many situations in which the amount of heat released to the fluid depends on the fluid properties and is therefore unknown.

We also need an equation of state. Since the temperature now appears in the formulation, we use an equation of the form $p = p(\rho, T)$ that, when linearized, takes the form

$$p' = \left(\frac{\partial p}{\partial \rho}\right)_{T_0} \rho' + \left(\frac{\partial p}{\partial T}\right)_{\rho_0} T' \qquad (5.10.4)$$

Making use of some of the thermodynamic identities developed in Section 1.3 and of the definition of β, we can write the coefficients of ρ' and T' as

$$\left(\frac{\partial p}{\partial \rho}\right)_{T_0} = \frac{c_{p0} - c_{v0}}{\beta_0^2 T_0} \qquad (5.10.5)$$

$$\left(\frac{\partial p}{\partial T}\right)_{\rho} = \frac{\rho_0(c_{p0} - c_{v0})}{\beta_0 T_0}$$

so that (5.10.4) can be written as

$$p' = \frac{(c_{p0} - c_{v0})}{\beta_0 T_0}\left(\frac{\rho'}{\beta_0} + \rho_0 T'\right) \qquad (5.10.6)$$

A more familiar form of this equation is obtained when the fluid is a perfect gas. In this special case, $\beta_0 = T_0^{-1}$, and the difference between specific heats is the universal gas constant R so that (5.10.6) reduces to $p' = R(T_0\rho' + \rho_0 T')$. This is the linearized form of the usual perfect-gas equation of state.

Equations (5.10.1) to (5.10.3), together with (5.10.6), can now be used to obtain a single equation for one of the variables appearing there, say p'. We first eliminate T' from (5.10.3) by making use of (5.10.6), and then use the continuity equation to eliminate $\partial\rho'/\partial t$ from the result. This yields

$$\frac{\beta_0 T_0}{\gamma - 1}\frac{\partial p'}{\partial t} + \frac{\rho_0 c_{p0}}{\beta_0}\nabla\cdot\mathbf{u} = \dot{q}(\mathbf{x}, t) \qquad (5.10.7)$$

where $\gamma = c_{p0}/c_{v0}$. Finally, we eliminate $\mathbf{u}$ by making use of (5.10.2):

$$\frac{\partial^2 p'}{\partial t^2} - c_0^2 \nabla^2 p' = \rho_0 c_0^2 \frac{\partial}{\partial t}\left(\frac{\beta_0 \dot{q}}{\rho_0 c_{p0}}\right) \qquad (5.10.8)$$

where we have used the identity $c_0^2 = c_{p0}(\gamma - 1)/\beta_0^2 T_0$. Equation (5.10.8) has exactly the same form as the equation that describes sound generation by means of mass addition [see Equation (5.2.8)]. The only difference is that instead of having, on the right-hand side, the time derivative of the volume flow rate, we have the time derivative of $(\beta_0/\rho_0 c_{p0})\dot{q}(\mathbf{x}, t)$.

In view of the remarks made earlier, this similarity is not surprising, and is useful to study sound waves produced by heat addition. Such a study will reveal some of the properties of the waves, such as frequency dependence, spatial variation, and so on. However, the magnitude of the acoustic intensity, which is proportional to the square of the amplitude of $\partial\dot{q}/\partial t$, cannot be estimated because the heat-release rate from a heater, say, to a fluid cannot be computed with ideal-fluid equations. The situation is entirely different from that encountered in monopole-addition problems. There, we found that the volume flow rate produced by body pulsations could be computed directly in terms of the velocity amplitude of the pulsations. In the present case, in order to compute the heat-release rate, we must include real-fluid effects. This will be done in Chapter 6.

Since the acoustic field produced by heat addition is given by (5.10.8), it follows, by comparison with (5.2.8), that p' is given by (5.2.38), with $(\beta_0/\rho_0 c_{p0})\dot{q}$ instead of Q. In terms of the velocity potential, we thus have

$$\phi(\mathbf{x}, t) = -\frac{\beta_0}{4\pi\rho_0 c_{p0}} \int_V \frac{\dot{q}\left(\mathbf{x}', t - \dfrac{|\mathbf{x}-\mathbf{x}'|}{c_0}\right)}{|\mathbf{x}-\mathbf{x}'|} dV(\mathbf{x}') \qquad (5.10.9)$$

If the distribution of heat sources is known, this equation gives the solution at points that do not coincide with the location of a source. The procedures used to obtain the solution are similar to those used in Section 5.3 to study sound emission by bodies pulsating with nonzero volume changes. In the present case, however, simulation by heat sources may not be straightforward (except in very simple situations, such as when an alternating electrical current flows along a thin, straight wire). For example, the sound emitted by unsteady flames may be difficult to simulate because the properties of unsteady flames are not well known.[2]

---

[2] The situation is similar to that described in Section 5.9 when discussing aerodynamic sound. In fact, flame unsteadiness is of a turbulent nature, also. The difference between aerodynamic noise and flame (or combustion) noise is that in the turbulent flame case we have both quadrupole emission (due to fluctuating stresses) and monopole emission (due to unsteady heat release). The strengths of these two sources may, of course, be connected, but it should be remembered that monopole radiation is usually more efficient than quadrupole radiation. Hence, the noise from unsteady flames should have the basic properties of a monopole field.

Let us return to (5.10.9). It gives the potential at x and t due to any known distribution of heat sources and can be written for specialized distributions, as in the case of mass sources. Thus, if the sources are limited to a surface, we have

$$\phi(\mathbf{x}, t) = -\frac{\beta_0}{4\pi\rho_0 c_{p0}} \int_{A(\mathbf{x}')} \frac{\dot{q}_A\left(\mathbf{x}', t - \frac{|\mathbf{x}-\mathbf{x}'|}{c_0}\right)}{|\mathbf{x}-\mathbf{x}'|} dA(\mathbf{x}') \qquad (5.10.10)$$

where $\dot{q}_A$ is the rate of heat release per unit surface area. Similarly, in the case of a line distribution, we may write

$$\phi(\mathbf{x}, t) = -\frac{\beta_0}{4\pi\rho_0 c_{p0}} \int_{l(\mathbf{x}')} \frac{\dot{q}_l\left(\mathbf{x}', t - \frac{|\mathbf{x}-\mathbf{x}'|}{c_0}\right)}{|\mathbf{x}-\mathbf{x}'|} dl(\mathbf{x}') \qquad (5.10.11)$$

where $\dot{q}_l$ is the rate of heat release per unit length. For the special case of a straight line of infinite extent, (5.10.11) may be written in the form given by (5.2.50). Thus, with $Q_l = \beta_0 \dot{q}_l / \rho_0 c_{p0}$, we have

$$\phi(r, t) = -\frac{\beta_0}{2\pi\rho_0 c_0} \int_{-\infty}^{t-r/c_0} \frac{\dot{q}_l(\eta)\, d\eta}{\sqrt{(t-\eta)^2 - r^2/c_0^2}} \qquad (5.10.12)$$

The corresponding acoustic pressure and velocity are [see Section 4.5, Equations (4.5.19) to (4.5.21)]

$$p' = \frac{\beta_0}{2\pi c_{p0}} \int_{-\infty}^{t-r/c_0} \frac{\ddot{q}_l(\eta)\, d\eta}{\sqrt{(t-\eta)^2 - r^2/c_0^2}} \qquad (5.10.13)$$

$$u_r = \frac{\beta_0}{2\pi\rho_0 c_{p0} r} \int_{-\infty}^{t-r/c_0} \frac{(t-\eta)\ddot{q}_l(\eta)\, d\eta}{\sqrt{(t-\eta)^2 - r^2/c_0^2}} \qquad (5.10.14)$$

The quantity $\ddot{q}_l$ in these equations represents the time derivative of the rate at which heat is added per unit length along the line source. Similarly, the equation for the potential due to a planar distribution of heat sources may be written in terms of its mass-source counterpart, Equation (5.2.43).

Because of the similarity with the equations that describe emission due to fluctuating mass addition, we will not give any examples of sound emission by unsteady heat release. However, the mechanism is clearly an important one. Thus, in addition to the already mentioned examples of combustion noise and noise due to joule resistance in an alternating-current conductor, we may mention the sound wave produced by the heat released during a lightning discharge, namely, thunder. Also, a wide variety of acoustic phenomena that occur when heat is released unsteadily in semienclosed regions could be mentioned. These include acoustic-combustion instability in several types of burners. An example of this instability is provided by the so-called "singing flame." This consists of a turbulent flame placed in a vertical tube open at both ends. When the flame is suitably located, the disturbances it produces are amplified by the tube, resulting, on occasion, in large-amplitude waves. The flames may, of course, be replaced by suitable heating elements. Tubes emitting sound due to heating elements inside them are known as Rijke tubes, if both ends of the tube are open, or as Sondhauss if only one end is open. The interested reader should consult some of the works cited in the bibliography.

PROBLEMS

5.10.1 A pulsed laser beam, passing through a real medium, deposits heat at a uniform rate of $\dot{q}_l$ per unit length along a straight line. The beam's width is very small and its duration is very short. Therefore, we may approximate the heat release by means of a line distribution of heat sources such that

$$\dot{q}_l(t) = \dot{q}_0 \delta(t)$$

where $\delta(t)$ is the Dirac delta. Making use of the definition of $\delta'(t)$ [see Equation (5.6.14)], show that the acoustic pressure at a distance r from the beam is

$$p'(r,t) = \frac{\beta_0 \dot{q}_0}{2\pi c_{p0}} \begin{cases} 0, & t - r/c_0 < 0 \\ \dfrac{t}{\sqrt{t^2 - r^2/c_0^2}}, & 0 < t - r/c_0 \end{cases}$$

5.10.2 Consider a high-voltage transmission line. Because of internal resistance, each unit length of the line dissipates energy at an instantaneous rate equal to

$$\dot{q}_l(t) = \frac{1}{\sigma} I_p^2 \cos 2\omega t$$

In this equation, I_p is the peak current, σ is the electrical conductivity of the line, and ω is the frequency. Show that in the far-field of the line, the potential is

$$\tilde{\phi} = \frac{\beta_0 I_p^2}{4\rho_0 c_{p0}\sigma} \sqrt{\frac{1}{\pi kr}} \; e^{2i(kr-\omega t + \pi/8)}$$

Show that the far-field intensity is

$$I_r = \frac{\rho_0 \omega}{8\pi r} \left(\frac{\beta_0 I_p^2}{\rho_0 \sigma c_{p0}} \right)^2$$

5.11 INTEGRAL FORMULATION FOR RADIATION

We conclude this chapter with a brief discussion of the general problem of radiation in the presence of boundaries. Situations that may arise in this regard include variations of those shown in Figure 5.11.1. Case (A) in the figure shows a surface distribution of sources, for example, pistons or membranes, radiating into a region that may be partially or totally enclosed by physical boundaries. The second case shows a wave falling on an object immersed in an infinite body of fluid. Strictly speaking, this is not a problem of radiation, but of scattering and diffraction. However, the problems are similar because the presence of the body results in radiation.

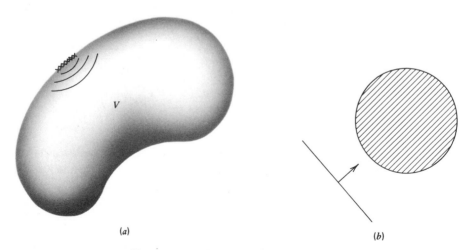

(a)   (b)

Figure 5.11.1 General radiation problems.

Some of the problems that were treated earlier display some of the features of the above situations. For example, the case of a plane wave falling on a rigid sphere, treated in Section 4.3, is a particularly simple example of the situation depicted by case (B) in Figure 5.11.1. Similarly, the emission of sound by a plane piston mounted in an infinite wall, has some of the elements of case (A) of that figure.

The common acoustical feature in cases (A) and (B) is the appearance of waves that are entirely due to the interaction of an incident wave with a boundary. Depending on the angle of incidence relative to the boundary's normal, these waves are referred to as scattered or as diffracted waves. For simplicity, we will refer to them as reflected waves. In the general case, the wavefronts of the reflected waves lack the symmetries of the plane, cylindrical, or spherical geometries we have thus far encountered. In fact, except for some isolated cases, the acoustical properties of the reflected waves cannot be obtained by the methods used earlier. A more fruitful approach is to use integral representations of the acoustical variables in terms of their values at the boundaries. Since, in general, these values are not known, unless the problem has already been solved, the representations are in fact integral equations; that is, the unknowns appear under one or more integrals. The method is based on the substitutions of boundaries by sources or dipoles whose strengths are determined by the boundary values of the acoustic variables, or on the introduction of other functions, chosen so that the boundary conditions are satisfied.

Before introducing the mathematical apparatus that is required to represent our acoustic variables, let us consider two simple situations dealing with reflection at a rigid, plane surface. Consider first the reflection of a plane wave, falling perpendicularly on an infinite plane. This situation was treated early in Chapter 3, where it was shown that the solution to the problem required that the normal gradient of the velocity potential should vanish on the plane. Since the general solution for one-dimensional propagation was known, all we had to do was to adjust the values of the arbitrary constants appearing in that solution. However, we saw in Problem 3.1.1 that the boundary condition could also be satisfied by considering the problem in terms of the interaction of the incident wave with its mirror image; that is, with a wave falling on the plane from the side opposite to that of the incident. In this approach, the solution in the region of interest was therefore given by the sum of the incident-wave potential, plus a solution of the wave equation for the region outside, chosen so as to satisfy the boundary condition.

Consider now the less trivial situation depicted in Figure 5.11.2. This shows a simple source located at a finite distance from an infinite rigid plane. While the potential for the waves emitted by the source is known, it is clear that the presence of the wall makes the problem difficult to solve. However, in analogy with the simple plane-wave reflection problem, it is also clear that the boundary

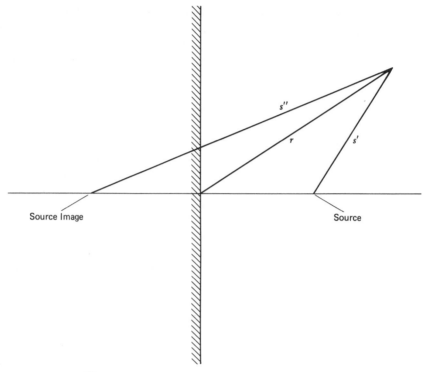

Figure 5.11.2 Simple source in front of a rigid plane.

condition on the plane will be satisfied by placing on the other side of the plane a mirror image of the actual source. Thus, with the geometry as indicated in Figure 5.11.2 we have, for a monochromatic source,

$$\phi(\mathbf{x}) = -\frac{1}{4\pi}\left(\frac{e^{iks'}}{s'} + \frac{e^{iks''}}{s''}\right) \tag{5.11.1}$$

where s' and s'' are the distances from the observation point to the source and to its image, respectively. It may be easily verified that this solution satisfies the boundary conditions. The procedure is an example of the *method of images*. This method may be used advantageously only when the geometries involved are of the simplest form.

Surface Integral Representation

We would like to obtain an integral representation for the velocity potential at an observation point $\mathbf{x}$ inside some region V. In general, the region may be

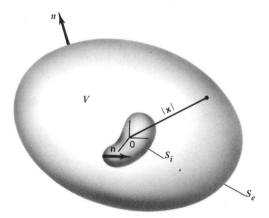

Figure 5.11.3 Region V bounded by surface $S = S_i + S_e$. Observation point x inside V.

bounded by internal and external surfaces. For example, the internal surface may be a vibrating body. To derive the desired result, we therefore consider the region V sketched in Figure 5.11.3. The volume is bounded by surfaces S_i and S_e, which, for the present application, may represent physical or mathematical surfaces. The first step in our derivation makes use of Green's theorem: let ϕ and ψ be two continuous functions that together with their first and second derivatives are finite in V. Then

$$\int_{S=S_i+S_e} \left[\phi \nabla \psi - \psi \nabla \phi \right] \cdot \mathbf{n} \, dS = \int_V \left[\phi \nabla^2 \psi - \psi \nabla^2 \phi \right] dV \qquad (5.11.2)$$

where $\mathbf{n}$ is the unit normal vector, *pointing out* of V. For the present application, we let ϕ be the velocity potential. For the function ψ, we select the potential due to a simple source:

$$\psi = -\frac{e^{ik|\mathbf{x}-\mathbf{x}'|}}{4\pi|\mathbf{x}-\mathbf{x}'|} \qquad (5.11.3)$$

Reference to (5.2.24) shows that this can also be written as $\psi = g(\mathbf{x}|\mathbf{x}')$ where g is the Green's function for the Helmholtz equation in unbounded media. Because of our choice for ϕ it follows that $\nabla^2 \phi = -k^2 \phi$. Similarly, from (5.2.18) it follows that

$$\nabla^2 \frac{e^{ik|\mathbf{x}-\mathbf{x}'|}}{|\mathbf{x}-\mathbf{x}'|} = -k^2 \frac{e^{ik|\mathbf{x}-\mathbf{x}'|}}{|\mathbf{x}-\mathbf{x}'|} - 4\pi\delta(\mathbf{x}-\mathbf{x}') \qquad (5.11.4)$$

Hence

$$\frac{1}{4\pi}\int_{S_i+S_e}\left[\phi\nabla\frac{e^{ik|\mathbf{x}-\mathbf{x}'|}}{|\mathbf{x}-\mathbf{x}'|}-\frac{e^{ik|\mathbf{x}-\mathbf{x}'|}}{|\mathbf{x}-\mathbf{x}'|}\nabla\phi\right]\cdot\mathbf{n}\,dS=-\int_V\phi\delta(\mathbf{x}-\mathbf{x}')\,dV \quad (5.11.5)$$

Because of the properties of the delta function, the volume integral is equal to $\phi(\mathbf{x})$, the potential at the observation point **x**. Thus

$$\phi(\mathbf{x})=\frac{1}{4\pi}\int_{S_i+S_e}\left[\mathbf{n}\cdot\nabla_{\mathbf{x}'}\phi\frac{e^{ik|\mathbf{x}-\mathbf{x}'|}}{|\mathbf{x}-\mathbf{x}'|}-\phi\mathbf{n}\cdot\nabla_{\mathbf{x}'}\frac{e^{ik|\mathbf{x}-\mathbf{x}'|}}{|\mathbf{x}-\mathbf{x}'|}\right]dS(\mathbf{x}') \quad (5.11.6)$$

The notation $\nabla_{\mathbf{x}'}$ means, as before, that the spatial derivative are with respect to the integration (primed) coordinates.

Equation (5.11.6) is the desired result. It expresses the potential ϕ at a point **x** inside V, in terms of surface integrals containing its values and those of its derivatives along the outward normal. Since, in general, these values are unknown, the representation is in fact an integral equation for ϕ. Further, the representation is not unique; that is, other suitably selected surface values of ϕ and $\mathbf{n}\cdot\nabla\phi=\partial\phi/\partial n$ will also give the same value of ϕ at **x**. This may be seen as follows. Suppose that we revert to (5.11.2) and consider a field point $\mathbf{x}=\boldsymbol{\xi}$ outside V (see Figure 5.11.4). Then the quantity $|\boldsymbol{\xi}-\mathbf{x}'|$ never vanishes and $\nabla^2 g(\boldsymbol{\xi}|\mathbf{x}')=-k^2 g(\boldsymbol{\xi}|\mathbf{x}')$ so that instead of (5.11.6) we obtain

$$0=\int_{S_i+S_e}\left[\mathbf{n}\cdot\nabla_{\mathbf{x}'}\phi\frac{e^{ik|\boldsymbol{\xi}-\mathbf{x}'|}}{4\pi|\boldsymbol{\xi}-\mathbf{x}'|}-\phi\mathbf{n}\cdot\nabla_{\mathbf{x}'}\frac{e^{ik|\boldsymbol{\xi}-\mathbf{x}'|}}{4\pi|\boldsymbol{\xi}-\mathbf{x}'|}\right]dS(\mathbf{x}') \quad (5.11.7)$$

If this is added to (5.11.6), we obtain another representation of $\phi(\mathbf{x})$ similar to

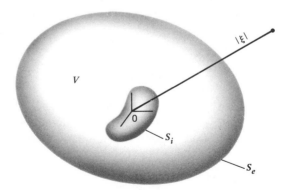

Figure 5.11.4 Region V bounded by surface S. Observation point outside V.

(5.11.6), but having different values of ϕ and $\partial\phi/\partial n$ on S. In fact, by varying the position of the point ξ external to V, we may obtain as many representations as we desire, each giving the value ϕ at $\mathbf{x}$.

The reason why our representation is not unique may be understood easily if we realize that (5.11.6) represents the potential ϕ in terms of a surface distribution of dipoles of strength ϕ, and a surface distribution of sources of strength $\partial\phi/\partial n$. Different combinations of these distributions may be used without altering the acoustic field in V.

Radiation Condition

In some applications, the region of interest is of infinite extent so that the external part of S, that is, S_e, is at "infinity." Physically it is clear that in a situation dealing with emission into an infinite region, we should get no reflected waves from infinity (unless there are sources there). Mathematically, we require that the contribution to $\phi(\mathbf{x})$ arising from the integral over S_e vanishes as that surface recedes to infinity. It is shown in Problem 5.11.2 that if we let S_e be a sphere of large radius R centered at $\mathbf{x}$, then the leading contribution to $\phi(\mathbf{x})$ arising from S_e is, as $R\to\infty$,

$$\frac{1}{4\pi}\int_{S_e}\left[\frac{\partial\phi}{\partial s}-ik\phi\right]\frac{e^{ikR}}{R}\,dS(\mathbf{x}') \tag{5.11.8}$$

where s is the coordinate along the normal to S_e. Since on S_e, dS is proportional to R^2, the requirement for the vanishing of the above contribution is that

$$\lim_{R\to\infty} R\left|\frac{\partial\phi}{\partial s}-ik\phi\right|_{s=R}=0 \tag{5.11.9}$$

This requirement is known as Sommerfeld's radiation condition. It may be stated in simpler, but more restrictive forms. For example, if for some constant K, $|\phi|<K/R$ as $R\to\infty$, then the above condition is met. Alternatively, we may assume, as we did in Section 5.2 [see Equation (5.2.62) and following discussion], that a small amount of dissipation exists that effectively reduces the contribution of (5.11.8) to zero.

Pulsating Sphere

As an example of the integral representation, we reconsider the emission of sound by pulsating sphere. Of course, this simple problem does not require such a representation, but the solution clarifies certain aspects of the formulation. The geometry is that shown earlier in Figure 4.2.1. For radial pulsations the field is centrally symmetric so that ϕ and $\partial\phi/\partial n$ are uniform over the sphere's surface. They may therefore be taken outside the integrals. Denoting their surface values by $\phi(a)$ and $\phi_n(a)$, respectively (but remembering that $\mathbf{n}$ represents a derivative along the outward normal), we obtain

$$\phi(r)=\phi_n(a)\int_{S_i}\frac{e^{ik|\mathbf{x}-\mathbf{x}'|}}{4\pi|\mathbf{x}-\mathbf{x}'|}\,dS(\mathbf{x}')-\phi(a)\int_{S_i}\mathbf{n}\cdot\nabla_{\mathbf{x}'}\frac{e^{ik|\mathbf{x}-\mathbf{x}'|}}{4\pi|\mathbf{x}-\mathbf{x}'|}\,dS(\mathbf{x}')$$

$$(5.11.10)$$

Thus, we have two uniform surface distributions; the first, a source distribution, is proportional to the volume flow-rate fluctuation produced by the sphere's radial motion. This part of the solution is consistent with our earlier statements; namely, that emission by a pulsating body is equivalent to the addition and subtraction of fluid. It may therefore appear somewhat puzzling that a dipole distribution—the second surface integral—appears in the solution. However, it is clear that one effect of an element of area on the sphere's surface is to exert a force to the fluid element in front of it. Thus, the surface distribution of dipoles, in which each dipole corresponds to a point force, simply represents the total force on the fluid. Does this mean that our previous result, given by (5.1.4), is incorrect because it contains no dipoles? The answer is: clearly no. The emission is monopole and this will be reflected by our solution when it is completed. There is in fact no paradox because, as pointed out earlier, different surface distributions in our integral representation will yield the same potential.

Another point of interest about our solution is that the source and dipole strengths are constant over the sphere's surface, a fact that simplifies the problem considerably. However, while the source strength is known—being in fact $\phi_n = -U$, the radial velocity of the surface—the dipole strength, $\phi(a)$, is unknown. But because of the simplicity of the problem we may eliminate it from (5.11.8). First, we write

$$\phi(r)=\phi_n(a)I_1(r)-\phi(a)I_2(r) \qquad (5.11.11)$$

where $I_1(r)$ and $I_2(r)$ are the surface integrals appearing in (5.11.10). Evaluating $\phi(r)$ at a by means of the above equation yields

$$\phi(a)=\frac{\phi_n(a)I_1(a)}{1+I_2(a)} \qquad (5.11.12)$$

so that

$$\phi(r)=-U\left[I_1(r)-\frac{I_1(a)}{1+I_2(a)}I_2(r)\right] \qquad (5.11.13)$$

It remains to evaluate the integrals $I_1(r)$ and $I_2(r)$. The first is considered in Problem 5.11.3 for both $r < a$ and $r > a$. In the present application we require it for points outside the sphere. Thus

$$I_1(r) = a \sin ka \frac{e^{ikr}}{kr} \qquad (5.11.14)$$

The second integral can be expressed as

$$I_2(r) = -\int_{S_i} \mathbf{n}_1 \cdot \nabla_{\mathbf{x}'} \frac{e^{ik|\mathbf{x}-\mathbf{x}'|}}{4\pi |\mathbf{x}-\mathbf{x}'|} \, dS(\mathbf{x}') \qquad (5.11.15)$$

where $\mathbf{n}_1 = -\mathbf{n}$ is a unit normal vector pointing into V. The integral may be changed into an integral over the volume bounded by S_i, that is, over the volume of the sphere. Thus,

$$I_2(r) = -\int_{V_s} \nabla^2_{\mathbf{x}'} \frac{e^{ik|\mathbf{x}-\mathbf{x}'|}}{4\pi |\mathbf{x}-\mathbf{x}'|} \, dV(\mathbf{x}') \qquad (5.11.16)$$

Since $|\mathbf{x}-\mathbf{x}'|$ never vanishes within V, we may write this as

$$I_2(r) = \frac{k^2}{4\pi} \int_{V_s} \frac{e^{ik|\mathbf{x}-\mathbf{x}'|}}{|\mathbf{x}-\mathbf{x}'|} \, dV(\mathbf{x}') \qquad (5.11.17)$$

This represents the potential due to a uniform volume distribution of simple sources of strength $-k^2$, and is another way of representing the dipole distributions of (5.11.6). The potential due to a spherical distribution of uniform sources was considered in Problem 5.2.1. The results of that problem show that

$$\frac{1}{4\pi} \int_{V_s} \frac{e^{ik|\mathbf{x}-\mathbf{x}'|}}{|\mathbf{x}-\mathbf{x}'|} \, dV(\mathbf{x}') = \frac{\sin ka - ka \cos ka}{k^2} \frac{e^{ikr}}{kr} \qquad (5.11.18)$$

so that

$$I_2(r) = (\sin ka - ka \cos ka) \frac{e^{ikr}}{kr} \qquad (5.11.19)$$

These values of $I_1(r)$ and $I_2(r)$ and the corresponding ones evaluated at $r = a$ yield, when used in (5.11.13),

$$\phi(r) = -Ua \sin ka \frac{e^{ikr}}{kr} \left(1 - \frac{\sin ka - ka \cos ka}{ka \exp(-ika) + \sin ka - ka \cos ka} \right) \qquad (5.11.20)$$

Since $\sin ka - ka \cos ka = -ka \exp(-ika) + (1 - ika)\sin ka$, this may be written as

$$\phi(r) = -Ua^2 \frac{e^{ik(r-a)}}{r(1-ika)} \qquad (5.11.21)$$

which is the result derived earlier by much simpler means.

Reduction to a Single Integral

Because of the arbitrariness of the surface values of ϕ and $\partial\phi/\partial n$, it is sometimes possible to obtain representation in terms of sources only or of dipoles only. To do this, we consider an auxiliary function ϕ' that is finite and continuous in the region V', external to V, where it satisfies $\nabla^2\phi' + k^2\phi' = 0$. Then ϕ' can also be expressed by an integral over S. However, for a point x inside V (and therefore exterior to V') the integral vanishes so that

$$0 = \frac{1}{4\pi} \int_S \left[\mathbf{n}_1 \cdot \nabla_{x'} \phi' \frac{e^{ik|x-x'|}}{|x-x'|} - \phi' \mathbf{n}_1 \cdot \nabla_{x'} \frac{e^{ik|x-x'|}}{|x-x'|} \right] dS(x') \qquad (5.11.22)$$

adding this to (5.11.6), we obtain

$$\phi(x) = \frac{1}{4\pi} \int_S \left(\frac{\partial\phi'}{\partial n_1'} - \frac{\partial\phi}{\partial n_1'} \right) \frac{e^{ik|x-x'|}}{|x-x'|} dS(x')$$

$$+ \frac{1}{4\pi} \int_S (\phi - \phi') \frac{\partial}{\partial n_1'} \frac{e^{ik|x-x'|}}{|x-x'|} dS(x') \qquad (5.11.23)$$

where $\partial/\partial n_1'$ represents derivation with respect to the primed coordinates, along the inward normal to V. Suppose, now, that we have a problem in which the prescribed boundary condition is in terms of the normal derivative $\partial\phi/\partial n$. Then we select ϕ' so that $\phi = \phi'$ on S. This gives

$$\phi(x) = \frac{1}{4\pi} \int_S \left(\frac{\partial\phi'}{\partial n_1'} - \frac{\partial\phi}{\partial n_1'} \right) \frac{e^{ik|x-x'|}}{|x-x'|} dS(x'), \quad \phi' = \phi \text{ on } S \qquad (5.11.24)$$

Similarly, if the prescribed boundary conditions are in terms of ϕ, we select ϕ' so that, on S, $\partial\phi/\partial n_1 + \partial\phi'/\partial n_1 = 0$. Then

$$\phi(x) = -\frac{1}{4\pi} \int_S (\phi - \phi') \frac{\partial}{\partial n_1'} \frac{e^{ik|x-x'|}}{|x-x'|} dS(x'), \qquad \frac{\partial\phi'}{\partial n_1} = -\frac{\partial\phi}{\partial n_1} \text{ on } S$$

$$(5.11.25)$$

The reduction of (5.11.6) to a single surface integral appears to simplify the problem considerably. In fact, the simplification is illusory, for we have replaced an integral having an unknown boundary condition with a function ϕ' that satisfies the wave equation together with prescribed conditions on S; a task that may be as difficult as that of solving the original problem for ϕ. There are situations, however, when some information about the auxiliary function ϕ' can be obtained without much effort, or when simple approximations to ϕ' can be used. For example, suppose that the surface S is a plane of infinite extent oscillating normally to itself with a small amplitude. Then $\partial\phi/\partial n_1$ is prescribed on S, and we might wish to find a solution ϕ' of the wave equation on the other side of the plane, which on the plane has a value equal to ϕ. Because of the symmetry about the plane, it is clear that if x is the coordinate along the normal into V, then $\phi'(-x, y, z) = \phi(x, y, z)$ so that $(\partial\phi'/\partial n_1 - \partial\phi/\partial n_1) = -2(\partial\phi/\partial n_1)$ and (5.11.24) becomes

$$\phi(x) = -\frac{1}{2\pi} \int_S \frac{\partial\phi}{\partial n_1} \frac{e^{ik|\mathbf{x}-\mathbf{x}'|}}{|\mathbf{x}-\mathbf{x}'|} \, dS(x') \tag{5.11.26}$$

This representation was used in Section 5.3 to obtain the field in front of a circular piston embedded in an inifinite wall.

Other simple geometries for which analytical results may be obtained with this procedure include the sphere and the cylinder. For more complicated geometries it is necessary to use approximate techniques involving numerical computations. It should be pointed out, however, that there are some distributions of ϕ or of $\partial\phi/\partial n$ for which the integral representations fail to give the required answer. These situations occur when the quantity k in those representations corresponds to one of the characteristic values of oscillations in the region enclosed by S. For a rigid enclosure, for example, these are prescribed by the condition $\partial\phi/\partial n = 0$ on S. For these values of k, the sources of sound have a frequency that matches one of the characteristic frequencies of the region, that is, we have resonance. Since the medium is considered ideal, we may obtain nonfinite solutions. Similarly, if k corresponds to one of the eigenvalues for the condition $\phi = 0$ on S, the representation also fails. The difficulties are discussed in Article 290 of Lamb's *Hydrodynamics* (Dover reprint, New York, 1945). A more recent discussion is given by Schenck [J. Acoust. Soc. Amer. **44**, 41–58 (1967)] who also presents an approximate numerical procedure for radiation problems that avoids those difficulties.

Reciprocity

Another important application of the extended Green's theorem, (5.11.2) is to the derivation of the reciprocity principle. Basically, this principle states that if

$\psi_A(\mathbf{x}_B)$ is the potential produced at location $\mathbf{x}_B$ by a source at $\mathbf{x}_A$, and if $\phi_B(\mathbf{x}_A)$ is the potential at x_A produced by a source at $\mathbf{x}_B$, then $\phi_B(\mathbf{x}_A)=\psi_A(\mathbf{x}_B)$, provided the sources have equal strength. This principle was mentioned in Section 5.2 with regard to unbounded media. We now derive it for a medium that is bounded by fixed and rigid boundaries. To do this, we first note that the potential ϕ_B due to a simple point source of strength Q_B located at $\mathbf{x}_B$ satisfies [see Equation (5.2.11)]

$$\nabla^2\phi_B + k^2\phi_B = Q_B\delta(\mathbf{x}_B) \tag{5.11.27}$$

Similarly, the potential $\psi_A(\mathbf{x}_B)$ due to a simple point source of strength Q_A located at $\mathbf{x}_A$ satisfies

$$\nabla^2\psi_A + k^2\psi_A = Q_A\delta(x_A) \tag{5.11.28}$$

If the internal and external boundaries are rigid and fixed (or if the internal boundary is fixed and rigid, and the external boundary extends to infinity), the normal derivatives of ϕ_B and ψ_A vanish on S so that the left-hand side of (5.11.2) vanishes. The right-hand side gives

$$Q_A\int_V\phi_B\delta(\mathbf{x}_A)\,dV = Q_B\int_V\psi_A\delta(\mathbf{x}_B)\,dV \tag{5.11.29}$$

Because of the properties of the delta function this becomes

$$Q_A\phi_B(\mathbf{x}_A) = Q_B\psi_A(\mathbf{x}_B) \tag{5.11.30}$$

Hence, if the two strengths are equal

$$\phi_B(\mathbf{x}_A) = \psi_A(\mathbf{x}_B) \tag{5.11.31}$$

as was to be shown. It should be added that the same result holds in more general conditions than those assumed here. In particular, it applies even if the fluid in the region does not have uniform mean properties, and if some of the boundaries are not rigid.

PROBLEMS

5.11.1 Consider a simple source of strength Q placed at a distance H from an infinite wall. Determine the far-field intensity, and sketch the angular variations of the intensity for the cases $ka\ll1$ and $ka\gg1$. Compare the total power emitted by the source in front of the plane with the power emitted by a simple source in an unbounded region.

5.11.2 Consider the contribution to $\phi(x)$ in (5.11.6) originating from S_e as S_e recedes to infinity. Let S_e be a sphere of radius R centered at x. Show that in the limit as $R \to \infty$, the leading contribution is given by (5.11.8).

5.11.3 Show that

$$\int_S \frac{e^{iks}}{s} dS(x') = \frac{a}{kr} \begin{cases} \sin ka \, e^{ikr}, & r > a \\ \sin kr \, e^{ika}, & r < a \end{cases}$$

where $s = |x - x'|$ and S is a sphere of radius a.

5.11.4 Suppose that in the pulsating-sphere example, k is equal to one of the characteristic values of the interior problem $r < a$, for the boundary condition $\phi(a) = 0$, that is, a value of k satisfying $\sin ka = 0$. Then $I_1(r)$ in (5.11.14) is identically zero. Determine $\phi(r)$.

5.11.5 Reconsider Problem 5.11.4, this time in terms of the single-integral representation (5.11.24). We know that $\phi = \exp(ikr)/r$. The auxiliary function ϕ' that is required in (5.11.24) satisfies $\nabla^2 \phi' + k^2 \phi' = 0$ inside the sphere $r < a$ and is equal to for $r = a$. Thus

$$\phi' = \frac{1}{\sin ka} \frac{\sin kr}{r}$$

Show that (5.11.24) gives the required answer for both $r < a$ and $r > a$, except when k is a characteristic value of the interior problem

5.11.6 Suppose that the volume V in Figure 5.11.3 contained a distribution of simple sources of strength $Q(x)$. Determine $\phi(x)$ in terms of surface and volume integrals.

5.11.7 Consider the region between two infinite, parallel planes. One of the planes is oscillating harmonically in a direction perpendicular to itself. The other plane, a distance L apart, is not moving. Assuming that both planes are rigid, determine the potential inside the region using the integral representation method.

SUGGESTED REFERENCES

Doak, P. E. "Analysis of Internally Generated Sound in Continuous Materials. 2. A Critical Review of the Conceptual Adequacy and Physical Scope of Existing Theories of Aerodynamic Noise, with Special Reference to Supersonic Jet Noise."

Ffowcs-Williams, J. E. "Aeroacoustics." This article presents a critical review of recent research in aeroacoustics, including combustion noise and diffusion-generated noise, and a wide variety of flow-noise problems.

Jones, H. "The Mechanics of Vibrating Flames in Tubes." This paper proposes a mechanism for flame vibration. Its Introduction gives a concise review of earlier work.

Lighthill, M. J. "The Bakerian Lecture, 1961: Sound Generated Aerodynamically." This article presents a clear introduction to aerodynamic sound, and reviews the most important work done since the theory was introduced in 1952.

Sommerfeld, A. *Partial Differential Equations of Physics*. Sections 25–28 of this excellent book deal with the Helmholtz equation and its solution by Green's functions.

Tyndall, J. *Sound*. This classical book is "a course of eight lectures delivered at the Royal Institution of Great Britain." Most of Lecture VI relates to "sounding flames."

CHAPTER SIX
SOUND ABSORPTION

In the previous chapter, we studied some of the basic mechanisms that are responsible for the emission of sound. In this chapter, we study those mechanisms responsible for the opposite effects; that is, for the absorption of sound. These mechanisms were ignored in Chapters 2–5 on the grounds that their effects were "small." However, no quantitative criterion was given that could be used to determine, in any given situation, whether or not they could in fact be ignored. At present, we would like to study these effects to determine when they can be neglected and to determine how they affect sound-wave propagation when they cannot be neglected. Another reason for studying these effects is that a clear understanding of their mode of action is important in eliminating or reducing undesirable sounds.

At the macroscopic level, the mechanisms responsible for sound absorption are the irreversible transfers of momentum and heat that occur due to viscosity and thermal conductivity, respectively. These quantities were set equal to zero in Chapter 2, a step that led to our study of acoustic waves in ideal fluids. We will now study their effects on the waves, assuming, however, that the other terms that were also neglected there—namely, the nonlinear terms—continue to be small compared to the retained terms. Whereas before it was sufficient that the nonlinear terms should be small compared to the linear, nondissipative terms, it is now necessary that they also be small compared to the dissipative terms. However, as we shall see, dissipation is also often small. Therefore, it is clear that this requirement may sometimes break down.

6.1 LINEARIZED DISSIPATIVE EQUATIONS

Our study of sound absorption is based on the full equations of motion, which include the effects of viscosity and heat conductivity. These equations were derived in Chapter 1, and are given in full in Section 1.8. If we linearize them, retaining only first-order dissipative terms, we obtain

$$\frac{\partial \rho'}{\partial t} + \rho_0 \nabla \cdot \mathbf{u} = 0 \tag{6.1.1}$$

$$\rho_0 \frac{\partial \mathbf{u}}{\partial t} + \nabla p' = \mu_0 \nabla^2 \mathbf{u} + \left(\mu_{v0} + \tfrac{1}{3}\mu_0\right)\nabla(\nabla \cdot \mathbf{u}) \tag{6.1.2}$$

$$\rho_0 T_0 \frac{DS'}{Dt} = \rho_0 c_{p0} \frac{DT'}{Dt} - T_0 \beta_0 \frac{Dp'}{Dt} = k_0 \nabla^2 T' \tag{6.1.3}$$

where c_p is the specific heat at constant pressure, β is the coefficient of thermal expansion, k is the coefficient of thermal conductivity, and μ and μ_v are the shear and the expansive coefficients of viscosity, respectively. The subscript zero in these quantities means that they are all evaluated at the ambient conditions p_0, ρ_0.

The equation of state, $p = p(\rho, S)$, becomes, in linearized form,

$$p' = c_0^2 \rho' + \left(\frac{\partial p}{\partial S}\right)_\rho S' \tag{6.1.4}$$

A useful form of this equation, which can be used together with (6.1.3), is

$$\frac{Dp'}{Dt} = c_0^2 \frac{D\rho'}{Dt} + \left(\frac{\partial p}{\partial S}\right)_\rho \frac{DS'}{Dt} \tag{6.1.5}$$

Obviously, in linearized acoustics, the substantive derivative D/Dt and the time derivative $\partial/\partial t$ are equal since the difference between them is of second order. However, in writing (6.1.5), we emphasize that the derivative applies to a given particle in motion.

Similarly, if instead of ρ and S we had used ρ and T as independent variables in the equation of state, we would have

$$p' = c_{T0}^2 \rho' + \rho_0 \beta_0 c_{T_0}^2 T' \tag{6.1.6}$$

where

$$c_{T0} = \sqrt{(\partial p/\partial \rho)_{T_0}} \tag{6.1.7}$$

is a quantity with the dimensions of a velocity and is known as the isothermal speed of sound. It is related to the adiabatic speed of sound c_0 by means of

$$c_0^2 = \gamma c_{T0}^2 \tag{6.1.8}$$

where $\gamma = c_{p0}/c_{v0}$.

Equations (6.1.1) to (6.1.3), together with (6.1.4) or (6.1.6), form a system of seven equations for the seven unknowns $u_1, u_2, u_3, p', \rho', T'$, and S'.

Physical Considerations

Equations (6.1.3) and (6.1.5) clearly show that when viscosity and heat conductivity are taken into account, the fluid pressure is no longer a function of the density alone. From the acoustic point of view, this means that the acoustic pressure and density are no longer in phase so that the pressure variations lag those of the density. These lags lead to acoustic-energy dissipation, as can be seen from simple thermodynamic arguments. Consider a fluid particle, oscillating back and forth about some mean position owing to the passage of an acoustic wave. Because of this motion, its pressure, density, and entropy (and also its temperature) vary, and the changes of these quantities are related by (6.1.5). Now, suppose that k, μ, and μ_v were identically zero. Then the entropy of the fluid particle would be constant, and the thermodynamic state of the fluid particle near the equilibrium position would be represented by a straight line bisecting the $c^2\rho$, p plane, as sketched in Figure 6.1.1. Since we are considering a given fluid particle, the mass is constant. Therefore, the coordinate ρc^2 is inversely proportional to the volume of the particle so that the figure is basically the pressure-volume diagram of the sequence of equilibrium states that the fluid particle goes through, owing to the passage of the wave. Thus, the pressure and volume of the particle go through cyclic variations, and the straight line merely

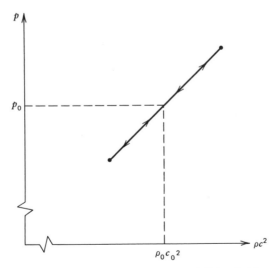

Figure 6.1.1 Pressure-volume diagram of acoustic "cycle" for fluid particle in ideal-fluid case.

represents the closed thermodynamic cycle. Since the area enclosed by the cycle is zero, it follows that the net work done by the wave on the fluid particle is also zero. Further, by the first law of thermodynamics, the net amount of heat given to the particle is also zero. We therefore conclude that the energy of the wave is conserved; that is, there is no dissipation of acoustic energy.

Consider now a case when μ, μ_v, and k are not all zero. Then S' is not zero, and the cycle becomes a three-dimensional curve in the $p, c^2\rho, S$ system of coordinates. Let us look at the $p, \rho c^2$ projection of that curve. As ρ increases, p also increases, but the increase is smaller than that of ρ owing to the term proportional to S' in (6.1.5). The opposite trend is true when ρ decreases so that during any one cycle (period), the increasing and the decreasing branches do not coincide. This is shown schematically in Figure 6.1.2 (for some cycle other than the initial one, where the curve would have to pass through the ambient equilibrium state $p_0, \rho_0 c_0^2$). The net result of this is that now the sound wave performs a net amount of work on the fluid particle at the expense of the acoustic energy. This energy is, of course, not lost because it increases the mean internal energy of the particle. However, as far as the sound wave is concerned, this energy transfer represents a loss. That is, the energy of the acoustic wave (not of the fluid as a whole) must decrease. This loss must obviously result in a decay of the amplitude of the wave as it travels in the fluid; that is, the wave is attenuated.

Thus, if a lag between pressure and density appears, we should expect some energy dissipation. The converse is not necessarily true. If in the linearized

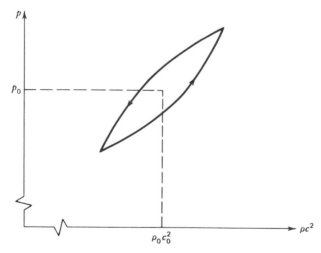

Figure 6.1.2 Pressure-volume diagram for acoustic cycle in a real fluid.

equations there is no apparent lag between these quantities, we may not necessarily conclude that the propagation is dissipationless. Consider, for example, the case of a viscous, nonheat-conducting fluid. Then (6.1.3) shows that the entropy of a fluid particle is a constant so that from (6.1.5) we have $p' = c_0^2 \rho'$ (i.e., no lags). However, we know that because of viscosity, there has to be some dissipation, and a closer study of the system of equations will show that this is indeed the case. The explanation of this apparent anomaly is that the lag between p' and ρ' still exists, but has been dropped because it is of second order of magnitude, as the entropy-variations equation shows. In other words, in a viscous, nonheat-conducting fluid, the variations of entropy due to the passage of a sound wave would be of second order so that approximately $p' = c_0^2 \rho'$. The dissipative effects of viscosity would, in that case, be taken care of by the linearized Navier–Stokes equation.

A better understanding of this exceptional case ($\mu \neq 0$, $k = 0$) is obtained by realizing that the linearized entropy equation is basically an equation for the energy of the acoustic wave. This energy is, as we know, a second-order quantity. Therefore, as far as energy calculations are concerned, one has to retain the second-order, viscous-dissipation function $\Phi = 2\mu(e_{ij}e_{ij} - \frac{1}{3}\Delta\delta_{ij})/\rho$, which was dropped when linearizing (1.6.10). These calculations provide, as we will see in Section 6.4, another means of studying acoustic-energy dissipation in fluids.

We will begin our study of dissipation by considering the effects of finite viscosity and heat conductivity on plane, one-dimensional waves, as they provide one of the simplest conditions for studying attenuation. However, since in this type of wave all the fluid fluctuations (i.e., pressure, velocity, etc.) depend only on one spatial coordinate, the waves must be unbounded. If we had boundaries present, the wave would not be purely one dimensional, except in the case when the normal to the boundary is along the direction of propagation. Consider, for example, the case when the direction of propagation is parallel to some solid boundary. Because of viscous effects, the relative velocity between the fluid and the wall is zero at the wall. Far from the wall, however, this relative velocity is not zero. Clearly, then, the motion cannot be considered purely one dimensional as the velocity must depend not only on the distance along the boundary but also on the distance normal to the boundary. There are some cases when such a motion can be considered approximately one dimensional. These situations, which will be studied later in some detail, occur when the distance over which the velocity changes from zero at the wall to its nearly uniform value away from the wall is small compared with other distances of importance in any given case. At present, however, we limit ourselves to studying attenuation in the absence of boundaries.

6.2   ATTENUATION DUE TO VISCOUS EFFECTS

We study, first, the propagation of a plane, one-dimensional wave in a viscous fluid that is, however, nonheat-conducting. This somewhat hypothetical situation is used because of its simplicity, and because it is the exceptional case ($k=0$, $\mu \neq 0$) referred to at the end of the last section.

When $k=0$, the entropy of a fluid element is, as explained earlier, conserved in the linear approximation so that from (6.1.5) we have

$$\frac{\partial \rho'}{\partial t} = \frac{1}{c_0^2} \frac{\partial p'}{\partial t} \tag{6.2.1}$$

The equation of continuity becomes, for the one-dimensional motion $\mathbf{u} = [u(x), 0, 0]$,

$$\frac{\partial p'}{\partial t} + \rho_0 c_0^2 \frac{\partial u}{\partial x} = 0 \tag{6.2.2}$$

Finally, the nonzero component of the linearized Navier–Stokes equation is

$$\rho_0 \frac{\partial u}{\partial t} + \frac{\partial p'}{\partial x} = \frac{4}{3} \mu_0' \frac{\partial^2 u}{\partial x^2} \tag{6.2.3}$$

where we have introduced the abbreviation

$$\mu_0' = \mu_0 \left(1 + \frac{3}{4} \frac{\mu_{v0}}{\mu_0} \right) \tag{6.2.4}$$

We can eliminate either p' or u from these equations. Eliminating p' first, we obtain

$$\frac{\partial^2 u}{\partial t^2} = c_0^2 \left(1 + \frac{4}{3} \frac{\nu_0'}{c_0^2} \frac{\partial}{\partial t} \right) \frac{\partial^2 u}{\partial x^2} \tag{6.2.5}$$

where $\nu_0' = \mu_0'/\rho_0$. Similarly, if u is eliminated, we find that p' satisfies the same equation as u:

$$\frac{\partial^2 p'}{\partial t^2} = c_0^2 \left(1 + \frac{4\nu_0'}{3c_0^2} \frac{\partial}{\partial t} \right) \frac{\partial^2 p'}{\partial x^2} \tag{6.2.6}$$

These equations reduce to the usual wave equation when $\nu_0' = 0$ and then have a general solution given by (2.4.4). It is clear that for nonzero ν_0', (2.4.4) will not work. A solution can now be obtained in terms of Fourier transforms. However, as a first step toward that solution, we consider the important case of monochromatic time dependence, where we seek solutions of the form

$$p' = \text{Re}\left[\tilde{P}(x)e^{-i\omega t} \right] \tag{6.2.7}$$

and obtain, after substituting into (6.2.6),

$$\tilde{P}''(x) + K^2 \tilde{P}(x) = 0 \tag{6.2.8}$$

where

$$K^2 = \frac{(\omega/c_0)^2}{1 - i\frac{4}{3}(\omega\nu_0'/c_0^2)} \tag{6.2.9}$$

The solution of (6.2.8) for positive-going waves is therefore

$$\tilde{p}' = \tilde{A}e^{i(Kx-\omega t)} \tag{6.2.10}$$

This is of the general form given by the first term of (2.5.7), except that now the effective wave number is complex:

$$K = k_1 + ik_2 \tag{6.2.11}$$

where k_1 and k_2 are real and positive (otherwise the amplitude would grow exponentially with x).

In terms of k_1 and k_2, the solution becomes (with $\tilde{A} = Ae^{i\theta}$)

$$p'(x, t) = Ae^{-k_2 x}\cos(k_1 x - \omega t + \theta) \tag{6.2.12}$$

This solution exhibits some interesting features. First, the amplitude of the wave is not constant, but decays exponentially with x. The decay of a wave's amplitude (and of its energy) is sometimes referred to as the *attenuation* of sound, and the quantity k_2, usually written as α, is called the (amplitude) *attenuation coefficient*. Since we can say that the wave is "damped" as it moves in space, k_2 is also referred to as the spatial damping coefficient. In either case, it is important to differentiate between the amplitude-attenuation coefficient and the energy-attenuation coefficient. The latter is also related to k_2 because the acoustic energy is proportional to the square of the waves' amplitude. Therefore, it is clear that the energy of a wave will decay, but with an exponential constant —the energy-attenuation coefficient—equal to $2k_2$. Later we will give yet another definition of the energy-attenuation coefficient. At present, we will limit our discussion to the amplitude coefficient, and adopt the widely used notation for it by writing α instead of k_2.

The second feature of our solution is that it predicts dispersion; that is, it predicts propagation with a frequency-dependent phase velocity. This is given by

$$c = \omega/k_1 \qquad (6.2.13)$$

where $k_1(\omega)$ is to be obtained from (6.2.9). Third, the profile of the wave does not change (other than a reduction of its amplitude). This result is applicable only to monochromatic waves. Any other profile—that is, one having several Fourier components—will be changed because each component will be propagated with its own velocity $c(\omega)$.

To summarize, the most notable features of plane-wave propagation in a viscous fluid are as follows: (a) There is dispersion and (b) there is attenuation. Both effects are described by the real and imaginary parts, respectively, of a complex wave number K. These statements can also be made when the sources of dissipation are different from viscosity; for example, when they are due to heat transfer. The main differences are, of course, the actual values of k_1 and k_2. In the viscosity case, these quantities are to be obtained from (6.2.9). Writing α instead of k_2 so that $K = k_1 + i\alpha$, we write that equation as

$$\left(\frac{c_0}{\omega}\right)^2 \left(k_1^2 - \alpha^2 + 2ik_1\alpha\right) = \frac{1}{1 - i\frac{4}{3}\left(\omega\nu_0'/c_0^2\right)} \qquad (6.2.14)$$

The real and imaginary parts of this equation constitute two equations from which k_1 and α can be obtained. To save writing, we introduce the following quantities:

$$\bar{\alpha} = \alpha c_0/\omega \qquad (6.2.15)$$

Since $c_0/\omega = \lambda/2\pi$, where λ is the wavelength, the quantity $\bar{\alpha} = 2\pi\alpha\lambda$ is sometimes referred to as the attenuation per wavelength, although that term applies more correctly to $\alpha\lambda$:

$$\bar{\beta} = k_1 c_0/\omega \qquad (6.2.16)$$

This is simply the ratio of the phase velocity in an ideal medium c_0, to the actual velocity given by (6.2.13):

$$\tau_\nu = \frac{4}{3} \frac{\nu_0'}{c_0^2} \qquad (6.2.17)$$

The meaning of this quantity can be best understood in terms of molecular quantities. We defer until later in this section a discussion based on such quantities. At present, we consider, instead, the nondimensional group

$$\omega\tau_\nu = \frac{4}{3}\frac{\omega\nu_0'}{c_0^2} \tag{6.2.18}$$

The quantity $\sqrt{\omega\nu_0'}$ has the dimensions of a velocity, and in a sense is proportional to velocity with which excess momentum is transmitted in an oscillatory flow due to the diffusive effects of viscosity. This can be seen by considering the situation described in Problem 6.2.1. Of course, excess longitudinal momentum can also be transmitted by means of ideal acoustic waves. Therefore, the quantity $\omega\tau_\nu$ measures the speed for viscous transfer to momentum relative to the speed with which momentum is transmitted by wave motion in an ideal fluid; that is, relative to c_0. The former type of propagation is dissipative, whereas the second is not. Hence, we can expect that the larger $\omega\tau_\nu$ is, the larger the attenuation should be.

In view of these remarks, the quantity τ_ν, which has the dimensions of time, can be considered as the time scale for viscous effects. It is therefore referred to as the *viscous relaxation time*. Another way of looking at $\omega\tau_\nu$ is as the ratio of the viscous time scale to the period of the waves.

We now return to (6.2.14). Using the definitions given above for $\bar{\alpha}, \bar{\beta}$, and τ_ν, we obtain, after equating real and imaginary parts on both sides of that equation,

$$\bar{\beta}^2 - \bar{\alpha}^2 = \frac{1}{1 + \omega^2\tau_\nu^2} \tag{6.2.19}$$

$$2\bar{\alpha}\bar{\beta} = \frac{\omega\tau_\nu}{1 + \omega^2\tau_\nu^2} \tag{6.2.20}$$

The values of τ_ν for many fluids of interest are such that, unless ω is exceedingly large, $\omega\tau_\nu$ is very small so that the term $(\omega\tau_\nu)^2$ can be neglected compared to unity. In fact, as will be seen later, there are fluids for which the range of validity of (6.2.19) is limited to small values of $\omega\tau_\nu$. However, in order to see the full implications of the above equations, we first obtain from them exact expressions for $\bar{\alpha}$ and $\bar{\beta}$. To do this, we first combine them in the form

$$\bar{\alpha}^2 + 2\bar{\alpha}\bar{\beta}/\omega\tau_\nu - \bar{\beta}^2 = 0$$

Solving for $\bar{\alpha}$, we obtain

$$\bar{\alpha} = \frac{\bar{\beta}}{\omega\tau_\nu}\left(\pm\sqrt{1 + \omega^2\tau_\nu^2} - 1\right) \tag{6.2.21}$$

We know that $\bar{\alpha}$ and $\bar{\beta}$ are both positive and real. Therefore, we take the positive sign in the radical. Finally, substitution for $\bar{\beta}$ from (6.2.20) yields

$$\left(\frac{\alpha c_0}{\omega}\right)^2 = \frac{1}{2}\left[\frac{1}{\sqrt{1+\omega^2\tau_\nu^2}} - \frac{1}{1+\omega^2\tau_\nu^2}\right] \qquad (6.2.22)$$

Similarly, solving for $\bar{\beta} = (k_1 c_0/\omega)$, we obtain

$$\left(\frac{k_1 c_0}{\omega}\right)^2 = \frac{1}{2}\left[\frac{1}{\sqrt{1+\omega^2\tau_\nu^2}} + \frac{1}{1+\omega^2\tau_\nu^2}\right] \qquad (6.2.23)$$

The variations of $\bar{\alpha}$ and $\bar{\beta}$ can be better appreciated by plotting $\bar{\alpha}^2$ and $\bar{\beta}^2$ versus $\omega\tau_\nu$, as in Figure 6.2.1. The salient feature of the attenuation-per-wavelength curve is its maximum at $\omega\tau_\nu = \sqrt{3}$. This maximum exists because at that frequency the period of the waves is basically equal to the viscous relaxation time τ_ν. In other words, viscosity effects are then "optimized." At either side of this maximum, $\bar{\alpha}$ approaches zero. However, whereas for $\omega\tau_\nu \to 0$, $\bar{\alpha} \to 0$ means that the dissipation is negligible, the same interpretation does not follow when

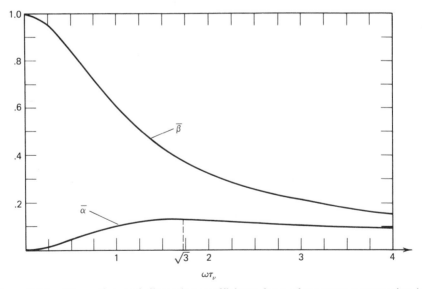

Figure 6.2.1 Attenuation and dispersion coefficients for a plane wave propagating in a viscous, nonheat-conducting fluid.

$\omega\tau_\nu\to\infty$. Thus, when $\omega\tau_\nu\to 0$, (6.2.23) shows that $k_1\to\omega/c_0$. Therefore (since for a given fluid τ_ν is fixed), $\omega\tau_\nu\to 0$ implies that the wavelength is finite so that the vanishing of $\bar\alpha$ implies that α itself approaches zero; that is, there is no dissipation. On the other hand, when $\omega\tau_\nu\to\infty$, the effective wavelength approaches zero, so that the vanishing of $\bar\alpha$ merely means that owing to the smallness of the wavelength λ, the wave decays little in a distance equal to λ. There is, nonetheless, significant dissipation, and this is measured by the attenuation coefficient α. In the limit $\omega\tau_\nu\to\infty$, α is given by

$$\alpha_\infty \approx \frac{\omega/c_0}{\sqrt{2\omega\tau_\nu}} = \sqrt{\frac{3}{8}\frac{\omega}{\nu_0'}} \qquad (6.2.24)$$

Contrary to the opposite limit, α is now finite and increases with frequency.

The other curve shown in the figure, namely, that for $(k_1 c_0/\omega)^2 = (c_0/c)^2$, shows that the actual phase velocity c is larger when viscosity is taken into account then when it is not. Similarly, the dispersion relation (6.2.23) may be used to show that the group velocity $c_g = (dk_1/d\omega)^{-1}$ is also larger than c_0 (see Problem 6.2.2). These results may appear somewhat puzzling, as one would expect viscosity to "slow down" the waves. However, the "slowing-down" concept applies only to macroscopic fluid velocities, and reference to (6.2.12) shows that, in fact, the fluid velocity has a smaller amplitude when viscosity is taken into account. On the other hand, the speed of sound is the speed with which momentum fluctuations are transmitted in a fluid by wave motion. This process depends, of course, on molecular collisions to which the slowing-down concept does not apply.

The increase of the value of the speed of propagation when viscosity is included in the formulation can be explained in terms of quantity $\sqrt{\omega\nu_0'}$. This quantity was interpreted as the velocity with which momentum is transferred owing to the diffusive effects of viscosity. Thus, when $\sqrt{\omega\nu_0'}$ is not negligible compared to c_0 (i.e., when $\omega\tau_\nu$ is finite), one effect of μ_0' is to increase the speed with which momentum is being propagated. In fact, for $\omega\tau_\nu\to\infty$, the phase and group velocities are, respectively,

$$c_\infty \approx \sqrt{\tfrac{8}{3}\omega\nu_0'}\,, \qquad c_{g_\infty} = 2c_\infty$$

values that correspond to the equivalent velocities for viscous diffusion (see Problem 6.2.1).

We now reconsider (6.2.22) and (6.2.23) for the range of values that the group $\omega\tau_\nu$ attains in situations of interest. To determine this range, we need numerical values of τ_ν. However, except for the monatomic-gas case, this quantity cannot

Table 6.2.1   Values of Viscous Relaxation Time and Related Quantities at $T = 295°K$ and 1Atm

Gas	ν_0 (cm^2/sec)	c_0 (cm/sec)	τ_ν (sec)	$f_t = 1/2\pi\tau_\nu$ (Hz)	λ_t/l
Ar	0.1158	31,900	1.52×10^{-10}	1.05×10^9	~5
He	1.0136	96,500	1.45×10^{-10}	1.10×10^9	~5
Ne	0.3219	43,500	2.27×10^{-10}	7.02×10^8	~5

be computed accurately because, for most fluids, the expansive viscosity μ_v is not known. In the monatomic-gas case, μ_v is identically zero, as explained in Chapter 1, and τ_ν can be computed from known values of μ_0 and c_0. Table 6.2.1 gives some of these quantities for three monatomic gases. The table also gives the value of the frequency for which $\omega\tau_\nu = 1$, and the ratio of the corresponding wavelength to the mean free path between molecular collisions l.

The table clearly shows that unless the frequency ω is exceedingly large, one can neglect in (6.2.19) and (6.2.20) terms of order $(\omega\tau_\nu)^2$. [If ω is so large that $(\omega\tau_\nu)^2$ is not small compared to unity, then one can use the full results given by (6.2.23). However, some caution is required when applying these equations because as Table 6.2.1 shows, when $\omega\tau_\nu$ is finite, the ratio of wavelength to mean free path l is not very large, as required for the continuum model to be applicable.] Therefore, retaining only those terms whose order of magnitude is $\omega\tau_\nu$ or larger, one obtains $\bar{\beta} \approx 1$ and $\bar{\alpha} = \frac{1}{2}\omega\tau_\nu$. In terms of c, c_g, and α, these give

$$\left. \begin{array}{l} c \approx c_0 \\ c_g \approx c_0 \\ \alpha \approx \dfrac{2}{3}\dfrac{\omega^2 \nu_0'}{c_0^3} \end{array} \right\} \tag{6.2.25}$$

Hence, in the limit $\omega\tau_\nu \to 0$, we find that the speed of propagation is the same as in a fluid devoid of viscosity, and that the amplitude of the wave decays very slowly, although this decay increases with frequency. Thus, after a distance x, the amplitude of a sinusoidal wave will become

$$A_0 e^{-\pi\omega\tau_\nu(x/\lambda)} \tag{6.2.26}$$

where A_0 is the initial amplitude. For example, the amplitude of a 1000-Hz wave, traveling in still air at uniform pressure and temperature, would become, in a distance equal to 1000 wavelengths (340m at STP), equal to $0.9968A_0$. This represents, for most purposes, a negligible decay.

Translational Relaxation Time

We digress a little to reconsider the physical meaning of the quantity τ_ν introduced earlier. The simplest case available for clarification is that of a monatomic gas for which $\mu_\nu = 0$, so that

$$\tau_\nu = \frac{4}{3} \frac{\mu_0}{\rho_0 c_0^2} \qquad (6.2.27)$$

Consider now a monatomic gas inside some adiabatic container in a state of complete thermodynamic equilibrium. At the molecular level, a state of equilibrium implies that each one of the three translational degrees of freedom of the gas has the same average kinetic energy. Suppose that one end of the container is fitted with a piston that at some time t moves a small distance into the container and then stops. If we confine our attention to a small volume in the vicinity of the piston, we will observe that a nonequilibrium state is developed because some of the molecules have gained additional x momentum, say, owing to collisions with the piston. These molecules will, in turn, collide with other molecules, and through each collision will lose some of the excess kinetic energy. After some collisions, a new state of equilibrium will set in that corresponds to the new macroscopic position of the piston. The time required for the establishment of the new equilibrium state depends on the nature of the collisional process, and on the magnitude of the departure from equilibrium. This time is called the *relaxation time*, and since only translational degrees of freedom are involved, we will refer to it as the translational relaxation time and will denote it temporarily by t^*. We would like to show that τ_ν is basically this relaxation time. To do this, we will use some known results from the kinetic theory of gases. We first assume that t^* depends on the number of collisions in the simplest way:

$$t^* = a/n$$

where n represents the number of collisions that a molecule experiences with other molecules per unit time, and a is a dimensionless constant of proportionality. The collision frequency (number of collisions per unit time) depends on the number N of molecules per unit volume, their average speed $\bar{c}$, and the effective cross-sectional area for collisions σ. Since in a unit time a molecule sweeps a collisional volume $\sigma \bar{c}$, it follows that

$$n = N \sigma \bar{c}$$

Therefore,

$$t^* = a/N \sigma \bar{c}$$

We know from the elementary kinetic theory of gases that $\bar{c}$ is of the order of the speed of sound c_0 in the gas. Also, the kinematic viscosity of a gas is of the order of $\bar{c}/N\sigma$:

$$\nu_0 = b\bar{c}/N\sigma$$

where b is a constant. Combining these estimates, we have

$$t^* \approx d\nu_0/c_0^2$$

where d is yet another constant. Comparison with (6.2.27) shows that τ_ν and t^* are equivalent. Thus, the quantity τ_ν is a measure of the time required to establish equilibrium of the translational (or external) degrees of freedom. Hence, it may be termed the *translational relaxation time*. Similarly, the frequency for which $\omega\tau_\nu = 1$,

$$f_t = 1/2\pi\tau_\nu$$

may be called the *translational relaxation frequency*. This frequency can be used to discuss the shape of the attenuation curve shown in Figure 6.2.1. Consider a plane sound wave propagating in a monatomic gas. If the frequency f of the wave is small compared to f_t, then the departures from equilibrium are small so that basically no dissipation takes place. As the frequency increases, the departures from equilibrium (and, therefore, the dissipation) increase, but provided $f < f_t$, the time required for the readjustment of equilibrium is smaller than the period of the wave, and at least for a fraction of each cycle the departures from equilibrium are very small. Therefore, if $f < f_t$, the dissipation per cycle is below its maximum possible value. This maximum occurs in the vicinity of $f = f_t$,[1] because then the time lag between the readjustment of equilibrium and the changes produced by the wave is at a maximum. This is the "optimum" condition to dissipation referred to earlier. Similarly, when $f > f_t$, the differences decrease so that the dissipation per cycle is smaller than at $f = f_t$.

A measure of the dissipation per unit cycle is the attenuation per wavelength $\alpha\lambda$,[2] and our discussion shows that this quantity should have a maximum in the vicinity of the translational relaxation frequency f_t. This is, in fact, the trend that (6.2.22) predicts. However, as stated earlier, those results are based on a continuum model for the fluid, and this is not strictly applicable when the quantity $\omega\tau_t = f/f_t$ is finite.

---

[1] If τ_t were exactly the time required to reestablish equilibrium and not just a measure of it, the maximum would occur at $f = f_t$.

[2] Since we are discussing dissipation per unit cycle, the suitable parameter here is βT, where β is the temporal damping coefficient and T is the period of the wave. However, $\beta = \alpha c_0$ and $T = 1/f$. Hence, $\beta T \equiv \alpha\lambda$.

If instead of a monatomic gas we have a polyatomic gas (say N_2), then in addition to translational motion, the molecules also rotate and (if the temperature is high enough) vibrate. Now, just as we have a translational relaxation time and frequency, we also have rotational and vibrational relaxation times and frequencies. These degrees of freedom take longer times to readjust to new conditions. Therefore, their relaxation frequency is smaller than that for the translational modes. If these relaxation frequencies were sufficiently far apart from each other, one would expect (from the above discussion for the translational modes) that the curve for $\alpha\lambda$ would have local maxima in the vicinity of each relaxation frequency. It is clear that these curves cannot be obtained from the Navier–Stokes equations because attenuation by molecular processes depends entirely on the molecular structure of the fluid, and this is ignored in the derivation of the equation. In principle, the expansive viscosity μ_v is supposed to take into account small departures from local equilibrium caused by the internal degrees of freedom. However, as pointed out in Section 1.4, very little is known about μ_v.

Except for a brief discussion of μ_v for diatomic gases in Section 6.3, this concludes our discussion of molecular relaxation. The reader interested in that subject should consult the excellent books by Hertzfeld and Litovitz and Bhatia listed in the bibliography. These books deal with the subject at length.

PROBLEMS

6.2.1 Consider a situation for which the pressure gradient in (6.2.3) vanishes. Show that (6.2.3) will then possess monochromatic solutions representing attenuated waves. Show that the attenuation coefficient α_∞ is

$$\alpha_\infty = \sqrt{3\omega/8\nu_0'}$$

Show, further, that the phase and group velocities are, respectively,

$$c = \sqrt{\tfrac{8}{3}\omega\nu_0'}\,, \qquad c_g = 2c$$

6.2.2 Equation (6.2.23) is a dispersion relation for k_1; that is, it is of the form $k_1 = k_1(\omega)$. Show that the small and large frequency limits for the group velocity are, respectively,

$$c_g/c_0 \simeq 1 + \tfrac{9}{8}\omega^2\tau_\nu^2, \quad \omega\tau_\nu \to 0$$

$$c_g/c_0 \simeq \sqrt{8\omega\tau_\nu}\,, \quad \omega\tau_\nu \to \infty$$

Compare these limits to the corresponding limits for the phase velocity.

6.3   ATTENUATION IN A VISCOUS HEAT-CONDUCTING FLUID

We now consider the more realistic case of wave propagation in a fluid that is both viscous and heat-conducting. The main difference from the previous case is that now the entropy changes of a fluid element are of first order, being given, in fact, by a term that is of first order in the variations of temperature (i.e., the term $k\nabla^2 T'$ in Equation 6.2.3). This implies that now p' is no longer a function of the density alone. The working equations are therefore

$$\frac{\partial \rho'}{\partial t} + \rho_0 \frac{\partial u}{\partial x} = 0 \tag{6.3.1}$$

$$\rho_0 \frac{\partial u}{\partial t} + \frac{\partial p'}{\partial x} = \frac{4}{3}\mu'_0 \frac{\partial^2 u}{\partial x^2} \tag{6.3.2}$$

$$\rho_0 c_{p0} \frac{\partial T'}{\partial t} - T_0 \beta_0 \frac{\partial p'}{\partial t} = k_0 \frac{\partial^2 T'}{\partial x^2} \tag{6.3.3}$$

and

$$p' = c_{T0}^2 \rho' + \rho_0 \beta_0 c_{T0}^2 T' \tag{6.3.4}$$

To study this linear, but coupled, system of equations, it is convenient to introduce, instead of u, p', ρ', T', x, and t, new nondimensional variables. We have, of course, a large variety of dimensional parameters that can be used to nondimensionalize these quantities. For example, the fluid velocity and pressure can be made nondimensional by dividing them by c_0 and p_0, respectively. However, it is usually more convenient to introduce variables such that each nondimensional quantity has a magnitude of order unity. This makes the bookkeeping much easier. Thus, if the maximum velocity amplitude in the wave is u_0, it is clear that

$$v = u/u_0 \tag{6.3.5}$$

is a quantity of order unity. For the other quantities, it is not generally clear which combinations of the physical parameters can be used to obtain variables of order unity. In the present case, however, we can use the information we already have about plane waves. Thus, in the viscous-attenuation case, we found that plane, one-dimensional waves are attenuated very little in a distance equal to one wavelength. This means that over distances of that order, the wave is not affected much by viscous effects. If we assume a priori that heat conduction will have similarly small effects on the wave, we can use the results from plane, one-dimensional waves in ideal fluids to obtain new variables of order unity.

Those results were given in Section 2.4. We found there that for a given fluid velocity u, the acoustic pressure in a progressive wave was given by $p' = \rho_0 c_0 u$. Thus, the order of magnitude of p' is $\rho_0 c_0 u_0$ so that the quantity

$$P = \frac{p'}{\rho_0 c_0 u_0} \qquad (6.3.6)$$

is of order unity. We do the same thing for temperature and density, and introduce the nondimensional variables

$$\theta = \frac{T'}{c_0 \beta_0 T_0 u_0 / c_{p0}} \qquad (6.3.7)$$

$$\delta = \frac{\rho'}{\rho_0 u_0 / c_0} \qquad (6.3.8)$$

instead of T' and ρ', respectively. Finally, if a typical frequency is ω and the corresponding wavelength is ω/c_0, the nondimensional distance and time variables

$$\left. \begin{array}{l} x' = (\omega/c_0)x \\ t' = \omega t \end{array} \right\} \qquad (6.3.9)$$

will define length and time scales; that is, distances and times over which the unknown variables change significantly. This completes the selection of our new variables. Substituting these into equations of continuity and momentum yields

$$\frac{\partial \delta}{\partial t'} + \frac{\partial v}{\partial x'} = 0 \qquad (6.3.10)$$

$$\frac{\partial v}{\partial t'} + \frac{\partial P}{\partial x'} = \omega \tau_\nu \frac{\partial^2 v}{\partial x'^2} \qquad (6.3.11)$$

where, as before, $\tau_\nu = \frac{4}{3}(\nu_0'/c_0^2)$. Similarly, the energy equation becomes

$$\frac{\partial \theta}{\partial t'} - \frac{\partial P}{\partial t} = \omega \tau_\kappa \frac{\partial^2 \theta}{\partial x'^2} \qquad (6.3.12)$$

where

$$\tau_\kappa = \frac{k_0/\rho_0 c_{p0}}{c_0^2} = \frac{\kappa_0}{c_0^2} \qquad (6.3.13)$$

is a quantity with the dimensions of time and is for heat conduction what τ_ν is for momentum. We can therefore refer to it as the thermal relaxation time. The quantity κ is known as the thermal diffusivity. For gases, $\kappa_0/c_0^2 \simeq 0(10^{-10}$ sec). The viscous and thermal relaxation times are related through the Prandtl number $Pr = \mu_0' c_{p0}/k_0$:

$$\tau_\kappa = \frac{3}{4Pr}\tau_\nu$$

Finally, with nondimensional variables the equation of state becomes

$$P = \left(\frac{c_{T0}}{c_0}\right)^2 \delta + \frac{T_0 \beta_0^2 c_{T0}^2}{c_{p0}}\theta \tag{6.3.14}$$

This equation can be written in a simpler form by recalling the general thermodynamic relation for $c_p - c_v$:

$$c_p - c_v = T\beta^2 \left(\frac{\partial p}{\partial \rho}\right)_T \tag{6.3.15}$$

The partial derivative appearing here is just c_T^2. Therefore, denoting, as before, the ratio c_{p0}/c_{v0} by γ and substituting into (6.3.15), we obtain

$$\gamma P = \delta + (\gamma - 1)\theta \tag{6.3.16}$$

Equations (6.3.10)—(6.3.12) and (6.3.16) form our final system of equations. The advantage of using the nondimensional variables defined earlier is now clear. Every one of the variables v, P, δ, and θ and their derivatives appearing in these equations has a magnitude of order unity. Therefore, the coefficients of these quantities in the system of equations tell us how important each term is when the time and length scale are those defined by (6.3.9). For example, (6.3.11) states that the term $\omega \tau_\nu (\partial^2 v/\partial x'^2)$ is of order $\omega \tau_\nu$, whereas the other terms are of order unity. Therefore, when $\omega \tau_\nu$ is very small, we may expect the viscous effects to be small also. This is, in fact, what our more detailed analysis of plane waves in a viscous fluid told us. Similar remarks apply to heat conduction, where the quantity $\omega \tau_\kappa$ represents the effects of heat conduction in a plane, one-dimensional acoustic wave.

Similar ideas apply to the relative contributions to the pressure variations as described by (6.3.16). This equation states that the contributions to P arising from density and temperature changes are not of the same order. Thus, in a fluid for which $\gamma - 1 \ll 1$, the contributions arising from temperature changes are negligible compared to those arising from density changes.

In the present case, where we have both viscosity and heat conductivity, we would like to retain all the terms appearing in our system of equations. The main consequence of this is that now we cannot obtain a single equation *of low order* for any one of the unknowns. While an equation of higher order can be obtained by suitable manipulation of our equations, and a solution to that equation can be obtained for monochromatic waves, we use an alternative procedure to study the effects of viscosity and heat conduction on sinusoidal waves. Thus, in view of our previous results, we expect that a sinusoidal wave will be attenuated, and will be propagated with a velocity different from c_0. However, we do not expect its profile to change. Therefore, we may expect that the effects of viscosity and heat conductivity can be taken into account by assuming a monochromatic-type solution, where ρ, δ, θ, and v all depend on distance and on time as

$$e^{i(K'x - \omega t)} = e^{i(Kx' - t')} \tag{6.3.17}$$

where $K = c_0 K'/\omega$ is a nondimensional complex wave number that, if known, gives the phase velocity and the attenuation coefficient by means of $c = c_0/\text{Re}(K)$ and $\alpha = (\omega/c_0)\text{Im}(K)$, respectively. Using the type of solution specified by (6.3.17), the equation of continuity becomes

$$\delta = Kv \tag{6.3.18}$$

Now, K is real only for $\tau_\nu = \tau_\kappa = 0$, in which case it is given by $K = 1$, and this implies nonattenuated propagation with speed c_0. For nonzero τ_ν or τ_k, K is complex, and (6.3.18) states that the density and velocity are out of phase.

The problem now is to find K. In order to do this, we must make sure that the assumed forms for δ, P, v, and θ are, in fact, a solution of our equations. Substituting $\delta = Kv$ in the equation of state, we have

$$Kv - \gamma P + (\gamma - 1)\theta = 0 \tag{6.3.19}$$

The other two equations we need are (6.3.11) and (6.3.12):

$$\frac{\partial v}{\partial t'} + \frac{\partial P}{\partial x'} - \omega \tau_\nu \frac{\partial^2 v}{\partial x'^2} = 0 \tag{6.3.20}$$

$$\frac{\partial \theta}{\partial t'} - \frac{\partial P}{\partial t'} - \omega \tau_\kappa \frac{\partial^2 \theta}{\partial x'^2} = 0 \tag{6.3.21}$$

Substituting the assumed form for our solution, we find, using matrix notation,

$$
\begin{bmatrix}
K & -\gamma & \gamma-1 \\
-i\left(1+i\omega\tau_\nu K^2\right) & iK & 0 \\
0 & i & -i\left(1+i\omega\tau_\kappa K^2\right)
\end{bmatrix}
\begin{bmatrix}
v \\
P \\
\theta
\end{bmatrix}=0
\qquad (6.3.22)
$$

As we know from the theory of linear algebra, a nontrivial solution to this system exists only if the determinant of the coefficient matrix vanishes. Expanding that determinant, we obtain

$$
K^2\left(1+i\omega\tau_\kappa K^2\right)-\left(1+i\gamma\omega\tau_\kappa K^2\right)\left(1+i\omega\tau_\nu K^2\right)=0
\qquad (6.3.23)
$$

This is a dispersion relation for the nondimensional wave number K. The equation is quadratic in K^2, and possesses two physically realistic roots. One of the roots represents highly attenuated waves, and is obtained in Problem 6.3.1. The other represents slightly attenuated waves, and can be obtained as follows. From (6.3.23), we can write

$$
K^2=\frac{\left(1+i\gamma\omega\tau_\kappa K^2\right)\left(1+i\omega\tau_\nu K^2\right)}{\left(1+i\omega\tau_\kappa K^2\right)}
\qquad (6.3.24)
$$

As stated earlier, the ideal-fluid result is $K=1$. Since we expect that viscosity and heat conductivity have a small effect on the waves, we put $K=1$ on the right-hand side of this equation and obtain

$$
K^2\approx\frac{\left(1+i\gamma\omega\tau_\kappa\right)\left(1+i\omega\tau_\nu\right)}{\left(1+i\omega\tau_\kappa\right)}
\qquad (6.3.25)
$$

This approximate result prescribes the speed of propagation and the attenuation coefficient as a function of $\omega\tau_\nu$ and $\omega\tau_\kappa$. Consider, first, the case when there is no heat conduction. Then,

$$
K\approx1+\tfrac{1}{2}i\omega\tau_\nu
\qquad (6.3.26)
$$

This is the result obtained in Section 6.2. It represents propagation with a viscous attenuation coefficient

$$
\alpha_\nu=\frac{1}{2}\frac{\omega^2\tau_\nu}{c_0}
\qquad (6.3.27)
$$

and with speed c_0. On the other hand, if no viscous effects are included,

$$K^2 \approx \frac{1 + \gamma \omega^2 \tau_\kappa^2 + i(\gamma - 1)\omega\tau_\kappa}{1 + \omega^2 \tau_\kappa^2} \qquad (6.3.28)$$

This has some interesting properties. We discuss the two extreme limits $\omega\tau_\kappa \ll 1$ and $\omega\tau_\kappa \gg 1$ separately.

$\omega\tau_\kappa \ll 1$.   Here, we obtain

$$K \approx 1 + \tfrac{1}{2}i(\gamma - 1)\omega\tau_\kappa \qquad (6.3.29)$$

This represents propagation with speed c_0 and with thermal attenuation coefficient

$$\alpha_\kappa = \tfrac{1}{2}(\gamma - 1)\frac{\omega^2 \tau_\kappa}{c_0} \qquad (6.3.30)$$

$\omega\tau_\kappa \gg 1$.   This case gives

$$K \approx \frac{c_0}{c_{T0}} + i\frac{\gamma - 1}{2\sqrt{\gamma}\ \omega\tau_\kappa} \qquad (6.3.31)$$

where we have used $c_0^2 = \gamma c_{T0}^2$. This wave number represents propagation with speed c_{T0}, and with attenuation coefficient $(\gamma - 1)/2\sqrt{\gamma}\ \omega\tau_\kappa$.

Since the speed of propagation is given in this limit by c_{T0} [which is obtained by evaluating $(\partial p / \partial \rho)$ at constant temperature], it follows that the temperature of the fluid remain nearly constant. This can also be seen by remembering that $\sqrt{\omega\tau_v}$ is the velocity with which momentum is propagated owing to the diffusive effects of viscosity. Similarly, the quantity $\sqrt{\omega\tau_\kappa}$ gives the velocity with which heat is conducted away from a fluid element. In the limit $\omega\tau_\kappa \gg 1$ this velocity is large so that a volume element cannot store any heat and, therefore, its temperature remains constant. The same conclusion follows from our equations. Thus, since in this case $v = 0$, (6.3.20) yields

$$v = KP \qquad (6.3.32)$$

so that (6.3.19) becomes

$$(\gamma - 1)\theta = (\gamma - K^2)P \qquad (6.3.33)$$

But for $\omega\tau_\kappa \gg 1$, K^2 is, from (6.3.28), given by

$$K^2 = \gamma + i\frac{\gamma-1}{\omega\tau_\kappa} \qquad (6.3.34)$$

and so

$$(\gamma-1)\theta \approx i\frac{\gamma-1}{\omega\tau_\kappa}P \qquad (6.3.35)$$

Thus, as $\omega\tau_\kappa \to \infty$, we obtain $\theta = 0$.

The quantity c_{T0} is called the isothermal speed of sound and is related to the adiabatic speed by means of $c_0^2 = \gamma c_{T0}^2$. Since, as we know from thermodynamics, $\gamma \geqslant 1$, it follows that c_{T0} is always smaller than c_0 (except for the case $\gamma = 1$).

We now return to (6.3.25) and consider the case when both τ_ν and τ_κ are small but finite. This is the most important case because it applies to waves of moderate frequency propagating in many different fluids, especially in air. Thus, for small values of $\omega\tau_\nu$ and $\omega\tau_\kappa$, (6.3.25) yields

$$K = 1 + i\tfrac{1}{2}\left[\omega\tau_\nu + (\gamma-1)\omega\tau_\kappa\right] \qquad (6.3.36)$$

Thus, the dimensional wave number is just ω/c_0 so that the waves are propagated with speed c_0. Similarly, the attenuation per wavelength $\alpha c_0/\omega$ is given by

$$\bar{\alpha} = \alpha c_0/\omega = \tfrac{1}{2}\omega\tau_\nu + \tfrac{1}{2}(\gamma-1)\omega\tau_\kappa \qquad (6.3.37)$$

We see from this equation that the effects of viscosity and heat conductivity are, for this case, separable. That is, the attenuation coefficient is given by the sum of a viscous and a heat-conduction contribution:

$$\alpha = \alpha_\nu + \alpha_\kappa \qquad (6.3.38)$$

This is important because it shows that for small values of $\omega\tau_\nu$ and $\omega\tau_\kappa$, viscous and heat-conduction effects are uncoupled, so that the energy dissipated by each of these irreversible mechanisms can be computed independently.

Comparison with Experimental Data

In view of our definitions of τ_ν and τ_κ, we can write (6.3.37) as

$$\alpha\lambda = \pi\frac{\omega\mu_0}{\rho_0 c_0^2}\left(\frac{4}{3} + \frac{\mu_{\nu 0}}{\mu_0} + \frac{\gamma-1}{Pr}\right) \qquad (6.3.39)$$

where $\Pr = \mu_0 c_{p0} / k_0$ is the ambient Prandtl number of the fluid. The attenuation coefficient α given by (6.3.39), with $\mu_v = 0$, was first derived by Kirchhoff in 1868. Earlier, Stokes (1845) had derived the contribution to α arising from shear viscosity. Therefore, the quantity α_c defined by

$$\alpha_c = \frac{\omega^2 \mu_0}{2 \rho_0 c_0^3} \left(\frac{4}{3} + \frac{\gamma - 1}{\Pr} \right) \qquad (6.3.40)$$

is knows as the *Stokes–Kirchhoff attenuation* coefficient. It is also known as the *classical absorption* coefficient.

Sound absorption in gases and liquids have been subject to several carefully performed experimental studies using a wide range of frequencies and pressures. The data obtained in these experiments, as well as the experimental techniques used, are reviewed at length by Markham, Beyer, and Lindsay (1953). Here we merely point out some of the main features of the attenuation data that were obtained in gases. First, we rewrite (6.3.40) in a form suitable for comparison with the experiments. We know that the predicted attenuation is small under normal conditions. However, (6.3.40) shows that α_c is inversely proportional to the density. Therefore, in order to increase absorption, most experiments with gases are conducted at low pressures; at these pressures, gases can be considered perfect, so that $\rho_0 c_0^2 = \gamma p_0$ and we may write (6.3.40) as

$$\frac{\alpha_c c_0 p_0}{\omega^2} = \frac{\mu_0}{2\gamma} \left(\frac{4}{3} + \frac{\gamma - 1}{\Pr} \right) \qquad (6.3.41)$$

The right-hand side of this equation is a function of the temperature alone. Hence, if one plots $\alpha_c c_0 p_0 / \omega$ versus frequency or versus mean pressure, the result should be (for a given temperature) invariable. Stated differently, the attenuation per wavelength $\alpha_c \lambda (= \alpha_c c_0 / f)$ should correlate, for a given temperature, with f / p_0. Experimental attenuation data in gases are therefore usually given by plotting $\alpha \lambda$ versus f / p_0 or versus some similar group.

Monatomic Gases. For monatomic gases, the expansive viscosity μ_v is zero, as there are no molecular degrees of freedom other than the translational. In that case, (6.3.39) reduces to the classical attenuation formula. Therefore, provided no other losses occur (such as those due to heat radiation), the theoretical prediction for α_c should be confirmed experimentally.

This is much easier said than done, for measuring such small attenuation is very difficult. For example, if the attenuation measurements are made in a tube by means of any of the three methods outlined in Section 3.6, wall effects come into play. However, as we shall see later, the attenuation due to wall effects is significantly larger than that due to the gas alone so that the accuracy required

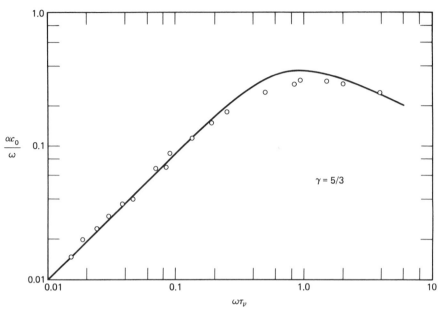

Figure 6.3.1 Comparison of the continuum theoretical prediction for attenuation in a monatomic gas with Greenspan's data for helium.

to measure α_c is very high.[3] To eliminate wall effects, α_c is sometimes measured in large-volume facilities so that the fluid may be considered unbounded. On the other hand, freely propagating waves are generally not plane so that the theoretical prediction for α, given by (6.3.39), does not apply to them. For example, the waves emitted by a vibrating piston in a wall are not plane near the piston. However, in the far-field of the piston, they are nearly so. This fact is used to measure attenuation of plane waves away from limiting boundaries. The measurements are made at large frequencies so that the "far-field" is relatively close to the emitter. Large frequencies have the additional advantage of increasing the magnitude of α_c.

Figure 6.3.1 shows attenuation data obtained by Greenspan (1956). In addition to the helium data shown in the figure, Greenspan also reported data for four other monatomic gases. The helium data shown here are typical.

---

[3] For example, at a frequency of 1600 Hz, the wall-attenuation coefficient for He at STP in a tube of 3-cm radius is of the order of 1.82×10^{-5}. If this coefficient could be determined with an accuracy of 99.99%, the uncertainty would still be larger than the value of α_c, which under the same conditions is $\alpha_c \approx 8 \times 10^{-10} \text{cm}^{-1}$.

Also shown in the figure is the classical prediction obtained from the full equation, (6.4.23), and for K (see Problems 6.3.2 and 6.3.3). The figure shows that agreement between theory and experiment is good, at least for values of $\omega \tau_\nu$ (or of $\omega \tau_\kappa$) that are not too large. Agreement for such values of $\omega \tau_\nu$ is not expected because the continuum model does not apply to them. We therefore conclude that within the limitations of the continuum model, the classical theory for attenuation and dispersion is verified for monatomic gases.

Diatomic Gases. The situation changes for diatomic gases because their molecules also have rotational and vibrational degrees of freedom. Depending on the temperature of the gas, these may or may not be excited. Thus, at room temperature, the rotational modes in a diatomic gas are fully excited, and contribute to the internal energy of the gas the full amount required by the classical equipartition-energy principle. However, the vibrational modes are nearly frozen. Only at much higher temperatures do these modes begin to get excited. For example, in N_2 the characteristic temperature for vibration is of the order of 3300°K. At temperatures much smaller than this, no vibrational modes are excited; and at temperatures much larger than this, the vibrational modes contribute their full amount to the internal energy of the gas. The same is true for the vibrational modes of other diatomic gases except, of course, that the corresponding characteristic temperatures vary from gas to gas. In any event, at room temperature we may ignore vibrational relaxation in diatomic gases. However, rotational relaxation cannot be ignored, as it will play some role under changing conditions. Thus, if we consider the contribution to the attenuation per wavelength arising from rotational relaxation, we may expect that it will have a maximum in the vicinity of the rotational relaxation frequency, and that it will tend to zero at both small and large frequencies. As explained earlier, this trend cannot be obtained from the continuum model. However, when the departures from equilibrium are small (i.e., when the sound-wave frequency is small compared to the rotational relaxation frequency), the departures can be taken into account by means of the expansive viscosity μ_v. Thus, if μ_v were known, (6.3.39) could be used to predict the total-attenuation coefficient. In practice, since μ_v is not known, the opposite procedure is used to determine it. That is, at frequencies well below the rotational relaxation frequency, the attenuation coefficient for a plane sound wave traveling in a diatomic gas is given by the sum of the classical attenuation coefficient α_c and a contribution arising from the expansive viscosity. Therefore, the attenuation due solely to rotational relaxation is given by

$$\alpha_e = \frac{\omega^2 \mu_v}{2\rho_0 c_0^3} \tag{6.3.42}$$

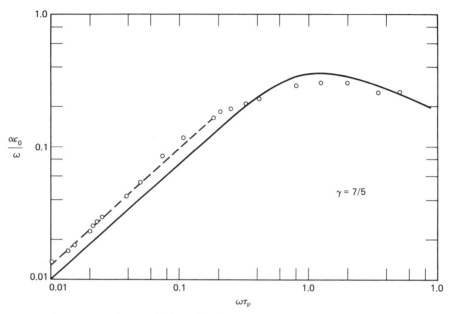

Figure 6.3.2 Attenuation coefficient for free propagation in nitrogen. O: Greenspan (1959).

where $\alpha_e = \alpha - \alpha_c$ is the "excess" attenuation; that is, the attenuation that exists above and beyond the classical attenuation. Thus, a measurement of the total attenuation in a diatomic gas, the use of the theoretical prediction for α_c, will yield the value μ_v in such a gas by means of the above equation.

There have been several careful experiments of attenuation in diatomic gases reported in the literature. Among the most recent is that of Greenspan, who measured attenuation and dispersion of sound in nitrogen, oxygen, and air at 11 MHz using a variety of pressures ranging from a few millimeters of Hg to atmospheric. Figure 6.3.2 shows his attenuation data for nitrogen. In the figure, the nondimensional attenuation coefficient $\alpha c_0 / \omega$ is plotted versus $\omega \tau_v$. Also shown is the theoretical prediction given by (6.3.40). The discrepancies at finite values of $\omega \tau_v$ are due to (6.3.39) not being valid then. On the other hand, we see that for $\omega \tau_v \ll 1$ the experimental and theoretical *trends* are in agreement, but that the "excess" attenuation is fairly constant over a large range of $\omega \tau_v$. To obtain the value of μ_v from these data, we use Greenspan's polynomial fit to his data for small values of $\omega \tau_v$.[4] He found that for $\omega \tau_v < 0.1$, the attenuation data were

---

[4]Instead of $\omega \tau_v$, Greenspan used the quantity $r = \rho_0 c_0^2 / \gamma \omega \mu_0$ as the independent variable. In terms of r, $\omega \tau_v$ is equal to $4/3r\gamma$.

Table 6.3.1 Coefficients a_1 and a_3 of Polynomial Fit of Greenspan's Data

	N_2	O_2	Air
a_1	0.9104	0.8569	0.8903
a_3	-6.52	-5.78	-8.02

well fitted by means of

$$\left(\frac{\alpha c_0}{\omega}\right)_E = \frac{3\gamma}{4}\left[a_1\omega\tau_\nu + a_3(\omega\tau_\nu)^3\right]$$

where the coefficients a_1 and a_3 are given in Table 6.3.1. In the limit $\omega\tau_\nu \to 0$, only the first term remains so that the fit to the experimental attenuation is given by $3\gamma a_1\omega\tau_\nu/4$. On the other hand, theory gives

$$\left(\frac{\alpha c_0}{\omega}\right)_T = \frac{3}{8}\omega\tau_\nu\left(\frac{4}{3} + \frac{\mu_{\nu 0}}{\mu_0} + \frac{\gamma-1}{Pr}\right)$$

Equating these two results, we obtain

$$\frac{\mu_{\nu 0}}{\mu_0} = 2\gamma a_1 - \tfrac{4}{3} - \frac{\gamma-1}{Pr}$$

For nitrogen at $300°K$, $Pr = 0.713$ and $\gamma = \tfrac{7}{5}$. These values, together with the value of a_1 from Table 6.3.1, give $\mu_{\nu 0} = 0.65\mu_0$. The corresponding values for pure oxygen and pure air are 0.52 and 0.60, respectively, at the same temperature.

Effects of Impurities

As stated above, these measurements were made with pure gases. Entirely different results for the total attenuation are obtained when impurities are present, even in small amounts. For example, at 2200 Hz, the attenuation in oxygen with a 2% hydrogen "impurity" is about 40% higher than that in pure oxygen. The reason for the increase is that collisions between oxygen and hydrogen molecules appear to be very efficient in producing vibrational relaxation in oxygen. The same is true of other hydrogen-bearing substances, water vapor in particular. The increased attenuation due to impurities is especially important when evaluating sound attenuation in atmospheric air, as this air contains significant amounts of water vapor. Figure 6.3.3 shows the effects of humidity on the attenuation coefficient α at various frequencies and at a fixed

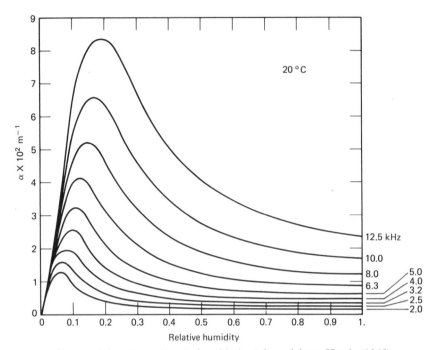

Figure 6.3.3 Attenuation in humid air. Adapted from Harris (1965).

temperature equal to 20°C. These curves are the result of attenuation measurements made by Harris (1966, 1975). The magnitude of the measured attenuation is significantly larger than in dry air. For example, at 12,500 Hz, the attenuation coefficient in dry air is found, from (6.3.40), to be equal to 0.00285 m^{-1}, whereas in air with 5% humidity it is 10 times as large. The main significance of this is that in a homogeneous atmosphere, humidity, and not viscosity or heat conductivity, determines how much a sound wave is attenuated.

Each of the curves shown in Figure 6.3.3 displays a maximum attenuation that depends, in addition to humidity, on frequency also. These maxima are due to vibrational relaxation of the oxygen molecules in air brought about by the addition of water vapor. As the concentration of water vapor is decreased, the maximum attenuation point is shifted toward lower frequencies. This trend is also obtained in other gases having hydrogen-bearing impurities. In fact, one of the techniques used to determine relaxation frequencies in polyatomic gases is to plot the maximum-absorption frequency against the impurity concentration. In most cases, the resulting plot appears to be a straight line, and when extrapolated to zero concentration, yields the desired relaxation frequency.

PROBLEMS

6.3.1 Equation (6.3.23) may be written as

$$K^4\left(\gamma\omega^2\tau_\nu\tau_\kappa + i\omega\tau_\kappa\right) + K^2\left[1 - i\omega\left(\tau_\nu + \gamma\tau_\kappa\right)\right] - 1 = 0$$

One root for K^2 corresponds to the slightly attenuated wave discussed in the text. A second root corresponds to a highly attenuated wave, in which case K^2 is very large. Show that the limiting form for K, when K^2 is large (but still with $\omega\tau_\nu$ and $\omega\tau_\kappa$ small), is

$$K = \frac{1+i}{\sqrt{2\omega\tau_\kappa}}$$

6.3.2 Because the quantity τ_κ is related to τ_ν by means of $\tau_\kappa = 3\tau_\nu/4\mathrm{Pr}$, the equation for K in Problem 6.3.1 shows that for a given fluid, K is a function of $\omega\tau_\nu$ alone. For gases, one of the two remaining parameters, γ and Pr, may be eliminated in favor of the other by means of Eucken's relation $\mathrm{Pr} = 4\gamma/(9\gamma - 5)$, which holds approximately well. Using these results for τ_κ and Pr, show that the equation for K can be written as

$$K^4\left(\omega^2\tau_\nu^2 + \frac{i}{\gamma}\omega\tau_\nu\right)\left(\frac{9\gamma - 5}{\frac{16}{3}}\right) + K^2\left[1 - i\omega\tau_\nu\left(1 + \frac{9\gamma - 5}{\frac{16}{3}}\right)\right] - 1 = 0$$

Show that when $\omega\tau_\nu \to 0$, this yields

$$K \simeq 1 + \frac{i}{2}\omega\tau_\nu\left[1 + \frac{3}{16\gamma}(\gamma - 1)(9\gamma - 5)\right]$$

and that this result corresponds to (6.3.36). Show, further, that when $\omega\tau_\nu \gg 1$, K is approximately given by

$$K = \sqrt{\frac{f(\gamma) + 1 + |f(\gamma) - 1|}{4f(\gamma)}} \ \frac{1+i}{\sqrt{\omega\tau_\nu}}$$

where $f(\gamma) = \frac{3}{16}(9\gamma - 5)$. For $\gamma \geqslant 1.15$, this reduces to $K = (1+i)/\sqrt{2\omega\tau_\nu}$. These limiting forms show that the attenuation per wavelength must have a maximum for some intermediate value of $\omega\tau_\nu$.

6.3.3 To obtain the roots of the fourth-order equation for K for all values of $\omega\tau_\nu$ in the previous problem, put $K = K_1 + iK_2$ in that equation. By separating real and imaginary parts of the resulting equation, obtain two coupled relationships for K_1 and K_2 from which these can be obtained numerically. Following this procedure, obtain $K_2 = \alpha c_0/\omega$ for $\gamma = \frac{5}{3}$ and compare your results with the solid line shown in Figure 6.3.1.

6.3.4 Consider a spherical wave propagating into a viscous but nonheat-conducting fluid. The fluid velocity is $\mathbf{u} = \mathbf{e}_r u_r(r, t)$, where $\mathbf{e}_r$ is a unit vector along r. Show that for this velocity field, the governing equation is

$$\frac{\partial^2 \mathbf{u}}{\partial t^2} = c_0^2 \left(1 + \frac{4}{3} \frac{\nu_0'}{c_0^2} \frac{\partial}{\partial t} \right) \nabla^2 \mathbf{u}$$

[*Hint*: $\nabla \times \mathbf{u} = 0$.]

6.3.5 For the velocity field of Problem 6.3.4, show that

$$\frac{\partial^2}{\partial t^2}(ru_r) = c_0^2 \left(1 + \frac{4}{3} \frac{\nu_0'}{c_0^2} \frac{\partial}{\partial t} \right) \left(\frac{\partial^2}{\partial r^2}(ru_r) - \frac{2u_r}{r} \right)$$

[*Hint*:   The r component of $\nabla^2 \mathbf{u}$ *is not* $\nabla^2 u_r$.]

6.3.6 Show that for monochromatic-time dependence, u_r, in the problem above, is given by

$$\tilde{u}_r = \frac{1}{r} J_0(Kr) e^{-i\omega t}$$

where the complex wave number $K = k \Big/ \sqrt{1 - 4i\omega\nu_0'/3c_0^2}$. For what values of kr does u_r decay exponentially? What is the corresponding attenuation coefficient?

6.4   ENERGY-DISSIPATION METHOD

Before considering the effects of boundaries on plane sound waves, we reconsider the free-propagation case treated in the previous section, this time using a simple method that is based on energy considerations. Suppose we compute the total acoustic energy in a given region: the moving region between two consecutive crests in a traveling wave, say, or the fixed region in which a standing wave exists. Denote this energy by E. If no dissipative effects are present, this energy will remain constant in time so that $dE/dt = 0$. However, if dissipation exists, then $E = E(t)$, and as time increases, the acoustic energy inside the region will decrease. We can then say that acoustic energy is being lost at a rate given by

$$\dot{E}_{\text{lost}} = dE/dt \geqslant 0 \tag{6.4.1}$$

If we can compute separately the time derivative of the energy and the rate at which energy is lost, we may be able to calculate attenuation coefficients without using the procedure presented in the previous section. Based on our previous discussions, we expect the decay of energy to be exponential:

$$E = E_0 e^{-2\beta t} \tag{6.4.2}$$

where E_0 is the value of E at $t = t_0$ and β $(= \alpha c_0)$ is the temporal damping coefficient. Using (6.4.2) in (6.4.1), we obtain

$$\dot{E}_{\text{lost}} = 2\alpha c_0 E \tag{6.4.3}$$

This equation holds at every instant. However, in many instances, one is interested in monochromatic waves for which time averages are more suitable than instantaneous values. Taking the time average of (6.4.3) and solving for α, we obtain

$$\alpha = \frac{\langle \dot{E}_{\text{lost}} \rangle}{2 c_0 \langle E \rangle} \tag{6.4.4}$$

This equation appears to be a suitable definition for α: in effect, if $\langle \dot{E}_{\text{lost}} \rangle$ and $\langle E \rangle$ are known, it can be used to compute the quantity. But $E(t)$ depends on α so that, in general, such an approach is questionable. However, consider the time average of E:

$$\langle E \rangle = \frac{1}{T} \int_0^T E_0 e^{-2\alpha c_0 t} \, dt = \frac{E_0}{2\alpha c_0 T} (1 - e^{-2\alpha c_0 T}) \tag{6.4.5}$$

If the energy changes little during the time required for the average (i.e., if $\beta T \ll 1$), then

$$\langle E \rangle = E_0 \tag{6.4.6}$$

Let us now consider the quantity $\langle \dot{E}_{\text{lost}} \rangle$. If it could also be expressed in terms of nonattenuated quantities, we may use (6.4.4) to obtain α. To show that this is indeed possible, we make use of an important concept from thermodynamics, that of *available energy*. The available energy of a system, in a given state, refers to the amount of work that the system is capable of producing when it is allowed to reach complete equilibrium with other systems. We know from thermodynamics that the work done by a system during a process depends on the details of the transformation, and that it is a maximum for a process for which the entropy change is zero. If the entropy change is not zero, some available energy is lost. In the acoustical context, the available energy is the total acoustical energy, and the lost available energy is our quantity E_{lost}. We also know from thermodynamics that if the total entropy increase is ΔS, the energy loss is $E_{lost} = T_0 \Delta S$. Therefore,

$$\dot{E}_{lost} = T_0 \dot{S}_T \qquad (6.4.7)$$

The quantity $\dot{S}_T$ is the rate at which the entropy of the fluid inside the volume V under consideration and that of its surroundings increases owing to irreversible mechanisms such as viscosity and heat conductivity. The entropy increase of the fluid in V is given by

$$\frac{d}{dt} \int \rho S \, dV$$

whereas that of the surroundings due only to a flow of entropy out of V is given by

$$\int_A \mathbf{J}_S \cdot \mathbf{n} \, dA$$

where J_s is the entropy flux vector (see Section 1.6). Therefore, (1.6.15) shows that

$$\dot{S}_T = \int_V \rho \sigma \, dV \qquad (6.4.8)$$

where $\rho\sigma$ gives the entropy production per unit volume and is given by (1.6.22). Substituting (6.4.8) into (6.4.7) and making use of (1.6.22), we obtain

$$\dot{E}_{lost} = T_0 \int \frac{1}{T} \left(\Phi + \frac{\mu_v}{\rho} \Delta^2 \right) dV + T_0 \int \frac{k}{T^2} (\nabla T)^2 \, dV \qquad (6.4.9)$$

where Φ is the viscous dissipation function defined by (1.6.9). For the acoustic case, it is sufficient to set $T = T_0$, $\rho = \rho_0$ in the denominators of each of the integrands in (6.4.9) and to consider the case of constant viscosities and heat conductivity. This gives

$$\dot{E}_{\text{lost}} = 2\mu_0 \int \left(e_{ij}e_{ij} - \frac{\Delta^2}{3} \right) dV + \mu_{v0} \int \Delta^2 \, dV + \frac{k_0}{T_0} \int (\nabla T)^2 \, dV \quad (6.4.10)$$

Making use of (1.5.15) and of the properties of the Kronecker delta, the following identity can be derived:

$$4e_{ij}e_{ij} - \tfrac{4}{3}\Delta^2 = \left(\frac{\partial u_i}{\partial x_j} + \frac{\partial u_j}{\partial x_i} - \tfrac{2}{3}\delta_{ij}\Delta \right)^2 \quad (6.4.11)$$

so that

$$\dot{E}_{\text{lost}} = \tfrac{1}{2}\mu_0 \int_V \left(\frac{\partial u_i}{\partial x_j} + \frac{\partial u_j}{\partial x_i} - \tfrac{2}{3}\delta_{ij}\Delta \right)^2 dV + \mu_{v0} \int_V \Delta^2 \, dV + \frac{k_0}{T_0} \int_V (\nabla T)^2 \, dV$$

$$(6.4.12)$$

This is the desired result. It gives the rate at which acoustic energy is being lost inside a volume V in terms of the squared spatial gradients of velocity and temperature. If these quantities were known, $\dot{E}_{\text{lost}}$ could be computed easily. It is clear, however, that even in the simplest, one-dimensional monochromatic case, those quantities are not known unless α is specified. Fortunately, as far as obtaining α from (6.4.4) is concerned, it is sufficient to use in (6.4.12) the nonattenuated results for the temperature and the pressure. The limitation imposed by the approximation is, again, that the attenuation in the region under consideration be small. Thus, if the region V over which $\langle E \rangle$ is being computed has a dimension of order L, it is necessary that $\alpha L \ll 1$. This is the requirement used before to obtain (6.4.6) because a wave travels a distance L in a time $T = L/c_0$ so that $\alpha L = \alpha c_0 T = \beta T \ll 1$. Therefore, when the attenuation is small in the above sense, all we have to do to obtain α is to substitute in (6.4.6) and (6.4.12) the unattenuated results and then use (6.4.4).

Unbounded Waves

As an example, we reconsider the attenuation of plane waves in unbounded media. If we place the x_1 axis of a coordinate system along the direction of propagation, then $\mathbf{u} = (u, 0, 0)$. Also, the nonattenuated velocity component u is a function of x and t only so that the only nonzero component of $\partial u_i / \partial x_j$ is $\partial u / \partial x$. Hence,

$$\left(\frac{\partial u_i}{\partial x_j} + \frac{\partial u_j}{\partial x_i} - \tfrac{2}{3}\delta_{ij}\Delta\right)^2 = 4\left(\frac{\partial u}{\partial x}\right)^2 - \frac{8}{3}\left(\frac{\partial u}{\partial x}\right)\Delta + \tfrac{4}{3}\Delta^2 = \frac{8}{3}\left(\frac{\partial u}{\partial x}\right)^2$$

Substitution into (6.4.12) yields

$$\dot{E}_{\text{lost}} = \tfrac{4}{3}\mu_0' \int_V \left(\frac{\partial u}{\partial x}\right)^2 dV + \frac{k_0}{T_0}\int_V \left(\frac{\partial T}{\partial x}\right)^2 dV \qquad (6.4.13)$$

where μ_0' is the quantity defined in (6.2.4). Consider the first term in (6.4.13). It gives the acoustic-energy loss due to viscous dissipation. Its value is obtained by using the nonattenuated value for u; that is, $u = \text{Re}\{u_0 \exp[i(kx - \omega t)]\}$. If we take, for example, the volume within two consecutive crests, the average energy dissipated by viscous effects per unit wave-front area of the wave is

$$\langle \dot{E}_{\text{vis}} \rangle = \tfrac{2}{3}\mu_0' \int_{x_0}^{x_0 + \lambda} \left(\frac{\partial \tilde{u}}{\partial x}\right)\left(\frac{\partial \tilde{u}}{\partial x}\right)^* dx = \tfrac{2}{3}\mu_0'\lambda(u_0 k)^2 \qquad (6.4.14)$$

The acoustical energy within the same region per unit wave-front area is

$$E_0 = \tfrac{1}{2}\rho_0 u_0^2 \lambda \qquad (6.4.15)$$

Using these results, one obtains from (6.4.4) the attenuation coefficient due to viscous effect alone:

$$\alpha_\nu = \tfrac{2}{3}\nu_0' k^2 / c_0 \qquad (6.4.16)$$

Since $k = \omega/c_0$, this is identical to (6.2.25)

Similarly, the second term in (6.4.13) results in the attenuation coefficient given by (6.3.29). It is seen that the present procedure leads to the results obtained previously by a different technique. However, the present method cannot be used to obtain attenuation coefficients if dissipation is large, nor does it yield the changes of the speed of sound. Nevertheless, it is clear that the method is quite powerful, and that it may be used in other, more complicated situations such as the one treated in the next section.

PROBLEM

6.4.1 Derive the attenuation coefficient due to thermal effects alone [i.e., Equation (6.3.29)] using the energy-balance procedure outlined in this section.

6.5 EFFECTS OF BOUNDARIES

We saw in Section 3.1 that when a plane sound wave, traveling in an *ideal* medium, meets a plane boundary, reflected and transmitted waves appear at the boundary. These additional waves are also plane, longitudinal waves, and propagate with the speed of sound. Furthermore, in the special case when the waves are being propagated in a direction parallel to the boundary, the propagation is not affected at all by the boundary.

The situation changes when real-fluid effects are taken into account. For example, nonzero viscosity results in zero-slip velocity between boundary and fluid. One effect produced by this no-slip condition is that plane-wave propagation, in a direction parallel to a boundary, can no longer be one-dimensional; that is, the fluid velocity component tangential to the wall will necessarily display a sheared profile, such as the one depicted in Figure 1.5.3. For monochromatic waves, these sheared profiles are not stationary; they oscillate back and forth with the main sound wave. They have amplitudes that at any given distance depend on position from the boundary but that change with time. These variations can be understood better in terms of waves moving away from the boundary. These waves have a speed that is quite different from the speed of the longitudinal waves, and have amplitudes that decay rapidly as they move into the fluid. Because the main effect of these waves is to reduce shear between contiguous layers of fluid, they are usually referred to as *shear waves*. Also, since their direction is along the local normal to the boundary, which in this example is parallel to the direction of propagation, they are sometimes called *transverse* waves. Name aside, it is clear that the resulting acoustic-energy dissipation in fluid layers near a boundary, while still possibly small, is nevertheless large when compared to that which would occur if no boundaries were present.

Similarly, if the fluid has a nonzero thermal conductivity, significant temperature gradients may exist in a direction perpendicular to the boundary. The main reason for these gradients is that in a sound wave the temperature is not constant, but fluctuates about a mean value. These fluctuations cannot be followed by real boundaries. For example, if the boundary is a wall having a large thermal conductivity (i.e., a metal wall), its temperature will remain basically constant even though heat transfer takes place. One of the results of a temperature gradient near a wall is the appearance of *thermal* waves, traveling away from the boundary in a direction along its normal. As we will see later, these waves are very similar to the viscous transverse waves mentioned above. Another thermal effect of some interest occurs when the temperature of a boundary in contact with a fluid changes in time. In addition to the thermal waves referred to earlier, the temperature fluctuations produce sound waves. This is an important sound-emission effect that does not occur in an ideal fluid and that will be considered later.

To study the effects of viscosity and thermal conductivity on sound waves traveling in the vicinity of a boundary, we will first consider related problems that illustrate these effects clearly.

Flow Induced by an Oscillating Plane

Suppose that we have a solid body immersed in a fluid and executing translatory oscillations of small amplitude. As we know from our discussion in Section 5.7, this motion will produce, far from the body, acoustic waves of small amplitude. The flow near the body will, in general, have normal and tangential velocity components relative to the body. On the body's surface, the normal velocity component is fixed by the requirement that there be no flow through the boundary. However, in the absence of viscosity, no requirement can be imposed on the tangential components so that, in general, there is a nonzero slip between the body and the fluid. For example, (5.5.18) shows that in the case of a solid sphere oscillating in a fluid, the relative tangential velocity $U_0 - \mathbf{U} \cdot \mathbf{e}_\theta$ between the fluid and the sphere is (for $ka \ll 1$) $\frac{3}{2} U_0 \sin \theta \cos \omega t$, where U_0 is the amplitude of the oscillatory sphere's velocity.

When viscosity effects are taken into account, the fluid in contact with the body can no longer slip over the body; instead, it adheres to it. This is not the only effect of viscosity, for in the same way that it precludes slip between fluid and solid, it also prevents complete slippage between contiguous layers of fluid. Therefore, the no-slip condition at a boundary will make the whole tangential-velocity profile significantly different from that which would exist if the fluid were inviscid. We would like to study the modifications induced by viscosity. These will, of course, depend on the geometry of the body, but there are several important features that are typical of problems with oscillatory boundaries. To illustrate some of these features, we consider a plane of infinite extent, oscillating in its own plane with a small velocity amplitude. The results will give us an idea of how shear waves are created by a body of *finite size*, at least locally, that is, in the vicinity of a given position on its boundary.

To study the motion over the infinite plane, we attach a rectangular system of coordinates to the plate in such a manner that the x axis is along the direction of oscillation, as sketched in Figure 6.5.1. Because of viscosity, the fluid above the plane will also move, but it is clear that the fluid velocity will have only one component, and this will be parallel to the velocity of the plane. Further, this velocity component cannot depend on distance along the plane so that $\mathbf{u} = [u(y, t), 0, 0]$. Therefore, $\nabla \cdot \mathbf{u} = 0$, and the continuity equation (6.1.1) yields $\rho' = 0$. Finally, because the motion does not depend on x, and $\nabla \cdot \mathbf{u} = 0$, (6.1.2) yields

$$\frac{\partial u}{\partial t} = \nu_0 \frac{\partial^2 u}{\partial y^2} \tag{6.5.1}$$

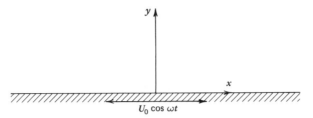

Figure 6.5.1 Plate oscillating in its own plane.

This is a diffusion equation. Therefore, if the plane starts to move at $t=0$ and imparts some momentum to the fluid in contact with it, we would expect this momentum to be diffused slowly into the fluid. At present, however, we are concerned only with the motion of the fluid after all transient effects have disappeared; that is, we would like to obtain the "steady-state" solution. Since the plane is oscillating as

$$U_p = U_0 \mathrm{Re}(e^{-i\omega t}) \tag{6.5.2}$$

we would also expect the fluid velocity to depend harmonically on time. Therefore, we set

$$u(y,t) = \mathrm{Re}\left[\tilde{U}(y)e^{-i\omega t}\right] \tag{6.5.3}$$

Substitution in (6.5.1) yields for $\tilde{U}(y)$

$$\tilde{U}''(y) + K^2 \tilde{U}(y) = 0 \tag{6.5.4}$$

where

$$K = (1+i)\sqrt{\omega/2\nu_0} \tag{6.5.5}$$

The solution for $\tilde{U}(y)$ is of the form $\exp(\pm iKy)$. Thus, with the notation

$$\delta_\nu = \sqrt{2\nu_0/\omega} \tag{6.5.6}$$

we have

$$\tilde{u}(y,t) = e^{-i\omega t}\left(Ae^{i(1+i)y/\delta_\nu} + Be^{-i(1+i)y/\delta_\nu}\right) \tag{6.5.7}$$

Since for $y \to \infty$ the velocity must be small, we must set $B = 0$. Also, at $y = 0$ the fluid velocity is equal to that of the plane, so that $A = U_0$ and

$$u(y, t) = U_0 e^{-(y/\delta_\nu)} \cos(\omega t - y/\delta_\nu) \qquad (6.5.8)$$

The fluid, therefore, also oscillates harmonically in time, but the oscillations lag those of the plane, and have very small amplitudes far from the plane. Figure 6.5.2 depicts relative-velocity profiles at various times during one oscillation. In the figure, time is measured from the point during a cycle when $u = U_0$ at $y = 0$. It is seen that for $\omega t = \pi/4$, the maximum fluid-velocity amplitude is at the plane $y = 0$. However, for $\omega t > \pi/4$, the maximum occurs for $y > 0$. In fact, its location in the fluid is given by

$$y_{max}(t) = (\omega t - \pi/4)\delta_\nu, \quad \omega t \geqslant \pi/4 \qquad (6.5.9)$$

Thus, one of the features of the oscillation, namely, the point of maximum fluid velocity, is seen to be moving into the fluid with velocity $\omega \delta_\nu = \sqrt{2\omega \nu_0}$.

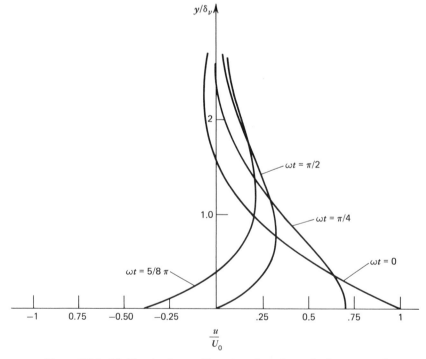

Figure 6.5.2 Fluid-velocity profiles at various times during one cycle.

Reference to (6.5.8) shows that this is simply the phase velocity so that $(1/\delta_\nu)$ can be thought of as the wave number. Therefore, the solution given by (6.5.8) represents a damped wave moving away from the plate in a direction perpendicular to the direction of oscillation. Because of the exponential decay, the effects produced by the plane are not significant far from the plane. In fact, at a distance equal to one wavelength, the amplitude of the fluid velocity oscillation has decreased to $1/e^{2\pi} \approx 0.0018$ of its amplitude at the plane. The distance δ_ν defined by (6.5.5) therefore gives an idea of how far the transverse waves penetrate into the fluid. It is called the *viscous penetration depth*[5] and, as we will see, plays an important role whenever there is a relative oscillatory flow near a boundary. In air at ambient conditions, for example, $\delta_\nu \approx 0.02$ cm at 100 Hz. The smallness of δ_ν compared to length scales typical of many sound propagation problems, such as propagation of relatively low frequency waves in tubes below cutoff, enabled us to use in Chapter 3 the ideal equations in our discussion of propagation in tubes. However, the effects of viscosity cannot be neglected very near a boundary, for then important effects may be left unaccounted for. In the present problem, for example, viscosity cannot be neglected at all, for if we did, the tangential force acting between the plane and the fluid would be zero. The net result of this would be that no fluid motion is induced by the plane. In reality, however, the force is nonzero even for small viscosity, as the following calculation shows.

In the present case, the tangential force of the plane is given by the x, y component of the stress tensor [see Equation (1.5.18)]:

$$\sigma_{xy} = \mu_0 \left(\frac{\partial u}{\partial y} \right)_{y=0} \tag{6.5.10}$$

Substitution from (6.5.8) yields

$$\sigma_{xy} = -\sqrt{2}\, \mu_0 U_0 \delta_\nu^{-1} \cos(\omega t + \pi/4) \tag{6.5.11}$$

This shows that for small δ_ν (i.e., when ω is large), the velocity gradient at $y=0$ is large so that σ_{xy} may not necessarily be small. We notice from this result that this force and the velocity of the plane are out of phase by an amount equal to $\pi/4$.

The rate at which the plane does work against this frictional force is, per unit surface area of plane,

$$\dot{W}_p = -\sigma_{xy} u(0, t)$$

---

[5] It is also referred to as the acoustic-boundary-layer thickness.

This has a nonvanishing time average. Thus, making use of (6.5.8) and (6.5.11) and taking the time average, we obtain, with $\langle \cos(\omega t + \pi/4)\cos \omega t \rangle = 1/2\sqrt{2}$,

$$\langle \dot{W}_p \rangle = \tfrac{1}{2}\mu_0 U_0^2 \delta_\nu^{-1} \tag{6.5.12}$$

What happens to this energy? Contrary to the cases dealt with in Chapter 5, these oscillations do not generate sound waves, for as we saw earlier, $\rho' = 0$. The answer is simply that this energy is dissipated in the fluid in the form of heat. That this is the case can be seen directly by computing the amount of energy dissipated in the fluid. This energy is given by (6.4.12). Since $\Delta = \nabla \cdot \mathbf{u} = 0$ and $\mathbf{u} = [u(y),0,0]$ in the present discussion, the first integral gives, when averaged,

$$\langle \dot{E}_{\text{lost}} \rangle_\nu = \mu_0 \int_V \left\langle \left(\frac{\partial u}{\partial y} \right)^2 \right\rangle dV \tag{6.5.13}$$

For a unit area of the plane, $dV = dy$, where the range of y extends to infinity. Therefore,

$$\langle \dot{E}_{\text{lost}} \rangle_\nu = \mu_0 \int_0^\infty \left\langle \left(\frac{\partial u}{\partial y} \right)^2 \right\rangle dy$$

Substituting from (6.5.8), we obtain

$$\langle \dot{E}_{\text{lost}} \rangle_\nu = \mu_0 U_0^2 \delta_\nu^{-2} \int_0^\infty e^{-2(y/\delta_\nu)} \, dy$$

$$= \tfrac{1}{2}\mu_0 U_0^2 \delta_\nu^{-1} \tag{6.5.14}$$

in agreement with (6.5.12). The point of this simple calculation is that, since the average energy lost can be calculated using two different procedures, it may be possible to use this approach to obtain, directly, quantities of interest such as the force acting on oscillating bodies in a viscous fluid. Later in this chapter we will, in fact, use this technique to obtain the force that acts on a rigid sphere.

Thermal Waves

We now study the effects of finite thermal conductivity near a boundary. To illustrate the similarity between these effects and those of viscosity, we consider a temperature-oscillation problem analogous to the viscous-oscillation problem treated above. Thus, suppose that we have a flat surface whose temperature is varying harmonically in time about a mean value T_0 according to

$$T_w = T_0 + \varepsilon \cos \omega t \tag{6.5.15}$$

The amplitude of the oscillation is assumed to be very small compared to T_0. The fluid in front of the flat surface will also develop temperature fluctuations T' of small amplitude. The variations of T' with time and position can be obtained from (6.1.1)–(6.1.3) together with (6.1.4) or (6.1.6). Consider, first, (6.1.3). Since T' can depend only on time and on distance normal to the plate, it can be written as

$$\rho_0 c_{p0} \frac{\partial T'}{\partial t} - \beta_0 T_0 \frac{\partial p'}{\partial t} = k_0 \frac{\partial^2 T'}{\partial y^2} \tag{6.5.16}$$

Therefore, for the case of harmonic oscillations, we expect that if

$$\left| \frac{p'}{T'} \right| \ll \frac{\rho_0 c_{p0}}{\beta_0 T_0} \tag{6.5.17}$$

the solution for T' would be of the form given by (6.5.8). Thus, with the thermal diffusivity k_0 instead of the viscous diffusivity ν_0,

$$T'(y,t) = \varepsilon e^{-(y/\delta_\kappa)} \cos(\omega t - y/\delta_\kappa) \tag{6.5.18}$$

where

$$\delta_\kappa = \sqrt{2\kappa_0/\omega} \tag{6.5.19}$$

The implications of the solution for T' are analogous to those of the solution obtained for u in the oscillating-plane problem. For example, the temperature fluctuations in the fluid result in an average energy dissipation rate per unit surface area of the heater given by

$$\langle \dot{E}_{\text{lost}} \rangle_\kappa = \tfrac{1}{2} \rho_0 c_{p0} (\varepsilon^2/T_0)(\omega \kappa_0)^{1/2} \tag{6.5.20}$$

This energy has to be supplied externally to the heater in order to sustain its temperature fluctuations. Further, because of the temperature gradient at the heater surface, there is heat transfer between the heater and the fluid. The heat-transfer rate per unit surface area of heater is

$$\dot{q} = k_0 (\partial T/\partial y)_{y=0} \tag{6.5.21}$$

Using (6.5.17) and the identity $k_0 = \rho_0 c_{p0} \kappa$, we obtain

$$\dot{q} = -\rho_0 c_{p0} \varepsilon (\omega \kappa)^{1/2} \cos(\omega t + \pi/4) \tag{6.5.22}$$

We notice that this rate is not in phase with the temperature fluctuation of the heater. This phase lag is responsible for the dissipation of energy.

The quantity δ_κ defined by (6.5.18) is, by analogy with δ_ν, called the *thermal penetration depth*. Its magnitude is, in general, different from the magnitude of δ_ν. In fact,

$$\frac{\delta_\nu}{\delta_\kappa} = \left(\frac{\nu_0}{\kappa_0}\right)^{1/2} \tag{6.5.23}$$

The ratio of the two diffusivities appearing in this equation is equal to the Prandtl number of the fluid:

$$\Pr = \frac{\nu_0}{\kappa_0}$$

$$= \frac{\mu_0 c_{p0}}{k_0} \tag{6.5.24}$$

and is a measure of the effects of viscosity relative to those of heat conductivity. In air at ambient conditions, for example, $\Pr = 0.73$, so that δ_ν and δ_κ are approximately equal.

The above solution for T' was obtained under the assumption that the term proportional to $\partial p'/\partial t$ could be neglected. As we will see later, nonzero pressure fluctuations imply the existence of a sound wave traveling away from the plate with speed approximately equal to c_0, and with an amplitude $|p'|$ that has an order of magnitude given by[6]

$$|p'| \sim \rho_0 c_0^2 \beta_0 \sqrt{\omega \kappa_0 / c_0^2} \; |T'|$$

[6]This can be shown as follows. Near the heater, the pressure remains approximately constant so that when the fluid temperature increases, its density decreases by an amount $\rho' = -\rho_0 \beta_0 T'$ (i.e., the fluid near the heater expands). Now, by the equation of continuity,

$$-i\omega\rho' = -\rho_0 |\partial v/\partial y| = i\omega\rho_0\beta_0 T'$$

$$\approx i\omega\rho_0\beta_0 e^{-i\omega t}$$

At a small distance δ_κ away, the fluid velocity is roughly

$$v \approx -i\omega\rho_0\varepsilon\delta_\kappa e^{-i\omega t}$$

Far from the heater, this becomes a wave:

$$v \approx \varepsilon\beta_0\omega\delta_\kappa e^{i(kx-\omega t-\pi/2)}$$

Therefore, in order that the solution for T' given by (6.5.18) be applicable, the condition (6.5.17) requires that

$$\left(\omega\kappa_0/c_0^2\right)^{1/2} \ll 1 \qquad (6.5.25)$$

The quantity $\omega\kappa/c_0^2$ is equal to the quantity $\omega\tau_\kappa$ introduced in Section 6.3. In view of (6.5.21) and (6.2.27), $\omega\tau_\kappa$ can be written as

$$\omega\tau_\kappa = \frac{3}{4\,\mathrm{Pr}}\,\omega\tau_\nu \qquad (6.5.26)$$

Unless the Prandtl number of the fluid is very small, our discussion in Section 6.2 shows that $\omega\kappa_0/c_0^2$ is small for all frequencies for which the Navier–Stokes equations apply. Hence, (6.5.17) is satisfied for a wide range of frequencies, and (6.5.18) is applicable.

When $\sqrt{\omega\tau_\kappa}$ is not entirely negligible, sound waves of small amplitude may appear. These sound waves are of some importance in certain applications and will be discussed later in some detail. Here we merely point out that in ideal fluids, such as those considered in Chapter 5, no sound waves could be emitted by this mechanism. On the other hand, the manner in which this mechanism generates waves in a real fluid is by the expansion and contraction of the fluid layers near the heating source. This is similar, for example, to sound generation by a pulsating body. Therefore, except for the mechanism that drives the fluid pulsations, sound waves produced by a heat source can be analyzed in terms of the mass-addition mechanism treated in Section 5.2, as our discussion in Section 5.10 shows.

PROBLEMS

6.5.1 Derive (6.5.9).

6.5.2 Consider a viscous fluid between two parallel rigid planes. One of the planes is stationary, and the other is executing translational oscillations with frequency ω in its own plane. Determine the fluid velocity between the two planes if their distance is h.

---

and since the wave is plane, $p' = \rho_0 c_0 v$, so that

$$|p'| \sim \rho_0 c_0 \beta_0 \omega \delta_\kappa \varepsilon = \rho_0 c_0^2 \beta_0 \left(2\omega\kappa/c_0^2\right)^{1/2} \varepsilon$$

This is another example of sound-wave generation by the unsteady-heat-release mechanism discussed earlier. This example will be treated in more detail in Section 6.13.

6.5.3   Consider a Helmholtz resonator having a neck of length ℓ and of circular cross section of radius a. The oscillations of the fluid in the neck result in viscous dissipation, which affect the value of the real part of the resonator's impedance. Provided $kl \ll 1$ (so that the fluid mass in the neck may be assumed to move as a whole) and that $a \gg \delta_\nu$ (so that the effects of the neck's curvature may be ignored), we may use (6.5.12) to compute the resistive part R'_ν, due to viscosity, of the resonator's specific impedance. Thus compute the total dissipation for a plane of "area" $2\pi a l$ to show that

$$R'_\nu = \frac{1}{\pi a^2} \rho_0 c_0 \left(\frac{l}{a} \right) \left(\frac{\omega \nu_0}{c_0^2} \right)^{1/2}$$

Compare this to the radiative part of the specific resistance, given by (5.3.46), for typical values of kl, ka, and of $(\omega \nu_0 / c_0^2)^{1/2}$.

6.6   ATTENUATION IN TUBES

The results derived in the previous section provide a simple means of obtaining a first-order approximation to the attenuation coefficient for sound waves propagating in tubes and in channels. This quantity plays an important role in many situations of interest. For example, the properties of sound-absorbing materials are studied experimentally in tubes using some of the techniques described in Section 3.6. Those techniques yield attenuation coefficients that are due not only to absorption in the material being tested, but also to dissipation at the tube's wall. Thus, in order to obtain a measure of the absorption in the material, one must independently determine the attenuation coefficient due to wall effects alone. While it is true that this coefficient can also be obtained experimentally, it is more convenient to determine it analytically. However, an accurate determination of this quantity requires the solution of a system of equations far more complicated than those treated so far. Later we will discuss that system in some detail. At present, we attempt to modify the results of the last section in order to obtain approximate expressions for the coefficient of attenuation of plane waves in tubes.

The first approximation we make is to assume that viscous and thermal effects can be calculated separately. The correctness of this assumption cannot be assessed unless the complete problem is solved. However, in view of our discussion in Section 6.3, we expect that when dissipation is small, the approximation may be fairly good, at least as far as the calculation of the attenuation coefficient is concerned. It is clear that in order to have small dissipation, it is necessary that the volume of fluid affected by viscous and thermal effects be small compared to the total volume of fluid in the tube. We know that wall effects extend to a distance of the order of δ_ν (or δ_κ) into the fluid. Therefore, if R is a measure of the lateral dimension of the tube and L of its length, the fluid volume affected by the wall is of the order of $\delta_\nu RL$ (or $\delta_\kappa RL$). On the other hand, the total volume of fluid is of the order of $R^2 L$. Thus, the condition that dissipation be small requires that $R \gg \delta_\nu$ and $R \gg \delta_\kappa$. Using the definitions of δ_ν and δ_κ, these can be written as

$$R \gg (2\nu_0/\omega)^{1/2}$$
$$R \gg (2\kappa_0/\omega)^{1/2}$$

(6.6.1)

Both conditions can be satisfied by considering only cases where the lateral dimension of the tube or channel is large.[7] For example, in the case of a two-dimensional channel, we would require that the distance between the walls of the channel be very large. In that case, we do not expect dissipation at one of the walls to be affected by the other. We may therefore consider a single wall in contact with an infinite body of fluid. The same situation applies to a circular tube. That is, provided the tube radius is sufficiently large, we may ignore the effects of wall curvature and again consider a plane wall. In this case, the amount of dissipation may be computed for a wall "depth" equal to $2\pi R$, whereas in the former it may be computed for a unit depth.

Let us now consider the effects of viscosity on the velocity of the fluid in a wave traveling in a direction parallel to the wall. From what has already been said, the velocity of the fluid will not be significantly affected by the wall. Far from the wall, we expect the wave to be plane, one dimensional, and slightly attenuated. Near the wall, however, we have a sheared velocity profile in which the velocity vanishes at the wall. As pointed out earlier, this profile can be obtained exactly only by solving the actual problem of propagation in tubes or channels, and this may be done using the formulation outlined in the next section. At present, however, we make use of a heuristic procedure to obtain an approximate profile that is adequate for obtaining first-order wall effects on sound waves. The procedure is simply to transform the velocity profile obtained in Section 6.5 by adding a uniform fluctuation to the velocity such that the boundary condition at the wall is satisfied. Thus, we consider a new fluid velocity u' defined by

$$u' = u_\infty - u$$

(6.6.2)

[7]Another condition has to be imposed because we have assumed plane-wave propagation. Reference to Section 4.4 shows that the condition is that the lateral dimension of the tube or of the channel should be smaller than the wavelength. Therefore, this lateral dimension cannot be arbitrarily large.

where $u_\infty = U_0 \cos(\omega t)$ is the uniform velocity fluctuation and u is given by (6.5.7). At the wall, $u = U_0 \cos(\omega t)$ so that this new velocity vanishes there, as required. Far from the wall, $u \to 0$ and $u' \to u_\infty$, and we merely have harmonic velocity fluctuations. This is what we would expect the velocity in the sound wave to do for a fixed value of x along the wall. There is, however, one major difference between the actual and heuristic velocity profiles: u' does not depend on distance along the wall, whereas the actual fluid velocity in a sound wave does. The reason for this difference is that while (6.6.2) satisfies the boundary conditions, it does not satisfy the equations of motion. In particular, u' as given by (6.6.2) requires the fluid to be incompressible since $\nabla \cdot \mathbf{u}' = \partial u'/\partial x = 0$. In the actual case, this derivative is not zero so that u' cannot describe exactly the actual fluid velocity in the wave. However, as far as dissipation at the wall is concerned, u' can be used provided that changes of the actual velocity along x be small compared to changes along other directions. Thus, if u represents the actual velocity, we impose, in addition to (6.6.1), the requirement that

$$\left| \frac{\partial u/\partial x}{\partial u/\partial y} \right| \ll 1$$

But $|\partial u/\partial x| \sim \omega |u|/c_0$ and $|\partial u/\partial y| \sim |u|/\delta_\nu$. Therefore, u' is also limited to values of ω such that $\omega \delta_\nu/c_0 \ll 1$, or using (6.5.6),

$$\left(\omega \nu_0 / c_0^2 \right)^{1/2} \ll 1 \tag{6.6.3}$$

This condition, together with the first of the conditions given by (6.6.1), must be satisfied in order that the fluid velocity at a fixed location x in a tube or in a channel be approximately given by (6.6.2). It should be noted that conditions (6.6.1) and (6.6.3) are not completely independent of one another.

With these approximations in mind, let us consider the velocity u'. It is explicitly given by

$$u' = U_0 \left[\cos(\omega t) - e^{-(y/\delta_\nu)} \cos(\omega t - y/\delta_\nu) \right] \tag{6.6.4}$$

While this differs from (6.5.8) only by a harmonic function of time, it has an interesting feature that does not appear in the oscillating-plane problem. To display this feature, we compute the root-mean-squared velocity from (6.6.4) and obtain

$$\frac{\langle u'^2 \rangle}{\langle u_\infty^2 \rangle} = 1 + e^{-2(y/\delta_\nu)} - 2e^{-(y/\delta_\nu)} \cos\left(\frac{y}{\delta_\nu} \right) \tag{6.6.5}$$

where u_∞ is the fluid velocity far from the wall. Figure 6.6.1 displays the variations of the normalized root-mean-squared velocity with distance from the wall. The most remarkable feature of this profile is that it shows a region in which, on the average, the fluid velocity is larger than the velocity far from the wall. The effect (called Richardson's annular effect) can be understood if we realize that the solution given by (6.6.4) is, in effect, the superposition of a transverse wave and a uniform oscillation. The transverse wave has, at $y=0$, a fluid velocity always equal and opposite to that of the uniform oscillation. For $y>0$, however, the fluid velocity in the transverse wave may exceed its values at $y=0$ and combine with the uniform velocity to produce, at some time during one cycle, velocities larger for some values of y than the value of the uniform, "free-stream" velocity. This is shown schematically in Figure 6.6.2a, where the two velocity distributions are plotted versus y/δ_ν for $\omega t=0$, and in Figure 6.6.2b, where the total velocity u' is shown at various times during half a cycle.

Since our derivation of u' was completely heuristic, it may be questioned whether this effect exists at all.[8] In Section 6.7 we will show that this is the case by considering the actual problem of sound propagation in a two-dimensional channel.

Let us return to the amplitude-attenuation coefficient α. Since, by assumption, dissipation is small, we use the procedure outlined in Section 6.4, in which α is defined by (6.4.4) in terms of the average rate at which acoustic energy is lost and of the unattenuated average energy of the wave.

The rate at which acoustic energy is lost due to viscous effects is given by (6.5.12), with the fluid velocity given by (6.5.8). Thus, for a surface of length λ and depth l of the wall, $dV=\lambda l\,dy$, so that upon integrating over y, we obtain

$$\langle \dot{E}_{\text{lost}} \rangle_\nu = \tfrac{1}{2}\mu_0 U_0^2 \delta_\nu^{-1} l \lambda \qquad (6.6.6)$$

The energy of the acoustic wave enclosed by the same region is

$$E_0 = \tfrac{1}{2}\rho_0 U_0^2 \lambda A_0 \qquad (6.6.7)$$

here A_0 is the area of the wave front. In this case of a circular tube of radius R, $l=2\pi R$ and $A_0 =\pi R^2$. Substitution of (6.6.6) and (6.6.7) into (6.4.4) yields the viscous part of the attenuation coefficient for propagation in a circular tube:

$$\alpha_\nu = \frac{\nu_0}{R\delta_\nu c_0} \qquad (6.6.8)$$

---

[8] For experimental evidence of this velocity overshoot, see O'Brien and Logan (1965).

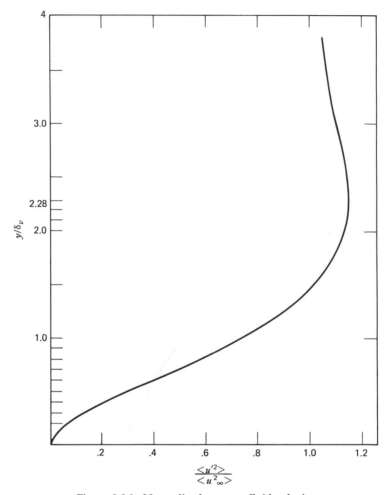

Figure 6.6.1 Normalized average fluid velocity.

Using the definition of δ_ν, this can be written as

$$\alpha_\nu R = \left(\omega \nu_0 / 2c_0^2 \right)^{1/2} \tag{6.6.9}$$

In view of the conditions imposed on $(\omega \nu_0 / c_0^2)^{1/2}$ by (6.6.3), this result implies that $\alpha_\nu R$ should be a small quantity. The other condition to be satisfied is given by (6.6.1). To see its implications, we rewrite (6.6.9) as

$$\alpha_\nu \lambda = \pi \left(2\nu_0 / \omega R^2 \right)^{1/2} \tag{6.6.10}$$

which, together with (6.6.1), shows that the attenuation per wavelength should also be small.

Consider now the dissipation due to temperature oscillations in the wave. If the wall is kept at constant temperature T_0, then arguments very similar to those used to obtain u' show that the local temperature variations in the fluid are approximately given by

$$T' = T_0 + \varepsilon\left[\cos(\omega t) - e^{-y/\delta_\kappa}\cos(\omega t - y/\delta_\kappa)\right] \qquad (6.6.11)$$

where ε is now the amplitude of the temperature in the wave, which in terms of U_0 is given by $\varepsilon = c_0\beta_0 T_0 U_0/c_{p0}$. The average acoustic-energy dissipation rate due to heat conduction is given by

$$\langle \dot{E}_{\text{lost}} \rangle_\kappa = \frac{k_0}{T_0} \int \left\langle \left(\frac{\partial T'}{\partial y}\right)^2 \right\rangle dV \qquad (6.6.12)$$

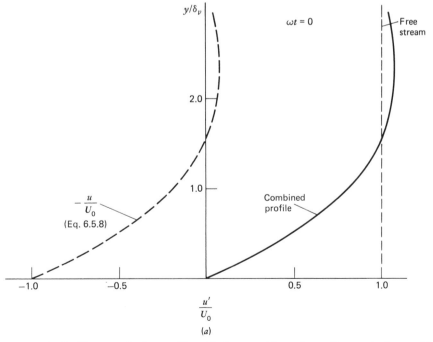

Figure 6.6.2 Oscillatory velocity profiles. (a) Superposition of a uniform velocity profile with the profile due to oscillating plate (see Figure 6.5.2); (b) fluid-velocity profiles at several times during one cycle.

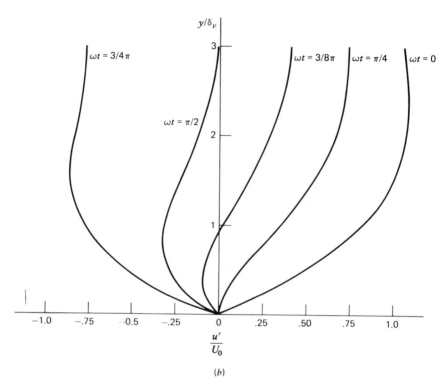

Figure 6.6.2   (continued)

where $dV = l\lambda\,dy$ and y extends to infinity. Substitution from (6.6.11) yields

$$\langle \dot{E}_{\text{lost}} \rangle_\kappa = \frac{1}{2}\frac{k_0}{T_0}\left(\frac{c_0\beta_0 T_0}{c_{p0}}\right)^2 U_0^2\frac{\lambda l}{\delta_\kappa}$$

$$= \frac{1}{2}\frac{k_0}{c_{p0}}\frac{\gamma-1}{\delta_\kappa}U_0^2\lambda l \qquad (6.6.13)$$

where the perfect gas result, $\beta_0 t_0 = 1$, has been used. The acoustic energy E_0 is still given by (6.6.7) so that the attenuation coefficient due to heat conduction at the walls of a circular tube is given by

$$\alpha_\kappa = \frac{\kappa_0}{R\delta_\kappa c_0}(\gamma-1) \qquad (6.6.14)$$

or, using (6.5.23) and (6.5.24),

$$\alpha_\kappa = \frac{\nu_0}{Rc_0\delta_\nu} \frac{\gamma-1}{Pr^{1/2}} \qquad (6.6.15)$$

This attenuation coefficient differs from that due to viscous effects only by the factor $(\gamma-1)Pr^{-1/2}$. For gases, the magnitude of this factor is of order one, so that thermal and viscous effects are equally important. Also, the limitations for the applicability of (6.6.15) are the same as those imposed on (6.6.8).

If no other energy-loss mechanisms exist in the tube, such as, for example, sound radiation through an opening in the tube, then the total-wall-attenuation coefficient is the sum of the viscous and thermal coefficients:

$$\alpha_w = \frac{1}{c_0}\left(\frac{\omega\nu_0}{2R^2}\right)^{1/2}\left(1+\frac{\gamma-1}{Pr^{1/2}}\right) \qquad (6.6.16)$$

This is known as the *Helmholtz–Kirchhoff wall-attenuation coefficient*, but it is sometimes referred to as the wide-tube, low-frequency approximation to the wall-attenuation coefficient. While this result is limited by the conditions (6.6.1) and (6.6.3) requiring that R be relatively large and ω be relatively small, respectively, it nevertheless shows that attenuation increases with frequency as $\omega^{1/2}$, as opposed to the attenuation coefficient for free propagation, which increases as ω^2. Also, as one would expect on physical grounds, attenuation increases as the tube radius is decreased. The $\omega^{1/2}$ variation arises simply because the quantity that determines how much attenuation will occur in a given distance is the ratio of viscous-penetration depth $(2\nu_0/\omega)^{1/2}$ to the tube radius. In fact, the attenuation per wavelength, given by

$$\alpha_w\lambda = \pi\left(\frac{\delta_\nu}{R}\right)\left(1+\frac{\gamma-1}{Pr^{1/2}}\right) \qquad (6.6.17)$$

shows this dependence clearly.

Comparison with Experimental Data

Because of their importance in many applications, sound-attenuation coefficients in tubes have been the subject of many experimental investigations. Most of the experiments were performed in rigid tubes by means of techniques such as those outlined in Section 3.6. For a given tube filled with a given fluid, these techniques yield the dependence of the attenuation coefficient on the frequency. For a tube with a different radius, the value of α_w will be different, but according to (6.6.16) the product $\alpha_w R$ is the same for two different tubes, provided they are filled with the same fluid, and the frequency is the same. From (6.6.16), $\alpha_w R$ is given by

$$\alpha_w R = \left(\frac{\omega \nu_0}{2c_0^2} \right)^{1/2} \left(1 + \frac{\gamma - 1}{\mathrm{Pr}^{1/2}} \right) \qquad (6.6.18)$$

In order to compare results obtained in tubes of different radii, we plot (see Figure 6.6.3) the quantity $\alpha_w R$ versus the nondimensional frequency $(\omega \nu_0 / 2c_0^2)^{1/2}$. The data shown in the figure were obtained with tubes filled with oxygen, nitrogen, and air. Although obtained with different gases, these data are comparable because for these gases the respective values of $(\gamma - 1) \mathrm{Pr}^{-1/2}$ are very much the same. Other wall-attenuation data may also be found in the literature. These data follow the same trend as that displayed by the data shown in the figure. That is, at low frequencies the agreement between theory and experiments is good, but as the frequency increases, the measured attenuation is larger than that predicted by theory, the difference increasing with frequency. Because of the conditions imposed on (6.6.16), such behavior might be expected. However, analytical studies of the actual problem of sound propagation in circular tubes indicate that for the range of parameters covered by the data shown in Figure 6.6.3, the approximate attenuation coefficient given by the Helmholtz–Kirchhoff formula is quite good. The observed discrepancies are, therefore, due to effects not included in the derivation of (6.6.16). For example, since most measurements are made in closed tubes, there will be some absorption at the reflected end, and this absorption is not accounted for in the simple theory. Other unaccounted effects include those of nonlinearity of the sound field, wall-surface roughness, sound radiation through imperfect seals, and gas-molecule adsorption at the walls of the tube. While it may be desirable to have a theoretical prediction that properly accounts for all dissipation effects in tubes, it is clear that its derivation would be rather complicated if not impossible. Fortunately, at moderate frequencies and in moderately wide tubes, the Helmholtz–Kirchhoff result is fairly accurate. For higher frequencies, some correlation may be required, and some authors have adopted the procedure of adding to the attenuation equation a term proportional to the frequency:

$$\alpha_w R = A \left(\omega_0 \nu_0 / 2c_0^2 \right)^{1/2} + B \left(\omega \nu_0 / 2c_0^2 \right) \qquad (6.6.19)$$

Here A and B are quantities that, in principle, must be determined experimentally, but if the Helmholtz–Kirchhoff formula is applicable at all, the measured value of A should agree closely with the coefficient of $(\omega \nu_0 / 2c_0^2)^{1/2}$ in (6.6.18).

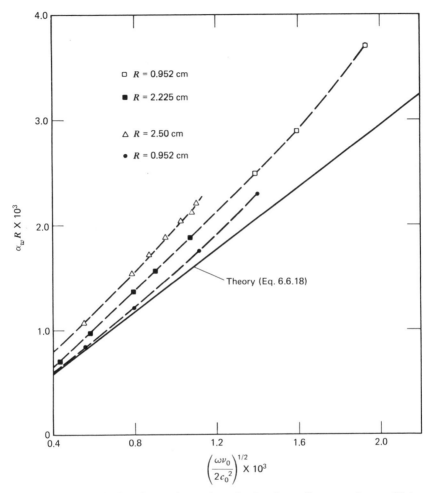

Figure 6.6.3 Theoretical and experimental results for the wall-attenuation coefficient. ● R. D. Fay (1940). △ J. G. Parker (1961). □, ■ S. Temkin and R. A. Dobbins (1966).

Thus, if the experimental results for $\alpha_w R$ are divided by $(\omega\nu_0/2c_0^2)^{1/2}$ and the results plotted versus that quantity, the data should nearly fall on a straight line whose intercept is A and whose slope is B. Figure 6.6.4 shows such a plot obtained from one of the sets of data shown in Figure 6.6.3. It is seen that, in fact, the data fall approximately on a straight line, which when extrapolated toward zero frequency yields $A = 1.52$, the theoretical value being 1.48. This agreement is an indication of the correctness of (6.6.18) for low frequencies. The slope of the straight line through the data in Figure 6.6.4 is $B=2.28\times10^2$ cm^{-1}, a value of the same order of magnitude as that derived from other data obtained under similar conditions.

Equation (6.6.19) may be thought of as an expansion of $\alpha_w R$ in terms of the parameter $(\omega\nu_0/2c_0^2)^{1/2}$, truncated after the second term. It is clear that higher-order terms may also be present. For example, dissipation in the bulk of the fluid leads to a contribution to the next term in the series. Since that dissipation is given by (6.3.40), its contribution to $\alpha_w R$ can be written as

$$\left(\frac{4}{3}+\frac{\mu_{v0}}{\mu_0}+\frac{\gamma-1}{\text{Pr}}\right)\left(\frac{\omega R^2}{\nu_0}\right)^{1/2}\left(\frac{\omega\nu_0}{2c_0}\right)^{3/2}$$

For the data shown in Figure 6.6.3, the maximum value of this contribution is of the order of 10^{-5}, and is negligible compared to the other two contributions.

The literature on the subject of sound attenuation in tubes is rather large. The interested reader may consult the paper by Tijdman (1975) referred to at the end of this chapter. In that paper, he reviews the theoretical works on the subject, and presents a numerical solution for the complex propagation constant in terms of the nondimensional groups $(\omega R^2/\nu_0)^{1/2}$ and $\omega R/c_0$. The maximum value of $\omega R/c_0$ for which results are presented in tabulated form is 0.5π. This value is

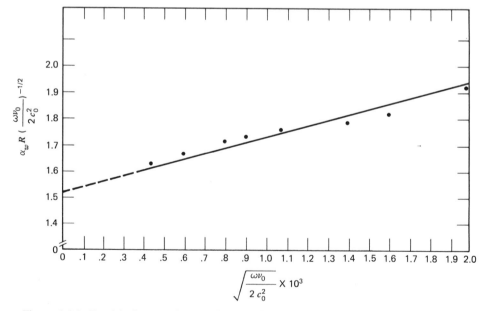

Figure 6.6.4 Empirical correction to the Helmholtz–Kirchhoff wall-attenuation coefficient.

just below that which corresponds to the lowest cutoff frequency in a circular tube.

The numerical solution presented by Tijdman is based on an analysis of the complete problem of sound propagation in tubes filled with viscous and heat-conducting fluids. In the next section, we introduce some techniques that are useful in dealing with such a problem and apply them to propagation in a channel. It will be seen how the approximate velocity distribution given by (6.6.4) follows from the complete set of equations.

6.7 BOUNDARY EFFECTS: VECTOR FORMULATION

In the last section, we used a heuristic argument to study some of the effects that arise when a sound wave traveling in a tube interacts, through the action of viscosity and heat conductivity, with the walls of the tube. In order to treat these interactions more rigorously and also to study similar problems not having such a simple geometry as the one treated in the last section, a more general formulation is needed. In this section, we first introduce such a formulation.

We begin, as we did in the free-propagation case, by considering fluids that are viscous but not heat-conducting. Later in this chapter, we will show how the effects of heat conductivity can be taken into account. The equations of motion for this case are

$$\frac{\partial \rho'}{\partial t} + \rho_0 \nabla \cdot \mathbf{u} = 0 \tag{6.7.1}$$

$$\rho_0 \frac{\partial \mathbf{u}}{\partial t} + \nabla p' = \mu_0 \nabla^2 \mathbf{u} + \left(\tfrac{1}{3}\mu_0 + \mu_{v0}\right) \nabla(\nabla \cdot \mathbf{u}) \tag{6.7.2}$$

and

$$p' = c_0^2 \rho' \tag{6.7.3}$$

It is clear that because of the no-slip condition, the fluid velocity may, in general, have more than one component, at least in the vicinity of boundaries, where the shear waves have larger amplitudes. Hence, the above system of equations will be far more complicated than that applicable to the free-propagation case. Several techniques can be used to obtain a solution to the present problem. One would be to eliminate, by cross-differentiation and substitution, all but one of the five variables $(u_1, u_2, u_3, p', \rho')$ appearing in the system. Another would be to assume wave-type solutions for all of the variables. This would result in a dispersion relation from which the propagation constants could be obtained, as was done in Section 6.3 for the free-propagation case. At present, however, we make use of yet another technique that is more general.

First, we combine (6.7.1)–(6.7.3) to obtain a single equation for the velocity vector **u**. This is easily done by substituting (6.5.3) into (6.7.1) and then eliminating p' by taking cross derivatives of the result and of (6.7.2). Thus,

$$\frac{\partial^2 \mathbf{u}}{\partial t^2} - c_0^2 \nabla(\nabla \cdot \mathbf{u}) = \nu_0 \nabla^2 \frac{\partial \mathbf{u}}{\partial t} + \frac{1}{\rho_0}\left(\mu_{v0} + \tfrac{1}{3}\mu_0\right)\nabla\left(\nabla \cdot \frac{\partial \mathbf{u}}{\partial t}\right) \qquad (6.7.4)$$

Further, since for any vector **A** we have

$$\nabla^2 \mathbf{A} = -\nabla \times (\nabla \times \mathbf{A}) + \nabla(\nabla \cdot \mathbf{A})$$

(6.7.4) can be written as

$$\frac{\partial^2 \mathbf{u}}{\partial t^2} - c_0^2\left(1 + \frac{4}{3}\frac{\nu_0'}{c_0^2}\frac{\partial}{\partial t}\right)\nabla(\nabla \cdot \mathbf{u}) = -\nu_0 \frac{\partial}{\partial t}\nabla \times (\nabla \times \mathbf{u}) \qquad (6.7.5)$$

This is still a linear equation for the velocity vector. We may therefore attempt to split **u** into several parts, each one of which satisfies a simpler equation than (6.7.5). Thus, we write $\mathbf{u} = \mathbf{u}_1 + \mathbf{u}_2 + \cdots$, and impose conditions on each one of these velocities, thereby allowing for some simplification.

Consider $\mathbf{u}_1$ first. Suppose that it is such that $\nabla \times \mathbf{u}_1 = 0$. Then it can be derived from a velocity potential by means of $\mathbf{u}_1 = \nabla\phi$, where, from (6.7.5), ϕ satisfies

$$\nabla\left[\frac{\partial^2 \phi}{\partial t^2} - c_0^2\left(1 + \frac{4}{3}\frac{\nu_0'}{c_0^2}\frac{\partial}{\partial t}\right)\nabla^2\phi\right] = 0 \qquad (6.7.6)$$

Without loss of generality, we can write

$$\frac{\partial^2 \phi}{\partial t^2} - c_0^2\left(1 + \frac{4}{3}\frac{\nu_0'}{c_0^2}\frac{\partial}{\partial t}\right)\nabla^2\phi = 0 \qquad (6.7.7)$$

The solutions to this equation represent, as we know, attenuated longitudinal waves. The velocity $\mathbf{u}_1 = \nabla\phi$ may therefore be referred to as the longitudinal part of the fluid velocity.

Consider now the opposite case; namely, consider a velocity field $\mathbf{u}_2$ such that $\nabla \cdot \mathbf{u}_2 = 0$ but with $\nabla \times \mathbf{u}_2 \neq 0$. In this case, we may put

$$\mathbf{u}_2 = \nabla \times \mathbf{B} \qquad (6.7.8)$$

The quantity **B** is called the vector potential. It is usually chosen such that $\nabla \cdot \mathbf{B} = 0$, because if **B** had a longitudinal part (i.e., if $\nabla \cdot \mathbf{B} \neq 0$), this part would be eliminated when taking the **curl** to obtain $\mathbf{u}_2$. Substitution of (6.7.8) into (6.7.5) and making use of $\nabla \cdot \mathbf{B} = 0$ yields

$$\nabla \times \frac{\partial}{\partial t}\left(\frac{\partial \mathbf{B}}{\partial t} - \nu_0 \nabla^2 \mathbf{B}\right) = 0 \tag{6.7.9}$$

Again, without loss of generality, this can be written as

$$\frac{\partial \mathbf{B}}{\partial t} = \nu_0 \nabla^2 \mathbf{B} \tag{6.7.10}$$

This is a vector-diffusion equation. Therefore, the quantity $\mathbf{B}$ represents diffusive waves that owe their existence to the effects of viscosity. The properties of these waves are basically the same as those of the waves discussed in Section 6.5; that is, they travel with a speed proportional to $\sqrt{\omega \nu_0}$ in a direction perpendicular to the local boundary normal and are damped rapidly as they move into the fluid.

Let us return to our discussion of the velocity vector $\mathbf{u}$. It is clear that the complete fluid velocity in an acoustic wave is given by the sum of the velocities $\mathbf{u}_1$ and $\mathbf{u}_2$ introduced above, for the only other possible addition would be a velocity $\mathbf{u}_3$, which is both solenoidal and irrotational (i.e., a velocity $\mathbf{u}_3$ such that $\nabla \cdot \mathbf{u}_3 = 0$ and $\nabla \times \mathbf{u}_3 = 0$). Such a velocity, however, applies only to incompressible fluids for which density changes are zero [see Equation (6.7.1)] so that $\mathbf{u}_3$ does not relate to acoustic waves. Thus, the acoustic velocity is, in general, given by

$$\mathbf{u} = \nabla \phi + \nabla \times \mathbf{B} \tag{6.7.11}$$

with ϕ and $\mathbf{B}$ given by (6.7.7) and (6.7.10), and $\mathbf{B}$ is such that $\nabla \cdot \mathbf{B} = 0$. In the important case of monochromatic waves, we may put

$$\phi = \mathrm{Re}\left[\tilde{\Phi}(\mathbf{x})e^{-i\omega t}\right] \tag{6.7.12}$$

$$\mathbf{B} = \mathrm{Re}\left[\tilde{\psi}(\mathbf{x})e^{-i\omega t}\right] \tag{6.7.13}$$

This yields

$$\nabla^2 \tilde{\Phi} + \frac{k^2}{1 - i\frac{4}{3}\left(\omega \nu_0'/c_0^2\right)} \tilde{\Phi} = 0 \tag{6.7.14}$$

and

$$\nabla^2 \tilde{\psi} + K^2 \tilde{\psi} = 0 \tag{6.7.15}$$

where

$$K^2 = i\omega/\nu_0 \tag{6.7.16}$$

and where $k = \omega/c_0$. It should be noted that the acoustic pressure is still given by the ideal result

$$\tilde{p}' = i\rho_0\omega\tilde{\phi} \tag{6.7.17}$$

as can be seen by substituting (6.5.11) and (6.7.3) into (6.7.1). Also, the velocity potential itself can sometimes be obtained from

$$\nabla^2\tilde{\Phi} + k^2\tilde{\Phi} = 0 \tag{6.7.18}$$

instead of (6.7.14). This approximation can be made when boundary dissipation is large compared to dissipation in the main body of fluid due to "classical" absorption.

Equations (6.7.14) and (6.7.15) can be used to study the effects of viscosity on monochromatic waves. Since (6.5.15) is a vector equation, it may appear that the number of unknowns, in the most general case, is four: the velocity potential and the three scalar components of $\tilde{\psi}$. In fact, there are only three unknowns, because the requirement that $\nabla \cdot \mathbf{B} = 0$ specifies one of the three components of $\mathbf{B}$ in terms of the other two. Thus, the general problem—that is, when $\mathbf{u}$ has three components—requires the solution of three Helmholtz equations, two of which (those for the two components of $\tilde{\psi}$) may be coupled. In many problems of interest, owing to symmetry of the boundaries or to some other simplifying factor, the number of independent spatial coordinates is reduced to two. In these cases, $\tilde{\psi}$ has only one nonzero component. For example, consider two-dimensional acoustic waves. Here $\mathbf{u} = (u, v, 0)$ so that $\nabla \times \mathbf{u}$ has only one component, which points in a direction perpendicular to the plane formed by u and v [i.e., $\nabla \times \mathbf{u} = \omega = (0, 0, \omega)$]. However, using (6.7.11) and remembering that $\nabla \cdot \mathbf{B} = 0$, we obtain

$$\omega = -\nabla^2 \mathbf{B} \tag{6.7.19}$$

But by (6.7.15), $\nabla^2\tilde{B} = K^2\tilde{B}$, which shows that $\mathbf{B}$, also, has only one nonzero component so that

$$\psi = (0, 0, \omega) \tag{6.7.20}$$

and (6.7.15) reduces to

$$\nabla^2\mathbf{e}_3\tilde{\psi} + K^2\mathbf{e}_3\tilde{\psi} = 0 \tag{6.7.21}$$

where $\mathbf{e}_3$ is a unit vector along the x_3 direction.

6.8 PROPAGATION IN A TWO-DIMENSIONAL CHANNEL

As an application of the formulation presented above, we consider propagation of a plane, monochromatic wave in a two-dimensional channel. The channel is assumed to be of infinite length and depth. We place a rectangular system of coordinates on the axis of the channel and align it so that the direction of propagation coincides with the x axis, with the y axis along the vertical. Under those conditions, the fluid velocity will have only two components, that is, $\mathbf{u} = (u, v, 0)$, each one being independent of z. They will, however, depend on x and y in such a manner that they vanish at the walls of the channel. The velocity potential also depends on x and y so that (6.7.14) written in this system of coordinates gives

$$\frac{\partial^2 \tilde{\phi}}{\partial x^2} + \frac{\partial^2 \tilde{\phi}}{\partial y^2} + \frac{k^2}{1 - i\frac{4}{3}\left(\omega v_0'/c_0^2\right)}\tilde{\phi} = 0 \qquad (6.8.1)$$

Consider now the equation for the vector potential $\boldsymbol{\psi}$. Since the present problem is two-dimensional, $\boldsymbol{\psi} = [0, 0, \psi(x, y)]$, so that (6.7.21) yields

$$\frac{\partial^2 \tilde{\psi}}{\partial x^2} + \frac{\partial^2 \tilde{\psi}}{\partial y^2} + K^2\tilde{\psi} = 0 \qquad (6.8.2)$$

where $K^2 = i\omega/v_0$. In terms of $\tilde{\phi}$ and $\tilde{\psi}$, the time-independent parts of the complex velocity components u and v are, respectively,

$$\tilde{U} = \frac{\partial \tilde{\Phi}}{\partial x} + \frac{\partial \tilde{\psi}}{\partial y} \qquad (6.8.3)$$

$$\tilde{V} = \frac{\partial \tilde{\Phi}}{\partial y} - \frac{\partial \tilde{\psi}}{\partial x} \qquad (6.8.4)$$

The solutions of (6.8.1) and (6.8.2) can be written as

$$\tilde{\Phi} = e^{iMx}\left(Ce^{imy} + De^{-imy}\right) \qquad (6.8.5)$$

$$\tilde{\psi} = e^{iNx}\left(ce^{iny} + de^{-iny}\right) \qquad (6.8.6)$$

where

$$M^2 + m^2 = k_c^2 \qquad (6.8.7)$$

$$N^2 + n^2 = K^2 \qquad (6.8.8)$$

and

$$k_c^2 = \frac{k^2}{1 - i\frac{4}{3}\left(\omega\nu_0'/c_0^2\right)} \tag{6.8.9}$$

It remains to evaluate the constants C, D, c, d, M, N, m, and n appearing in the solution. This can be done by means of (6.8.7) and (6.8.8) and the boundary conditions. These conditions require that

$$u(x, \pm h) = v(x, \pm h) = 0 \tag{6.8.10}$$

where $2h$ is the channel's height. Since the boundary conditions at the walls are independent of x, it is clear that we should have $M = N$. Further, since propagation is symmetric with respect to the channel axis, the x component of velocity u is even with respect to y; that is, $u(-y) = u(y)$. From (6.8.3), this requires that $\tilde{\Phi}(-y) = \tilde{\Phi}(y)$ and that $\partial \tilde{\psi}/\partial y$ be symmetric with respect to y. These requirements also follow from the antisymmetry of v and are satisfied by setting $D = C$ and $d = -c$. Thus, our solutions for $\tilde{\Phi}$ and $\tilde{\psi}$ can be written as

$$\tilde{\Phi} = 2Ae^{iMx}\cos(my) \tag{6.8.11}$$

$$\tilde{\psi} = 2iBe^{iMx}\sin(ny) \tag{6.8.12}$$

The velocity components $\tilde{U}$ and $\tilde{V}$ are then given by

$$\tilde{U} = 2i\left[AM\cos(my) + nB\cos(ny)\right]e^{iMx} \tag{6.8.13}$$

$$\tilde{V} = 2\left[MB\sin(ny) - mA\sin(my)\right]e^{iMx} \tag{6.8.14}$$

whereas the time-independent acoustic pressure $\tilde{P}$ is

$$\tilde{P} = 2i\rho_0\omega Ae^{iMx}\cos(my) \tag{6.8.15}$$

Applying the boundary conditions at $y = -h$ as prescribed by (6.8.10), we obtain

$$AM\cos(mh) + nB\cos(nh) = 0 \tag{6.8.16}$$

$$Am\sin(nh) - MB\sin(nh) = 0 \tag{6.8.17}$$

In order to obtain a nontrivial solution for A and B, it is necessary that the coefficient matrix have a vanishing determinant. This requirement yields

$$M^2\tan(nh) = -mn\tan(mh) \tag{6.8.18}$$

This equation, together with (6.8.7) and (6.8.8), specifies the value of M. However, it remains to solve the transcendental equation (6.8.18), a task that is not trivial.

Since the propagation constant M is the quantity that determines the attenuation coefficient and the phase velocity, we eliminate m and n in favor of M by means of (6.8.7) and (6.8.8) and obtain

$$M^2 \tan\left(\sqrt{K^2 - M^2}\, h\right) = -\sqrt{k_c^2 - M^2}\; \sqrt{K^2 - M^2}\; \tan\left(\sqrt{k_c^2 - M^2}\, h\right)$$

(6.8.19)

Our task is to obtain values of M that approximately satisfy this equation and also satisfy the requirement that only positive-going waves exist in the channel. This imposes the condition that M must have positive real and imaginary parts. To obtain such a solution, it is convenient to introduce suitable nondimensional parameters instead of M, K, h, and k_c. Since in the absence of viscous effects, $M = k$, we introduce, instead of M,

$$\beta = M/k \qquad (6.8.20)$$

Since viscous effects act over distances of the order of $\delta_\nu = (2\nu_0/\omega)^{1/2}$, we eliminate h in favor of H, where

$$H = h/\delta_\nu \qquad (6.8.21)$$

Another quantity that can be identified in (6.8.19) is the ratio of h to the unattenuated wavelength, or what is the same thing, the product

$$\Gamma = kh \qquad (6.8.22)$$

A third ratio of length scales is

$$\Theta = k\delta_\nu = \left(2\omega\nu_0/c_0^2\right)^{1/2} \qquad (6.8.23)$$

This quantity can be defined in terms of H and Γ by means of

$$\Theta = \Gamma/H \qquad (6.8.24)$$

The quantity $4\omega\nu_0/3c_0^2$ also appears in the definition of k_c in (6.8.9). It can be expressed in terms of the viscous attenuation per wavelength $\bar{\alpha}_\nu$, for the free-propagation case, by means of [see Equation (6.2.18)]

$$2\bar{\alpha}_\nu = \frac{4}{3}\frac{\omega\nu_0'}{c_0^2} \qquad (6.8.25)$$

This, in turn, can be expressed in terms of Θ by means of

$$2\bar{\alpha}_\nu = \tfrac{2}{3}\Theta^2\left(1 + \frac{3}{4}\frac{\mu_{v0}}{\mu_0}\right) \tag{6.8.26}$$

which shows that $\bar{\alpha}_\nu$ is of the order of Θ^2. With these definitions and with $K = \sqrt{2i}\,/\delta_\nu$, we can write (6.8.19) as

$$\Theta\beta^2\tan\left(\sqrt{2i}\,H\sqrt{1+(i/2)\beta^2\Theta^2}\,\right) = -\frac{\sqrt{2i}}{\sqrt{1-2i\bar{\alpha}_\nu}}\sqrt{1-\beta^2(1-2i\bar{\alpha}_\nu)}$$

$$\times\sqrt{1+(i/2)\beta^2\Theta^2}$$

$$\times\tan\left(\sqrt{1-\beta^2(1-2i\bar{\alpha}_\nu)}\,\frac{\Gamma}{\sqrt{1-2i\bar{\alpha}_\nu}}\right) \tag{6.8.27}$$

As we know from our discussion in Section 6.3, the quantity $\bar{\alpha}_\nu$ is very small. Therefore, if we anticipate that the attenuation in the channel due to wall effects is significantly larger than that due to free propagation, we can set $\bar{\alpha}_\nu = 0$ in (6.8.27). Since $\bar{\alpha}_\nu$ is proportional to Θ^2, it may be thought that we also have to set $\Theta = 0$. However, all we have to do is to consider the limiting case when quantities of order Θ^2 can be neglected compared to those of order Θ. Under these conditions, (6.8.27) can be written as

$$\Theta\beta^2\tan(\sqrt{2i}\,H) = -\sqrt{2i}\,\sqrt{1-\beta^2}\,\tan\left(\Gamma\sqrt{1-\beta^2}\,\right) \tag{6.8.28}$$

where we have also assumed that $(\Theta\beta)^2 \ll 1$. The roots of this equation are seen to be functions of the parameters H and Γ [because Θ can be eliminated by means of (6.8.24)] or, explicitly,

$$\beta = f\left[\left(\frac{\omega h}{2\nu_0}\right)^{1/2}, \frac{\omega h}{c_0}\right] \tag{6.8.29}$$

While it is clear that the roots cannot be obtained in closed form, (6.8.29) suggests that a numerical solution may be obtained in terms of the two parameters appearing there. Also, suitable asymptotic solutions of (6.8.28) may be obtained by letting these parameters attain limiting values of interest. Here we use only the latter approach, and consider the important cases when the channel is, in some sense, very "wide" or very "narrow." To obtain these limits, we first take advantage of the fact that Θ is small, and consider the implications of letting this parameter approach zero. If $\Theta \to 0$ and H is such that $\Theta H \ll 1$, then $\tan(\Gamma\sqrt{1-\beta^2}) \approx \Theta H \sqrt{1-\beta^2}$ so that (6.8.28) becomes

$$\beta^2 \tan(\sqrt{2i}\, H) \approx -\sqrt{2i}\,(1-\beta^2)H$$

This yields

$$\beta^2 = \frac{1}{1 - \dfrac{\tan(\sqrt{2i}\, H)}{\sqrt{2i}\, H}} \tag{6.8.30}$$

an approximation that is applicable provided that

$$\Theta \ll 1$$

$$H \ll 1/\Theta \tag{6.8.31}$$

Wide-Tube, Low-Frequency Approximation

Suppose that H is large (but still satisfies $\Theta H \ll 1$). Then, using the identity

$$\tan(\sqrt{2i}\, H) = i\frac{1 - e^{2i(1+i)H}}{1 + e^{2i(1+i)H}} \tag{6.8.32}$$

we obtain, for large H, $\tan(\sqrt{2i}\, H) \approx i$, so that

$$\frac{\tan(\sqrt{2i}\, H)}{\sqrt{2i}\, H} \approx \frac{i}{(1+i)H} \tag{6.8.33}$$

Therefore, β^2 is then approximately given by

$$\beta^2 = 1 + \frac{1+i}{2H} \tag{6.8.34}$$

This result predicts that the propagation constant is nearly equal to k, a result that was to be expected owing to the smallness of Θ. However, in order that this result be applicable, we must have $\Theta H \ll 1$. In terms of the original quantities, this requires that

$$(\omega h/c_0) \ll 1 \tag{6.8.35}$$

Thus, while we assumed H large (i.e., that $h \gg \delta_\nu$), (6.8.35) requires that the channel height be small compared to the wavelength. The result given by (6.6.18) therefore gives the low-frequency, wide-tube approximation to the propagation constant. Returning to (6.8.34), we have, with $M = \beta k$,

$$M \approx k\left(1 + \frac{1+i}{4}\frac{\delta_\nu}{h}\right) \tag{6.8.36}$$

The phase velocity of the waves in the channel, given by $\omega/\mathrm{Re}(M)$, is

$$c \approx c_0(1 - \delta_\nu/4h) = c_0\left(1 - \tfrac{1}{4}\sqrt{2\nu_0/\omega h^2}\right) \tag{6.8.37}$$

Thus, the phase velocity differs, in this case, little from c_0. Also, the waves are only slightly attenuated, the attenuation coefficient being [with $\alpha_{w\nu} = \mathrm{Im}(M)$]

$$\alpha_{w\nu} = \frac{\delta_\nu}{4h}k \tag{6.8.38}$$

or, using the definition of δ_ν,

$$\alpha_{w\nu}h = \frac{1}{2}\left(\frac{\omega\nu_0}{2c_0^2}\right)^{1/2} \tag{6.8.39}$$

Comparison with (6.6.9) shows that $\alpha_{w\nu}$ is basically the same thing as the viscous contribution to the tube-attenuation coefficient introduced there. The factor of $\tfrac{1}{2}$ appearing in (6.8.39) and not in (6.6.9) is due to the two-dimensionality of our channel.

It should be noticed that while the attenuation due to the walls is small, it is nevertheless large compared to that which would occur if no wall effects existed. In fact, the ratio of $\alpha_{w\nu}$ to the viscous part of the classical attenuation coefficient is, from (6.8.38) and (6.8.25), of the order of $1/H\Theta^2$, and this is large because $\Theta H \ll 1$ and $\Theta \ll 1$. This relative magnitude was anticipated when we eliminated $\bar{\alpha}_\nu$ from (6.8.26). The correctness of this anticipated result implies that when we are considering propagation over boundaries, we may use for k_c in (6.8.9) the approximation $k_c \approx k$, or what is the same thing, use (6.7.18) instead of (6.7.14) to discuss the behavior of the longitudinal waves when boundaries are present.

Now that we have an approximate value for β, applicable for $H \gg 1$ and for $\Theta H \ll 1$, let us return to the fluid velocity and pressure as given by (6.8.13)–(6.8.15). In order to compute these quantities, we need explicit expressions for m, n, and B. Using (6.8.7) with $k_c = k$ and with M^2 from (6.8.36), we obtain

$$m \approx \tfrac{1}{2}(1-i)(\delta_\nu/h)^{1/2}k \tag{6.8.40}$$

Similarly, the leading terms in n is

$$n \approx \frac{1+i}{\delta_\nu} \tag{6.8.41}$$

With these values of m and n, (6.8.16) yields

$$nB = -2AMe^{(i-1)(h/\delta_\nu)} \tag{6.8.42}$$

so that the x component of the velocity is given by

$$\tilde{u} = 2iAM\left[\cos(my) - 2e^{(i-1)(h/\delta_\nu)}\cos(ny)\right]e^{i(Mx-\omega t)} \tag{6.8.43}$$

Consider $\cos(my)$. In view of (6.8.40), it is given by

$$\cos(my) \approx \cos\left(\frac{1-i}{2}\frac{ky}{\sqrt{H}}\right)$$

But $ky \leqslant kh = \omega h/c_0$, and this is small compared to unity [see Equation (6.8.35)]. Thus, $\cos(my) \approx 1$. This implies that the acoustic pressure is constant across the channel.

The other function appearing in (6.8.43) is

$$\cos(ny) = \tfrac{1}{2}\left[e^{(i-1)(y/\delta_\nu)} + e^{-(i-1)(y/\delta_\nu)}\right] \tag{6.8.44}$$

and this does not have an expansion that is valid for all values of y. However, provided that $y \gg \delta_\nu$, we may set

$$\cos(ny) \approx \tfrac{1}{2}e^{-(i-1)(y/\delta_\nu)} \tag{6.8.45}$$

so that, then,

$$\tilde{u} \approx 2iAke^{-\alpha_{wr}x}\left[1 - e^{(i-1)(h-y)/\delta_\nu}\right]e^{i(kx-\omega t)} \tag{6.8.46}$$

On the other hand, if $y \ll \delta_\nu$, we have $\cos(ny) \approx 1 - i(y/\delta_\nu)^2$ so that near the axis of the channel we may take

$$\tilde{u} \approx 2iAke^{-\alpha_{wr}x}\left[1 - 2e^{(i-1)(h/\delta_\nu)}\right]e^{i(kx-\omega t)}$$

However, when y/δ_ν is large, the differences between this and (6.8.46) are very small. We may therefore use (6.8.46) as a first approximation to u throughout the channel provided, of course, that the conditions given by (6.8.31) are satisfied.

As suggested by (6.8.46), we use, instead of y, the variable $\eta = h - y$. This simply measures distances from the lower wall of the channel. Then (6.8.46) becomes

$$\tilde{u} = 2iAke^{-\alpha_{wr}x}\left[1 - e^{(i-1)(\eta/\delta_r)}\right]e^{i(kx-\omega t)} \tag{6.8.47}$$

This can be used to obtain the mean-squared velocity $\langle u^2 \rangle = \frac{1}{2}\operatorname{Re}(\tilde{u}\tilde{u}^*)$. Thus,

$$\frac{\langle u^2 \rangle}{\langle u_\infty^2 \rangle} = e^{-2\alpha_{wr}x}\left[1 - 2\cos\left(\frac{\eta}{\delta_r}\right)e^{-(\eta/\delta_r)} + e^{-2(\eta/\delta_r)}\right] \tag{6.8.48}$$

where $\tilde{u}_\infty = 2ikA\exp[i(kx - \omega t)]$ represents the velocity that would exist if no wall effects were present. Comparison of this result with (6.6.5) shows that the variations with vertical distance given by the two expressions are identical. Our present result also gives the manner in which this profile is attenuated as the wave is propagated in the channel. More importantly, it yields the conditions under which (6.6.5) apply. These are given by (6.8.31) so that (6.6.5) applies only to wide channels ($H \gg 1$) and low-frequency waves ($(\omega h/c_0) \ll 1$).

Narrow-Tube, Low-Frequency Approximation

We now consider the case when the channel's height $2h$ is small compared with the viscous penetration depth δ_r (i.e., when $H \ll 1$). The other parameter, Θ, remains small. Since $\Theta = \Gamma/H$, these assumptions require that $\Gamma = kh$ be very small. Under these conditions, (6.8.30) is still applicable. Therefore, when $H \ll 1$, we obtain

$$\beta^2 = 3i/2H^2 \tag{6.8.49}$$

which gives

$$M \approx \frac{\sqrt{3}}{2}\frac{1+i}{H}k \tag{6.8.50}$$

The quantities m^2 and n^2 are, from (6.8.7) and (6.8.8), given by

$$m^2 = k^2(1 - \beta^2) \approx -(k\beta)^2$$

$$n^2 = \frac{2i}{\delta_r^2}\left[1 + \frac{1}{2}i(\beta\Theta)^2\right] \approx 2i\delta_r^{-2} \tag{6.8.51}$$

Thus, the limiting values of m and n, when $\Theta \ll 1$ and $H \ll 1$, are

$$m \approx \frac{\sqrt{3}}{2}(1-i)\frac{k\delta_\nu}{h} \qquad (6.8.52)$$

$$n \approx \frac{1+i}{\delta_\nu} \qquad (6.8.53)$$

Notice that in (6.8.53) we dropped the quantity $(\beta\Theta)^2$. This was done in order to be consistent with the previous assumption made when deriving (6.8.26) from (6.8.19). In terms of Θ and H, this requires that $\Theta \ll H$.

Now, since $|my| < \Theta \ll 1$ and $|ny| < H \ll 1$, we can approximate $\cos(my)$ and $\cos(ny)$ by

$$\cos(my) \approx 1 - \tfrac{3}{8}(1-i)^2(ky)^2(\delta_\nu/h)^2$$

and

$$\cos(ny) \approx 1 - \tfrac{1}{2}(1+i)^2(y/\delta_\nu)^2$$

Therefore,

$$\frac{\cos(ny)}{\cos(nh)} \approx 1 - \frac{(1+i)^2}{2}\frac{y^2-h^2}{\delta_\nu^2} \qquad (6.8.54)$$

Since (6.8.49) is limited to values of $\Theta \ll H$, $\cos(my)$ can be set equal to unity so that the velocity in the channel

$$\tilde{u} = 2iAM\left[\cos(my) - \frac{\cos(mh)}{\cos(nh)}\cos(ny)\right]e^{i(Mx-\omega t)} \qquad (6.8.55)$$

can be written as

$$\tilde{u} = -2AM\frac{(y^2-h^2)}{\delta_\nu^2}e^{i(Mx-\omega t)} \qquad (6.8.56)$$

Thus, at any fixed location, the instantaneous-velocity profile across the channel is parabolic, the maximum value of the velocity occurring at the centerline and having a magnitude given by

$$|u(0,t)| = 2\sqrt{3}\, AkHe^{-\sqrt{3}\,kx/2H} \qquad (6.8.57)$$

In view of the smallness of H, it follows that $u(0, t)$ is very small when compared to the velocity that would exist there in the absence of viscosity (i.e., u_∞). Also, $u(0, t)$ decays rapidly with distance along the channel. In fact, in a distance x nearly equal to a fraction H of the wavelength, the relative amplitude has been reduced by an amount equal to $e^{2\pi} \approx 535$. Further, the phase velocity $c = \omega/\mathrm{Re}(M)$ is a small fraction of c_0.

The above results show clearly that sound waves are attenuated rapidly when they propagate in very narrow channels. The results are therefore of much interest in noise-reduction problems, for they show that any material having narrow pores and crevices will reduce sound-pressure levels when placed in a noisy environment.

It should be pointed out that in this section we have not taken into account the effects of heat conduction. These effects include not only additional dissipation, but also changes in the basic nature of the propagation problem. For example, when the channel is very narrow, the assumption of adiabatic propagation is grossly incorrect, because heat is easily conducted to and from the wall. In such a case the propagation is more nearly isothermic so that (6.7.3) is not applicable. The effects of finite heat conductivity near boundaries can be handled in a manner similar to that used for the viscous effects, as shown in Section 6.10.

6.9   SPHERE OSCILLATING IN A VISCOUS FLUID

As a second application of the formulation introduced in Section 6.7 to study the effects of viscosity near boundaries, we reconsider the problem of a rigid sphere oscillating in a fluid. The problem was treated in Section 5.7 for an inviscid fluid and was used to introduce dipole emission. Since, however, all fluids have finite viscosity, it is of some practical interest to study how viscosity modifies the inviscid results.

In the inviscid-fluid case, the fluid motion around the sphere was described by the scalar velocity potential $\tilde{\Phi}$. In the viscous-fluid case, we also require the vector potential $\tilde{\psi}$. These quantities satisfy (6.7.14) and (6.7.15), respectively. Since we expect boundary dissipation to be large compared to dissipation in the main body of fluid, we write (6.7.14) as

$$\nabla^2 \tilde{\Phi} + k^2 \tilde{\Phi} = 0 \qquad (6.9.1)$$

Because of the type of motion under consideration, it is convenient to use polar-spherical coordinates r, θ, φ. Since the flow is symmetric about the polar axis ($\theta = 0$), Φ is independent of the azimuthal angle φ. Therefore, for outgoing waves, Φ may be expressed as (see Section 4.3)

$$\tilde{\Phi} = \sum_{n=0}^{\infty} i^n (2n+1) A_n h_n^{(1)}(kr) P_n(\cos\theta) \qquad (6.9.2)$$

The equation for $\tilde{\psi}$ has, in general, three components. However, because of the symmetry of the fluid motion under consideration about the direction of motion, $\tilde{\psi}$ has only one component, and this is along the azimuth direction (i.e., $\psi = [0, 0, B(r, \theta)]$) so that the equation for $\tilde{\psi}$ becomes

$$\nabla^2 \mathbf{e}\tilde{B} + K^2 \mathbf{e}\tilde{B} = 0 \qquad (6.9.3)$$

where $\mathbf{e}$ is a unit vector along the azimuthal direction. Since this vector is not constant with respect to a cartesian system of coordinates, $\nabla^2 \mathbf{e}B \neq \mathbf{e}\nabla^2 B$. Instead,

$$\nabla^2 \mathbf{e}\tilde{B} = \mathbf{e}\left(\nabla^2 \tilde{B} - \frac{\tilde{B}}{r^2 \sin^2\theta} \right) \qquad (6.9.4)$$

where we have taken advantage of the fact that B also depends only on r and θ. Hence, (6.9.3) can be expressed in scalar form as

$$\nabla^2 \tilde{B} + \left(K^2 - \frac{1}{r^2 \sin^2\theta} \right) \tilde{B} = 0 \qquad (6.9.5)$$

The solution to this equation may also be obtained by separation of variables, and is given by

$$\tilde{B} = \sum_{n=0}^{\infty} i^n (2n+1) B_n h_n^{(1)}(Kr) P_n^1(\cos\theta) \qquad (6.9.6)$$

where P_n^1 is the associated Legendre polynomial of order unity and degree n and is defined as

$$P_n^1(\cos\theta) = -\sin\theta \frac{dP_n(\cos\theta)}{d(\cos\theta)} \qquad (6.9.7)$$

The coefficients A_n and B_n may be obtained from the boundary conditions. These require that the fluid velocity $\mathbf{u}$ be equal to the sphere velocity $\mathbf{U}_p$ on the sphere surface. If, as in Section 5.7, the sphere oscillates harmonically along the polar axis with amplitude U_{p0}, the radial and tangential velocity components must satisfy, on $r = a$, the following conditions:

$$u_r = U_{p0} \cos \theta$$

$$u_\theta = -U_{p0} \sin \theta \qquad (6.9.8)$$

The components of the fluid velocity may be obtained from ϕ and B by means of (6.7.11), which in the present system of coordinates gives

$$u_r = \frac{\partial \phi}{\partial r} + \frac{1}{r \sin \theta} \frac{\partial (B \sin \theta)}{\partial \theta} \qquad (6.9.9)$$

$$u_\theta = \frac{1}{r} \frac{\partial \phi}{\partial \theta} - \frac{1}{r} \frac{\partial (rB)}{\partial r} \qquad (6.9.10)$$

Dropping the superscript on $h_n^{(1)}$, we obtain for u_r,

$$u_r = \sum_{n=0}^{\infty} i^n (2n+1) k \left[A_n h_n'(kr) - n(n+1) \frac{B_n}{Kr} h_n(Kr) \right] P_n(\cos \theta) \qquad (6.9.11)$$

where the prime denotes differentiation with respect to the argument and where we have used the identity

$$\frac{dP_n^1}{d\theta} + \frac{\cos \theta}{\sin \theta} P_n^1(\cos \theta) = -n(n+1) P_n(\cos \theta) \qquad (6.9.12)$$

Similarly, making use of (6.9.8) and (6.9.9), we obtain

$$u_\theta = \sum_{n=0}^{\infty} i^n (2n+1) \frac{1}{r} \left\{ A_n h_n'(kr) - B_n \left[Krh_n'(Kr) + h_n(Kr) \right] \right\} P_n^1(\cos \theta)$$

$$(6.9.13)$$

Since $P_0^1(\cos \theta) = 0$ and $P_0(\cos \theta) = 1$, it follows that the $n=0$ terms in these results must be absent (i.e., $A_0 = B_0 = 0$). Further, since $P_1^1(\cos \theta) = -\sin \theta$ and $P_1(\cos \theta) = \cos \theta$, the boundary conditions give for $n=1$

$$U_{p0} = 3ik \left[A_1 h_1'(ka) - 2 \frac{B_1}{ka} h_1(Ka) \right] \qquad (6.9.14)$$

$$U_{p0} = (3i/a) \left\{ A_1 h_1(ka) - B_1 \left[Kah_1'(Ka) + h_1(Ka) \right] \right\} \qquad (6.9.15)$$

For $n>1$, the boundary conditions are

$$A_n h'_n(ka) - n(n+1)\frac{B_n}{ka} h_n(ka) = 0 \qquad (6.9.16)$$

$$A_n h_n(ka) - B_n[Kah_n(Ka) + h_n(Ka)] = 0 \qquad (6.9.17)$$

but these will not be needed for computing the intensity of the emitted waves because far from the sphere the $n=1$ term dominates. In fact, all we need for computing the far-field intensity is A_1, and this is, from (6.9.14) and (6.9.15), given by

$$A_1 = -\frac{U_{p0}}{3ik} \frac{\beta h'_1(\beta) - h_1(\beta)}{\begin{vmatrix} h'_1(b) & -(2/b)h_1(\beta) \\ h_1(b) & -[\beta h'_1(\beta) + h_1(\beta)] \end{vmatrix}} \qquad (6.9.18)$$

where we have introduced the notations

$$b = ka, \qquad \beta = Ka \qquad (6.9.19)$$

Now, from Section 5.7 we have

$$h'_1(z) = \frac{e^{iz}}{z^3} \left[2z + i(2 - z^2) \right] \qquad (6.9.20)$$

$$h_1(z) = -\frac{e^{iz}}{z^2}(z + i) \qquad (6.9.21)$$

so that (6.9.18) can be written as

$$-3ikA_1 = U_{p0} \frac{\left[3\beta + i(3 - \beta^2) \right] e^{-ib}/\beta^2}{\begin{vmatrix} \dfrac{1}{b^3}[2b + i(2 - b^2)] & \dfrac{2}{b\beta^2}(\beta + i) \\ -\dfrac{1}{b^2}(b + i) & -\dfrac{1}{\beta^2}[\beta + i(1 - \beta^2)] \end{vmatrix}} \qquad (6.9.22)$$

The determinant in the denominator can be written as

$$d = \frac{-1}{b^3\beta^2} \left\{ [2b + i(2 - b^2)][\beta + i(1 - \beta^2)] - 2(\beta + i)(b + i) \right\} \qquad (6.9.23)$$

or, after some rearrangement, as

$$d = \frac{1}{b^3\beta^2}\left[\beta^2(b^2-2)-b^2+i(\beta b^2+2b\beta^2)\right] \tag{6.9.24}$$

Hence,

$$-3ikA_1 = U_{p0}b^3e^{-ib}\frac{3\beta+3i-i\beta^2}{\beta^2(b^2-2)-b^2+i(\beta b^2+2b\beta^2)} \tag{6.9.25}$$

Since the quantity $\beta = Ka$ is complex, it is convenient to introduce the real quantity y defined by

$$y = \sqrt{\omega a^2/2\nu_0} \tag{6.9.26}$$

With this definition, we have

$$\beta^2 = 2iy^2, \qquad \beta = (1+i)y \tag{6.9.27}$$

The quantity y defined by (6.9.26) is simply the ratio of the sphere radius to the thickness of the oscillatory boundary layer (see Section 6.5).

In terms of y and of $b = ka$, (6.9.25) can be written as

$$3ikA_1 = U_{p0}b^3e^{-ib}\frac{3y+2y^2+3i(1+y)}{4by^2+b^2(1+y)-i\left[yb^2-2y^2(2-b^2)\right]} \tag{6.9.28}$$

The coefficient B_1 may also be obtained from (6.9.14) and (6.9.15). Following the same procedure used for A_1, we obtain

$$-3ikB_1 = U_{p0}\beta^2be^{-i\beta}\frac{3b+i(3-b^2)}{4by^2+b^2(1+y)-i\left[yb^2-2y^2(2-b^2)\right]} \tag{6.9.29}$$

These equations describe some of the effects of viscosity on the waves emitted by the oscillating sphere. When these effects do not exist, they reduce to our previous results. Thus, if the viscosity approaches zero, $y\to\infty$. In that limit, ($y\to\infty$), $B_1=0$, and (6.9.28) becomes

$$3ikA_1 = U_{p0}b^3e^{-ib}\frac{1}{2b+i(2-b^2)} \tag{6.9.30}$$

The velocity potential far from the sphere is given by the first term in (6.9.2):

$$\tilde{\Phi} = 3iA_1 h_1(kr)\cos\theta$$

Hence, using our result for A_1 and the explicit result for $H_1(kr)$, we obtain

$$\tilde{\Phi} = -U_{p0}b^3 \frac{1}{k^3 r^2} e^{ik(r-a)} \frac{kr+i}{2b+i(2-b^2)}\cos\theta$$

Rearranging, and restoring the time factor, we obtain

$$\tilde{\phi} = U_{p0}\left(\frac{a}{r}\right)^2 \cos\theta \frac{ikr-1}{2-b^2-2ib} e^{ik(r-a)-i\omega t}$$

This is the result previously obtained in Section 5.5.

We now return to the results for the viscous case and compute the far-field intensity. This is, as usual, given by

$$I = \tfrac{1}{2}\operatorname{Re}(\tilde{p}'^* \tilde{u}_r) \tag{6.9.31}$$

where $\tilde{p}' = i\rho_0\omega\tilde{\phi}$. As stated earlier, in the far-field the dominant contribution to our variables arises from the term with $n=1$. Now, $\tilde{u}_r$ contains contributions from both $\tilde{\phi}$ and $\tilde{B}$. However, the last contribution decreases as $1/r^2$, whereas that due to $\tilde{\phi}$ decreases as $1/r$. Hence, for sufficiently large distances, we will simply have

$$I = \tfrac{1}{2}\rho_0\omega \operatorname{Re}\left(-i\tilde{\phi}_1^* \frac{\partial\tilde{\phi}_1}{\partial r}\right) \tag{6.9.32}$$

where $\tilde{\phi}_1 = 3iAh_1(kr)\cos\theta$. Therefore,

$$I = \tfrac{9}{2}\rho_0\omega k \cos^2\theta |A_1|^2 \operatorname{Re}\left[-ih_1^*(kr)h_1'(kr)\right] \tag{6.9.33}$$

The real part of the quantity in brackets is simply $(kr)^{-2}$ so that

$$I = \tfrac{9}{2}\rho_0 c_0 \frac{\cos^2\theta}{r^2}|A_1|^2 \tag{6.9.34}$$

The $\cos^2\theta$ factor gives an angular distribution for the intensity that is typical of dipole emission; as expected, it is identical to that obtained in the inviscid case. However, as we will see below, the typical dipole dependence of the intensity on frequency is recovered only in the limit $y\to\infty$ [i.e., $(\delta_v/a)\to0$]. For finite values of y, the dependence differs from the typical ω^4 value. In particular, when $y \ll 1$, the intensity varies with frequency as ω^2, indicating that in spite of the angular dependence, the basic emission mechanism is monopole.

To show these effects, we need $|A_1|^2$. This may be obtained from (6.9.28), but the result will, in general, be too involved. We therefore consider the special but important case when $b = ka \ll 1$. Using the definitions of b and y, we may write

$$b = \sqrt{2\omega\nu_0/c_0^2}\, y \qquad (6.9.35)$$

This shows that we cannot set b equal to zero for arbitrary values of ω, ν, and c_0 while retaining y. However, unless the frequency is very high, the parameter $\sqrt{2\omega\nu_0}/c_0$ is very small. In the case of air it is equal to 1.26×10^{-3} at 1000 Hz. Therefore, whatever the value of y may be, we have $(b/y) \ll 1$, so that as $b \to 0$, we obtain from (6.9.28)

$$3ikA_1 = -\tfrac{1}{2}iU_{p0}b^3\left[1 + \frac{3}{2y} + \frac{3}{2}i\left(\frac{1}{y} + \frac{1}{y^2}\right)\right] \qquad (6.9.36)$$

which yields

$$|A_1|^2 = \tfrac{1}{36}U_{p0}^2\frac{b^6}{k^2}\left(1 + \frac{3}{y} + \frac{9}{2y^2} + \frac{9}{2y^3} + \frac{9}{4y^4}\right) \qquad (6.9.37)$$

Finally, substituting this into (6.9.34), we obtain

$$I = \tfrac{1}{8}\rho_0\omega^4\frac{U_{p0}^2}{c_0^3}\frac{a^6}{r^2}\left(1 + \frac{3}{y} + \frac{9}{2y^2} + \frac{9}{2y^3} + \frac{9}{4y^4}\right)\cos^2\theta \qquad (6.9.38)$$

The corresponding power is

$$\Pi = \frac{\pi}{6}\rho_0\omega^4\frac{U_{p0}^2}{c_0^3}a^6\left(1 + \frac{3}{y} + \frac{9}{2y^2} + \frac{9}{2y^3} + \frac{9}{4y^4}\right) \qquad (6.9.39)$$

In the limit $y \to \infty$, we recover the inviscid results applicable for $b \ll 1$ (see Section 5.7). Only in this limit does the intensity depend on ω as ω^4. For any other value, the exact dependence is modified by the quantity in parentheses, which is never smaller than one. Thus, the inviscid-fluid assumption results in emitted intensities that are always smaller than the actual. Further, when $y \ll 1$, we obtain, with $y = (\omega a^2/2\nu_0)^{1/2}$,

$$I = \frac{9}{8} \frac{1}{r^2 c_0^3} \rho_0 \nu_0^2 \omega^2 U_{p0}^2 a^2 \cos^2 \theta \tag{6.9.40}$$

This is basically monopole emission modified by an angular factor equal to $\cos^2 \theta$. It is not difficult to see why the sphere is, in this limit, emitting as a monopole, for when $\omega a^2/2\nu_0 \ll 1$, the radius of the sphere is very small compared with the viscous-penetration depth $\sqrt{2\nu_0/\omega}$. When $y \ll 1$, the effective radius of the sphere corresponds to the instantaneous thickness of the boundary layer, so that its value oscillates harmonically in time between a value equal to a and a value that is of the order of $a(1 + 1/y)$. The only difference between this emission and the basic pulsating-sphere emission is that in the present case the sphere is moving axially, a fact that accounts for the $\cos^2 \theta$ in the intensity.

Force on the Sphere

We now consider the total force on the oscillating sphere. In the inviscid-fluid case, this force could be separated into two types: one proportional to the sphere's velocity and another proportional to the sphere's acceleration. The former was responsible for the emission of sound, whereas the net effect of the second was to increase the net mass of the sphere. In the present case, viscous forces come into play and it is of interest to calculate them, as it is clear that they will modify the inviscid results significantly.

In order to simplify our calculations, we will consider only the case $ka \ll 1$, which basically corresponds to the case of an incompressible fluid, at least near the sphere. One could, therefore, make use of the incompressible-fluid equations to obtain the force on the sphere. However, since we have already computed the flow field around the sphere in the acoustic case, it is more convenient to use those results, even though they contain effects we are not interested in at present.

Because of the type of motion under consideration, the total force on the sphere will be along the axial direction. If $\sigma_{r\theta}$ and σ_{rr} are the tangential and normal components of the stress tensor, then the total force on the sphere will be given by

$$F = 2\pi a^2 \int_0^\pi (\sigma_{rr} \cos \theta - \sigma_{\theta r} \sin \theta)_{r=a} \sin \theta \, d\theta \tag{6.9.41}$$

where

$$\sigma_{rr} = -p + 2\mu_0 \left(\frac{\partial u_r}{\partial r} \right) - \tfrac{2}{3} \mu_0 \nabla \cdot \mathbf{u} \tag{6.9.42}$$

$$\sigma_{r\theta} = \mu_0 \left(\frac{\partial u_\theta}{\partial r} - \frac{u_\theta}{r} + \frac{1}{r} \frac{\partial u_r}{\partial \theta} \right) \tag{6.9.43}$$

In the present case, all the variables vary harmonically in time. Therefore, if $F = \mathrm{Re}[\tilde{F}_0 \exp(-i\omega t)]$,

$$\tilde{F}_0 = 2\pi a^2 \int_0^\pi \left[\left(-\tilde{p}' + 2\mu \frac{\partial \tilde{u}_r}{\partial r} - \tfrac{2}{3}\mu_0 \tilde{\Delta} \right) \cos\theta \right.$$

$$\left. - \mu_0 \left(\frac{\partial \tilde{u}_\theta}{\partial r} - \frac{\tilde{u}_\theta}{r} + \frac{1}{r} \frac{\partial \tilde{u}_r}{\partial\theta} \right) \sin\theta \right]_{r=a} \sin\theta \, d\theta \qquad (6.9.44)$$

where the dilatation $\tilde{\Delta} = \nabla \cdot \tilde{\mathbf{u}} = \nabla^2 \tilde{\phi} = -k^2 \tilde{\phi}$ appears, owing to compressibility effects, and where

$$\tilde{p}' = i\rho_0 \omega \tilde{\phi} \qquad (6.9.45)$$

Hence, the first and third terms in the first parentheses of (6.9.44) combine to give

$$\tilde{p}' + \tfrac{2}{3}\mu_0 \tilde{\Delta} = \tilde{p}'\left(1 + \tfrac{2}{3} i\omega\nu_0 / c_0^2 \right) \qquad (6.9.46)$$

As remarked earlier, the quantity $\omega\nu_0 / c_0^2$ is very small so that consistent with our previous approximations [i.e., Equation (6.7.18)], we can neglect the contribution to σ_{rr} arising from finite-compressibility effects.

The quantities $\tilde{\phi}$, $\tilde{u}_r$, and $\tilde{u}_\theta$ appearing in (6.9.44) and (6.9.45) are given by (6.9.2), (6.9.11), and (6.9.13), respectively. If we substitute these into the force equation, we obtain terms containing the integrals $\int_0^\pi P_n(\cos\theta)\cos\theta\sin\theta \, d\theta$ and $\int_0^\pi P_n^1(\cos\theta)\sin^2\theta \, d\theta$. Because of the orthogonality conditions satisfied by P_n and P_n^1, these integrals vanish unless $n=1$, in which case they are equal to $\tfrac{2}{3}$ and $-\tfrac{4}{3}$, respectively. Further, because of the boundary conditions, we have, for $n=1$,

$$\left(\partial \tilde{u}_r / \partial\theta - \tilde{u}_\theta \right)_{r=a} = 0 \qquad (6.9.47)$$

so that (6.9.44) gives, upon integration,

$$\tilde{F}_0 = \tfrac{4}{3}\pi a^2 \left(-i\rho_0 \omega \tilde{\Phi} + 2\mu_0 \frac{\partial \tilde{U}_r}{\partial r} \right)_{r=a} + \tfrac{8}{3}\pi a^2 \mu_0 \left(\frac{\partial \tilde{U}_\theta}{\partial r} \right)_{r=a} \qquad (6.9.48)$$

where the symbols $\tilde{\Phi}$, $\tilde{U}_r$, and $\tilde{U}_\theta$ represent the term with $n = 1$ of the corresponding quantities without the angular dependence:

$$\tilde{\Phi} = 3iA_1 h_1(kr) \tag{6.9.49}$$

$$\tilde{U}_r = 3ik\left[A_1 h_1'(kr) - 2\frac{B_1}{kr} h_1(Kr) \right] \tag{6.9.50}$$

$$\tilde{U}_\theta = 3i\frac{1}{r}\left\{ A_1 h_1(kr) - B_1\left[Krh_1'(Kr) + h_1(Kr) \right] \right\} \tag{6.9.51}$$

Substitution of these quantities into (6.9.48) yields, after some arrangement,

$$\tilde{F}_0 = 4\pi\rho_0\omega a^2 A_1 h_1(b) + 8\pi i\mu_0\left\{ A_1 b^2 h_1''(b) - 2B_1\left[\beta h_1'(\beta) - h_1(\beta) \right] \right\}$$

$$+ 8\pi i\mu_0\left\{ A_1\left[bh_1'(b) - h_1(b) \right] - B_1\left[\beta^2 h_1''(\beta) + \beta h_1'(\beta) - h_1(\beta) \right] \right\} \tag{6.9.52}$$

The quantity $h_1''(z)$ may be eliminated by means of

$$z^2 h_1''(z) = -2zh_1'(z) + (2 - z^2)h_1(z) \tag{6.9.53}$$

This yields

$$\tilde{F}_0 = 4\pi\rho_0\omega a^2 A_1 h_1(b) + 8\pi i\mu_0 A_1\left[(1 - b^2)h_1(b) - bh_1'(b) \right]$$

$$- 8\pi i\mu_0 B_1\left[\beta h_1'(\beta) - (1 + \beta^2)h_1(\beta) \right] \tag{6.9.54}$$

We now substitute the explicit expressions for h_1 and h_1' given by (6.19.20) and (6.9.21). This yields

$$\tilde{F}_0 = -\frac{A_1}{b^2} e^{ib}\left[4\pi\rho_0\omega a^2(b + i) - 8\pi i\mu_0(3b - b^3 + 3i - 2ib^2) \right]$$

$$- 8\pi i\mu_0\frac{B_1}{\beta^2} e^{i\beta}(3\beta + \beta^3 + 3i) \tag{6.9.55}$$

This equation, together with (6.9.28) and (6.9.29) for A_1 and B_1, yields the force of the sphere for arbitrary values of b and y. As stated earlier, we are presently interested only in the case $b \to 0$, for which (6.9.28) and (6.9.29) give

$$\frac{A_1}{b^2} e^{ib} \approx -\tfrac{1}{6} U_{p0} a \left[1 + \frac{3}{2y} + \frac{3i}{2}\left(\frac{1}{y} + \frac{1}{y^2} \right) \right] \tag{6.9.56}$$

$$\frac{B_1 e^{i\beta}}{\beta^2} \approx \frac{i}{4y^2} U_{p0} a \tag{6.9.57}$$

Using these approximations in (6.9.55), together with (6.9.26), yields for $b \to 0$

$$\tilde{F}_0 = \pi U_{p0} a \left(\tfrac{2}{3} \rho_0 a^2 i\omega - 4\mu_0 \right) \left[1 + \frac{3}{2y} + \frac{3i}{2}\left(\frac{1}{y} + \frac{1}{y^2} \right) \right]$$

$$+ 2\pi \mu_0 U_{p0} a \left[\frac{3}{y} - 2y + 3i\left(\frac{1}{y} + \frac{1}{y^2} + \frac{2}{3}y \right) \right] \tag{6.9.58}$$

Rearranging,

$$\tilde{F}_0 = \tfrac{2}{3}\pi \rho_0 a^3 i\omega U_{p0} \left[1 + \frac{3}{2y} + \frac{3i}{2}\left(\frac{1}{y} + \frac{1}{y^2} \right) \right] - 4\pi \mu_0 U_{p0} a (1 + y - iy)$$

$$\tag{6.9.59}$$

Separating real and imaginary quantities and using the fact that $\rho_0 \omega a^2 = 2\mu_0 y^2$, we obtain

$$\tilde{F}_0 = -6\pi \mu_0 U_{p0} a (1 + y) + \tfrac{4}{3}\pi \rho_0 a^3 \left(\frac{1}{2} + \frac{9}{4y} \right) i\omega U_{p0} \tag{6.9.60}$$

Finally, restoring the time factor and making use of the fact that $i\omega U_{p0} \exp(-i\omega t)$ $= -dU_p/dt$, we obtain

$$F = -6\pi \mu_0 U_p a (1 + y) - \tfrac{4}{3}\pi \rho_0 a^3 \left(\frac{1}{2} + \frac{9}{4y} \right) \frac{dU_p}{dt} \tag{6.9.61}$$

This is the desired result. It gives the force acting on a sphere oscillating, at a frequency ω, in a viscous fluid with velocity $U_p = U_{p0} \cos \omega t$. The result was first obtained by Stokes in 1845 and was later adapted to other situations. The form given by (6.9.61) may be compared with the inviscid result, which in this limit ($ka\rightarrow0$) is given by

$$F_{\text{inviscid}} \rightarrow \tfrac{2}{3}\pi\rho_0 a^3 \, dU_p/dt \qquad (6.9.62)$$

Hence, one of the effects of viscosity is to increase the added mass coefficient from $\tfrac{1}{2}$ to $\tfrac{1}{2}+9/4y$. This increase decreases with frequency and vanishes in the limit as $y\rightarrow\infty$. This is as it should be, because as ω increases, the thickness of the oscillating viscous layer decreases. Another effect of viscosity on the force is given by the first term in (6.9.61). This is proportional to the instantaneous velocity of the sphere and to the sphere radius. Its counterpart in inviscid-flow theory is proportional to $(ka)^3$ and is therefore negligible in the limit $ka\rightarrow0$.

Two special limits of (6.9.61) are of interest. One, $y\rightarrow0$, corresponding to very low frequencies, yields

$$F = -6\pi\mu_0 U_p a, \qquad y\ll1 \qquad (6.9.63)$$

This is the well-known Stokes law for drag. The second limit, $y\rightarrow\infty$, corresponds to the case when the viscous layer around the sphere is very thin. It yields the result

$$F = -6\pi U_p a^2\sqrt{\tfrac{1}{2}\rho_0\mu_0\omega} \; -\tfrac{2}{3}\pi\rho_0 a^3 \, dU_p/dt \qquad (6.9.64)$$

This is the correct limit, provided $b=ka$ can still be considered small. However, since b is given in terms of y by (6.9.35), it is clear that for sufficiently large y, the limit given by (6.9.64) will be incorrect. In that case, it becomes necessary to retain terms of order b or smaller in the formulation.

6.10 SPHERE IN A SOUND WAVE

The results of the preceding section may be used to study the effects of viscosity on the motion of a small sphere in a sound wave. The motion, which was treated in Section 4.3 for an inviscid fluid, is of some importance in a variety of areas, such as in acoustic coagulation of aerosols. In this and in other applications, the quantities of interest are the oscillatory amplitude of the sphere and the phase of their oscillations relative to those of the surrounding fluid. These quantities may be computed easily if the forces acting on the sphere are known. However, this is not the case, so that the forces have to be calculated also. This can be done by following a procedure similar to that used in the last section, but of a slightly more complicated nature because the sphere's velocity is not known. Fortunately, when the frequency of the wave is not exceedingly large, the results given by (6.9.61) may be adapted easily to the present problem because the flow field near the sphere is basically incompressible. Thus, the forces on the sphere (which are calculated in terms of the pressure fluctuations and of the velocity gradients evaluated on the sphere's surface) may be obtained from the incompressible results. Therefore, except for one difference that will be mentioned later, the force on a sphere oscillating in a fluid due to the passage of a sound wave is the same as the force that was calculated in the previous section, provided the instantaneous relative velocity between sphere and fluid is used instead of the sphere's velocity. In the previous example, the relative velocity was simply U_p because if the sphere had not been there, the fluid velocity would be zero. In the present case, the incident wave has a fluid velocity u along the direction of propagation. The sphere's velocity is also along the same direction and will be denoted by U_p so that the relative velocity becomes $U_p - u$.[9] Hence, the force on the sphere in an acoustic wave due to viscous effects is, from (6.9.60), given by

$$-6\pi\mu_0 a(1+y)(U_p-u) - \tfrac{2}{3}\mu_0\rho_0 a^3\left(1+\frac{9}{2y}\right)\frac{d(U_p-u)}{dt}$$

This is not, however, the total force acting on the sphere. An additional force must be included to take into account the acceleration of the frame of reference. This additional force is equivalent to a force per unit volume equal to $\rho_0(du/dt)$, as explained in Section 4.3. The need for this additional force may be seen by considering a situation when the sphere moves with the fluid, in which case the expression given above vanishes. Since the fluid is accelerating with velocity u, it is clear that a force $\rho_0 V_0(du/dt)$ must be acting on any volume V_0 of fluid. Thus, if $V_p = \tfrac{4}{3}\pi a^3$ represents the volume of the sphere, the total force on the sphere in a sound wave will be

$$F = \rho_0 V_p \frac{du}{dt} - 6\pi\mu_0 a(1+y)(U_p-u) - \tfrac{2}{3}\pi\rho_0 a^3\left(1+\frac{9}{2y}\right)\frac{d(U_p-u)}{dt} \qquad (6.10.1)$$

or, combining the first and third terms,

$$F = \tfrac{4}{3}\pi\rho_0 a^3\left(\frac{3}{2}+\frac{9}{4y}\right)\frac{du}{dt} - 6\pi\mu_0 a(1+y)(U_p-u)$$

$$-\tfrac{4}{3}\pi\rho_0 a^3\left(\frac{1}{2}+\frac{9}{4y}\right)\frac{dU_p}{dt} \qquad (6.10.2)$$

[9]Notice that the fluid velocity u as defined in the relative velocity above is the velocity that the fluid would have at the instantaneous location of the sphere if the sphere were absent.

This result applies only to a sphere in a monochromatic sound wave (more precisely, to a sphere in a fluid whose local velocity varies harmonically in time), but may be extended to other types of time dependence by means of Fourier-transform methods. Now we need it in its present form because the fluid velocity u varies harmonically in time:

$$u = \text{Re}\left(\tilde{u}_0 e^{-i\omega t}\right) \tag{6.10.3}$$

where the complex amplitude $\tilde{u}_0$ is, in general, a function of x [i.e., $\tilde{u}_0 = U_0 \exp(ikx)$]. However, if the amplitude of the oscillation of the sphere is small compared to the wavelength (a condition that must be satisfied if the oscillations are to be considered linear), we may set $\tilde{u}_0 = U_0$. Assuming that U_p is of the form

$$U_p = \text{Re}\left(\tilde{U}_{p0} e^{-i\omega t}\right) \tag{6.10.4}$$

we have, from (6.10.2),

$$-i\omega\tilde{U}_{p0} = -i\omega\left(\frac{\rho_0}{\rho_p}\right)\left(\frac{3}{2} + \frac{9}{4y}\right)U_0 - \frac{1+y}{\tau_d}\left(\tilde{U}_{p0} - U_0\right)$$

$$+ i\omega\left(\frac{\rho_0}{\rho_p}\right)\left(\frac{1}{2} + \frac{9}{4y}\right)\tilde{U}_{p0} \tag{6.10.5}$$

where

$$\tau_d = \frac{2}{9}\frac{a^2}{\nu_0(\rho_0/\rho_p)} \tag{6.10.6}$$

is a quantity with the dimensions of time. This quantity plays the role of a relaxation time for the motion of the sphere. Solving for the velocity amplitude ratio $\tilde{U}_{p0}/U_0$, we obtain

$$\frac{\tilde{U}_{p0}}{U_0} = \frac{1 + y - i\omega\tau_d\delta\left(\frac{3}{2} + 9/4y\right)}{1 + y - i\omega\tau_d\left[1 + \delta\left(\frac{1}{2} + 9/4y\right)\right]} \tag{6.10.7}$$

where $\delta = (\rho_0/\rho_p)$. Making use of the identity

$$\omega\tau_d = \frac{4}{9}\frac{y^2}{\delta} \tag{6.10.8}$$

(6.10.7) may be written as

$$\frac{\tilde{U}_{p0}}{U_0} = \frac{1+y-i\left(\frac{2}{3}y^2+y\right)}{1+y-i\left[\omega\tau_d\left(1+\frac{1}{2}\delta\right)+y\right]} \tag{6.10.9}$$

Before considering this equation in full, we obtain from it a special, limiting form that is of interest. This limiting form refers to small, high-density spheres in a low-density fluid; for example, a small solid sphere in a gas. For this case, $\delta \ll 1$, so that

$$\frac{\tilde{U}_{p0}}{U_0} = \frac{1+y-i\left(\frac{2}{3}y^2+y\right)}{1+y-i(\omega\tau_d+y)} \tag{6.10.10}$$

Further, if the sphere radius is very small relative to the viscous-penetration thickness $\delta_\nu = \sqrt{2\nu_0/\omega}$, then $y \ll 1$ and

$$\frac{\tilde{U}_{p0}}{U_0} = \frac{1-iy}{1-i(\omega\tau_d+y)} \tag{6.10.11}$$

Particularly simple forms may be obtained from (6.10.11) if either $\omega\tau_d \ll y$ or $y \ll \omega\tau_d$. When $\omega\tau_d \ll y$, we simply have $\tilde{U}_{p0} = U_0$ for all y such that $y \ll 1$. On the other hand, when $\omega\tau_d \gg y$,

$$\frac{\tilde{U}_{p0}}{U_0} = \frac{1}{1-i\omega\tau_d} \tag{6.10.12}$$

This remarkably simple result may be derived easily by assuming that the force on the sphere is given by Stokes's law[10]:

$$F = 6\pi\mu_0 a\left(U_p - u\right)$$

The derivation of (6.10.12) tells us under what conditions this law can be applied to a sphere in an acoustic wave. First, since compressibility effects were neglected in obtaining (6.9.60), we require that

$$ka \ll 1 \tag{6.10.13}$$

---

[10] For this force, the equation of motion of the sphere may be written as $dU_p/dt = -(U_p-u)/\tau_d$. This shows that τ_d is the relaxation time for the motion, for if a sphere having a constant velocity $U(0)$ is placed at $t=0$ in a stagnant fluid, the velocity at a time $t=\tau_d$ will have relaxed to a value equal to a fraction $1/e$ of $U(0)$.

Second, the quantities δ and y were assumed small. Therefore,

$$\delta = \rho_0 / \rho_p \ll 1 \qquad (6.10.14)$$

$$y = \sqrt{\omega a^2 / 2\nu_0} \ll 1 \qquad (6.10.15)$$

the relative magnitude of these parameters being given by

$$y \ll \omega \tau_d \qquad (6.10.16)$$

Because $y^2 \ll y$ when $y \ll 1$, and $y^2 = \frac{9}{4} \omega \tau_d \delta$, (6.10.16) may also be written as

$$y \ll \omega \tau_d \ll y / \delta$$

A numerical example will show that these conditions are not as restrictive as they appear to be. Thus, if a five-micron-radius sphere with a density equal to that of water is exposed to a 1000-Hz sound wave in air at standard temperature and pressure, then $ka = 9.1 \times 10^{-5}$, $\delta = 1.2 \times 10^{-3}$, $y = 7.15 \times 10^{-2}$, $\delta_\nu / y = 1.66 \times 10^{-2}$, and $\omega \tau_d = 1.95$.

Returning to (6.10.9), we write it in the form

$$U_p = |\tilde{U}_{p0}| \cos(\omega t - \eta) \qquad (6.10.17)$$

where, from (6.10.12),

$$\frac{|\tilde{U}_{p0}|}{U_0} = \frac{1}{\sqrt{1 + \omega^2 \tau_d^2}} \qquad (6.10.18)$$

and where

$$\tan \eta = \omega \tau_d \qquad (6.10.19)$$

Also of interest is the displacement amplitude of the sphere, X_p. Thus, if $X_p = \text{Re}[\tilde{X}_p \exp(-i\omega t)]$, then

$$\frac{|\tilde{X}_p|}{U_0 / \omega} = \frac{1}{\sqrt{1 + \omega^2 \tau_d^2}} \qquad (6.10.20)$$

where U_0/ω represents the displacement amplitude of the gas. Equation (6.10.20) shows that when $\omega\tau_d \to 0$, the sphere's amplitude is equal to that of the gas, but is negligible when $\omega\tau_d \gg 1$. Thus, for a given frequency (and therefore a fixed gas amplitude of oscillation), the sphere's amplitude decreases as the sphere's diameter increases. If a sound wave is propagated in a gas containing several spheres having different sizes, the smaller ones will oscillate with larger amplitudes. The differential motion between the spheres may therefore produce collisions between different-sized spheres.

Evidence for the validity of (6.10.20) is given in Figure 6.10.1, where displacement measurements made with small liquid droplets in nitrogen are shown versus droplet radius. The experiments were performed at a single frequency (4850 Hz) using different sphere radii. The maximum values of our parameters ka, δ, and y were 3.6×10^{-4}, 1×10^{-3}, and 0.127, respectively. Therefore, the requirements stipulated by (6.10.13) to (6.10.15) were approximately met by the experimental conditions.

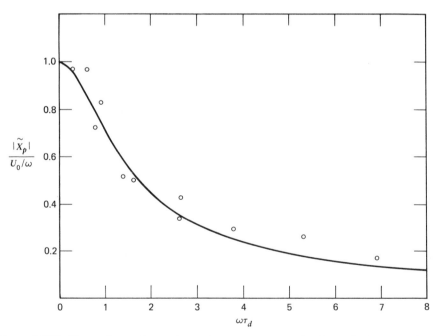

Figure 6.10.1 Comparison of experimental measurements of sphere displacement in a sound wave with the theoretical prediction given by Equation (6.10.20). O: Gucker and Doyle (1956).

For other values of y and δ, the more complete results given by (6.10.9) must be used. Writing that equation at $U_p = |\tilde{U}_{p0}|\exp(i\eta)$, we obtain

$$\frac{|\tilde{U}_{p0}|}{U_0} = 3\delta\sqrt{\frac{4y^4 + 12y^3 + 18y^2 + 18y + 9}{4(2+\delta)^2 y^4 + 36\delta y^3(2+\delta) + 81\delta^2(2y^2 + 2y + 1)}}$$

$$(6.10.21)$$

$$\tan\eta = \frac{12(y+1)y^2(1-\delta)}{4y^4(2+\delta) + 12y^3(1+2\delta) + 27\delta(2y^2 + 2y + 1)} \qquad (6.10.22)$$

Later we will compare these results to the simple ones given earlier, but first we note that when the sphere is of the same density as the fluid, we obtain $U_{p0} = U_0$

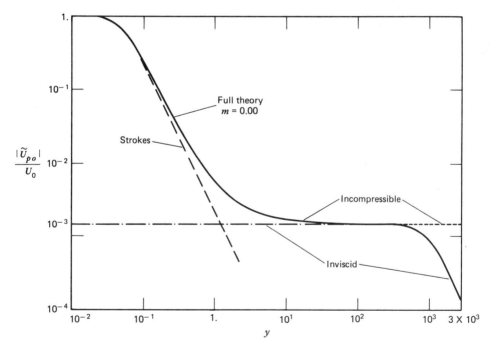

Figure 6.10.2 Velocity-magnitude ratio. (———) Full results given by Equation (6.10.24) for $m = 0.001$. (— · · —) Simple results obtained by using Stokes's law. (-------) Incompressible results. (— · —) Inviscid results.

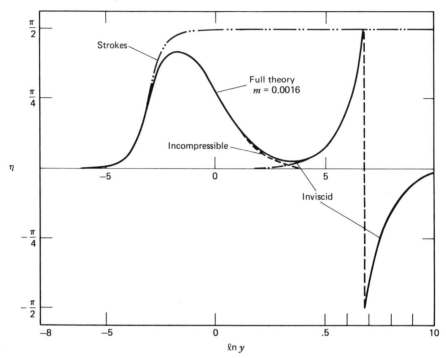

Figure 6.10.3 Phase angle for oscillations of a sphere in a sound wave. (———) Full results given by Equation (6.10.15). (— · · —) Simple results. (------) Incompressible results. (— · —) Inviscid results.

and $\eta = 0$. That is, there is no relative motion between sphere and fluid, a result that was expected. Further, when $y \gg 1$ we obtain $\eta = 0$ and

$$\frac{|\tilde{U}_{p0}|}{U_0} = \frac{3\delta}{2+\delta}. \tag{6.10.23}$$

These are the inviscid results (see Section 4.3), applicable in the limit $b \to 0$.

To see the effects of finite values of y on the sphere's velocity, we plot in Figures 6.10.2 and 6.10.3 the quantities $|\tilde{U}_{p0}|/U_0$ and η versus y for the case $\delta \ll 1$. Three types of results are shown in these figures: the simple results given by (6.10.18) and (6.10.19), the incompressible results given by (6.10.21) and (6.10.22), and the results of more complete calculations in which the effects of compressibility (i.e., finite values of ka) were retained. In the notation of this section, the more complete results may be expressed as[11]

$$\frac{|\tilde{U}_{p0}|}{U_0} = 3\delta\sqrt{\frac{F^2 + G^2}{H^2 + I^2}} \tag{6.10.24}$$

$$\eta = \tan^{-1}\frac{GH - IF}{FH + IG} \tag{6.10.25}$$

where

$$F = 2y^2 + 3y + (b/y)^2(1+y) \tag{6.10.26}$$

$$G = 3(1+y) - 2b^2 - b^2/y \tag{6.10.27}$$

$$H = 2y^2(b^2 - 2 - \delta) + y\left[b^2(1+2\delta) - 9\delta(b+1)\right]$$

$$+ 9\delta b\left(\tfrac{2}{9}b^2 - 1\right) + 3\delta(b^2/y)(b-1) - 3\delta(b/y)^2 \tag{6.10.28}$$

$$I = 2y^2b(2+\delta) + y\left[9\delta(b-1) + (1+2\delta)b^2\right] + (1+4\delta)b^2 - 9\delta$$

$$+ 3\delta\frac{b^2}{y}(b+1) + 3\delta b\left(\frac{b}{y}\right)^2 \tag{6.10.29}$$

For a given value of δ, these results depend on b and y. However, because of their definition, these two quantities are related by means of

$$y = bc_0/\sqrt{2\omega\nu_0} \tag{6.10.30}$$

This relationship clearly shows that for a given frequency and radius, the inviscid results ($\nu_0 = 0$) should be obtained by holding b fixed and letting $y \to \infty$. It also shows that the incompressible results (which correspond to $c_0 \to \infty$) may be obtained by holding y fixed and letting $b \to 0$. It is therefore clear that the more complete results can be plotted versus y (or versus b) with the quantity

$$m = \sqrt{2\omega\nu_0}/c_0 \tag{6.10.31}$$

---

[11]For details of the computations, see Temkin and Leung (1976).

appearing as a parameter. This quantity is the ratio of the velocity of propagation of transverse viscous waves [see the discussion following Equation (6.5.9)] to the ideal speed of sound. It may also be expressed in terms of the translational relaxation time $\tau_\nu = 4\nu_0/3c_0^2$. As we know from Section 6.2, the quantity $\omega\tau_\nu$ should be small if the continuum formulation is to be applicable. Therefore, the quantity m also should be small.

Figures 6.10.2 and 6.10.3 show plots of $|\tilde{U}_{p0}/U_0|$ and of η versus y for two values of m that apply to the important case of liquid or solid spheres in a gas. Also shown are the inviscid results given by (4.3.56) and (4.3.57) and the simple results obtained by using Stokes's law. The curves with $m=0$ correspond to the incompressible results given by (6.10.21) and (6.10.22). It is seen that for m small but finite, the full theory predicts results that join smoothly with the inviscid results when y is large, and with the incompressible results otherwise. The figures indicate clearly the regions of validity of the inviscid and incompressible results. For example, it is evident that the inviscid results are incorrect, except when y is large. On the other hand, the incompressible results provide a fairly good approximation for values of y ranging from very small to moderately large. Such a wide range is often more than is required in practical situations. Finally, the simple results given by Stokes's law have limited validity, but provide a useful approximation to the more complete incompressible results when both y and δ are small.

PROBLEMS

6.10.1 From (6.10.9) it follows that when $y \ll \omega\tau_d \ll 1$,

$$\frac{\tilde{U}_{p0}}{U_0} = \frac{1}{1 - i\omega\tau_d\left(1 + \frac{1}{2}\delta\right)}$$

For the case $\delta \gg 1$, compare the velocity magnitude and phase predicted by this equation with those predicted by (6.10.21) and (6.10.25).

6.10.2 One possible application of the results of this section is in the determination of the size of very small particles of known density. Using the equation given in Problem 6.10.1, show that if the real and imaginary parts of U_p are known (from measurements at a frequency ω, say), it is possible to determine a. What type of experimental setup would you use to measure U_p?

Another possible application of the results given in this section is in the measurement of fluid velocities in a sound wave. How can this be accomplished?

6.10.3 The velocity U_p of a sphere translating unsteadily in a viscous, incompressible fluid is described by the following equation:

$$m_p \frac{dU_p}{dt} = \delta m_p \frac{du}{dt} + \tfrac{1}{2}\delta m_p \frac{d(u-U_p)}{dt} + 6\pi\mu_0 a(u-U_p)$$

$$+ 6a^2(\pi\mu_0\rho_0)^{1/2} \int_{-\infty}^{t} \frac{d(u-U_p)}{dz} \frac{dz}{\sqrt{t-z}}$$

This is known as the B–B–O Equation (Basset–Boussinesq–Oseen). The last term is called the "history" term, as it depends on all the past history of the motion. For monochromatic time dependence, this equation may be solved easily. Show that if $u = U_0 \, \mathrm{Re}[\exp(-i\omega t)]$ and $U_p = \mathrm{Re}[\tilde{U}_{p0}\exp(-i\omega t)]$,

$$\frac{\tilde{U}_{p0}}{U_0} = \frac{1 + \tfrac{3}{2}(\delta\omega\tau_d)^{1/2} - i\left[\tfrac{3}{2}\delta\omega\tau_d + \tfrac{3}{2}(\omega\tau_d\delta)^{1/2}\right]}{1 + \tfrac{3}{2}(\delta\omega\tau_d)^{1/2} - i\left[\left(1 + \tfrac{1}{2}\delta\right)\omega\tau_d + \tfrac{3}{2}(\omega\tau_d\delta)^{1/2}\right]}$$

Show that this corresponds to (6.10.7).

6.10.4 Using the full results given by (6.10.24) to (6.10.29), plot $|\tilde{U}_{p0}|/U_0$ and η versus y for the case $\delta = 1000$ and $m = 0.001$. Show that the full results join the inviscid and the incompressible results.

6.11 ATTENUATION AND DISPERSION IN A DILUTE SUSPENSION

In the previous section, we studied the motion of a rigid sphere that is induced by a monochromatic sound wave. The main result of that study was an equation for the velocity of the sphere in the wave. It is clear, however, that in addition to the motion of the sphere, there are other effects resulting from the sphere-wave interaction. For example, as we saw in Chapter 5, the sphere's translational oscillation results in sound emission. In addition, there is scattering of the incident wave due to partial blockage of the incident beam. These effects remove energy from the incoming wave at a rate that is of interest. In the case when there is only one small sphere in the path of the wave, the energy removed is clearly insignificant. However, if there are many such spheres, their total effect on the wave is, within some limits, additive and may therefore be noticeable. For example, an atmospheric cloud, having as many as a few hundred droplets per cubic centimeter and having overall dimensions of the order of a few thousand meters, can significantly attenuate sound waves. This attenuation is, in general, frequency dependent; that is, some frequencies are more easily attenuated than others, and it is desirable to determine conditions for which this type of attenuation is most significant. To do this, we study sound attenuation due to a suspension of spherical particles in a carrier fluid.

The problem is fairly complicated because the results for the one-sphere problem generally cannot be used in a situation where there are many spheres. Furthermore, in addition to the losses induced by viscosity, there are losses due to heat transfer from the sound wave to the spheres and due to evaporation and condensation if the spheres are liquid. These types of losses are clearly coupled, which increases the complexity of the problem. In this section, we give a simple derivation of the attenuation coefficient for propagation through a dilute suspension of rigid, spherical particles in a gas. We also give, without derivation, the corresponding result for the phase velocity in such a suspension. For a more complete derivation of these results, the article by Temkin and Dobbins (1966a) could be consulted.

Attenuation

In a dilute suspension, the volume occupied by the spheres is small compared to the volume occupied by the carrier fluid. In such a suspension, we may assume that the total energy removal is the sum of the energies removed by each sphere, because then the distances between spheres is necessarily large compared to their radius. For the same reason, we anticipate that the attenuation per wavelength will be relatively small. This will enable us to use the approximate procedure introduced in Section 6.4 to obtain attenuation coefficients. Thus, if $\dot{e}$ is the energy-removal rate due to one sphere, then the energy dissipation rate per unit volume is

$$\dot{E} = n\dot{e} \qquad (6.11.1)$$

where n is the number of spheres per unit volume. To compute $\dot{e}$, we make use of the fact that for very small spheres, the energy-loss rate due to scattering is very small (see Section 4.3) so that it may be neglected when compared to viscous losses. These, in turn, may be computed easily. Thus, if the velocity of the sphere relative to that of the fluid is $U_p - u$ and the viscous force on the sphere is F_p, then the power that must be spent instantaneously to maintain the motion is $F_p(U_p - u)$. Hence, the average energy-dissipation rate per unit volume is

$$\langle \dot{E} \rangle = n \langle F_p(U_p - u) \rangle \qquad (6.11.2)$$

Now, if we are considering a plane wave, whose average energy per unit volume is simply $E_0 = \frac{1}{2}\rho_0 U_0^2$, where U_0 is the fluid's velocity amplitude, then (6.4.4) gives for the *amplitude* attenuation coefficient

$$\alpha = n \frac{\langle F_p(U_p - u) \rangle}{\rho_0 c_0 U_0^2}$$

$$= n \frac{\operatorname{Re}\left[\tilde{F}_p(\tilde{U}_p - \tilde{u})^* \right]}{2\rho_0 c_0 U_0^2} \tag{6.11.3}$$

The amplitudes of the complex force of the sphere's velocity are given by (6.10.2) and (6.10.7) for arbitrary values of y and δ. However, to simplify the calculations, we will consider only the case when both y and δ are very small. This corresponds to heavy spheres in a gas and to relatively low frequencies and, as discussed in the previous section, applies to many important practical situations. Under those conditions, the force on the sphere and the relative sphere velocity are

$$\tilde{F}_p = 6\pi\mu_0 a \left(\tilde{U}_p - \tilde{u} \right) \tag{6.11.4}$$

$$\tilde{U}_p - \tilde{u} = U_0 \frac{i\omega\tau_d}{1 - i\omega\tau_d} \tag{6.11.5}$$

Substitution of these two equations into (6.11.3) yields

$$\alpha = \frac{3\pi\nu_0 na}{c_0} \frac{\omega^2\tau_d^2}{1 + \omega^2\tau_d^2} \tag{6.11.6}$$

This is the desired result, but it is convenient to recast it in terms of other quantities such as $\bar{\alpha} = \alpha c_0 / \omega$, the attenuation per wavelength, and of the quantity

$$C_m = \tfrac{4}{3}\pi a^3 n\rho_p / \rho_0 \tag{6.11.7}$$

where ρ_p is the density of the sphere. The quantity C_m thus defined is the mass concentration of the suspension. It is the ratio of the mass of the spheres per unit volume of fluid to the mass of fluid in a unit volume of fluid. Sometimes a mass loading ratio is used instead of C_m. The mass loading ratio is defined as the mass of the spheres in a unit volume of the suspension. For a dilute suspension, such as that under consideration, the two quantities are equivalent.

In terms of $\bar{\alpha}$ and of C_m, one has for $\bar{\alpha}$, on using (6.10.6),

$$\bar{\alpha} = \tfrac{1}{2} C_m \frac{\omega \tau_d}{1 + \omega^2 \tau_d^2} \tag{6.11.8}$$

This result is reminiscent of attenuation due to internal relaxation (see Section 6.2). This similarity is not surprising because attenuation by suspended particles is, in fact, caused by the inability of the particles to follow the changes in the fluid that are induced by the sound wave; that is, it is caused by relaxation.

Equation (6.11.7) shows that the attenuation per wavelength has a maximum at $\omega \tau_d = 1$. Thus, sound waves having frequencies of the order of τ_d^{-1} are damped the most by the suspension. On the other hand, even for those frequencies, the attenuation per wavelength is relatively small. For example, in a cumulus cloud having a liquid content of, say, $1 g/m^3$, the value of C_m is of the order of 10^{-3}. Therefore, a sound wave having the optimum frequency must travel approximately 1000 wavelengths before its amplitude decays to a fraction $e^{-\pi} = 0.043$ of its original value.

The frequency for maximum attenuation per wavelength is given by the condition that $\omega \tau_d = 1$. It is therefore given by

$$f_0 = \frac{1}{4\pi} \frac{9 \nu_0 \delta}{a^2} \tag{6.11.9}$$

A suspension of liquid droplets in air, having droplets with a mean radius equal to 5×10^{-4} cm, would have an optimum frequency of about 525 Hz. Waves having lower or higher frequencies would, in one wavelength, be attenuated less than those having the optimum frequency. Of course, at the higher frequency range, the absolute attenuation is higher than at the optimum value.

It should be added that when other dissipation mechanisms are taken into account, the general features of the attenuation curves (e.g., the location of the most optimum frequency) are necessarily changed. For instance, if we had taken heat conductivity into account, additional attenuation would take place owing to heat transfer from the fluid to the sphere. In the limit when this transfer is due only to conduction, the contribution to $\bar{\alpha}$ is

$$\frac{1}{2}(\gamma - 1) \frac{C}{c_{p0}} \frac{\omega \tau_t}{1 + \omega^2 \tau_t^2} C_m$$

where C is the specific heat of the sphere and τ_t is given by

$$\tau_t = \frac{1}{3} \frac{a^2}{\kappa \delta} \frac{C}{c_{p0}} \tag{6.11.10}$$

where κ is the terminal diffusivity of the fluid. The quantity τ_t plays the role of a relaxation time for heat transfer to the sphere due to conduction.[12] Thus, the attenuation coefficient per wavelength due to viscous and heat-conduction effects is given by

$$\frac{\bar{\alpha}}{C_m} = \frac{1}{2}\frac{\omega\tau_d}{1+\omega^2\tau_d^2} + \frac{1}{2}(\gamma-1)\frac{C}{c_{p0}}\frac{\omega\tau_t}{1+\omega^2\tau_t^2} \qquad (6.11.11)$$

In view of the definitions of τ_t and τ_d, we have

$$\tau_t = \frac{3}{2}\frac{C}{c_{p0}}\mathrm{Pr}\,\tau_d \qquad (6.11.12)$$

where

$$\mathrm{Pr} = \mu c_{p0}/k_0 \qquad (6.11.13)$$

Equation (6.11.12) enables us to plot $\bar{\alpha}/C_m$ against a single relaxation time, and Figure 6.11.1 shows a plot for the case of water droplets in air. The maximum of the heat-conduction contribution, which occurs for $\omega\tau_t = 1$, appears to the left of $\omega\tau_d = 1$ simply because in this case $\tau_t < \tau_d$. Also, the maximum of the heat-conduction contribution is larger than that due to viscosity because the fraction $(\gamma-1)(C/c_{p0})$ is greater than unity. Other combinations of properties may change these relative positions and magnitudes.

Figure 6.11.2 shows the result of experimental measurements of sound attenuation in a droplet cloud made of oleic acid droplets in nitrogen ($C/c_{p0} = 0.55$). The bar appearing over τ_d is used to indicate that for each test, this quantity is based on a mean droplet radius. Also shown in the figure is the analytical result obtained by integrating $\bar{\alpha}(a)$, as given by (6.11.11), over a distribution of droplet sizes. (For details, see Temkin and Dobbins, 1966b.)

Dispersion

A sound wave traveling in a fluid medium containing suspended particles will have a phase velocity that is generally different from that in the pure fluid. The difference occurs mainly because the mechanical and thermal properties of the carrier fluid are changed by the particles. Furthermore, these changes are frequency dependent so that the phase velocity in a suspension depends on frequency. That is, propagation is dispersive.

---

[12]Under those conditions, the temperature of the sphere obeys $m_p C\, dT_p/dt = 4\pi a k_0(T_p - T)$, where T is the temperature of the fluid around the sphere and k_0 is the fluid's thermal conductivity. This is a relaxation equation, with $m_p C/4\pi a k_0 = \tau_t$ appearing as a relaxation time.

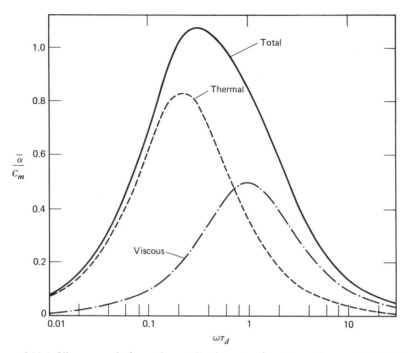

Figure 6.11.1 Viscous and thermal contributions to the attenuation coefficient for a dilute suspension.

Before we write the equations that predict the changes of phase velocity with frequency in a dilute suspension of rigid particles in a gas, let us consider the low-frequency limit. Here, the velocity and temperature fluctuations of the particles are able to follow the respective fluctuations in the gas so that a state of thermodynamic equilibrium exists at all times. It is therefore possible to obtain, then, a limiting form for the phase velocity by means of simple thermodynamic arguments. Thus, assume that the gas is a perfect gas having a specific heat ratio γ. Further, since the suspension is dilute, we may assume that the mean pressure in the gas is not affected by the particles. Then we can write, for the equilibrium speed of sound in the suspension, the perfect-gas result

$$\left(c_s^2\right)_{eq} = \Gamma_s p_0 / \rho_s \tag{6.11.14}$$

where Γ_s and ρ_s are the specific-heat ratio and density of the suspension, respectively. Consider, first, the suspension's density. Since the suspension is

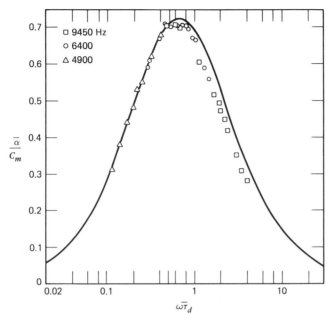

Figure 6.11.2 Comparison of experimental and theoretical results for the attenuation coefficient in a dilute suspension of nonevaporating droplets in a gas.

dilute, a unit volume of suspension is equal to a unit volume of the gas alone, so that

$$\rho_s = \rho_0 + \rho_0 C_m \qquad (6.11.15)$$

To obtain Γ_s, we consider the specific heats at constant pressure c_{ps} and at constant volume c_{vs} for the suspension. Again, because the suspension is dilute, we have

$$\rho_s c_{ps} = \rho_0 c_{p0} + \rho_0 C_m C \qquad (6.11.16)$$

$$\rho_s c_{vs} = \rho_0 c_{v0} + \rho_0 C_m C \qquad (6.11.17)$$

where we have assumed that for particles there is no difference between their specific heat at constant pressure and at constant volume. These equations give

$$\Gamma_s = \gamma \frac{1 + C_m C / c_{p0}}{1 + \gamma C_m C / c_{p0}} \qquad (6.11.18)$$

Substituting (6.11.15) and (6.11.18) into (6.11.14) yields

$$\left(\frac{c_s}{c_0}\right)^2_{eq} = \frac{1 + C_m C / c_{p0}}{(1 + \gamma C_m C / c_{p0})(1 + C_m)} \tag{6.11.19}$$

where $c_0 = \sqrt{\gamma p_0 / \rho_0}$ is the equilibrium speed in the pure gas. Since $\gamma \geqslant 1$, this result shows that for all values of C_m the equilibrium speed of sound in the suspension is smaller than c_0. Consider (6.11.19) for the common case when $C_m \ll 1$:

$$\left(\frac{c_0}{c_s}\right)^2_{eq} = 1 + C_m + \frac{(\gamma - 1)C_m C}{c_{p0}}$$

This shows that there are two contributions to the change of speed: the first, equal to C_m, is due to the change of density; the second, equal to $(\gamma - 1)C_m C / c_{p0}$, is clearly due to the change of thermal properties. Now, as the frequency increases, the equilibrium between particles and gas is destroyed because the particles cannot follow completely the fluctuations of temperature and velocity in the gas. Eventually, a limit is reached where the particles' temperature and

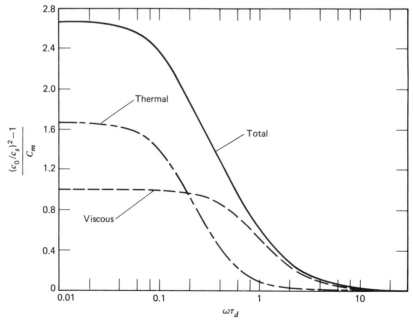

Figure 6.11.3 Dispersion of sound in a dilute suspension of heavy spheres in air.

velocity do not change. In this "frozen" state, the particles do not alter the properties of the gas, and we should then expect that $c_s = c_0$.

For the case when the particles' drag obeys Stokes's law, and the heat transfer is due to conduction, the changes for speed of sound with frequency are given by (for details, see Temkin and Dobbins, 1966a)

$$\left(\frac{c_0}{c_s}\right)^2 = 1 + C_m \frac{1}{1 + \omega^2 \tau_d^2} + C_m \frac{(\gamma - 1)C/c_{p0}}{1 + \omega^2 \tau_t^2} \qquad (6.11.20)$$

As anticipated, this shows that as $\omega\tau_d \to \infty$, $c_s \to c_0$. Also, as $\omega\tau_d$ (and also $\omega\tau_t$) becomes very small, we recover the equilibrium limit given by (6.11.19). Figure 6.11.3 shows the variations of $[(c_0/c_s)^2 - 1]C_m^{-1}$ with $\omega\tau_d$.

Figure 6.11.4 shows measurements of sound dispersion made under conditions similar to those for the attenuation measurements described earlier in this section. Also shown in the figure is the analytical prediction obtained by integrating (6.11.10) over a distribution of droplet sizes.

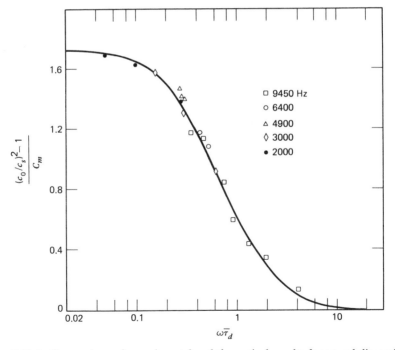

Figure 6.11.4 Comparison of experimental and theoretical results for sound dispersion in a dilute suspension of nonevaporating droplets in a gas.

Measurements of Particle Size

As an application of these results, we briefly indicate how measurements of attenuation and dispersion in a suspension may be used to obtain the particle size and concentration in the suspension. For simplicity, we will consider only monodisperse suspensions, although the method can also be used for polydisperse suspensions (for details, see Dobbins and Temkin, 1967).

As the basis of the method we use (6.11.12) and (6.11.20). These give the attenuation and dispersion coefficients in the suspension. Figure 6.11.5 shows the ratio of a dispersion coefficient $\bar{\beta} = [(c_0/c_s)^2 - 1]$ to $\bar{\alpha}$, as given by those equations for the case $\gamma = 1.40$, $C/c_{p0} = 4.1$, $\mathrm{Pr} = 0.71$, as a function of $\omega\tau_d$. Since $\sqrt{\omega\tau_d}$ is proportional to the particle radius, the figure shows that measurements of $\bar{\alpha}$ and $\bar{\beta}$ at the same frequency may, together with the figure, be used to determine the radius of the particle. Further, once this radius is known, the mass fraction C_m can be determined from either (6.11.11) or (6.11.20). Another procedure that may be used to determine C_m is to measure $\bar{\beta}$ in the limit $\omega\tau_d \to 0$.

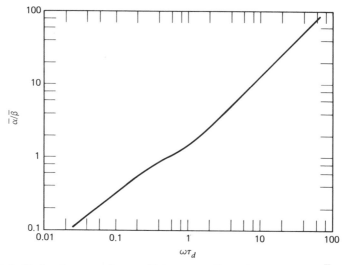

Figure 6.11.5 Ratio of attenuation coefficient $\bar{\alpha}$ to dispersion coefficient $\bar{\beta}$ in a suspension.

PROBLEMS

6.11.1 Show that the temperature change T_p' of a small sphere imbedded in a medium whose temperature varies according to $T' = \varepsilon \cos \omega t$ is

$$\tilde{T}_p' = \frac{\tilde{T}'}{1 - i\omega\tau_t}$$

6.11.2 Consider the quantity

$$C_m \langle U_p / u \rangle$$

where $U_p = \text{Re}(\tilde{U}_p)$ and $u = U_0 \cos \omega t$. Using the simple results for $\tilde{U}_p$ given by (6.10.12), show that the above quantity corresponds to the second term in (6.11.20) for $(c_0/c_s)^2$.

6.11.3 Using the results of Problem 6.11.1, show that the quantity

$$(\gamma - 1) C_m (C/c_{p0}) \langle T_p' / T' \rangle$$

corresponds to the last term in (6.11.20).

6.11.4 Problems 6.11.2 and 6.11.3 show that

$$\left(\frac{c_0}{c_s} \right)^2 = 1 + C_m \left\langle \frac{U_p}{u} \right\rangle + (\gamma - 1) C_m \frac{C}{c_{p0}} \left\langle \frac{T_p'}{T'} \right\rangle$$

Can you provide physical backing for this equation?

6.11.5 Consider the attenuation coefficient $\bar{\alpha}$ in a suspension, at two different frequencies, ω_1 and $\omega_2 = 2\omega_1$. For simplicity put $\tau_t = 0$. Plot the ratio of $\bar{\alpha}_1/\bar{\alpha}_2$ versus $\omega_1 \tau_d$. Show that, within certain limits, the plot may be used together with measurements of $\bar{\alpha}_1$ and $\bar{\alpha}_2$ to obtain particle size and concentration in a suspension.

6.12 BOUNDARY VISCOUS AND THERMAL EFFECTS

When both viscosity and heat conductivity are retained, the linearized equations of continuity, motion, energy, and state are

$$\frac{\partial \rho'}{\partial t} + \rho_0 \nabla \cdot \mathbf{u} = 0 \tag{6.12.1}$$

$$\rho_0 \frac{\partial \mathbf{u}}{\partial t} + \nabla p' = \tfrac{4}{3} \mu_0' \nabla (\nabla \cdot \mathbf{u}) - \mu_0 \nabla \times (\nabla \times \mathbf{u}) \tag{6.12.2}$$

$$\rho_0 c_{p0} \frac{\partial T'}{\partial t} - \beta_0 T_0 \frac{\partial p'}{\partial t} = k_0 \nabla^2 T' \tag{6.12.3}$$

$$p' = c_{T0}^2 \rho' + \rho_0 (c_{p0} - c_{v0}) T' / \beta_0 T_0 \tag{6.12.4}$$

In the small-dissipation case, the contributions of viscous and thermal effects to the total dissipation are separable, and can be computed in terms of the velocity and temperature gradients. It is therefore convenient to obtain, from the above system of equations, single equations for the temperature and the velocity alone. Using (6.12.1) and (6.12.4), we eliminate p' from (6.12.3) and obtain

$$\frac{\partial T'}{\partial t} + \frac{(\gamma - 1)}{\beta_0} \nabla \cdot \mathbf{u} = \gamma \kappa \nabla^2 T' \tag{6.12.5}$$

where we have used the identity $\beta_0 c_{T0}^2 = (c_{p0} - c_{v0})/\beta_0 T_0$. Similarly, we eliminate p' from (6.12.2) by taking a time derivative of that equation and using (6.12.1) and (6.12.4). The final result is

$$\frac{\partial^2 \mathbf{u}}{\partial t^2} - c_{T0}^2 \nabla(\nabla \cdot \mathbf{u}) + \beta_0 c_{T0}^2 \nabla \frac{\partial T'}{\partial t} = \tfrac{4}{3} \nu_0' \nabla \left(\nabla \cdot \frac{\partial \mathbf{u}}{\partial t} \right) - \nu_0 \nabla \times \left(\nabla \times \frac{\partial \mathbf{u}}{\partial t} \right) \tag{6.12.6}$$

For the monochromatic case, these equations become

$$\nabla^2 T' + \frac{i\omega}{\gamma \kappa_0} T' = \frac{(\gamma - 1)}{\gamma \beta_0 \kappa_0} \nabla \cdot \mathbf{u} \tag{6.12.7}$$

$$\left(c_{T0}^2 - \tfrac{4}{3} \nu_0' i\omega \right) \nabla(\nabla \cdot \mathbf{u}) + \omega^2 \mathbf{u} + i\omega \nu_0 \nabla \times (\nabla \times \mathbf{u}) = -i\beta_0 c_{T0}^2 \omega \nabla T' \tag{6.12.8}$$

Proceeding as in the viscous case (see Section 6.7), we split $\mathbf{u}$ into a longitudinal and a transverse part by writing

$$\mathbf{u} = \nabla \tilde{\phi} + \nabla \times \tilde{\mathbf{B}} \tag{6.12.9}$$

where, as before, $\nabla \cdot \mathbf{B} = 0$. Substitution of this into (6.12.7) gives

$$\nabla^2 \tilde{T}' + \frac{i\omega}{\gamma \kappa_0} \tilde{T}' = \frac{\gamma - 1}{\gamma \beta_0 \kappa_0} \nabla^2 \tilde{\phi} \tag{6.12.10}$$

showing clearly that T' does not depend on $\mathbf{B}$. Similarly, substitution in (6.12.8) yields, with $\nabla \cdot \mathbf{B} = 0$,

$$\nabla \left[\left(c_{T0}^2 - \tfrac{4}{3} \nu_0' i\omega \right) \nabla^2 \tilde{\phi} + \omega^2 \tilde{\phi} + i\beta_0 c_{T0}^2 \omega \tilde{T}' \right] = -\nabla \times \left[\omega^2 \tilde{\mathbf{B}} - i\omega \nu_0 \nabla^2 \tilde{\mathbf{B}} \right] \tag{6.12.11}$$

As in Section 6.7, the quantities in square brackets can be set equal to zero, so that

$$\nabla^2\tilde{\mathbf{B}}+(i\omega/\nu_0)\tilde{\mathbf{B}}=0 \qquad (6.12.12)$$

$$(1-2i\gamma\bar{\alpha}_\nu)\nabla^2\tilde{\phi}+\gamma(\omega/c_0)^2\tilde{\phi}+i\beta_0\omega\tilde{T}'=0 \qquad (6.12.13)$$

The first equation is the same as (6.7.15), and therefore describes transverse waves of viscous origin. The second equation involves both $\tilde{\phi}$ and $\tilde{T}'$, but $\tilde{T}'$ can be eliminated by means of (6.12.10). Thus, if we first apply to (6.12.13) the Laplacian operator ∇^2 and then use (6.12.10) for $\nabla^2\tilde{T}'$, we obtain

$$(1-2i\gamma\bar{\alpha}_\nu)\nabla^4\tilde{\phi}+\left[\gamma\left(\frac{\omega}{c_0}\right)^2+i\frac{\omega}{\kappa_0}\frac{\gamma-1}{\gamma}\right]\nabla^2\tilde{\phi}+\frac{\omega^2\beta_0}{\gamma\kappa_0}\tilde{T}'=0 \quad (6.12.14)$$

We now substitute $\tilde{T}'$ from (6.12.13) and obtain

$$(1-2i\gamma\bar{\alpha}_\nu)\nabla^4\tilde{\phi}+\left[\gamma\left(\frac{\omega}{c_0}\right)^2+\frac{i\omega}{\kappa_0}(1-2i\bar{\alpha}_\nu)\right]\nabla^2\tilde{\phi}+i\left(\frac{\omega}{c_0}\right)^2\left(\frac{\omega}{\kappa_0}\right)\tilde{\phi}=0$$

$$(6.12.15)$$

Further, since [see Equation (6.3.30)]

$$\frac{\omega\kappa_0}{c_0^2}=\frac{2\bar{\alpha}_\kappa}{\gamma-1} \qquad (6.12.16)$$

we can write (6.12.15) as

$$(1-2i\gamma\bar{\alpha}_\nu)2\bar{\alpha}_\kappa\nabla^4\tilde{\phi}+i(\gamma-1)\left(\frac{\omega}{c_0}\right)^2\left[1-2i\left(\bar{\alpha}_\nu+\frac{\gamma}{\gamma-1}\bar{\alpha}_\kappa\right)\right]\nabla^2\tilde{\phi}$$

$$+i(\gamma-1)\left(\frac{\omega}{c_0}\right)^4\tilde{\phi}=0 \quad (6.12.17)$$

This fourth-order differential equation can, by suitable choice of constants K_1^2 and K_2^2, be written in the form

$$(\nabla^2+K_1^2)(\nabla^2+K_2^2)\tilde{\phi}=0 \qquad (6.12.18)$$

As will be shown below, K_1 and K_2 are never equal. Therefore, a solution $\tilde{\phi}$ can be obtained by adding up solutions to

$$\left(\nabla^2 + K_1^2\right)\tilde{\phi}_1 = 0 \tag{6.12.19}$$

and

$$\left(\nabla^2 + K_2^2\right)\tilde{\phi}_2 = 0 \tag{6.12.20}$$

Thus, instead of a fourth-order equation, we have two second-order Helmholtz equations of the same type encountered earlier. To obtain K_1 and K_2, we first write (6.12.17) as

$$\left(A\nabla^2 + B^2\right)\left(\nabla^2 + C^2\right)\tilde{\phi} = 0 \tag{6.12.21}$$

where the quantities A, B, and C can be obtained by comparing (6.12.17) and (6.12.21). Thus,

$$AC^2 + B^2 = i(\gamma - 1)\left(\frac{\omega}{c_0}\right)^2\left[1 - 2i\left(\bar{\alpha}_\nu + \frac{\gamma}{\gamma - 1}\bar{\alpha}_\kappa\right)\right] \tag{6.12.22}$$

$$B^2C^2 = i(\gamma - 1)(\omega/c_0)^4 \tag{6.12.23}$$

and

$$A = 2\bar{\alpha}_\kappa(1 - 2i\gamma\bar{\alpha}_\nu) \tag{6.12.24}$$

Substituting for C^2 from (6.12.23) into (6.12.22) and solving for B^2, we obtain

$$2B^2 = i\left(\frac{\omega}{c_0}\right)^2(\gamma - 1)\left[1 - 2i\left(\bar{\alpha}_\nu + \frac{\gamma}{\gamma - 1}\bar{\alpha}_\kappa\right)\right.$$

$$\left. + \sqrt{\left[1 - 2i\left(\bar{\alpha}_\nu + \frac{\gamma}{\gamma - 1}\bar{\alpha}_\kappa\right)\right]^2 + 4i\bar{\alpha}_\kappa\frac{(1 - 2i\gamma\bar{\alpha}_\nu)}{\gamma - 1}}\right] \tag{6.12.25}$$

Remembering that $\bar{\alpha}_\nu$ and $\bar{\alpha}_\kappa$ are very small quantities, we may write (6.12.25) as

$$B^2 = i\left(\frac{\omega}{c_0}\right)^2(\gamma - 1)\left[1 - 2i(\bar{\alpha}_\nu + \bar{\alpha}_\kappa)\right] \tag{6.12.26}$$

and

$$C^2 = \left(\frac{\omega}{c_0}\right)^2 \left[1 + 2i(\bar{\alpha}_\nu + \bar{\alpha}_\kappa)\right] \qquad (6.12.27)$$

Therefore, K_1^2 and K_2^2 in (6.12.18) are given by

$$K_1^2 = \frac{i}{2\bar{\alpha}_\kappa}(\gamma - 1)\left(\frac{\omega}{c_0}\right)^2 \left[1 + 2i[(\gamma - 1)\bar{\alpha}_\nu - \bar{\alpha}_\kappa]\right] \qquad (6.12.28)$$

and

$$K_2^2 = (\omega/c_0)^2 \left[1 + 2i(\bar{\alpha}_\nu + \bar{\alpha}_\kappa)\right] \qquad (6.12.29)$$

To see the significance of these results, we first make use of the smallness of $\bar{\alpha}_\nu$ and $\bar{\alpha}_\kappa$ to write

$$K_2 \approx (\omega/c_0)\left[1 + i(\bar{\alpha}_\nu + \bar{\alpha}_\kappa)\right] \qquad (6.12.30)$$

Since the imaginary part of this propagation constant is very small, one part of the solution represents longitudinal waves, slightly attenuated in the main body of fluid by the effects of viscosity and heat conductivity. In fact, the propagation constant as given above is identical to that derived in Section 6.3 [see Equation (6.3.36)].

Similarly, the quantity K_1 can, with the aid of (6.5.19) and (6.12.16), be written as

$$K_1 \approx (1 + i)/\delta_\kappa \qquad (6.12.31)$$

This shows that $\tilde{\phi}_1$ in (6.12.19) also represents longitudinal waves, but this time the waves are highly attenuated and the mechanism responsible for the rapid decay is irreversible heat conduction. These waves can be recognized as the thermal waves introduced in Section 6.5. It is clear that they can have amplitudes significantly different from zero only in the vicinity of boundaries or of regions where heat is added and removed from the fluid. Thus, away from these locations, only the wave associated with K_2 exists. This behavior was used in Section 6.3 to obtain one root for the propagation constant in the problem treated there. The other root, mentioned in Problem 6.3.1, corresponds to (6.12.31).

We have therefore seen that when viscosity and heat conductivity are taken into account, the monochromatic solution involves three different types of waves:

1. The longitudinal, slightly attenuated wave ϕ_2,
2. The longitudinal, highly attenuated wave ϕ_1,
3. The transverse, highly attenuated wave **B**.

The last two waves are important only in the vicinity of the regions where they are produced, and result in rates of acoustic-energy dissipation much larger than those associated with the first type. Therefore, in the vicinity of boundaries, one can safely neglect dissipation in the first type of wave so that $\tilde{\phi}_2$ can be obtained from the ideal Helmholtz equation

$$\nabla^2\tilde{\phi}_2 + k^2\tilde{\phi}_2 = 0 \tag{6.12.32}$$

where $k = \omega/c_0$. The other two quantities can be obtained from

$$\nabla^2\tilde{\phi}_1 + (i\omega/\kappa)\tilde{\phi}_1 = 0 \tag{6.12.33}$$

and

$$\nabla^2\tilde{\mathbf{B}} + (i\omega/\nu_0)\tilde{\mathbf{B}} = 0 \tag{6.12.34}$$

The complete solution can then be obtained from these and from whatever boundary conditions are to be imposed. For example, at a fixed boundary, $\mathbf{u} = 0$, so that ϕ_1, ϕ_2, and **B** must be such that

$$\mathbf{u} = \nabla(\phi_1 + \phi_2) + \nabla \times \mathbf{B} \tag{6.12.35}$$

is zero there.

6.13 WAVES EMITTED BY PLANE HEATER

As an application of the results obtained above, we reconsider the plane-heater problem treated at the end of Section 6.5. There we studied the effects of a periodic heat addition on the assumption that the pressure fluctuations could be neglected. This enabled us to study thermal waves alone. However, it was pointed out that owing to heat addition, pressure waves would be produced. We now study how these waves are emitted.

As in Section 6.5, we assume that the heater is a plane surface coinciding with the xz plane and whose temperature is given by

$$T' = T_0 + \varepsilon\cos(\omega t) \tag{6.13.1}$$

where T_0 is the mean temperature and $\varepsilon \ll T_0$ is the amplitude of the heater's temperature fluctuations. It is clear that the fluid velocity $\mathbf{u}$ will have, at most, one component and that this component is along y (i.e., $\mathbf{u} = [0, v(y, t), 0]$). Accordingly, $\nabla \times \mathbf{u} = 0$, so that $\mathbf{B} = 0$. Hence, in order to solve this problem, we have only to obtain $\tilde{\phi}_1$ and $\tilde{\phi}_2$ from (6.12.33) and (6.12.34). It may be thought that since all we want is the acoustic wave (i.e., $\tilde{\phi}_2$), and since $\tilde{\phi}_1$ decays very rapidly with distance from the heater, we could ignore $\tilde{\phi}_1$ altogether. However, the amplitude of $\tilde{\phi}_2$ is determined by the boundary condition at the heater. This condition relates $\tilde{\phi}_2$ to $\tilde{\phi}_1$ so that one cannot neglect $\tilde{\phi}_1$. Hence, we must consider both (6.12.33) and (6.13.34), which in the present case give

$$\frac{d^2\tilde{\phi}_2}{dy^2} + \left(\frac{\omega}{c_0}\right)^2 \tilde{\phi}_2 = 0 \tag{6.13.2}$$

$$\frac{d^2\tilde{\phi}_1}{dy^2} + \left(\frac{i\omega}{\kappa}\right)\tilde{\phi}_1 = 0 \tag{6.13.3}$$

The fluid in front of the heater is unbounded. Therefore, the solutions of (6.13.2) and (6.13.3) are

$$\tilde{\phi}_1 = A e^{i(1+i)y/\delta_\kappa} \tag{6.13.4}$$

and

$$\tilde{\phi}_2 = B e^{iky} \tag{6.13.5}$$

where $(1+i)/\delta_\kappa = \sqrt{i\omega/\kappa_0} = K_1$ and $k = \omega/c_0$.

At present, we are interested mainly in the amplitude B of the acoustic wave. To obtain this quantity, we apply the boundary condition given by (6.12.35):

$$\left(\frac{d\tilde{\phi}_1}{dy}\right)_0 + \left(\frac{d\tilde{\phi}_2}{dy}\right)_0 = 0 \tag{6.13.6}$$

This yields

$$B = -A K_1 / k \tag{6.13.7}$$

This shows that the amplitude of the acoustic wave is proportional to the amplitude of the thermal wave which, in turn, can be obtained from the amplitude of the temperature fluctuations at the heater's surface. Thus, since $\bar{\alpha}_v$

is very small, we have from (6.12.13)

$$\nabla^2 \tilde{\phi} + \gamma (\omega/c_0)^2 \tilde{\phi} = -i\beta_0 \omega \tilde{T}' \tag{6.13.8}$$

Comparison of this equation with those derived in the previous chapter [see, for example, Equation (5.2.7)] shows that the temperature fluctuations act as a source of sound. In fact, the solution of (5.2.35) can be used to obtain $\tilde{\phi}$. However, since all that is required at present are the constants A and B, it is simpler to evaluate (6.13.8) at $y=0$ and obtain

$$\left[\frac{d^2(\tilde{\phi}_1 + \tilde{\phi}_2)}{dy^2} + \gamma \left(\frac{\omega}{c_0} \right)^2 (\tilde{\phi}_1 + \tilde{\phi}_2) \right]_{y=0} = -i\beta_0 \omega \varepsilon \tag{6.13.9}$$

Making use of (6.13.2)–(6.13.5), this can be written as

$$-AK_1^2 - Bk^2 + \gamma k^2 (A + B) = -i\beta_0 \omega \varepsilon$$

or using (6.13.7) to eliminate A,

$$B = \varepsilon \frac{i\beta_0 \omega/k^2}{\gamma(k/K_1 - 1) - (K_1/k - 1)} \tag{6.13.10}$$

The ratio k/K_1 can be written in terms of the quantity δ_κ as

$$\frac{k}{K_1} = \frac{1-i}{2} k\delta_\kappa$$

We know that $k\delta_\kappa$ should, in general, be small, so that the limiting form of B is

$$B = -\frac{1+i}{\sqrt{2}} \beta_0 c_0 \left(\frac{\kappa_0}{\omega} \right)^{1/2} \varepsilon \tag{6.13.11}$$

The fluid velocity in the acoustic wave is then

$$\tilde{v} = \frac{1-i}{\sqrt{2}} \beta_0 \varepsilon (\omega\kappa_0)^{1/2} e^{i(ky - \omega t)} \tag{6.13.12}$$

and since we are dealing with a plane wave, $p' = \rho_0 c_0 v$, so that the intensity is found to be

$$I = \tfrac{1}{2} \rho_0 \omega \kappa c_0 \beta_0^2 \varepsilon^2 \tag{6.13.13}$$

Energy Considerations

Equation (6.13.13) gives the acoustic intensity, due to a unit surface area of the heater, in terms of the amplitude of the heater's temperature fluctuation ε. It is of interest to study the relationship between this acoustic intensity and the rate at which heat is transferred from the heater to the fluid. To do this, we first compute the heat-transfer rate. This is given by

$$\dot{q} = k_0 (\partial T / \partial y)_{y=0} \qquad (6.13.14)$$

where $T = T_0 + \tilde{T}'$ and $\tilde{T}'$ is given by (6.13.8), or

$$\tilde{T}' = \frac{i}{\beta_0 \omega} \left[\nabla^2 (\tilde{\phi}_1 + \tilde{\phi}_2) + \gamma k^2 (\tilde{\phi}_1 + \tilde{\phi}_2) \right] \qquad (6.13.15)$$

Making use of (6.13.2) and (6.13.3), this can be written as

$$\tilde{T}' = \frac{iK_1^2}{\beta_0 \omega} \left[(\gamma - 1) \left(\frac{k}{K_1} \right)^2 \tilde{\phi}_2 - \left(1 - \gamma \frac{k^2}{K_1^2} \right) \tilde{\phi}_1 \right] \qquad (6.13.16)$$

where $k = \omega / c_0$ and $K_1 = (1 + i)/\delta_\kappa$ [see Equation (6.12.31)]. Using these definitions of k and K_1 and the values of $\tilde{\phi}_1$ and $\tilde{\phi}_2$ given by (6.13.4) and (6.13.5), we have

$$\tilde{T}' = \frac{A}{\beta_0 \kappa} \left(\frac{i}{2} (\gamma - 1)(k\delta_\kappa)^2 \left(\frac{B}{A} \right) e^{iky} + \left[1 + \frac{i\gamma}{2} (k\delta_\kappa)^2 \right] e^{iK_1 y} \right) e^{-i\omega t} \qquad (6.13.17)$$

or, on using (6.13.7),

$$\tilde{T}' = \frac{A}{\beta_0 \kappa} \left(\frac{1-i}{2} (\gamma - 1)(k\delta_\kappa) e^{iky} + \left[1 + \frac{i\gamma}{2} (k\delta_\kappa)^2 \right] e^{iK_1 y} \right) e^{-i\omega t} \qquad (6.13.18)$$

The amplitude A can be obtained from the condition that at $y = 0$, $\tilde{T}' = \varepsilon e^{-i\omega t}$, or from the already found value of B. In either case, the result is

$$\frac{A}{\beta_0 \kappa} = \frac{\varepsilon}{1 + \frac{1}{2}(\gamma - 1)k\delta_\kappa - (i/2)(\gamma - 1)\left[1 - \gamma k \delta_\kappa (\gamma - 1)^{-1} \right] k \delta_\kappa} \qquad (6.13.19)$$

Remembering that $k\delta_\kappa \ll 1$, this can be approximated by

$$\frac{A}{\beta_0 \kappa} = \frac{\varepsilon}{1 + \dfrac{\gamma-1}{2} k\delta_\kappa (1-i)}$$ (6.13.20)

To the same order of approximation, the temperature fluctuation in the fluid can be written as

$$\tilde{T}' \approx \varepsilon e^{i(K_1 y - \omega t)} + \frac{\varepsilon}{2}(1-i)(\gamma - 1)k\delta_\kappa \left[e^{i(ky-\omega t)} - e^{i(K_1 y - \omega t)} \right]$$ (6.13.21)

We have written this result in this form so that we may easily identify the meaning of each term. For example, the first term can be recognized as the temperature fluctuation in the incompressible fluid that was derived in Section 6.5 [see Equation (6.5.18)]. The other term is the first-order contribution arising from the effects of compressibility. These effects are of order $k\delta_\kappa$ relative to the incompressible results, and are made of two parts: (a) a plane, undamped acoustic wave and (b) a thermal wave whose amplitude at the heater is equal to that of the acoustic wave, and which is negligible far from the heater. Thus, sufficiently far from the heater, the only temperature fluctuations are acoustic.

Let us return to the heat-transfer rate. Using (6.13.21) we have for $(\partial \tilde{T}'/\partial y)_{y=0}$

$$\left(\frac{\partial \tilde{T}'}{\partial y} \right)_{y=0} = i\varepsilon K_1 \left[1 - \frac{i-1}{2}(\gamma-1)k\delta_\kappa \left(1 - \frac{k}{K_1} \right) \right] e^{-i\omega t}$$ (6.13.22)

With $K_1 = (1+i)/\delta_\kappa$ and $(1-i) = \sqrt{2}\, e^{-i\pi/4}$, this becomes

$$\left(\frac{\partial \tilde{T}'}{\partial y} \right)_{y=0} = -\sqrt{2}\, \frac{\varepsilon}{\delta_\kappa} e^{-i(\omega t + \pi/4)} + (\gamma-1)k\varepsilon e^{-i(\omega t + \pi/2)}$$ (6.13.23)

Hence, the heat transfer per unit surface area of heater is

$$\frac{\dot{q}}{\rho_0 c_{p0}} = -\varepsilon(\omega\kappa_0)^{1/2}\cos\left(\omega t + \frac{\pi}{4} \right) + \varepsilon(\gamma-1)\left(\frac{\omega\kappa}{c_0} \right)\sin(\omega t)$$ (6.13.24)

where we have used the identities $k_0 = \rho_0 c_{p0}\kappa_0$ and $\delta_\kappa = (2\kappa_0/\omega)^{1/2}$. Comparison of this with (6.5.22) shows that the first term is the heat-transfer rate in the case of an incompressible fluid. The second term, while also proportional to ε, has a much smaller magnitude because $\omega\kappa_0 \ll c_0$.

We would like to show that only the heat-addition rate given by the first term in (6.13.24) is responsible for the emitted intensity. To do this, we may integrate

(5.10.10) over the heater's surface, using for the source strength $\dot{q}_A$, the amount prescribed by the first term in (6.13.24):

$$\dot{q}_A(t) = \text{Re}\left[-\rho_0 c_{p0} \varepsilon (\omega \kappa_0)^{1/2} e^{-i(\omega t + \pi/4)} \right] \tag{6.13.25}$$

If we use polar coordinates on the heater surface, $dA(x') = r' \, d\theta', dr'$, then (5.10.10) can be written as

$$\tilde{\phi}(y, t) = -\frac{1}{2\pi} B_0 \varepsilon (\omega \kappa_0)^{1/2} e^{-i(\omega t + \pi/4)} \int_0^{2\pi} \int_0^{\infty} \frac{r' e^{ik\sqrt{y^2 + r'^2}}}{\sqrt{y^2 + r'^2}} \, dr'$$

$$\tag{6.13.26}$$

where we have replaced the factor $\frac{1}{4}$ in (5.10.10) with the factor $\frac{1}{2}$ because we are considering only one side of the heater (see Section 5.2). The area integral is very similar to that discussed in that same section. Its value is simply $(2\pi/ik)e^{iky}$ so that

$$\tilde{\phi}(y, t) = \frac{i}{k} B_0 \varepsilon (\omega \kappa_0)^{1/2} e^{i(ky - \omega t - \pi/4)} \tag{6.13.27}$$

The corresponding velocity is then

$$v = -B_0 \varepsilon (\omega \kappa_0)^{1/2} e^{i(ky - \omega t - \pi/4)} \tag{6.13.28}$$

As expected, this is identical to the acoustic velocity given by (6.13.12). While the agreement is not surprising, it does show that heat addition is the mechanism responsible for the emission. Also, it has enabled us to identify which of the various terms in the heat transfer rate equation leads to the first-order acoustic effects.

Another question that arises is the relative efficiency of heat transfer as a sound-production mechanism. This efficiency may be measured by the ratio of average acoustic output per unit area of heater; that is, the intensity to the average energy input. Thus,

$$\eta_a = \frac{I}{\langle \dot{E}_{\text{input}} \rangle} \tag{6.13.29}$$

The intensity is given by (6.13.13). The required average energy input was obtained in Section 6.5. The discussion given there shows that in order to sustain the specified temperature fluctuation of the heater, an amount of energy equal to

$$\langle \dot{E}_n \rangle = \tfrac{1}{2} \rho_0 c_{p0} (\varepsilon^2/T_0)(\omega \kappa_0)^{1/2} \tag{6.13.30}$$

has to be supplied on the average to each unit surface area of the heater. That result was obtained by integrating the quantity $(\partial T/\partial y)^2$ over the complete body of fluid. Since that computation did not include the emitted sound waves, we will rederive it by applying to the heater the concept of available energy loss introduced earlier in this chapter. This is associated with the total average rate of entropy increase $\langle \dot{S}_T \rangle$. Since the heater returns every cycle to the same state, its entropy remains constant, and the entropy changes are due to the fluid alone. These are given by

$$\langle \dot{S}_T \rangle = \langle \dot{q}/T_H \rangle \tag{6.13.31}$$

where $\dot{q}$ is to be obtained from (6.13.24) and where T_H is the temperature at which the heat transfer occurs. Consider, first, the contribution to $\langle \dot{S}_T \rangle$ arising from the second term in (6.13.24). This contribution is proportional to

$$\int_0^{2\pi/\omega} \frac{\sin(\omega t)}{T_0 + \varepsilon \cos \omega t} dt$$

and this integral is identically zero. Hence,

$$\langle \dot{S}_T \rangle = -\rho_0 c_{p0} \left(\frac{\varepsilon}{T_0} \right) (\omega \kappa_0)^{1/2} \frac{1}{2\pi/\omega} \int_0^{2\pi/\omega} \frac{\cos(\omega t + \pi/4)}{T_0(1 + (\varepsilon/T_0)\cos \omega t)} dt$$

$$\tag{6.13.32}$$

Since ε/T_0 is a small quantity, (6.13.32) can be written as

$$\langle \dot{S}_T \rangle = -\rho_0 c_{p0} \left(\frac{\varepsilon}{T_0} \right) (\omega \kappa_0)^{1/2} \left(\frac{\omega}{2\pi} \right) \int_0^{2\pi/\omega} \left[1 - \frac{\varepsilon}{T_0} \cos(\omega t) \right] \times \cos\left(\omega t + \frac{\pi}{4} \right) dt$$

$$\tag{6.13.33}$$

The value of this integral is $(-\varepsilon/T_0)(2\pi/\omega)(\tfrac{1}{2}\sqrt{2})$. Thus,

$$\langle \dot{S} \rangle = \tfrac{1}{2} \rho_0 c_{p0} (\varepsilon/T_0)^2 (\omega \kappa_0/2)^{1/2} \tag{6.13.34}$$

The related energy $T_0 \langle \dot{S} \rangle$ is the energy that has to be supplied, and is

$$T_0 \langle \dot{S}_T \rangle = \tfrac{1}{2} \rho_0 c_{p0} (\varepsilon^2 / T_0)(\omega \kappa_0 / 2)^{1/2} \tag{6.13.35}$$

This is identical to the energy rate specified by (6.13.30). The thermo-acoustic efficiency η_a is therefore given by

$$\eta_a = \left(\frac{\omega \kappa_0}{2} \right)^{1/2} \frac{\beta_0^2 T_0 c_0}{c_p} \tag{6.13.36}$$

or, using the identity $\beta_0^2 T_0 c_0^2 = c_{p0}(\gamma - 1)$, by

$$\eta_a = (\gamma - 1)(\omega \kappa_0 / 2 c_0^2)^{1/2} \tag{6.13.37}$$

The quantity $(\omega \kappa_0 / 2 c_0^2)^{1/2}$ is, as we have seen, very small. Therefore, the conversion efficiency is also very small. However, the amounts of heat released unsteadily can, on occasion, be rather large, so that then the emitted intensity may be significant in spite of the smallness of η_a.

SUGGESTED REFERENCES

Chu, B.-T. "Pressure Wave Generated by Addition of Heat in a Gaseous Medium." This was the first work to point out the analogy between sound emission by heat release and by mass addition.

Lighthill, M. J. "Viscosity Effects in Sound Waves of Finite Amplitude." Sections 2–4 give a lucid description of attenuation in gases.

Markham, J. J., Beyer, R. T., and Lindsay, R. B. "Absorption of Sound in Fluids." This paper presents a detailed discussion of absorption in fluids. Chapter III discusses the experimental methods used, and Chapters IV and V cover the experimental results as of 1951.

Rott, N. "Thermoacoustics." This work presents a general review of recent work on thermoacoustic phenomena, including sound emission by heated surfaces and heat generation by finite amplitude waves.

Tijdman, H. "On the Propagation of Sound Waves in Cylindrical Tubes."

BIBLIOGRAPHY

This bibliography is divided into three sections: (a) books dealing with fairly general subjects, (b) specialized books, and (c) review and research articles. The lists include the works listed in the text, as well as a few others that touch on some of the topics treated in the text.

GENERAL BOOKS

Acoustics

Kinsler, L. E., and E. R. Frey. *Fundamentals of Acoustics*, Second Edition (Wiley, New York, 1962).

Meyer, E., and E.-G. Neumann. *Physical and Applied Acoustics* (Academic Press, New York, 1972).

Morse, P. M. *Vibration and Sound* (McGraw-Hill, New York, 1948).

Morse, P. M., and K. U. Ingard. *Theoretical Acoustics* (McGraw-Hill, New York, 1968).

Rayleigh, Lord. *The Theory of Sound*, Vols. I and II, Second Edition, 1896. (Dover, New York, 1945).

Skudrzyk, E. *The Foundations of Acoustics* (Springer-Verlag, New York, 1971).

Stephens, R. W. B., and A. E. Bate. *Acoustics and Vibrational Physics* (E. Arnold, London, 1966).

Stewart, G. W., and R. B. Lindsay. *Acoustics: A Text on Theory and Applications* (van Nostrand, Princeton, N.J., 1930).

Tyndall, J. *Sound* (D. Appleton, New York, 1867).

Wood, A. *Acoustics*, Second Edition (Blackie and Son, London, 1960). Also, Dover, New York, 1966.

Fluid Dynamics

Batchelor, G. K. *An Introduction to Fluid Dynamics* (Cambridge University Press, Cambridge, 1967).

Currie, I.G. *Fundamental Mechanics of Fluids* (McGraw-Hill, New York, 1974).

Lamb, H., *Dynamical Theory of Sound*, Second Edition (Dover, N.Y., 1960).

——*Hydrodynamics*, Sixth Edition (Dover, N.Y., 1945).

Landau, L. D., and E. M. Lifshitz. *Fluid Mechanics* (Pergamon Press, London, 1959).

Yih, C.-S. *Fluid Mechanics* (McGraw-Hill, New York, 1969).

Mathematical Tables

Abramowitz, M., and I. Stegun. *Handbook of Mathematical Functions*. National Bureau of Standards, Applied Mathematics Series 55, 1964. Available from Dover.

Gradshteyn, I. S., and I. M. Ryzhik. *Table of Integrals, Series and Products* (Academic Press, New York, 1965).

Mathematics

Hildebrand, P. M. *Advanced Calculus for Applications*, Second Edition (Prentice Hall, Englewood, N.J., 1976).

Jeffreys, H. *Cartesian Tensors* (Cambridge University Press, Cambridge, 1961).

Morse, P. M., and H. Feshbach Vols. I and II. *Methods of Theoretical Physics* (McGraw-Hill, New York, 1953).

Sommerfeld, A. *Partial Differential Equations in Physics* (Academic Press, New York, 1964).

Thermodynamics

Callen, H. B. *Thermodynamics* (Wiley, New York, 1962).

Kestin, J. *A Course in Thermodynamics*, Vols. I and II, Revised Printing (McGraw-Hill, New York, 1979).

Zemansky, M. *Heat and Thermodynamics*, Fourth Edition (McGraw-Hill, New York, 1957).

SPECIALIZED BOOKS

Acoustical Techniques and Measurements

Beranek, L. *Acoustic Measurements* (Wiley, New York, 1949).

Bergmann, L. *Ultrasonics* (Wiley, New York, 1938).

Hueter, T. F., and R. H. Bolt. *Sonics* (Wiley, New York, 1955).

Aerodynamic Noise; Acoustics in Moving Media

Blokhintsev, D. I. *Acoustics of a Nonhomogeneous Moving Medium* (National Advisory Committee for Aeronautics, Technical Memorandum 1399, February 1956).

Goldstein, M. E. *Aeroacoustics* (McGraw-Hill, New York, 1977).

Richards, E. J., and D. J. Mead, Eds. *Noise and Acoustic Fatigue in Aeronautics* (Wiley, London, 1968).

Gas Dynamics

Liepmann, H. W., and A. Roshko. *Elements of Gas Dynamics* (Wiley, New York, 1957).

Thompson, P. A. *Compressible Fluid Dynamics* (McGraw-Hill, New York, 1972).

Vincenti, W. G., and C. H. Kruger. *Physical Gas Dynamics* (Wiley, New York, 1965).

Molecular Relaxation

Bhatia, A. B. *Ultrasonic Absorption* (Oxford University Press, Oxford, 1967).

Herzfeld, K. F., and T. A. Litovitz. *Absorption and Dispersion of Ultrasonic Waves* (Academic Press, New York, 1959).

Noise Control

Beranek, L., Ed. *Noise and Vibration Control* (McGraw-Hill, New York, 1971).

Harris, C. M., Ed. *Handbook of Noise Control*, Second Edition (McGraw-Hill, New York, 1979).

Nonlinear Acoustics

Beyer, R. T. *Nonlinear Acoustics* (Department of the Navy, Sea Systems Command, Washington, D.C., 1974).

Rudenko, O. V., and S. I. Soluyan. *Theoretical Foundations of Nonlinear Acoustics* (Consultants Bureau, New York, 1977).

Wave Motion

Lighthill, J. *Waves in Fluids* (Cambridge University Press, Cambridge, 1978).

Pain, H. J. *The Physics of Vibrations and Waves* (Wiley, London, 1968).

Tolstoy, I. *Wave Propagation* (McGraw-Hill, New York, 1973).

Towne, D. H. *Wave Phenomena* (Addison-Wesley, Reading, Mass., 1967).

Witham, G. B. *Linear and Nonlinear Waves* (Wiley, New York, 1974).

REVIEW AND RESEARCH ARTICLES

Absorption in Fluids

Diatomic Gases

Boitnott, C. A., and R. C. Warden, Jr. "Shock-Tube Measurements of Rotational Relaxation in Hydrogen," *Phys. Fluids*, **14**, 2312–2316 (1971).

Greenspan, M. "Rotational Relaxation in Nitrogen, Oxygen, and Air," *J. Acoust. Soc. Amer.*, **31**, 155–160 (1959).

Hanson, F. B., T. F. Morse, and L. Sirovich, "Kinetic Description of the Propagation of Plane Sound Waves in Diatomic Gas," *Phys. Fluids*, **12**, 84–95 (1969).

Monchik, L. "Sound Dispersion in Diatomic Gases at High Frequency," in *Molecular Relaxation Processes*, M. Davis, Ed. (Academic Press, 1966). pp. 257–275.

Parker, J. G., C. E. Adams, and R. M. Stavseth. "Absorption of Sound in Argon, Nitrogen and Oxygen at Low Pressures," *J. Acoust. Soc. Amer.*, **25**, 263–269 (1953).

Parker, J. G., and R. H. Swope. "Vibrational Relaxation Times of Oxygen in the Temperature Range 100°–200°C," *J. Acoust. Soc. Amer.*, **37**, 718–723 (1965).

Effects of Humidity

Bass, H., and F. D. Shields. "Absorption of Sound in Air," *J. Acoust. Soc. Amer.*, **62**, 571–576 (1972).

Harris, C. M. "Absorption of Sound in Air versus Humidity and Temperature," *J. Acoust. Soc. Amer.*, **40**, 148–159 (1966).

——"Effects of Humidity on the Velocity of Sound in Air," *J. Acoust. Soc. Amer.*, **49**, 890–893 (1971).

——"Normalized Curve of Molecular Absorption versus Humidity," *J. Acoust. Soc. Amer.*, **57**, 241–242 (1975).

Effects of Impurities

Henderson, M. C., K. F., Herzfeld, J. Bry, R. Coakley, and G. Carrsere. "Thermal Relaxation in Nitrogen with Wet Carbon Dioxide as Impurity," *J. Acoust. Soc. Amer.*, **45**, 109–114 (1969).

Parker, J. G. "Effect of Several Light Molecules on the Vibrational Relaxation Time of Oxygen," *J. Chem. Phys.*, **34**, 1763–1772 (1961).

Gas Mixtures

Angona, F. A. "The Absorption of Sound in Gas Mixtures," *J. Acoust. Soc. Amer.*, **25**, 1116–1122 (1953).

Shields, F. D., and G. P. Carney. "Sound Absorption in Pure D_2S, CO_2/D_2S Mixtures," *J. Acoust. Soc. Amer.*, **47**, 1269–1273 (1970).

Wight, H. M. "Vibrational Relaxation in $N_2O–H_2O$ and $N_2O–D_2O$ Mixtures," *J. Acoust. Soc. Amer.*, **28**, 459–461 (1956).

General Reviews

Herzfeld, K. F. "Fifty Years of Physical Ultrasonics," *J. Acoust. Soc. Amer.*, **39**, 813–825 (1966).

Lighthill, M. J. "Viscosity Effects in Sound Waves of Finite Amplitude," in *Surveys in Mechanics*, G. K. Batchelor and R. M. Davis, Eds. (Cambridge University Press, Cambridge 1956). pp. 250–351.

Markham, J. J., R. T. Beyer, and R. B. Lindsay. "Absorption of Sound in Fluids," *Revs. Mod. Phys.*, **23**, 353–411 (1951).

Monatomic Gases

Greenspan, M. "Propagation of Sound in Five Monatomic Gases," *J. Acoust. Soc. Amer.*, **28**, 644–648 (1956).

Schotter, R. "Rarefied Gas Acoustics in the Noble Gases," *Phys. Fluids*, **17**, 1163–1168 (1974).

Acoustic Interferometer (impedance tube)

Lippert, W. K. R. "The Practical Representation of Standing Waves in an Acoustic Impedance Tube," *Acoustica*, **3**, 153–160 (1953).

Scott, R. A. "An Apparatus for Accurate Measurement of the Acoustic Impedance of Sound-Absorbing Materials," *Proc. Phys. Soc.*, **58**, 153–264 (1946). (The most significant portions of this article may be found in L. Beranek, *Acoustic Measurements*, listed under Section 2: Specialized Books, Acoustical techniques, and Measurements.)

Temkin, S., and R. A. Dobbins. "Measurements of Attenuation and Dispersion of Sound by an Aerosol," *J. Acoust. Soc. Amer.*, **40**, 1016–1024 (1966).

Aerodynamic Sound

Curle, N. "The Influence of Solid Boundaries upon Aerodynamic Sound," *Proc. Roy. Soc.*, Series A, **231**, 505–514 (1955).

Doak, P. E. "Analysis of Internally Generated Sound in Continuous Materials. 2. A Critical Review of the Conceptual Adequacy and Physical Scope of Existing Theories of Aerodynamic Noise, with Special Reference to Supersonic Jet Noise," *J. Sound Vib.*, **25**, 263–335 (1972).

Ffowcs Williams, J. E., "Aeroacoustics," *Ann. Rev. Fluid Mech.*, **9**, 447–468 (1977).

——"Hydrodynamic Noise," *Ann. Rev. Fluid Mech.*, **1**, 197–222 (1969).

Lighthill, M. J. "On Sound Generated Aerodynamically. I. General Theory," *Proc. Roy. Soc.*, Series A, **211**, 565–587 (1952).

——"On Sound Generated Aerodynamically. II. Turbulence as a Source of Sound," *Proc. Roy. Soc.*, Series A, **222**, 1–32 (1954).

——The Bakerian Lecture, 1961, "Sound Generated Aerodynamically," *Proc. Roy. Soc.*, Series A, **267**, 117–182 (1962).

Reethof, G. "Turbulence-Generated Noise in Pipe Flow," *Ann. Rev. Fluid Mech.*, **10**,333–367 (1978).

Ribner, H. S. "Jets and Noise," *Can. Aero. Space J.*, **14**, 282–298 (1968).

Sears, W. R. "Aerodynamics, Noise and the Sonic Boom," *Amer. Inst. Aero. Astro. J.*, **7**, 577–586 (1969).

Attenuation in Suspensions

Allegra, J. R., and S. A. Hawley. "Attenuation and Dispersion of Sound in Suspensions and Emulsions: Theory and Experiments," *J. Acoust. Soc. Amer.*, **51**, 1545–1564 (1972).

Batchelor, G. K. "Compression Waves in a Suspension of Gas Bubbles," *Fluid Dynamics Trans.*, **4**, 425–445 (1969).

Carstensen, E. L., and L. I. Foldy. "Propagation of Sound through a Liquid Containing Bubbles," *J. Acoust. Soc. Amer.*, **19**, 481–501 (1957).

Cole, III, J. E., and R. A. Dobbins. "Propagation of Sound through Atmospheric Fog," *J. Atmos. Sci.*, **27**, 426–434 (1970).

Crespo, A. "Sound and Shock Waves in Liquid Containing Bubbles," *Phys. Fluids*, **12**, 2274–2282 (1969).

Epstein, P. S., and R. R. Carhart. "The Absorption of Sound in Suspensions and Emulsions. I. Water Fog in Air," *J. Acoust. Soc. Amer.*, **25**, 553–565 (1953).

Marble, F. E. "Dusty Gases," *Ann. Rev. Fluid Mech.*, **2**, 397–446 (1967).

Marble, F. E., and D. C. Wooten. "Sound Attenuation in a Condensing Vapor," *Phys. Fluids*, **13**, 2657–2664 (1970).

Noordzij, L., and L. van Wijngaarden. "Relaxation Effects, Caused by Relative Motion, on Shock Waves in Gas-Bubble Liquid Mixtures," *J. Fluid Mech.*, **66**, 115–143 (1974).

Rudinger, G. "Wave Propagation in Suspensions of Solid Particles in Gas Flow," *Appl. Mech. Revs.*, **26**, 273–279 (1973).

Temkin, S., and R. A. Dobbins. "Attenuation and Dispersion of Sound by Particulate Relaxation Processes," *J. Acoust. Soc. Amer.*, **40**, 317–324 (1966).

——"Measurements of Attenuation and Dispersion of Sound by an Aerosol," *J. Acoust. Soc. Amer.*, **40**, 1016–1024 (1966).

van Wijngaarden, L. "One-Dimensional Flow of Liquids Containing Small Gas Bubbles," *Ann. Rev. Fluid Mech.*, **4**, 369–396 (1972).

Zink, J. W., and L. P. Delsasso. "Attenuation and Dispersion of Sound by Solid Particles Suspended in a Gas," *J. Acoust. Soc. Amer.*, **30**, 765–771 (1958).

Emission by Heat Release

Arnold, J. S. "Generation of Combustion Noise," *J. Acoust. Soc. Amer.*, **52**, 5–12 (1972).

Babcock, W. R., K. L. Baker, and A. C. Cattaneo. "Musical Flames," *Nature*, **216**, 676–678 (1967).

Carrier, G. "The Mechanics of the Rijke Tube," *Quart. Appl. Math.*, **12**, 383–395 (1955).

Chu, B.-T. "Analysis of a Self-sustained Thermally Driven Nonlinear Vibration," *Phys. Fluids*, **6**, 1638–1644 (1963).

——"Pressure Waves Generated by Addition of Heat in a Gaseous Medium," National Advisory Committee for Aeronautics, Technical Note 3411, 1955.

Feldman, Jr., K. T. "Review of the Literature on Sondhauss Thermoacoustic Phenomena," *J. Sound Vib.* **7**, 71–82 (1968).

——"Review of the Literature on Rijke Thermoacoustic Phenomena," *J. Sound Vib.*, **7**, 83–89 (1968).

Jones, H. "The Mechanics of Vibrating Flames in Tubes," *Proc. Roy. Soc.*, Series A, **353**, 459–473 (1977).

Kempton, A. J. "Heat Diffusion as a Source of Aerodynamic Sound," *J. Fluid Mech.*, **78**, 1–31 (1976).

Putnam, A. A., and W. R. Dennis. "Survey of Organ-Pipe Oscillations in Combustion Systems," *J. Acoust. Soc. Amer.*, **28**, 246–259 (1956).

Rosencwaig, A., and A. Gersho. "Photoacoustic Effect with Solids: A Theoretical Treatment," *Science*, **190**, 556–557 (1975).

Rott, N. "Recent Research on Thermoacoustic Oscillations," *J. Acoust. Soc. Amer.*, **65**, 541 (1979).

——"Thermoacoustics" in *Advances in Applied Mechanics*, Vol. 20, C.-S. Yih, Ed. (Academic Press, New York, 1980). pp. 135–175.

Sinai, Y. L. "The Generation of Combustion Noise by Chemical Inhomogeneities in Steady, Low-Mach-Number Duct Flow," *J. Fluid Mech.*, **99**, 383–397.

Strahle, W. C. "On Combustion Generated Noise," *J. Fluid Mech.*, **49**, 399–414 (1971).

Temkin, S. "A Model for Thunder Based on Heat Addition," *J. Sound Vib.*, **52**, 401–414 (1977).

Whitehead, D. S., and D. K. Holger. "Excitation of Nonlinear Acoustic Resonances by Unsteady Heat Input," *J. Acoust. Soc. Amer.*, **65**, 324–335 (1979).

Filters and Resonators

Davis, Jr., D. D. "Acoustical Filters and Mufflers," in *Handbook of Noise Control*, C. M. Harris, Ed., Second Edition (McGraw-Hill, New York, 1979).

Embelton, T. F. W. "Mufflers," in *Noise and Vibration Control*, L. L. Beranek, Ed. (McGraw-Hill, New York, 1971), Chapter 12.

Garrett, S., and D. S. Stat. "Peruvian Whistling Bottles," *J Acoust. Soc. Amer.*, **62**, 449–453 (1977).

Ingard, U. "On the Theory of Design of Acoustic Resonators," *J. Acoust. Soc. Amer.*, **25**, 1037–1061 (1953).

Panton, R., and J. M. Miller. "Resonant Frequencies of Cylindrical Helmholtz Resonators," *J. Acoust. Soc. Amer.*, **57**, 1533–1535 (1975).

Tang, P. K., and W. A. Sirignano. "Theory of a Generalized Helmholtz Resonator," *J. Sound Vib.*, **26**, 247–262 (1973).

Group and Phase Velocities

Lighthill, M. J. "Group Velocity," *J. Inst. Maths Applics*, **1**, 1–28 (1965).

Thau, S. A. "Linear Dispersive Waves," in *Nonlinear Waves*, S. Leibovich and A. R. Seebas, Eds. (Cornell University Press, Ithaca, N.Y., 1974), pp. 44–81.

Horns

Eisner, E. "Complete Solutions of the 'Webster' Horn Equation," *J. Acoust. Soc. Amer.*, **41**, 1126–1146 (1966).

Lesser, M. B., and D. G. Creighton. "Physical Acoustics and the Method of Matched Asymptotic Expansions," in *Physical Acoustics*, Vol. X, W. P. Mason and R. N. Thurston, Eds. (Academic Press, New York, 1975), pp. 69–149.

Lesser, M. B., and J. A. Lewis. "Applications of Matched Asymptotic Expansion Methods to Acoustics. I. The Webster Horn Equation and the Stepped Duct," *J. Acoust. Soc. Amer.*, **51**, 1664–1669 (1971).

Pyle, Jr., R. W. "Effective Length of Horns," *J. Acoust. Soc. Amer.*, **57**, 1309–1317 (1975).

Miscellaneous Topics

Farrel, W. E., D. P. McKenzie, and R. L. Parker. "On the Note Emitted from a Mug While Mixing Instant Coffee," *Proc. Camb. Phil. Soc.*, **65**, 365–367 (1969).

Few, A. A. "Thunder," *Sci. Amer.* (July 1975).

Gold, T., and S. Soter. "Brontides: Natural Explosive Noises," *Science*, **204**, 371–375 (1979).

Lindsay, R. B. "The Story of Acoustics," *J. Acoust. Soc. Amer.*, **39**, 629–644 (1965).

Takahara, H. "Sounding Mechanism of Singing Sand," *J. Acoust. Soc. Amer.*, **53**, 634–639 (1973).

Westfall, R. S. "Newton and the Fudge Factor," *Science*, **179**, 751–758 (1973).

Nonlinear Acoustics

Bayer, R. T. "Nonlinear Acoustics," *Amer. J. Phys.*, **41**, 1060–1067 (1973).

Bjørnø, L. "Nonlinear Acoustics," in *Acoustics and Vibration Progress*, Vol. 2, R. W. B. Stephens and H. G. Leventhall, Eds. (Chapman and Hall, London, 1976), pp. 101–198.

Blackstock, D. T. "Nonlinear Acoustics (Theoretical)," in Chapter 3n of *American Institute of Physics Handbook*, D. E. Grey, Ed. (McGraw-Hill, New York, 1972), pp. 183–205.

——"Propagation of Plane Sound Waves of Finite Amplitude in Nondissipative Fluids," *J. Acoust. Soc. Amer.*, **34**, 9–30 (1962).

Creighton, D. C. "Model Equations in Nonlinear Acoustics," *Ann. Rev. Fluid Mech.*, **11**, 11–33 (1979).

Westervelt, P. J. "The Status and Future of Nonlinear Acoustics," *J. Acoust. Soc. Amer.*, **57**, 1352–1356 (1975).

Nonlinear Resonance in Tubes

Brocher, E. "Oscillatory Flows in Ducts: A Report on Euromech 73," *J. Fluid Mech.*, **79**, 113–126 (1977).

Chester, W. "Resonant Oscillations in Closed Tubes," *J. Fluid Mech.*, **18**, 44–66 (1964).

Cruikshank, Jr., D. B. "Experimental Investigation of Finite-Amplitude Acoustic Oscillations in a Closed Tube," *J. Acoust. Soc. Amer.*, **52**, 1024–1036 (1972).

Disselhorst, J. H. M., and L. Van Wijngaarden. "Flow in the Exit of Open Pipes During Acoustic Resonance," *J. Fluid Mech.*, **99**, 293–319.

Jimenez, J. "Nonlinear Gas Oscillations in Pipes. Part 1. Theory," *J. Fluid Mech.*, **59**, 23–46 (1973)(1980).

Keller, J. J. "Resonant Oscillations in Closed Tubes: The Solution of Chester's Equations," *J. Fluid Mech.*, **77**, 279–304 (1976).

Seymour, B. R., and M. P. Mortell. "Nonlinear Resonant Oscillations in Open Tubes," *J. Fluid Mech.*, **60**, 733–749 (1973).

Sturtevant, B. "Nonlinear Gas Oscillations in Pipes. Part 2. Experiment," *J. Fluid Mech.*, **63**, 97–120 (1974).

Temkin, S. "Nonlinear Gas Oscillations in a Resonant Tube," *Phys. Fluids*, **11**, 960–963 (1968).

Radiation from Pistons

Burnett, D. S., and W. W. Soroka. "Tables of Rectangular Piston Radiation Impedance Functions, with Application to Sound Transmission Loss through Deep Apertures," *J. Acoust. Soc. Amer.*, **51**, 1618–1623 (1972).

Neubauer, W. G. "Radiated Field of a Rectangular Piston," *J. Acoust. Soc. Amer.*, **37**, 671–672 (1965).

Rogers, P. H. "Acoustic Field of Circular Plane Piston in Limits of Short Wavelength or Large Radius," *J. Acoust. Soc. Amer.*, **52**, 865–870 (1972).

Stepanishen, P. R. "Asymptotic Behavior of the Acoustic Nearfield of a Circular Piston," *J. Acoust. Soc. Amer.*, **59**, 749–754 (1976).

Zemanek, J. "Beam Behavior of the Nearfield of a Vibrating Piston," *J. Acoust. Soc. Amer.*, **49**, 181–191 (1971).

Reflection from Porous Boundaries

Rayleigh, Lord. "On Resonant Reflection of Sound from a Perforated Wall," in *Scientific Papers*, Vol. VI (Dover, New York, 1964), pp. 662–669. (See, also, article No. 351 of Rayleigh's *The Theory of Sound*.)

Roetman, E. L., and R. P. Kochhar. "Reflection of Acoustical Waves at Porous Boundaries," *J. Acoust. Soc. Amer.*, **59**, 1057–1064 (1976).

Scattering by Spheres

Davis, C. M., L. R. Dragonette, and L. Flax. "Acoustic Scattering from Silicone Rubber Cylinders and Spheres," *J. Acoust. Soc. Amer.*, **63**, 1694–1698 (1978).

George, J., and H. Überall. "Approximate Methods to Describe the Reflections from Cylinders and Spheres with Complex Impedance," *J. Acoust. Soc. Amer.*, **65**, 15–24 (1979).

Hickling, R., and N. M. Wang. "Scattering of Sound by a Rigid Movable Sphere," *J. Acoust. Soc. Amer.*, **39**, 276–279 (1966).

Johnson, R. K. "Sound Scattering from a Fluid Sphere Revisited," *J. Acoust. Soc. Amer.*, **61**, 275–377 (1977).

Neubauer, W. G., R. H. Vogt, and L. R. Dragonette. "Acoustic Reflection from Elastic Spheres. I. Steady-State Signals," *J. Acoust. Soc. Amer.*, **55**, 1123–1129 (1974).

Size Measurements in Suspensions

Dobbins, R. A., and S. Temkin. "Acoustical Measurements of Aerosol Particle Size and Concentration," *J. Colloid Interface Sci.*, **25**, 329–333 (1967).

Medwin, H. "Acoustical Determination of Bubble-Size Spectra," *J. Acoust. Soc. Amer.*, **62**, 1041–1044 (1977).

Small Spheres in Sound Waves

Gucker, F. T., and G. J. Doyle. "The Amplitude of Vibration of Aerosol Droplets in a Sonic Field," *J. Phys. Chem.*, **60**, 989–996 (1956).

Taylor, K. J. "Absolute Measurement of Acoustic Particle Velocity," *J. Acoust. Soc. Amer.*, **59**, 691–694 (1976).

Temkin, S., and C.-M. Leung. "On the Velocity of a Rigid Sphere in a Sound Wave," *J. Sound Vib.*, **49**, 75–92 (1976).

Speed of Sound in Water

Del Grosso, V. A. "New Equation for the Speed of Sound in Natural Waters (with comparisons to other equations)," *J. Acoust. Soc. Amer.*, **56**, 1084–1091 (1974).

Del Grosso, V. A. "Sound Speed in Pure Water and Sea Water," *J. Acoust. Soc. Amer.*, **47**, 947–949 (1969).

Del Grosso, V. A., and C. W. Mader. "Speed of Sound in Pure Water," *J. Acoust. Soc. Amer.*, **52**, 1442–1446 (1972).

Lovett, J. R. "Merged Seawater Sound-Speed Equations," *J. Acoust. Soc. Amer.*, **63**, 1713–1719 (1978).

Medwin, H. "Speed of Sound in Water: A Simple Equation for Realistic Parameters," *J. Acoust. Soc. Amer.*, **58**, 1318–1319 (1975).

Millero, F. J., and T. Kubinski. "Speed of Sound in Seawater as a Function of Temperature and Salinity," *J. Acoust. Soc. Amer.*, **57**, 312–319 (1975).

Transverse Modes in Tubes

Hartig, H. E., and C. E. Swanson. "Transverse Acoustic Waves in Rigid Tubes," *Phys. Rev.*, **54**, 618–626 (1938).

Morfey, C. L. "Rotating Pressure Patterns in Ducts: Their Generation and Transmission," *J. Sound Vib.*, **1**, 60–87 (1964).

Snow, D. J., and M. V. Lowson. "Attenuation of Spiral Modes in a Circular and Annular Lined Duct," *J. Sound Vib.*, **25**, 465–477 (1972).

Zorumski, W. E., and J. P. Mason. "Multiple Eigenvalues of Sound-Absorbing Circular and Annular Ducts," *J. Acoust. Soc. Amer.*, **55**, 1158–1165 (1974).

Viscous and Thermal Effects in Tubes

Fay, R. D. "Attenuation of Sound in Tubes," *J. Acoust. Soc. Amer.*, **12**, 62–67 (1940).

Monkewitz, P. A. "The Linearized Treatment of General Forced Oscillations in Tubes," *J. Fluid Mech.*, **91**, 357–397 (1979).

O'Brien, V., and F. E. Logan. "Velocity Overshoot within the Boundary Layer in Laminar Pulsating Flow," *Phys. Fluids*, **9**, 214–215 (1965).

Parker, J. G. "Effects of Adsorption on Acoustic Boundary-Layer Losses," *J. Chem. Phys.*, **36**, 1547–1554 (1962).

—— "Effects of Several Light Molecules on the Vibrational Relaxation Time of Oxygen," *J. Chem. Phys.*, **34**, 1763–1772 (1961).

Rott, N. "Damped and Thermally Driven Acoustic Oscillations in Wide and Narrow Tubes," *J. Appl. Math. Phys.* (ZAMP), **20**, 230–243 (1969).

Shields, F. D., K. P. Lee, and W. J. Wiley. "Numerical Solution for Sound Velocity and Adsorption in Cylindrical Tubes," *J. Acoust. Soc. Amer.*, **37**, 724–729 (1965).

Temkin, S., and R. A. Dobbins. "Measurements of Attenuation and Dispersion of Sound by an Aerosol," *J. Acoust. Soc. Amer.*, **46**, 1016–1024 (1966).

Tijdman, H. "On the Propagation of Sound Waves in Cylindrical Tubes," *J. Sound Vib.*, **39**, 1–33 (1975).

APPENDIX A
USEFUL FORMULAS FROM VECTOR ANALYSIS

In the following formulas, **A**, **B**, **C**, and **D** represent vector functions; ϕ and ψ represent scalar functions.

General Relations

$$\mathbf{A} \cdot \mathbf{B} \times \mathbf{C} = \mathbf{B} \cdot \mathbf{C} \times \mathbf{A} - \mathbf{C} \cdot \mathbf{A} \times \mathbf{B}$$

$$\mathbf{A} \times \mathbf{B} \times \mathbf{C} = (\mathbf{A} \cdot \mathbf{C})\mathbf{B} - (\mathbf{A} \cdot \mathbf{B})\mathbf{C}$$

$$(\mathbf{A} \times \mathbf{B}) \cdot (\mathbf{C} \times \mathbf{D}) = (\mathbf{A} \cdot \mathbf{C})(\mathbf{B} \cdot \mathbf{D}) - (\mathbf{A} \cdot \mathbf{D})(\mathbf{B} \cdot \mathbf{C})$$

$$(\mathbf{A} \times \mathbf{B}) \times (\mathbf{C} \times \mathbf{D}) = (\mathbf{A} \times \mathbf{B} \cdot \mathbf{D})\mathbf{C} - (\mathbf{A} \times \mathbf{B} \cdot \mathbf{C})\mathbf{D}$$

$$\nabla \times \nabla \phi = 0$$

$$\nabla \cdot \nabla \times \mathbf{B} = 0$$

$$\nabla(\phi + \psi) = \nabla\phi + \nabla\psi$$

$$\nabla(\phi\psi) = \psi\nabla\phi + \phi\nabla\psi$$

$$\nabla \cdot (\mathbf{A} + \mathbf{B}) = \nabla \cdot \mathbf{A} + \nabla \cdot \mathbf{B}$$

$$\nabla \times (\mathbf{A} + \mathbf{B}) = \nabla \times \mathbf{A} + \nabla \times \mathbf{B}$$

$$\nabla \cdot (\phi\mathbf{A}) = \mathbf{A} \cdot \nabla\phi + \phi\nabla \cdot \mathbf{A}$$

$$\nabla \times (\phi \mathbf{A}) = \nabla \phi \times \mathbf{A} + \phi \nabla \times \mathbf{A}$$

$$\nabla (\mathbf{A} \cdot \mathbf{B}) = (\mathbf{A} \cdot \nabla) \mathbf{B} + (\mathbf{B} \cdot \nabla) \mathbf{A} + \mathbf{A} \times (\nabla \times \mathbf{B}) + \mathbf{B} \times (\nabla \times \mathbf{A})$$

$$\nabla \cdot (\mathbf{A} \times \mathbf{B}) = \mathbf{B} \cdot \nabla \times \mathbf{A} - \mathbf{A} \cdot \nabla \times \mathbf{B}$$

$$\nabla \times (\mathbf{A} \times \mathbf{B}) = \mathbf{A} \nabla \cdot \mathbf{B} - \mathbf{B} \nabla \cdot \mathbf{A} + (\mathbf{B} \cdot \nabla) \mathbf{A} - (\mathbf{A} \cdot \nabla) \mathbf{B}$$

$$\nabla \times (\nabla \times \mathbf{A}) = \nabla \nabla \cdot \mathbf{A} - \nabla^2 \mathbf{A}$$

Special Relations

If $\mathbf{x} = \mathbf{i}x + \mathbf{j}y + \mathbf{k}z$ in the position vector of a point (x, y, z) and $r = |\mathbf{x}|$, then:

$$\nabla \cdot \mathbf{x} = 3$$

$$\nabla \times \mathbf{x} = 0$$

$$\nabla r = \mathbf{x}/r$$

$$\nabla (1/r) = -\mathbf{x}/r^3$$

$$(\mathbf{A} \cdot \nabla) \mathbf{x} = \mathbf{A}$$

Integral Relations

In the following formulas, V is a volume bounded by a closed surface S, having a unit normal vector $\mathbf{n}$ directed outward from V:

$$\int_S \phi \mathbf{n} \, dS = \int_V \nabla \phi \, dV$$

$$\int_S \mathbf{A} \cdot \mathbf{n} \, dS = \int_V \nabla \cdot \mathbf{A} \, dV$$

$$\int_S \mathbf{n} \times \mathbf{A} \, dS = \int_V \nabla \times \mathbf{A} \, dV$$

Let S be an open surface with unit normal $\mathbf{n}$, bounded by the closed contour C having a line element $d\mathbf{l}$. Then

$$\oint_C \phi \, d\mathbf{l} = \int_S \mathbf{n} \times \nabla \phi \, dS$$

$$\oint_C \mathbf{A} \cdot d\mathbf{l} = \int_S \mathbf{n} \times (\nabla \times \mathbf{A}) \, dS$$

APPENDIX B

EXPLICIT EXPRESSIONS FOR SOME VECTOR AND TENSOR QUANTITIES IN SPECIAL COORDINATE SYSTEMS

The following tables give explicit formulas for some of the vector and tensor quantities used in the text, for the particular systems of cartesian, spherical, cylindrical, and polar coordinates. These systems are emphasized in the text, but are not the only ones in which the Helmholtz equation is separable. For a complete list of such systems, see "Table of Separable Coordinates in Three Dimensions," in Chapter 5 of *Methods of Theoretical Physics*, by P. M. Morse and H. Feshbach (McGraw-Hill, New York, 1953), page 655.

Table B.1 Cartesian Coordinates

Coordinate symbols	x	y	z
Line elements	dx	dy	dz
Orthonormal vectors along coordinate axes	$\mathbf{i}$	$\mathbf{j}$	$\mathbf{k}$
Velocity components	u	v	w
Components of position vector $\mathbf{x}$	x	y	z
Components of $\nabla\phi$	$\partial\phi/\partial x$	$\partial\phi/\partial y$	$\partial\phi/\partial z$
Components of $\nabla \times \mathbf{B}$, $\mathbf{B}=(B_1, B_2, B_3)$	$\left(\dfrac{\partial B_2}{\partial z} - \dfrac{\partial B_3}{\partial y}\right)$	$\left(\dfrac{\partial B_1}{\partial z} - \dfrac{\partial B_3}{\partial x}\right)$	$\left(\dfrac{\partial B_2}{\partial x} - \dfrac{\partial B_1}{\partial y}\right)$

Table B.1—*Continued*

Components of $\nabla^2 \mathbf{B}$	$\partial^2 B_1/\partial x^2$	$\partial^2 B_2/\partial y^2$	$\partial^2 B_3/\partial z^2$

Special Operators

$$\nabla \cdot \mathbf{f} = \frac{\partial f_1}{\partial x} + \frac{\partial f_2}{\partial y} + \frac{\partial f_3}{\partial z}$$

$$\nabla^2 \phi = \frac{\partial^2 \phi}{\partial x^2} + \frac{\partial^2 \phi}{\partial y^2} + \frac{\partial^2 \phi}{\partial z^2}$$

Components of rate-of-strain tensor

$$e_{xx} = \frac{\partial u}{\partial x}, \qquad e_{yy} = \frac{\partial v}{\partial y}, \qquad e_{zz} = \frac{\partial w}{\partial z}$$

$$e_{xy} = e_{yx} = \frac{1}{2}\left(\frac{\partial u}{\partial y} + \frac{\partial v}{\partial x}\right), \qquad e_{xz} = e_{zx} = \frac{1}{2}\left(\frac{\partial u}{\partial z} + \frac{\partial w}{\partial x}\right), \qquad e_{yz} = e_{zy} = \frac{1}{2}\left(\frac{\partial v}{\partial z} + \frac{\partial w}{\partial y}\right)$$

Table B.2 Spherical-Polar Coordinates

Coordinate symbols	r	θ	φ
Line elements	dr	$r\,d\theta$	$r\sin\theta\,d\varphi$
Orthonormal vectors along coordinate axes	$\mathbf{e}_r$	$\mathbf{e}_\theta$	$\mathbf{e}_\varphi$
Orthonormal vectors referred to cartesian coordinates (with $\varphi=0$ along positive x axis)	$\mathbf{i}\cos\varphi\sin\theta+\mathbf{j}\sin\varphi$ $\times\sin\theta+\mathbf{k}\cos\theta$	$\mathbf{i}\cos\varphi\cos\theta+\mathbf{j}\sin\varphi$ $\times\cos\theta-\mathbf{k}\sin\theta$	$-\mathbf{i}\sin\varphi+\mathbf{j}\cos\varphi$
Velocity components	u_r	u_θ	u_φ
Components of $\nabla\phi$	$\dfrac{\partial\phi}{\partial r}$	$\dfrac{1}{r}\dfrac{\partial\phi}{\partial\theta}$	$\dfrac{1}{r\sin\theta}\dfrac{\partial\phi}{\partial\varphi}$
Components of $\nabla\times\mathbf{B}$,	$\dfrac{1}{r\sin\theta}\left[\dfrac{\partial(B_\varphi\sin\theta)}{\partial\theta}-\dfrac{\partial B_\theta}{\partial\varphi}\right]$	$\dfrac{1}{r}\left[\dfrac{1}{\sin\theta}\dfrac{\partial B_r}{\partial\varphi}-\dfrac{\partial(rB_\varphi)}{\partial r}\right]$	$\dfrac{1}{r}\left[\dfrac{\partial(rB_\theta)}{\partial r}-\dfrac{\partial B_r}{\partial\theta}\right]$

$\mathbf{B}=(B_r,B_\theta,B_\varphi)$

Components of $\nabla^2\mathbf{B}$

$$\mathbf{e}_r\cdot\nabla^2\mathbf{B}=\nabla^2 B_r-\frac{2B_r}{r^2}-\frac{2}{r^2\sin\theta}\frac{\partial(B_\theta\sin\theta)}{\partial\theta}-\frac{2}{r^2\sin\theta}\frac{\partial B_\varphi}{\partial\varphi}$$

$$\mathbf{e}_\theta\cdot\nabla^2\mathbf{B}=\nabla^2 B_\theta+\frac{2}{r^2}\frac{\partial B_r}{\partial\theta}-\frac{B_\theta}{r^2\sin^2\theta}-\frac{2\cos\theta}{r^2\sin^2\theta}\frac{\partial B_\varphi}{\partial\varphi}$$

$$\mathbf{e}_\varphi\cdot\nabla^2\mathbf{B}=\nabla^2 B_\varphi+\frac{2}{r^2\sin\theta}\frac{\partial B_r}{\partial\varphi}+\frac{2\cos\theta}{r^2\sin^2\theta}\frac{\partial B_\theta}{\partial\varphi}-\frac{B_\varphi}{r^2\sin^2\theta}$$

Table B.2—Continued

Special Operators

$$\nabla \cdot \mathbf{B} = \frac{1}{r^2}\frac{\partial(r^2 B_r)}{\partial r} + \frac{1}{r\sin\theta}\frac{\partial(\sin\theta\, B_\theta)}{\partial\theta} + \frac{1}{r\sin\theta}\frac{\partial B_\varphi}{\partial\varphi}$$

$$\nabla^2\phi = \frac{1}{r^2}\frac{\partial}{\partial r}\left(r^2\frac{\partial\phi}{\partial r}\right) + \frac{1}{r^2\sin\theta}\frac{\partial}{\partial\theta}\left(\sin\theta\frac{\partial\phi}{\partial\theta}\right) + \frac{1}{r^2\sin^2\theta}\frac{\partial^2\phi}{\partial\varphi^2}$$

Components of rate-of-strain tensor

$$e_{rr} = \frac{\partial u_r}{\partial\theta}, \qquad e_{\theta\theta} = \frac{1}{r}\frac{\partial u_\theta}{\partial\theta} + \frac{u_r}{r}, \qquad e_{\varphi\varphi} = \frac{1}{r\sin\theta}\frac{\partial u_\varphi}{\partial\varphi} + \frac{u_r}{r} + \frac{u_\theta\cot\theta}{r}$$

$$e_{\theta\varphi} = e_{\varphi\theta} = \frac{\sin\theta}{2r}\frac{\partial}{\partial\theta}\left(\frac{u_\varphi}{\sin\theta}\right) + \frac{1}{2r\sin\theta}\frac{\partial u_\theta}{\partial\varphi}$$

$$e_{\varphi r} = e_{r\varphi} = \frac{1}{2r\sin\theta}\frac{\partial u_r}{\partial\varphi} + \frac{r}{2}\frac{\partial}{\partial r}\left(\frac{u_\theta}{r}\right) + \frac{r}{2}\frac{\partial}{\partial r}\left(\frac{u_\varphi}{r}\right)$$

$$e_{r\theta} = e_{\theta r}\frac{r}{2}\frac{\partial}{\partial r}\left(\frac{u_\theta}{r}\right) + \frac{1}{2r}\frac{\partial u_r}{\partial\theta}$$

Table B.3 Cylindrical Coordinates

Coordinate symbols	x	σ	φ
Line elements	dx	$d\sigma$	$\sigma\, d\varphi$
Orthonormal vectors along coordinate axes	$\mathbf{i}$	$\mathbf{e}_\sigma$	$\mathbf{e}_\varphi$
Orthonormal vectors referred to cartesian coordinates	$\mathbf{i}$	$\mathbf{i}\cos\varphi + \mathbf{j}\sin\varphi$	$-\mathbf{i}\sin\varphi + \mathbf{j}\cos\varphi$
Velocity components	u	v	w
Components of $\nabla\phi$	$\dfrac{\partial\phi}{\partial x}$	$\dfrac{\partial\theta}{\partial\sigma}$	$\dfrac{1}{\sigma}\dfrac{\partial\phi}{\partial\varphi}$

Components of $\nabla\times\mathbf{B}$,

$$\left[\frac{1}{\sigma}\frac{\partial(\sigma B_\varphi)}{\partial\sigma} - \frac{1}{\sigma}\frac{\partial B_\sigma}{\partial\varphi}\right] \qquad \left(\frac{1}{\sigma}\frac{\partial B_x}{\partial\varphi} - \frac{\partial B_\varphi}{\partial x}\right) \qquad \left(\frac{\partial B_\sigma}{\partial x} - \frac{\partial B_x}{\partial\sigma}\right)$$

$$\mathbf{B} = (B_x, B_\sigma, B_\varphi)$$

Components of $\nabla^2\mathbf{B}$

$$\nabla^2 B_x \qquad \nabla^2 B_\sigma - \frac{B_\sigma}{\sigma^2} - \frac{2}{\sigma^2}\frac{\partial B_\varphi}{\partial\varphi} \qquad \nabla^2 B_\varphi + \frac{2}{\sigma^2}\frac{\partial B_\sigma}{\partial\varphi} - \frac{B_\varphi}{\sigma^2}$$

Special Operators

$$\nabla\cdot\mathbf{B} = \frac{\partial B_x}{\partial x} + \frac{1}{\sigma}\frac{\partial(\sigma B_\sigma)}{\partial\sigma} + \frac{1}{\sigma}\frac{\partial B_\varphi}{\partial\varphi}$$

$$\nabla^2\phi = \frac{\partial^2\phi}{\partial x^2} + \frac{1}{\sigma}\frac{\partial}{\partial\sigma}\left(\sigma\frac{\partial\phi}{\partial\sigma}\right) + \frac{1}{\sigma^2}\frac{\partial^2\phi}{\partial\varphi^2}$$

Components of rate-of-strain tensor

$$e_{xx} = \frac{\partial u}{\partial x}, \qquad e_{\sigma\sigma} = \frac{\partial v}{\partial\sigma}, \qquad e_{\varphi\varphi} = \frac{1}{\sigma}\left(\frac{\partial w}{\partial\varphi} + v\right)$$

$$e_{\sigma\varphi} = e_{\varphi\sigma} = \frac{\sigma}{2}\frac{\partial}{\partial\sigma}\left(\frac{w}{\sigma}\right) + \frac{1}{2\sigma}\frac{\partial v}{\partial\varphi}, \qquad e_{\varphi x} = e_{x\varphi} = \frac{1}{2\sigma}\frac{\partial u}{\partial\varphi} + \frac{1}{2}\frac{\partial w}{\partial x}, \qquad e_{x\sigma} = e_{\sigma x} = \frac{1}{2}\frac{\partial v}{\partial x} + \frac{1}{2}\frac{\partial u}{\partial\sigma}$$

Table B.4 Polar Coordinates

	r	θ
Coordinate symbols	r	θ
Line elements	dr	$r\,d\theta$
Orthonormal vectors along coordinate axis	$\mathbf{e}_r$	$\mathbf{e}_\theta$
Orthonormal vectors referred to cartesian coordinates	$\mathbf{i}\cos\theta+\mathbf{j}\sin\theta$	$-\mathbf{i}\sin\theta+\mathbf{j}\cos\theta$
Velocity components	u_r	u_θ
Components of $\nabla\phi$	$\dfrac{\partial\phi}{\partial r}$	$\dfrac{1}{r}\dfrac{\partial\phi}{\partial\theta}$
Components of $\nabla^2\mathbf{B}$,	$\nabla^2 B_r-\dfrac{B_r}{r^2}-\dfrac{2}{r^2}\dfrac{\partial B_\theta}{\partial\theta}$	$\nabla^2 B_\theta+\dfrac{2}{r^2}\dfrac{\partial B_r}{\partial\theta}-\dfrac{B_\theta}{r^2}$

$\mathbf{B}=(B_r,\,B_\theta)$

Special Operators

$$\nabla\cdot\mathbf{B}=\frac{1}{r}\frac{\partial(rB_r)}{\partial r}+\frac{1}{r}\frac{\partial B_\theta}{\partial\theta}$$

$$\nabla\times\mathbf{B}=\mathbf{e}_r\times\mathbf{e}_\theta\left[\frac{1}{r}\frac{\partial(rB_\theta)}{\partial r}-\frac{1}{r}\frac{\partial B_r}{\partial\theta}\right]$$

$$\nabla^2\phi=\frac{1}{r}\frac{\partial}{\partial r}\left(r\frac{\partial\phi}{\partial r}\right)+\frac{1}{r^2}\frac{\partial^2\phi}{\partial r^2}$$

Components of rate-of-strain tensor

$$e_{rr}=\frac{\partial u_r}{\partial r},\qquad e_{\theta\theta}=\frac{1}{r}\frac{\partial u_\theta}{\partial\theta}+\frac{u_r}{r},\qquad e_{r\theta}=2_{\partial r}\frac{r}{2}\frac{\partial}{\partial r}\left(\frac{u_\theta}{r}\right)+\frac{1}{2r}\frac{\partial u_r}{\partial\theta}$$

APPENDIX C

SOME PROPERTIES OF THE BESSEL FUNCTIONS

The standard Bessel differential equation is

$$\frac{d^2f}{dz^2} + \frac{1}{z}\frac{df}{dz} + \left(1 - \frac{\nu^2}{z^2}\right)f = 0$$

where ν may have arbitrary values. For nonintegral values of ν, the general solution of this equation may be written as

$$f = AJ_\nu(z) + BJ_{-\nu}(z)$$

where $J_{\pm\nu}(z)$ are the Bessel functions of the first kind. The general solution may also be written as

$$f = aJ_\nu(z) + bY_\nu(z), \quad \nu \text{ arbitrary}$$

where $Y_\nu(z)$ is the Bessel function of the second kind. The solution may also be expressed in terms of the Bessel functions of the third kind, $H_\nu^{(1)}$ and $H_\nu^{(2)}$ (also known as the Hankel functions) by means of

$$H_\nu^{(1)}(z) = J_\nu(z) + iY_\nu(z)$$

$$H_\nu^{(2)}(z) = J_\nu(z) - iY_\nu(z)$$

These are independent for every value of ν.

Expressions for Small Arguments

When ν is fixed and $z \to 0$, the limiting forms of $J_\nu(z)$ and $Y_\nu(z)$ are

$$J_\nu(z) \simeq \left(\tfrac{1}{2}z\right)^\nu / \Gamma(\nu+1), \quad \nu \neq -1, -2, \ldots$$

where Γ is the gamma function. For n equal to a positive integer or zero, $\Gamma(n+1) = n!$, with $0! = 1$. (For other values of the gamma function, see Chapter 6 of Abramowitz and Stegun.)

$$Y_\nu \simeq -(1/\pi)\Gamma(\nu)\left(\frac{1}{2}z\right)^{-\nu}, \quad \operatorname{Re}(\nu) > 0$$

$$Y_0 \simeq (2/\pi)\ln z$$

Asymptotic Expressions for Large Arguments

When ν is fixed and $|z| \to \infty$, the limiting forms of $J_\nu(z)$ and $Y_\nu(z)$ are

$$J_\nu(z) \simeq \sqrt{2/\pi z}\ \cos\left(z - \tfrac{1}{2}\nu\pi - \tfrac{1}{4}\pi\right), \quad |\arg z| < \pi$$

$$Y_\nu(z) \simeq \sqrt{2/\pi z}\ \sin\left(z - \tfrac{1}{2}\nu\pi - \tfrac{1}{4}\pi\right), \quad |\arg z| < \pi$$

Asymptotic Expressions for Large Orders of ν

When z is fixed and $\nu \to \infty$,

$$J_\nu(z) \simeq \frac{1}{\sqrt{2\pi\nu}}\left(\frac{ez}{2\nu}\right)^\nu$$

$$Y_\nu(z) \simeq \sqrt{\frac{2}{\pi\nu}}\left(\frac{ez}{2\nu}\right)^{-\nu}$$

where $e = 2.71828\ldots$.

Recurrence Relations

The following expressions apply to $J_{\pm\nu}$, Y_ν, $H_\nu^{(1)}$, and $H_\nu^{(2)}$:

$$f_\nu'(z) = f_{\nu-1}(z) - \frac{\nu}{z}f_\nu(z)$$

$$f_{\nu+1}(z) = \frac{2\nu}{z}f_\nu(z) - f_{\nu-1}(z)$$

Wronskians

$$W[J_\nu(z), Y_\nu(z)] = J_{\nu+1}(z)Y_\nu(z) - J_\nu(z)Y_{\nu+1}(z) = 2/(\pi z)$$

$$W[H_\nu^{(1)}(z), H_\nu^{(2)}(z)] = H_{\nu+1}^{(1)}(z)H_\nu^{(2)}(z) - H_\nu^{(1)}(z)H_{\nu+1}^{(2)}(z) = -4i/(\pi z)$$

Further information about these functions may be found in Chapter 9 of Abramowitz and Stegun.

Orthogonality Conditions

Suppose that the Bessel function J_p satisfies any of the following boundary conditions:

(1) $J_p(\alpha_n L) = 0$

(2) $J_p'(\alpha_n L) = 0$

(3) $J_p'(\alpha_n L) = -k J_p(\alpha_n L)$

Then,

$$\int_0^L x J_p(\alpha_m x) J_p(\alpha_n x)\, dx = 0$$

for $m \neq n$. When $m = n$, the integral is given by

$$\int_0^L x J_p^2(\alpha_n x)\, dx = \frac{L^2}{2} J_{p+1}^2(\alpha_n L), \quad \text{if } J_p(\alpha_n L) = 0$$

$$= \frac{\alpha_n^2 L^2 - p^2}{2\alpha_n^2} J_p^2(\alpha_n L), \quad \text{if } J_p'(\alpha_n L) = 0$$

$$= \frac{\alpha_n^2 L^2 - p^2 + k^2 L^2}{2\alpha_n^2} J_p^2(\alpha_n L), \quad \text{if } J_p'(\alpha_n L) = -k J_p(\alpha_n L)$$

APPENDIX D

SOME PROPERTIES OF THE SPHERICAL BESSEL FUNCTIONS

For n equal to an integer, the solution to

$$f''(z) + \frac{2}{z} f'(z) + \left[1 - \frac{n(n+1)}{z} \right] f(z) = 0$$

may be written in terms of the *spherical Bessel functions of the first kind*:

$$j_n(z) = \sqrt{\pi/2z}\ J_{n+1/2}(z)$$

the spherical Bessel functions of the second kind:

$$y_n(z) = \sqrt{\pi/2z}\ Y_{n+1/2}(z)$$

and the spherical Bessel functions of the third kind:

$$h_n^{(1)}(z) = j_n(z) + i y_n(z) = \sqrt{\pi/2z}\ H_{n+1/2}^{(1)}(z)$$

$$h_n^{(2)}(z) = j_n(z) - i y_n(z) = \sqrt{\pi/2z}\ H_{n+1/2}^{(2)}(z)$$

The pairs j_n, y_n and $h_n^{(1)}$, $h_n^{(2)}$ are linearly independent.

Explicit Expressions

When the order n is small, the spherical Bessel functions may be expressed in terms of elementary functions:

$n=0$

$$j_0(z) = \frac{\sin z}{z}$$

$$y_0(z) = -\frac{\cos z}{z}$$

$$h_0^{(1)}(z) = -\frac{i}{z}e^{iz}, \qquad h_0^{(2)}(z) = \frac{i}{z}e^{-iz}$$

$n=1$

$$j_1(z) = \frac{\sin z}{z^2} - \frac{\cos z}{z}$$

$$y_1(z) = -\frac{\cos z}{z^2} - \frac{\sin z}{z^2}$$

$$h_1^{(1)}(z) = -\frac{1}{z}\left(1 + \frac{i}{z}\right)e^{iz}, \qquad h_1^{(2)}(z) = -\frac{1}{z}\left(1 - \frac{i}{z}\right)e^{-iz}$$

Recurrence Relations

For $n \geq 2$, the spherical Bessel functions, and their derivatives, may be obtained from the following relations, applicable to j_n, y_n, $h_n^{(1)}$, and $h_n^{(2)}$:

$$f_{n+1}(z) = \frac{2n+1}{z}f_n(z) - f_{n-1}(z)$$

$$f_n'(z) = f_{n-1}(z) - \frac{n+1}{2}f_n(z)$$

Wronskians

$$W\left[j_n(z), y_n(z)\right] = j_{n+1}(z)y_n(z) - j_n(z)y_{n+1}(z) = 1/z^2$$

$$W\left[h_n^{(1)}(z), h_n^{(2)}(z)\right] = h_{n+1}^{(1)}(z)h_n^{(2)}(z) - h_n^{(1)}(z)h_{n+1}^{(2)}(z) = -2i/z^2$$

Expressions for Small Arguments

When $z \to 0$, the leading terms in the series expansion of $j_n(z)$ and $y_n(z)$ are

$$j_n(z) \simeq \frac{z^n}{1 \cdot 3 \cdot 5 \cdots (2n+1)} \left[1 - \frac{z^2/2}{1!(2n+3)} + \cdots \right]$$

$$y_n(z) \simeq - \frac{1 \cdot 3 \cdot 5 \cdots (2n-1)}{z^{n+1}} \left[1 - \frac{z^2/2}{1!(1-2n)} - \cdots \right]$$

For the cases $n=0$ and $n=1$, these give

$$j_0(z) \simeq 1 - z^2/6, \qquad j_1(z) \simeq \tfrac{1}{3} z - z^3/30$$

$$y_0(z) \simeq -1/z + z/2, \qquad y_1(z) \simeq -1/z^2 - \tfrac{1}{2}$$

Expressions in Terms of Modulus and Phase

The spherical Bessel functions of any order n may be expressed in terms of their modulus and phase as follows:

$$j_n(z) = \sqrt{\pi/2z} \; M_{n+1/2}(z) \cos \theta_{n+1/2}(z)$$

$$y_n(z) = \sqrt{\pi/2z} \; M_{n+1/2}(z) \sin \theta_{n+1/2}(z)$$

$$h_n^{(1)} = \sqrt{\pi/2z} \; M_{n+1/2}(z) e^{i\theta_{n+1/2}(z)}, \qquad h_n^{(2)}(z) = \sqrt{\pi/2z} \; M_{n+1/2}(z) e^{-i\theta_{n+1/2}(z)}$$

Similarly,

$$j_n'(z) = \sqrt{\pi/2z} \; N_{n+1/2}(z) \cos \varphi_{n+1/2}(z)$$

$$y_n'(z) = \sqrt{\pi/2z} \; N_{n+1/2}(z) \sin \varphi_{n+1/2}(z)$$

$$h_n^{(1)'}(z) = \sqrt{\pi/2z} \; N_{n+1/2}(z) e^{i\varphi_{n+1/2}(z)},$$

$$h_n^{(2)'}(z) = \sqrt{\pi/2z} \; N_{n+1/2}(z) e^{-i\varphi_{n+1/2}(z)}$$

[The phase angle $\varphi_{n+1/2}(z)$ was denoted by δ_n' in Section 4.3.] Tabulated values of $M_{n+1/2}$, $N_{n+1/2}$, $\theta_{n+1/2}$, and $\varphi_{n+1/2}$ may be found in Chapters 9 and 10 of Abramowitz and Stegun.

Asymptotic Expansions for Large Arguments

For n fixed and x large, positive, and real, and with the notation $\mu = (2n+1)^2$, the leading terms in the asymptotic expressions of $M_{n+1/2}$, $\theta_{n+1/2}$, $N_{n+1/2}$, and $\varphi_{n+1/2}$ are

$$M^2_{n+1/2} \approx \frac{2}{\pi x}\left[1 + \frac{\mu-1}{8x^2} + \cdots\right]$$

$$\theta_{n+1/2} \approx x - \frac{n+1}{2}\pi + \frac{\mu-1}{8x} + \cdots$$

$$N^2_{n+1/2} \approx \frac{2}{\pi x}\left[1 - \frac{\mu-3}{8x^2} - \cdots\right]$$

$$\varphi_{n+1/2} \approx x - \tfrac{1}{2}n\pi + \frac{\mu+3}{8x} + \cdots$$

In the particular cases of $n=0$ and $n=1$, the explicit expressions for the spherical Bessel functions show that for $x \to \infty$,

$$j_0(x) = \frac{1}{x}\sin x, \qquad j_0'(x) \simeq \frac{1}{x}\cos x$$

$$y_0(x) = -\frac{1}{x}\cos x, \qquad y_0'(x) \simeq \frac{1}{x}\sin x$$

$$h_0^{(1)}(x) = -\frac{i}{x}e^{ix}, \qquad h_0^{(1)\prime}(x) \simeq -\frac{1}{x}e^{ix}$$

$$j_1(x) \simeq -\frac{1}{x}\cos x, \qquad j_1'(x) \simeq \frac{1}{x}\cos x$$

$$y_1(x) \simeq -\frac{1}{x}\sin x, \qquad y_1'(x) \simeq -\frac{1}{x}\cos x$$

$$h_1^{(1)}(x) \simeq -\frac{1}{x}e^{ix}, \qquad h_1^{(1)\prime}(x) \simeq -\frac{i}{x}e^{ix}$$

APPENDIX E
LEGENDRE POLYNOMIALS

With $x = \cos\theta$, (4.3.7) may be written as

$$(1-x^2)\frac{d^2\Theta}{dx^2} - 2x\frac{d\Theta}{dx} + n(n+1)\Theta = 0$$

The solutions to this equation are the Legendre polynomials of the first and second kinds $P_n(x)$ and $Q_n(x)$, respectively. Explicit forms for low values of n are:

$n = 0$

$$P_0(x) = 1, \qquad Q_0(x) = \tfrac{1}{2}\ln\left(\frac{1+x}{1-x}\right)$$

$n = 1$

$$P_1(x) = x, \qquad Q_1(x) = \frac{x}{2}\ln\left(\frac{1+x}{1-x}\right) - 1$$

$n = 2$

$$P_2(x) = \tfrac{1}{2}(3x^2 - 1), \qquad Q_2(x) = \frac{3x^2 - 1}{4}\ln\left(\frac{1+x}{1-x}\right) - \frac{3x}{2}$$

Recurrence Relations for $P_n(x)$

$$(n+1)P_{n+1}(x) = (2n+1)xP_n(x) - nP_{n-1}(x)$$

$$(x^2-1)\frac{dP_n(x)}{dx} = nxP_n(x) - nP_{n-1}(x)$$

These recurrence relations are also satisfied by Q_n.

Special Relations and Values

$$P_n(\theta) = 1$$

$$P_{-n-1}(x) = P_n(x)$$

$$P_n(-x) = P_n(x)$$

Associated Legendre Polynomials

These are defined by

$$P_n^m(x) = (-1)^m(1-x^2)^{m/2}\frac{d^mP_n(x)}{dx^m}$$

$$Q_n^m(x) = (-1)^m(1-x^2)^{m/2}\frac{d^mQ_n}{dx^m}$$

Thus,

$$P_0^0 = 0, \qquad P_0^1 = 0$$

$$P_1^0 = 1, \qquad P_1^1 = -(1-x^2)^{1/2}$$

AUTHOR INDEX

SUBJECT INDEX